Relationships of Natural Enemies and Non-Prey Foods

Progress in Biological Control

Volume 7

Published:

Volume 1
H.M.T. Hokkanen and A.E. Hajek (eds.)
Environmental Impacts of Microbial Insecticides – Need and Methods for Risk Assessment. 2004
ISBN: 978-1-4020-0813-9

Volume 2
J. Eilenberg and H.M.T. Hokkanen (eds.)
An Ecological and Societal Approach to Biological Control. 2007
ISBN 978-1-4020-4320-8

Volume 3
J. Brodeur and G. Boivin (eds.):
Trophic and Guild Interactions in Biological Control. 2006
ISBN 978-1-4020-4766-4

Volume 4
J. Gould, K. Hoelmer and J. Goolsby (eds.):
Classical Biological Control of Bemisia tabaci in the United States. 2008
ISBN 978-1-4020-6739-6

Volume 5
J. Romeis, A.M. Shelton and G. Kennedy (eds.):
Integration of Insect-Resistant Genetically Modified Crops within IPM Programs. 2008
HB ISBN 978-1-4020-8372-3; PB ISBN 978-1-4020-8459-1

Volume 6
A.E. Hajek, T.R. Glare and M. O'Callaghan (eds.):
Use of Microbes for Control and Eradication of Invasive Arthropods. 2008
ISBN: 978-1-4020-8559-8

For other titles published in this series, go to
www.springer.com/series/6417

Jonathan G. Lundgren

Relationships of Natural Enemies and Non-Prey Foods

Springer

Jonathan G. Lundgren
North Central Agricultural Research Laboratory
USDA-ARS
2923 Medary Avenue
Brookings, SD, 57006
USA

Cover pictures: The photo credits should read (clockwise from top left):
Amara carinata (LeConte), a common granivorous carabid in the Midwest and Great Plains of North America (photo by Jonathan Lundgren).
A mixture of seeds commonly encountered in cropland (photo by Jonathan Lundgren).
A fourth instar *Coleomegilla maculata* DeGeer, consuming maize pollen (photo by Michael Jeffords, Illinois Natural History Survey).
Adult *Coleomegilla maculata* DeGeer, foraging on the tassel of maize for pollen. This species is one of the most widespread and abundant ladybeetles in North American cropland, and is highly omnivorous on pollen, fungus, and sugar sources (Photo by Michael Jeffords, Illinois Natural History Survey).

ISBN 978-1-4020-9234-3 e-ISBN 978-1-4020-9235-0

Library of Congress Control Number: 2008936830

Published by Springer Science+Business Media B.V.

Printed on acid-free paper

springer.com

This book is dedicated above all to

Jenna
My successes are brighter and failures are bearable because of you

R. N. W.
For the wisdom that gave me a wide berth to explore what I felt was important, and the good sense to know when to bring me back to reality. These are traits found only in the best of mentors

Progress in Biological Control

Series Preface

Biological control of pests, weeds, and plant and animal diseases utilising their natural antagonists is a well-established and rapidly evolving field of science. Despite its stunning successes world-wide and a steadily growing number of applications, biological control has remained grossly underexploited. Its untapped potential, however, represents the best hope to providing lasting, environmentally sound, and socially acceptable pest management. Such techniques are urgently needed for the control of an increasing number of problem pests affecting agriculture and forestry, and to suppress invasive organisms which threaten natural habitats and global biodiversity.

Based on the positive features of biological control, such as its target specificity and the lack of negative impacts on humans, it is the prime candidate in the search for reducing dependency on chemical pesticides. Replacement of chemical control by biological control – even partially as in many IPM programs – has important positive but so far neglected socio-economic, humanitarian, environmental and ethical implications. Change from chemical to biological control substantially contributes to the conservation of natural resources, and results in a considerable reduction of environmental pollution. It eliminates human exposure to toxic pesticides, improves sustainability of production systems, and enhances biodiversity. Public demand for finding solutions based on biological control is the main driving force in the increasing utilisation of natural enemies for controlling noxious organisms.

This book series is intended to accelerate these developments through exploring the progress made within the various aspects of biological control, and via documenting these advances to the benefit of fellow scientists, students, public officials, policymakers, and the public at large. Each of the books in this series is expected to provide a comprehensive, authoritative synthesis of the topic, likely to stand the test of time.

Heikki M.T. Hokkanen, Series Editor

Foreword

Feeding on Non-Prey Resources by Natural Enemies

Moshe Coll

Reports on the consumption of non-prey food sources, particularly plant materials, by predators and parasitoids are common throughout the literature (reviewed recently by Naranjo and Gibson 1996, Coll 1998a, Coll and Guershon, 2002). Predators belonging to a variety of orders and families are known to feed on pollen and nectar, and adult parasitoids acquire nutrients from honeydew and floral and extrafloral nectar. A recent publication by Wäckers et al. (2005) discusses the provisioning of plant resources to natural enemies from the perspective of the plant, exploring the evolutionary possibility that plants enhance their defenses by recruiting enemies to food sources. The present volume, in contrast, presents primarily the enemies' perspective, and as such is the first comprehensive review of the nutritional importance of non-prey foods for insect predators and parasitoids.

Although the ecological significance of feeding on non-prey foods has long been underappreciated, attempts have been made to manipulate nectar and pollen availability in crop fields in order to enhance levels of biological pest control by natural enemies (van Emden, 1965; Hagen, 1986; Coll, 1998a). The importance of non-prey foods for the management of pest populations is also discussed in the book. To place our view of interactions between prey, predators and supplemental foods in a historical context, I will briefly review developments in our understanding of trophic interactions in ecological systems, from consumer-resource relationships, through interactions in linear food chains with three species, to more complex direct and indirect effects in community modules with closed loops of omnivorous interactions. Finally, I will touch upon more recent research on trophic interactions of greater complexity, and discuss the need to place omnivorous feeding habits in spatial, evolutionary and conservation biology contexts.

Moshe Coll
Department of Entomology, The Hebrew University of Jerusalem,
P.O. Box 12, Rehovot 76100, Israel
email: coll@agri.huji.ac.il

1 From Simple Interactions in Linear Food-Chains to Omnivorous Trophic Loops

Ecologists have traditionally focused on feeding by organisms on food items at the trophic level immediately below their own. This perspective is not indicative of a belief that interactions between two trophic levels occur in isolation, but rather expresses the hope that a simplistic view of ecological systems would yield in-depth understanding of underlying processes (Begon et al. 1996a). This approach led trophic research in two general directions, one dealing with animal–plant interactions, and the other focusing on interactions between predators and their prey.

Studies of animal-plant relationships allowed for detailed investigation of the effects of plant chemistry, morphology, and structure on herbivores, and resulted in the development of numerous testable hypotheses. Although many major advances in our understanding of insect–plant interactions date back some 50 years, with notable contributions by Fraenkel (1959), Ehrlich and Raven (1964), Feeny (1975, 1976) and Rhoades (1979), ecological investigations of predator–prey and host–parasitoid interactions were initiated even earlier (Lotka, 1924; Volterra, 1926; Nicholson, 1933; Nicholson and Bailey, 1935). It should be noted that the effect of predators and parasitoids on herbivore populations was in fact appreciated more than 2,300 years ago, when natural enemies were used for biological pest control (DeBach and Rosen, 1991). It therefore seems that agricultural use of natural enemies both preceded and stimulated the ecological investigations of predator–prey interactions that gained momentum during the 20th century. Much effort over the last 50 years was therefore aimed at identifying regulators of insect populations and was focused on inter-trophic level interactions, be they herbivore-plant or predator-prey associations (Hairston et al., 1960; Hassell, 1978, 1985; Strong et al., 1984).

After the late 1970's, trophic interactions between consumers and their food sources were placed in a three-trophic level context (Fig 1a) (Campbell and Duffey, 1979; Lawton and McNeill, 1979; Price et al., 1980; Schultz, 1983). These early studies were focused on direct adverse effects of plant defenses on natural enemies, and on positive indirect influences of plants that acted to increase enemy-induced mortality by slowing herbivore development (Campbell and Duffey, 1979; Lawton and McNeill, 1979). While the traditional view, in which ecological communities are composed of distinct trophic levels, is in evidence in both early and more recent entomological studies (Pierce et al., 1912; Pearson and Dyer, 2006), later studies also focus on other trophic configurations in modules with three species. These modules include an enemy that attacks two herbivores (Fig. 1b), and a single prey which is attacked by two enemies (Fig. 1c).

More recently, ecologists have addressed interactions of greater complexity in three-species community modules (e.g., Polis, 1991; Polis and Holt, 1992; Coll and Guershon, 2002; Finke and Denno, 2004). The widely accepted view of communities as consisting of species occupying three functionally discrete trophic levels, as proposed by Hairston et al. (1960), has been replaced by a recognition of the importance of modules containing omnivorous species, defined as consumers that feed at more

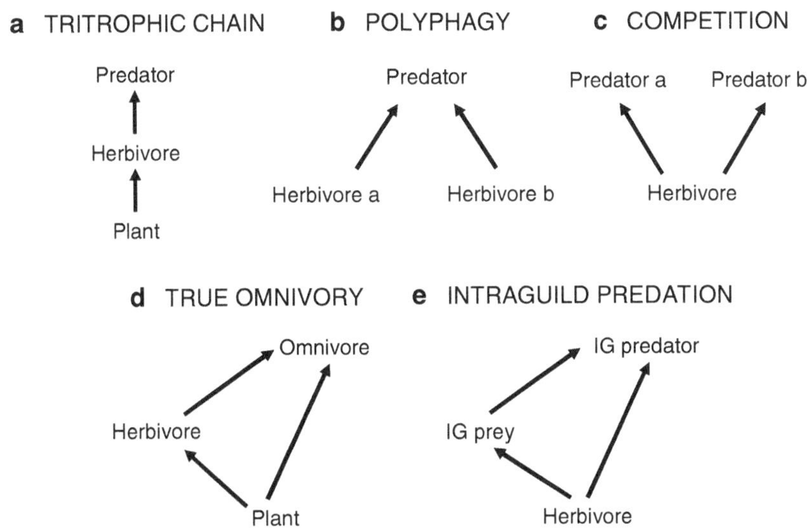

Fig. 1 Diagrammatic representation of possible direct trophic relations in community modules composed of three species. Feeding also on the plant by the predator in module (a) will result in true omnivory (d); two prey species with a shared predator (b) may exhibit apparent competition; and a predator feeding on the second predator in module (c) will lead to intraguild predation (e) of the intraguild (IG) prey by the IG predator. Both true omnivory (d) and intraguild predation (e) are cases of trophic omnivory, whereby the omnivore and IG predator feed at more than one trophic level

than one trophic level (Pimm and Strong 1978). These modules, which act to blur trophic organization, include closed loops wherein one species, be it the herbivore in true omnivory (Fig. 1d) or the intraguild (IG) prey in the case of IG predation (Fig. 1e), is not only consumed by top predators (the omnivore and IG predator), but also competes with them for shared food sources (plant and herbivore).

Indeed, examination of ecology textbooks[1] reveals an exponential increase in the attention devoted to omnivory over the last 40 years (Fig. 2); yet, data show that until the early 1990's, the term omnivory appeared in these books only to describe the nature of animal feeding habits, much like herbivory and carnivory. This use of the term omnivory dates back to ancient Greece. Aristotle, who is regarded as the founder of the biological sciences, recognized that terrestrial animals differ in their diets: they can be carnivorous, graminivorous, omnivorous, or "special" (e.g., nectivorous)

[1] The indexes of 27 authored, general ecology and insect ecology textbooks were examined for the number of pages that include the words 'omnivory', 'omnivore' or 'omnivorous' in the volume. The number of pages with the terms was divided by the total number of text pages in the volume to obtain the "attention level". Then, the context in which the terms appear in the text was classified as "feeding habit" (much like herbivory and carnivory) or "trophic organization".

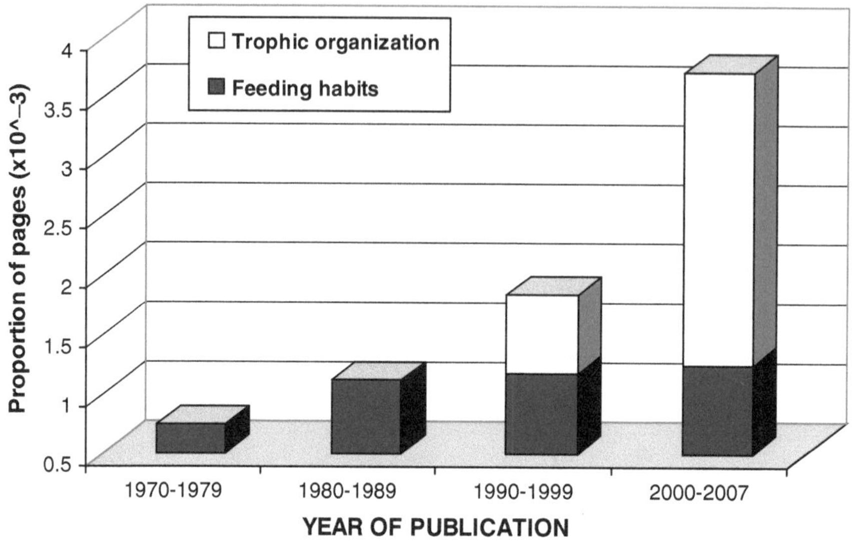

Fig. 2 Change over time in the attention devoted to omnivory in ecological textbooks. Shown is the proportion of pages in authored general ecology (n = 22) and insect ecology (n = 5) textbooks[2] in which the terms 'omnivory', 'omnivore' or 'omnivorous' appear, as indicated by the books' subject indexes. Each mention of the terms in the text was then classified as based on 'feeding habits' or 'trophic organization' according context

(outlined by Allee et al., 1949: 15). Discussion of trophic omnivory and its significance for community structure and function has appeared in textbooks only very recently, with a four-fold increase in attention between the 1990s and 2000s (Fig. 2). A similar situation came to light when the glossary definitions of 'omnivory' or 'omnivore' found in these books were compared (n = 14). Six of the glossaries did not define 'omnivory' at all; the trophic definition of omnivory appeared only after 1990; and only one glossary (Ricklefs, 1990) lists both definitions. I argue that the changes appearing in the definitions of these terms act to confuse trophic and true omnivory, thus obscuring peculiarities of the latter. A case in point is the definition used by Begon et al. (2006b) of omnivory as 'feeding on prey from more than one trophic level' that excludes true omnivory altogether.

The present volume deals primarily with true omnivory; it is a timely compilation of our understanding of the nutritional, ecological, and agricultural significance of feeding on non-prey foods, be they fungi, or plant- or herbivore-produced materials.

---

[2] Begon et al. (1986, 1990, 1996b, 2006); Chapin et al. (2002); Chapman and Reiss (1992); Colinvaux (1993); Dodson et al. (1998); Krebs (1972, 1978, 1994, 2001); McIntosh (1985); McNaughton and Wolf (1979); Odum (1971, 1983); Price (1975, 1984, 1997); Ricklefs (1973, 1990, 1997); Ricklefs and Miller (2000); Schowalter (2006); Smith (1974); Speight et al. (1999); and Townsend et al. (2003).

This comprehensive contribution, together with other publications which have appeared over the last 12 years (Alomar and Wiedenmann, 1996; Jervis and Kidd, 1996; Schaefer, 1997; Coll, 1998b; Coll and Guershon, 2002; Wäckers et al., 2005), encompasses many aspects of plant-feeding by natural enemies which need not be repeated in this foreword. I will instead focus in the remainder of this chapter on several little-studied implications of true omnivory.

2 Some Ecological Implications of True Omnivory

The dramatic change in our view of the structure of animal communities, from discrete to more diffused trophic organization, presents ecologists with new research challenges. I briefly discuss some of these challenges in this section.

2.1 Synergistic Nutritional Effects on True Omnivores

Omnivorous predators are assumed to switch between prey and plant feeding (Cohen, 1996; Naranjo and Gibson, 1996; Coll, 1998b; Agrawal et al., 1999; Coll and Guershon, 2002), but little is known about the nutritional relations between these vastly different food sources. If prey and plant-based foods are in principle nutritionally equivalent, more herbivores should be consumed by omnivores when plant quality decreases (Agrawal et al., 1999; Eubanks and Denno, 2000, Janssen et al., 2003). Plant materials, on the other hand, may provide some essential nutrients that facilitate prey consumption. It has recently been reported that consumption of plant-derived water facilitates prey feeding in *Dicyphus hesperus* (Gillespie and McGregor, 2000; Sinia et al., 2004). Plant tissue may also provide nutrients that are not available in prey; in this case, omnivores can be expected to switch between plant- and prey-feeding to supplement their dietary needs. Exploring the nutritional interactions between plant and prey foods would help us predict short and long term effects of omnivores on herbivore populations.

2.2 Foraging Behavior by True Omnivores

Foraging strategy often affects the diet of consumers; sit-and-wait predators, for example, may be able to consume mobile but not sessile prey (see discussion in Rosenheim and Corbett (2003)). If this is the case, then we may expect true omnivores to be less mobile than pure carnivores, as they may feed on plants when plant and prey foods are nutritionally equivalent. This proposition has not been tested for true omnivores, although resolving the issue may help both to predict the ability of omnivores to suppress various prey species that differ in their mobility, and to inform us about the susceptibility of omnivores to their own predators: it has been proposed, for example, that true omnivores are more likely to serve as IG predators than as IG prey (see discussion below).

Another important, yet little explored topic is the effect of prey and plant foods on patch dynamics of true omnivores. It is often assumed that omnivores, because they may sustain themselves on plant materials, are less likely than pure carnivores to leave the habitat when prey becomes scarce. Yet few studies have actually addressed this issue in depth (but see Eubanks and Denno, 1999). Van Laerhoven et al. (2006) showed that both plant and prey foods influence the length of time individual bugs remain on a given host plant. Investigating specific contributions of prey and plant resources to omnivore behavior in habitat patches is particularly challenging because plant resources, unlike prey, are not usually depleted by the omnivore, and the plant defines the prey's habitat. Foraging omnivores may therefore encounter both food types simultaneously; finding one type is likely to be dependant on finding the other. This dependence is likely to be asymmetrical: although finding the plant may not always result in encountering prey, the reverse is more likely to be true. Finally, nutritional constraints are often assumed to be the primary, if not the only, cause for food mixing by omnivores. Singer and Bernays (2003) pointed out that other considerations, such as toxin dilution and predator avoidance, may also lead to omnivorous feeding habits.

2.3 Competition and Cannibalism by True Omnivores

In food webs with omnivorous trophic loops, the omnivore also competes with its prey for shared food sources. It has been predicted that these competitive interactions will destabilize such three-species modules and make omnivory rare in nature (Pimm and Lawton, 1978). Theoretical studies indicate that IGP systems will persist only if the IG prey is superior to the IG predator as an exploiter of the herbivore (see discussion in Rosenheim and Harmon, 2006 and Janssen et al., 2006). This prediction is also plausible for three-species modules which include true omnivores. With a few exceptions, the nature of competitive interactions between true omnivores and their prey nonetheless remains in the realm of theory.

Contrary to predictions for IGP systems, Coll and Izraylevich (1997) showed that the true omnivorous bug *Orius insidiosus* displaces its thrips prey from preferred feeding sites on plants. The study, however, did not compare host plant utilization by the omnivore and by its prey. Instead, a heuristic mathematical model showed that a decrease in plant palatability for the omnivore tends to stabilize the system (Coll and Izraylevich, 1997). Taken together, these results support predictions formulated for IGP systems and suggest that three-species modules with true omnivores may persist on well-defended plants that adversely affect the omnivore but not the herbivore. This may result in a higher prevalence of true omnivores in systems with more monophagous than polyphagous herbivore prey, and in natural rather than managed ecosystems. These predictions await testing.

True omnivory may also reduce cannibalism because of the availability of plant-based foods (Coll and Guershon 2002). Recent studies indicate that the presence of plant materials, much like prey, does indeed reduce the intensity of cannibalism by an omnivorous bug (Leon-Beck and Coll, 2007), and that plant characteristics have an important effect on cannibalism in another true omnivore (Laycock et al., 2006). The presence of pollen also reduced cannibalism in field populations of a true omnivorous coccinellid (Cottrell and Yeargan, 1998). Yet, true omnivores may exhibit intraspecific competition for plant resources. Groenteman et al. (2006) showed, for example, that *Orius albidipennis* females guard preferred oviposition sites on cotton leaves against conspecific females, and that this behavior is more pronounced on nitrogen-rich than on nitrogen-poor plants. Likewise, proportionately fewer eggs were deposited by *O. albidipennis* at preferred sites in the presence of two intraguild predator species than in their absence (Groenteman, 2004).

2.4 True Omnivory and Population Dynamics

Much progress has been made in recent years in our understanding of dynamic properties of omnivore populations. Most of this progress, however, involves omnivores that feed on herbivorous and carnivorous prey (discussed recently by Janssen et al., 2006; Rosenheim and Harmon, 2006; Denno and Finke, 2006). Relatively little is known about the population dynamics of true omnivores (Coll and Izraylevich, 1997; Lalonde et al., 1999; Gillespie and Roitberg, 2006). Three unique features of these modules call for more theoretical explorations. First, works to date have treated plant-based foods as non-depletable resources. This is hardly the case in natural systems, where availability of these foods often varies over time and space and therefore is likely to change the intensity of omnivore–prey interactions. The ways in which these changes in resource availability affect the dynamics of the system remain to be studied.

The second unique feature of these systems involves the nature of the competitive interactions between true omnivores and their prey, whether characterized by exploitation (scramble) or by interference (contest) competition. Exploitation competition may occur when availability of resources such as nectar is inversely related to omnivore density, whereas interference competition has been reported when the omnivore displaces its prey from preferred feeding sites (Coll and Izraylevich, 1997). These two types of competition may have different effects on the behavior of community modules with true omnivory.

The third important feature is the differential suitability of various foods for the omnivore. The multifaceted difference in nutritional properties of plant and prey food sources often leads to differential contributions to the survival, development and fecundity of the omnivore (Coll, 1998b). For this reason, modeling of age-structured populations is expected to yield different predictions because of the diverse effects diet mixing has on various fitness traits of the omnivore.

2.5 True Omnivory and Intraguild Predation

It has been hypothesized that true omnivores will sustain themselves on plant resources when prey density is low, rather than being forced to leave the habitat or starve, as is the case for pure carnivores (Coll, 1998b; Coll and Guershon, 2002; Sabelis and van Rijn, 2006). Therefore, plant-feeding omnivores are expected to remain in the habitat and prevent subsequent rapid build-ups of herbivore populations (Coll, 1998b; Eubanks and Denno, 2000; van Rijn et al., 2002). This unique characteristic of omnivorous natural enemies is particularly desirable for biological control early in the growing season, when true omnivores can colonize fields before pests become abundant, and for regulation of pest populations exhibiting transient declines during the season. Yet when pests are scarce and true omnivores are expected to express their special advantage, the intensity of intraguild predation (IGP) also increases (Polis et al., 1989; Gillespie and Quiring, 1992; Lucas et al., 1998; Rosenheim, 2001). It is important to realize, however, that the two phenomena – true omnivory and IGP – are in many cases expressions of the single underlying fact that predators have broad diets. Many, though not all, predators whose diets are wide enough to include both prey and plant-based foods, will consume both herbivorous and carnivorous prey (Arim and Marquet, 2004). Thus, many omnivores are also IG predators, and communities that are rich in true omnivores will contain many IG predators as well. It therefore may be inappropriate to treat true omnivory and IGP as separate ecological phenomena, as has been done in the past.

In systems that include true omnivores engaged in IGP, supplementation by pollen and other plant foods is expected to have two counteracting short term effects on prey populations. First, pollen-feeding by the two predators will release prey populations from predation because factors such as gut fullness are important determinants of predator attack rate (Sabelis, 1990). This would result in greater abundance of prey in the presence of pollen. In contrast, supplementation by pollen will lessen the disruptive effect of IGP on prey suppression, i.e., it will reduce predation on the IG prey by the IG predator, which should lead to lower prey densities. A recent study on a system containing two true omnivores that are engaged in IGP confirmed these predictions by showing that both the IG prey and IG predator consumed significantly more prey in the absence of pollen than in its presence. Likewise, fewer IG prey were consumed by the IG predator in the presence of pollen than in its absence. Thus, results show that trophic interactions are weakened in the short term by supplementation with plant-based foods.

It is much harder to predict how such short term effects influence the behavior of these systems in the long run. The various ways in which plant quality affects IGP by true omnivores have been explored recently by Gillespie and Roitberg (2006). They conclude that understanding the influence of plants on IGP is important because they are likely to mediate IGP by true omnivores. Assessment of IGP occurrence in 113 food webs suggests that true omnivorous species are less likely than expected to be IG prey in nature, and more likely than expected to be IG predators (Arim and Marquet, 2004). These results suggest that the ability of IG predators to feed on non-prey foods may relax predation on and competition with

IG prey, thus allowing such IGP systems to persist. Also, true omnivorous IG prey are likely to be excluded by IG predators and thus be under-represented in food webs, if true omnivores are inferior predators relative to pure carnivores, as suggested by Coll and Guershon (2002). This issue is being explored recently also in theoretically and empirically studies of the effect of food supplements on the dynamics of community modules with IGP (Daugherty et al., 2007).

2.6 Spatial Dynamics of True Omnivore Populations

The consumption of prey and plant-derived foods by true omnivores may be separated not only in time (e.g., life-history omnivory; Polis and Strong, 1996), but also in space. If true omnivores and their prey respond differently to spatial variation in the availability of plant and prey food, the intensity of trophic interactions is expected to vary spatially, which may allow the system to persist over a larger spatial scale. This may be the case for the omnivorous bug *Anthocoris nemoralis*, which moves between tree species in Mediterranean woods to feed on pollen and different psylla species (Shaltiel and Coll, 2004). Similar disjunct distributions of omnivores and their foods can occur on a smaller scale. Recently showed that the spatial dynamics generated when true omnivores and their prey track food sources differently on the plant, and possibly when prey alter their distribution to escape predation, lead to site-specific configurations of interacting populations. Specifically, omnivorous bugs and mites congregated on pollen-bearing flowers, whereas their thrips prey colonized the fruits. The intensity of resulting trophic interactions was weakened by the heterogeneous distribution of plant and prey foods. Similar ideas were put forth by Tilman (1982), who argued that competing species may persist in a region if the supplies of different resources vary in space. Asynchronous spatial dynamics in the availability of plant and prey foods thus may enable true omnivores and their prey to coexist in heterogeneous areas. Further empirical and theoretical exploration of spatial aspects of omnivory is likely to enhance our understanding of the stability properties of these systems.

2.7 Evolutionary Transition to True Omnivory

Little is known about the adaptive advantages and disadvantages of omnivory and about constraints to the evolution of true omnivory (but see Diehl (2003) and Roitberg et al. (2005)). Studies of morphological, physiological, and behavioral traits associated with true omnivory are scant (see discussions by Coll and Guershon, 2002 and Eubanks et al., 2003), and the evolutionary path to omnivory has been the subject of extensive debate. This debate is well documented for the Heteroptera, an order encompassing diverse feeding habits and many true omnivorous species (Cobben, 1978; Sweet, 1979; Schuh, 1986; Wheeler, 2001; Eubanks et al., 2003). Investigating the evolutionary consequences of the ability to feed on both plants and prey provides a unique opportunity to gain valuable insight into the speciation processes driven by food diversification.

2.8 True Omnivory, Conservation Biology and Global Climate Changes

Our new view of the structure and function of ecological communities as possessing diffused trophic organization, should also change our decisions concerning both conservation and the impacts of global climate changes on ecological systems. I bring three examples for such considerations. The first, which deals with invasion biology, suggests that true omnivores may pose the highest risk of invading new areas (Berkvens et al., 2008). This can be expected if polyphagous consumers are more successful invaders than specialist ones (Vázquez, 2005). If this prediction holds true, our efforts in curbing biological invasions should be directed accordingly. The second example addresses the difficulty of identifying species that warrant protection in communities with complex food webs (i.e., with many omnivores). The concept of 'keystone interaction' may be useful in such cases, as it refers to those pair-wise interactions whose disturbance may alter the function of the whole community (Eubanks and Styrsky, 2006). The last example applies to the implications of true omnivory for predicting the impact of global changes on ecological systems. Until recently, the effect of atmosphere enrichment with CO_2 has been assessed mostly for plants and, to a lesser degree, for herbivorous arthropods. Much less is known about the flow-on effects of elevated CO_2 on the performance of natural enemies, and nothing was known until recently about its direct and indirect effects on true omnivores (Coll and Hughes, 2008). A recent study showed that elevated CO_2 may (1) benefit the omnivore indirectly by slowing prey development and thus increasing its vulnerability to predation, but (2) hamper omnivore development because of reduced nitrogen content of plant foliage on which it feeds (Coll and Hughes, 2008). These examples illustrate some of the dramatic ways in which our ability to make predictions and take effective action may be altered by considering the omnivorous feeding habits of consumers in ecological communities.

3 Closing Remarks

This single-authored book provides a comprehensive review of the basic ecological and applied significance of feeding by predators and parasitoids on non-prey foods, a topic that was largely ignored until recently. The author brings to the volume valuable experience in the field. It is my hope that the present volume as a whole, and my thoughts in the second part of the foreword, will encourage further exploration of the poorly understood implications of omnivorous feeding habits for the function of ecological communities and the management of pest populations.

Acknowledgements I thank Ruth-Ann Yonah for valuable help with manuscript preparation. My recent work on omnivory has been supported by funding from the Israel Science Foundation, funded by the Academy of Science and Humanities (ISF grant 574/07), and by the US-Israel Binational Agricultural Research & Development Fund (BARD grant TB-8027-04)

References

Agrawal, A.A., Kobayashi, C. & Thaler, J.S. (1999) Influence of prey availability and induced host-plant resistance on omnivory by western flower thrips. Ecology, 80, 518–523.

Allee, W.C., Emerson, A.E., Park, O., Park, T., & Schmidt, K.P. (1949) Principles of Animal Ecology. W.B. Saunders, Philadelphia, PA, 837 pp.

Alomar, O. & Wiedenmann, R.N. (eds.) (1996) Zoophytophagous Heteroptera: Implications for Life History and Integrated Pest Management. Entomological Society of America, Lanham, MA.

Arim, M. & Marquet, P.A. (2004) Intraguild predation: a widespread interaction related to species biology. Ecology Letters, 7, 557–564.

Begon, M., Harper, M.J. & Townsend, C.R. (1986) Ecology: Individuals, Populations, and Communities. Sinaure Associates, Sunderland, MA.

Begon, M., Harper, M.J. & Townsend, C.R. (1990) Ecology: Individuals, Populations, and Communities (4th edition). Blackwell, Oxford.

Begon, M., Browers, R.G., Sait, S.M., & Thompson, D.J. (1996a) Population Dynamics Beyond Two Species: Hosts, Parasitoids and Pathogens, pp. 115–126, In Frontiers of Population Ecology (eds. Floyd RB, Sheppard AW & De Barro PJ) CSIRO, Melbourne.

Begon, M., Harper, J.L. & Townsend, C.R. (1996b) Ecology: Individuals, Populations and Communities. (3rd edition) Blackwell, Oxford.

Begon, M., Harper, J.L. & Townsend, C.R. (2006) Ecology: From Ecosystems to Individuals (4th edition). Blackwell, Oxford.

Berkvens, N., Bonte, J., Berkvens, D., Deforce, K., Tirry, L. & De Clercq, P. (2008) Pollen as an alternative food for *Harmonia axyridis*. BioControl, 53, 201–210.

Campbell, B.C. & Duffey, S.S. (1979) Tomatime and parasitic wasps: potential incompatibility of plant abiosis with biological control. Science, 205, 700–702.

Chapman, J.L. & Reiss, M.J. (1992) Ecology: Principles and Applications. Cambridge University Press, Cambridge, 294 pp.

Chapin, F.S., Matson, P.A. & Mooney, H.A. (2002) Principles of Terrestrial Ecosystems. Springer, New York.

Cobben, R.H. (1978) Evolutionary trends in Heteroptera. Part II. Mouthpart-structures and feeding strategies. Mededelingen Landbouwhogeschool, Wageningen, 407 p.

Cohen, A.C. (1996) Plant feeding by predatory Heteroptera: evolutionary and adaptational aspects of trophic switching. In Zoophytophagous Heteroptera: Implications for Life History and Integrated Pest Management (eds. O. Alomar & R.N. Wiedenmann), pp. 1–17. Entomological Society of America, Lanham, MA.

Colinvaux, P. (1993) Ecology 2. Wiley, New York, 688 pp.

Coll, M. (1998a) Parasitoids in diversified intercropped systems. In Enhancing Biological Control: Habitat Management to Promote Natural Enemies of Agricultural Pests. (eds. C.H. Pickett & R. Bugg), pp. 85–120. University of California Press, Berkeley, CA.

Coll, M. (1998b) Living and feeding on plants in predatory Heteroptera. In Predatory Heteroptera: Their Ecology and Use in Biological Control. (eds. M. Coll & J.R. Ruberson), pp. 89–130. Entomological Society of America, Lanham, MA.

Coll, M. & Guershon, M. (2002) Omnivory in terrestrial arthropods: Mixing plant and prey diets. Annual Review of Entomology, 47, 267–297.

Coll, M. & Hughes, L. (2008) Effects of elevated CO2 on an insect omnivore: A test for nutritional effects mediated by host plants and prey. Agriculture, Ecosystems and Environment, 123, 271–279.

Coll, M. & Izraylevich, S. (1997) When predators also feed on plants: effects of competition and plant quality on omnivore-prey population dynamics. Annals of the Entomological Society of America, 90, 155–161.

Cottrell, T.E. & Yeargan, K.V. (1998) Effect of pollen on *Coleomegilla maculata* (Coleoptera: Coccinellidae) population density, predation, and cannibalism in sweet corn. Environmental Entomology, 27, 1402–1410.

Daugherty, M.P., Harmon, J.P. & Briggs, C.J. (2007) Trophic supplements to intraguild predation. Oikos, 116, 662–677.

DeBach, P. & Rosen, D. (1991) Biological Control by Natural Enemies. 2nd ed. Cambridge Univiversity Press, Cambridge.

Denno, R.F. & Finke, D.L. (2006) Multiple predator interactions and food web connectance: Implications for biological control. In Trophic and Guild Interactions in Biological Control (eds. J. Brodeur & G. Boivin), pp. 45–70. Springer, Dordrecht.

Diehl, S. (2003) The evolution and maintenance of omnivory: Dynamic constraints and the role of food quality. Ecology, 84, 2557–2567

Dodson, S., Allen, T.F.H., Carpenter, S.R., Ives, A.R., Jeanne, R.L., Kitchell, J.F., Langston, N.E. & Turner, M.G. (1998) Ecology. Oxford University Press, Oxford, UK, 434 pp.

Ehrlich, P.R. & Raven, P.H. (1964) Butterflies and plants: a study in coevolution. Evolution, 18, 586–608.

Eubanks, M. D. & Denno, R. F. (1999) The ecological consequences of variation in plants and prey for an omnivorous insect. Ecology, 80, 1253–1266.

Eubanks, M.D. & Denno, R.F. (2000) Host plants mediate omnivore–herbivore interactions and influence prey suppression. Ecology, 81, 936–947.

Eubanks, M.D., & Styrsky, J.D. (2006) Ant-Hemipteran mutualisms: keystone interactions that alter food web dynamics and influence plant fitness, In Trophic and Guild Interactions in Biological Control (eds. J. Brodeur & G. Boivin), pp.1–20. Springer, Dordrecht.

Eubanks, M.D., Styrsky, J.D. and Denno, R.F. (2003) The evolution of omnivory in heteropteran insects. Ecology, 84, 2549–2556.

Feeny, P.P. (1975) Biochemical coevolution between plants and their insect herbivores. In Coevolution of Animals and Plants (eds. L.E. Gilbert & P.H. Raven), pp. 3–19. University of Texas Press, Austin.

Feeny, P.P. (1976) Plant apparency and chemical defense. Recent Advances in Phytochemistry, 10, 1–40.

Finke, D.L. & Denno, R.F. (2004) Predator diversity dampens trophic cascades. Nature, 429, 407–410.

Fraenkel, G.S. (1959) The raison d'être of secondary plant substances. Science, 129, 1466–1470.

Gillespie, D.R. & McGregor, R.R. (2000) The functions of plant feeding in the omnivorous predator *Dicyphus hesperus*: water places limits on predation. Ecological Entomology, 25, 380–386.

Gillespie, D.R. & Quiring, D.J.M. (1992) Competition between *Orius tristicolor* (White) (Hemiptera: Anthocoridae) and *Amblyseius cucumeris* (Oudemans) (Acari: Phytoseiidae) feeding on *Frankliniella occidentalis* (Pergande) (Thysanoptera: Thripidae). Canadian Entomologist, 124, 1123–1128.

Gillespie, D.R. & Roitberg, B.D. (2006) Inter-guild influences on intra-guild predation in plant-feeding omnivores. In Trophic and Guild Interactions in Biological Control (eds. J. Brodeur & G. Boivin), pp. 71–100. Springer, Dordrecht.

Groenteman R. (2004) Intraspecific and intraguild effects on egg deposition by *Orius albidipennis* Reuter. M.Sc. thesis, Hebrew University of Jerusalem, Rehovot.

Groenteman, R., Guershon, M. & Coll, M. (2006) Effects of leaf nitrogen content on oviposition site selection, offspring performance, and intraspecific interactions in an omnivorous bug. Ecological Entomology, 31, 155–161.

Hagen, K.S. (1986) Ecosystem analysis: plant cultivars (HPR), entomophagous species and food supplements. In Interactions of Plant Resistance and Parasitoids and Predators of Insects (eds. D.J. Boethel & R.D. Eikenbary), pp. 151–97. Halsted, New York.

Hairston, N.G., Smith, F.E. & Slobodkin, L.B. (1960) Community structure, population control, and competition. American Naturalist, 94, 421–425.

Hassell, M.P. (1978) The Dynamics of Arthropod Predator-Prey Systems. Princeton University Press, Princeton, NJ.

Hassell, M.P. (1985) Insect natural enemies as regulating factors. Journal of Animal Ecology, 54, 323–334.

Janssen, A., Willemse, E. & van der Hammen, T. (2003) Poor host plant quality causes omnivore to consume predator eggs. Journal of Animal Ecology, 72, 478–483.

Janssen, A., Monserrat, M., HilleRisLambers, R., de Roos, A.M., Pallini, A. & Sabelis, M.W. (2006) Intraguild predation usually does not disrupt biological control. In Trophic and Guild Interactions in Biological Control (eds. J. Brodeur & G. Boivin), pp. 21–44. Springer, Dordrecht.

Jervis, M.A. & Kidd, N.A.C. (1996) Phytophagy. In Insect Natural Enemies: Practical Approach to Their Study and Evaluation (eds. M.A. Jervis & N.A.C. Kidd), pp. 375–394. Chapman & Hall, London.

Krebs, C.J. (1972) Ecology: The Experimental Analysis of Distribution and Abundance. Harper and Row, New York, 694 pp.

Krebs C.J. (1978) Ecology: The Experimental Analysis of Distribution and Abundance. (2nd edition). Harper and Row, New York, 677 pp.

Krebs C.J. (1994) Ecology: The Experimental Analysis of Distribution and Abundance. (4th Edition). Harper Collins, New York, 801 pp.

Krebs, C.J. (2001) Ecology: The Experimental Analysis of Distribution and Abundance. (5th Edition). Benjamin Cummings, New York, 695 pp.

Lalonde, R.G., McGregor, R.R., Gillespie, D.R., Roitberg, B.D. (1999) Plant-feeding by arthropod predators contributes to the stability of predator-prey population dynamics. Oikos, 87, 603–608.

Lawton, J.H. & McNeill, S. (1979) Between the devil and the deep blue sea: on the problems of being an herbivore. In Population Dynamics. Symposium of the British Ecological Society (eds. R.M. Anderson, B.D. Turner & L.R. Taylor), pp. 223–244. Blackwell, Oxford.

Laycock, A., Camm, E., & van Laerhoven, S. (2006) Cannibalism in a zoophytophagous omnivore is mediated by prey availability and plant substrate. Journal of Insect Behavior, 19, 219–229.

Leon-Beck, M. & Coll, M. (2007) Plant and prey consumption cause a similar reduction in cammibalism by an omnivorous bug. Journal of Insect Behavior, 20, 67–76.

Lotka, A.J. (1924) Elements of Physical Biology. Williams & Wilkins, Baltimore, MD.

Lucas, E., Coderre, D. & Brodeur, J. (1998) Intraguild predation among aphid predators: characterization and influence of extraguild prey density. Ecology, 79, 1094–1092.

McIntosh, R.P. (1985) The Background of Ecology: Concept and Theory. Cambridge University Press, Cambridge, 377 pp.

McNaughton, S.J. & Wolf, L.L. (1979) General Ecology (2nd Edition). Holt, Rinehart & Winston, New York.

Naranjo, S. E., & Gibson, R. L. (1996) Phytophagy in predaceous Heteroptera: Effects on life-history and population dynamics. In Zoophytophagous Heteroptera: Implications for Life History and Integrated Pest Management (eds. O. Alomar & R.N. Wiedenmann), pp. 57–93. Thomas Say Symposium Proceedings, Entomological Society of America, Lanham, MA.

Nicholson, A.J. (1933) The balance of animal populations. Journal of Animal Ecology, 2, 132–178.
Nicholson, A.J. & Bailey, V.A. (1935) The balance of animal populations. Part I. Proceedings of the Zoological Society of London, 3, 551–598.
Odum, E.P. (1971) Fundamentals of Ecology (3rd Edition). W.B. Saunders, Philadelphia, PA, 574 pp.
Odum, E.P. (1983) Basic Ecology. Saunders College, Philadelphia, PA, 613 pp.
Pearson, C.V. & Dyer, L.A. (2006) Trophic diversity in two grassland ecosystems, 11 pp. Journal of Insect Science, 6, 25, available online: insectscience.org/6.25 Journal of Insect Science.
Pierce, W.D., Cushman, R.A., Hood, C.E. & Hunter W.D. (1912) The insect enemies of the cot-ton-boll-weevil. USDA Bureau of Entomology Bulletin, 100, 1–99.
Pimm, S.L. & Lawton, J.H. (1978) On feeding on more than one trophic level. Nature, 275, 542–544.
Polis, G.A. (1991) Complex trophic interactions in deserts: an empirical critique of food-web theory. American Naturalist, 138, 123–155.
Polis, G.A. & Holt, R.D. (1992) Intraguild predation: the dynamics of complex trophic interactions. TREE, 7, 151–154.
Polis, G.A. & Strong, D.R. (1996) Food web complexity and community dynamics. American Naturalist, 147, 813–846.
Polis, G.A., Myers, C.A. & Holt, R.D. (1989) The ecology and evolution of intraguild predation: potential competitors that eat each other. Annual Review of Ecology and Systematics, 20, 297–330.
Price, P.W. (1975) Insect Ecology. Wiley, New York, 514 pp.
Price, P.W. (1984) Insect Ecology (2nd Edition). Wiley, New York.
Price, P.W. (1997) Insect Ecology (3rd Edition). Wiley, New York, 874 pp.
Price, P.W., Bouton, C.E., Gross, P., McPheron, B.A., Thompson, J.N. & Weis, A.E. (1980) Interactions among three trophic levels: Influence of plants on interactions between insect herbivores and natural enemies. Annual Review of Ecology and Systematics, 11, 41–65.
Rhoades, D.F. (1979) Evolution of plant chemical defense against herbivores. In Herbivores: Their Interaction with Secondary Plant Metabolites (eds. G.A. Rosenthal & D.H. Janzen). Academic, New York, pp. 3–54.
Ricklefs, R.E. (1973) Ecology (1st Edition). Chiron Press, Newton, MA.
Ricklefs, R.E. (1990) Ecology (3rd Edition) W.H. Freeman, New York.
Ricklefs, R.E. (1997) The Economy of Nature (4th Edition). W.H. Freeman, New York.
Ricklefs, R.E. & Miller, G.L. (2000) Ecology (4th Edition). W.H. Freeman, New York, 822 pp.
Roitberg, B.D., Gillespie, D.R., Quiring, D.M.J., Alma, C.R., Jenner, W.H., Perry, J., Peterson, J.H., Salomon, M., and VanLaerhoven, S. (2005) The cost of being an omnivore: mandible wear from plant feeding in a true bug. Naturwissenschaften, 92, 431–434.
Rosenheim, J.A. (2001) Source-sink dynamics for a generalist insect predator in a habitat with strong higher-order predation. Ecological Monographs, 71, 93–116.
Rosenheim, J.A. & Corbett, A. (2003) Omnivory and the indeterminacy of predator function: Can a knowledge of foraging behavior help? Ecology, 84, 2538–2548.
Rosenheim, J.A. & Harmon, J.P. (2006) The influence of intraguild predation on the suppression of a shared prey population: an empirical reassessment. In Trophic and Guild Interactions in Biological Control (eds. J. Brodeur & G. Boivin), pp.1–20. Springer, Dordrecht.
Sabelis, M.W. (1990) How to analyze prey preference when prey density varies? A new method to discriminate between effects of gut fullness and prey type composition. Oecologia, 82, 289–298.
Sabelis, M.W. & van Rijn, P.C.J. (2006) When does alternative food promote biological pest control? IOBC/wprs Bulletin, 29, 195–200.
Schaefer, C.W. (1997) The origin of secondary carnivory from herbivory in Heteroptera (Hemiptera). In Ecology and Evolution of Plant-Feeding Insects in Natural and Man-Made environments (ed. A. Raman), pp. 229–239. International Science Publications, New Delhi.

Schowalter, T.D. (2006) Insect Ecology: An Ecosystem Approach (2nd Edition). Elsevier, Amsterdam.
Schuh, R.T. (1986) The influence of cladistics on Heteropteran classification. Annual Review of Entomology, 31, 67–93.
Schultz, J.C. (1983) Impact of variable plant chemical defenses on insect susceptibility to parasites, predators and diseases. In Plant Resistance to Insects (ed. P.A. Hedin), pp. 37–55, American Chemical Society, Washington, DC.
Shaltiel, L. & Coll, M. (2004) Reduction of pear psylla damage by the predatory bug *Anthocoris nemoralis* (Heteroptera: Anthocoridae): the importance of orchard colonization time and neighboring vegetation. Biocontrol Science and Technology, 14, 811–821.
Singer, M.S. & Bernays, E.A. (2003) Understanding omnivory needs a behavioral perspective. Ecology, 84, 2532–2537.
Sinia, A., Roitberg, B., McGregor, R.R., & Gillespie, D.R. (2004) Prey feeding increases water stress in the omnivorous predator *Dicyphus hesperus*. Entomologia Experimentalis et Applicata, 110, 243–248.
Smith, R.L. (1974) Ecology and Field Biology (2nd Edition). Harper & Row, New York, 850 pp.
Speight, M.R., Hunter, M.D., Watt, A.D. & Southwood, R. (1999) Ecology of Insects: Concepts and Applications. Wiley-Blackwell, New York, 360 pp.
Strong, D.R., Lawton, J.H. & Southwood, R. (1984) Insects on Plants: Community Patterns and Mechanisms. Harvard University Press, Cambridge, 313 pp.
Sweet, M.H. (1979) On the original feeding habits of the Hemiptera (Insecta). Annals of the Entomological Society of America, 72, 575–579.
Tilman, D. (1982) Resource Competition and Community Structure. Princeton University Press, Princeton, NJ.
Townsend, C.R., Harper, M. & Harper, J.L. (2003) Essentials of Ecology (2nd edition). Blackwell, Oxford, 554 pp.
van Emden, H.F. (1965) The role of uncultivated land in the biology of crop pests and beneficial insects. Scientific Horticulture, 17, 121–136.
Van Laerhoven, S.L., Gillespie, D.R. & Roitberg, B.D. (2006) Patch retention time in an omnivore, *Dicyphus hesperus* is dependent on both host plant and prey type. Journal of Insect Behavior, 19, 613–621.
van Rijn, P.C.J., van Houten, Y.M. & Sabelis, M.A. (2002) How plants benefit from providing food to predators even when it is also edible to herbivores. Ecology, 83, 2664–2679.
Vázquez, D.P. (2005) Exploring the relationship between niche breadth and invasion success. In Conceptual Ecology and Invasions Biology, (eds. M.W. Cadotte, S.M. McMahon and T. Fukami), pp. 317–332. Springer. Dordrecht, The Netherlands, 505 p.
Volterra, V. (1926) Variazioni e fluttuazioni del numero d'individui in specie animali conviventi. Mem. Acad. Lincei Roma, 2, 31–113.
Wäckers, F.L., van Rijn, P.C.J. & Bruin, J. (eds.) (2005) Plant-Provided Food for Carnivorous Insects: A Protective Mutualism and its Applications. Cambridge University Press, Cambridge, 356 pp.
Wheeler, A.G. (2001) Biology of the Plant Bugs (Heteroptera: Miridae). Cornell University Press, Ithaca, New York.

Preface

A growing proportion of the research devoted to top-down regulation of herbivore communities has been devoted to the complexity that is present in nearly all food webs. Much attention has been given to intraguild interactions among predators and parasitoids, and the importance of omnivory by these same organisms to both intra- and interguild interactions is too often ignored or trivialized. The current state of knowledge regarding the use of non-prey foods by natural enemies is diffuse. An extensive body of literature exists for glucophagy by parasitoids, and granivory by omnivorous epigeal predators (ants and carabids) is also well developed although less well synthesized. Pollinivory and mycophagy, although occasionally discussed within the context of natural enemy nutritional ecology, is relatively poorly understood. This book represents the first attempt to congeal these disparate sources of the literature to illustrate just how pervasive omnivory is within higher trophic levels, and to highlight the evolutionary interactions that have helped to shape both entomophagous arthropods and the non-prey foods themselves.

The book arose from a series of rejections. Initially, an extensive (perhaps too extensive) review of pollinivory by natural enemies was prepared (what essentially became Section II), and I tried shopping it around to several journals. Both my youth and the page limitations of the journals contributed to several rejections. I contacted Heikki Hokannen (then editor-in-chief of BioControl), who again explained that the article was simply too long. But rather than leaving it there, he presented that if I wanted to expand the idea, he would consider it as a title for the Progress in Biological Control series that Springer had entrusted him with editing. After some negotiations, I committed to a 250 page expansion of the review article, but in completing the first 250 pages, I realized that the book was only half finished. Although a bit later than hoped, the current volume is what eventually materialized, thanks to the patience of all those involved.

The goal of the book was to summarize much of the literature pertaining to the fascinating relationships that natural enemies have with non-prey foods. In initial preparations, it became very clear that to include omnivory on vegetative plant tissues would become grossly unmanageable, and so I decided to focus on fungi and microorganisms, sugar sources, pollen, and seeds as the major food categories addressed by the book. At every opportunity, I have attempted to give credit to my forebears that have provided synthesis to particular topics dealt with in this book as

I attempted to provide broader synthesis of the topic of omnivory and the influences of non-prey foods on the dynamics and ecological functions of higher trophic levels. It is my hope that this book will inspire additional research on this topic, and will help to elevate the level of attention that is given to omnivory by naturalists and applied scientists alike.

Brookings, South Dakota Jonathan G. Lundgren

Acknowledgements

This was a large project, and it could not have been completed without the help and patience of many people. First, many evenings and weekends over the past 3 years have been devoted to this book at the expense of time with my family, so I offer my greatest thanks to my wife (Jenna) and children (Gabrielle and Ian) for their sacrifice. Next, I am extremely indebted to Heikki Hokkanen for giving me the opportunity to prepare this volume for the Progress in Biological Control series. These opportunities do not often come to young scientists, and I appreciate his faith in my abilities.

For their extremely helpful and thought-provoking comments on drafts of selected portions of this manuscript, my thanks go out to Mickey Eubanks, George Heimpel, Heikki Hokkanen, Pavel Saska, Paul van Rijn, Felix Wäckers, Don Weber, Paula Westerman, and Livy Williams. Moshe Coll provides a particularly thought-inspiring foreword on the current state of knowledge and possible futures looming for omnivory and biological control.

The species names discussed in the book were extracted from literature dating back 130 years, and many have been revised. Many thanks to the following systematists and databases that were used to provide the most update nomenclature for the discussed taxa.

Sternorrhyncha: Colin Favret (SEL, Beltsville, MD) and the Scalenet website (http://www.sel.barc.usda.gov/scalenet/query.htm);

Plants: Gary Larson (SD State University) and the Plants Database maintained by the USDA (http://plants.usda.gov/);

Parasitoid Hymenoptera: Jim Whitfield (University of Illinois, Urbana, IL) and the Nomina Insecta Nearctica database: (http://www.nearctica.com/nomina/main.htm);

Phytoseiidae: Hans Klompen (Ohio State University) and De Moraes, G. J., McMurtry, J. A., Denmark, H. A. Campos, C. B. 2004. A revised catalogue of the mite family Phytoseiidae. Zootaxa, 434, 1–494;

Diptera: Evenhuis, N. L., Pape, T., Pont, A. C. & Thompson, F. C. (eds.). 2008. Biosystematic Database of World Diptera. (http://www.diptera.org/biosys.htm). Accessed on 8 June 2008);

Formicidae: Bolton, B., Alpert, G., Ward, P. S., and Naskrecki, P. 2007. Bolton's Catalogue of Ants of the World 1758–2005. Harvard University Press; CD format.

Heteroptera: Dr. Henry Thomas (SEL, USDA-ARS, Beltsville, MD).

Carabidae: Kip Will (University of California, Berkeley, CA) and Lorenz, W. 2005. Systematic List of Extant Ground Beetles of the World, 2nd Edition. Published by the author, Tutzing, Germany. 530 pp.

Neuroptera: Oswald, J. D. (chief editor). Lacewing Digital Library. Lacewing Digital Library module. http://lacewing.tamu.edu/. Accessed on 29 July 2008.

Other species were checked to the best of my ability using various on-line resources.

The USDA-ARS allowed me to devote time and federal resources to researching this book (mention of any proprietary products does not constitute endorsement on behalf of the USDA). Staffs at the libraries of the University of Illinois, South Dakota State University, and the National Agriculture Library provided considerable help to locate rare or old resources. My thanks also goes out to Springer, especially Zuzana Bernhardt, Ilse Hensen, Ineke Ravesloot, and their colleagues who helped with the logistics that advanced this project to its fruition and for their patience when the project's completion dragged on longer than anyone had hoped or anticipated. Finally, this book builds upon the ideas and research put forward by my colleagues (living and dead); I hope that this volume does service unto them and inspires additional research that widens our perspective on how predators and parasitoids function within complex food webs.

Contents

Chapter 1
The Functions of Non-Prey Foods in the Diets of Entomophagous Species

While found in almost every conceivable situation, while our naturalists count the species in their cabinets by the thousands, it would be difficult to point out a single species, the food habits of which we fully understand, when both the larvae and imago state are taken under consideration.

Webster, 1881

For too long the entomologist has centered his attention on the beneficial parasitic stage of the parasite, often failing to appreciate the equal importance of the remainder of its life history.

Wolcott, 1942

Nutrition is at the heart of biology even in its broadest sense. We are what we eat, after all, and the quest for something to eat drives the behavior and anatomy of most organisms. Biologists reinforce this premise, since the ecological placement of an organism is often trophic in nature. Herbivores eat plants and thereby influence vegetational communities; detritivores consume dead material, thereby facilitating nutrient cycling; entomophagous species are key to the top-down regulation of insect communities. These broad trophic designations are important to understanding how food webs work, but the trophic placement of arthropods is seldom so simplistic. Facultative consumption of different types of food outside an organism's normal trophic designation offers flexibility to an organism and partially fuels the evolution of new species, but it also adds a shroud of complexity to traditional understanding of how organisms interact. Indeed, the intricate nutritional ecology of an organism has striking implications for where it ultimately fits in a food web, and ignoring key dietary components of an organism will quickly disrupt the predictability of where, when and how this organism functions within a community.

Applied entomology inherently relies on predicting what insects are going to do and when. Stakeholders want entomologists to provide a reliable solution to their pest problems. Action thresholds that are devised to predict spatio-temporal

J.G. Lundgren, *Relationships of Natural Enemies and Non-Prey Foods*,
Progress in Biological Control 7,

Published by Springer Science+Business Media B.V.

occurrence of pest outbreaks are an example of this; they tell pest managers what to do and when and how to do it. But biology on its best day is a complicated science that frequently defies generalization. It could be argued that the more biologically intensive a solution is for managing pests, the less predictable is the outcome. Still, society calls for environmentally friendly, ecologically minded pest management options that are sustainable, and seeing the societal and environmental benefits, many scientists want to give it to them.

Biological control of pests is one such biologically intensive pest management solution that suffers from a lack of predictability as one of its major hindrances. There are a number of factors responsible when the outcomes of biological control are erratic (i.e., intraguild interactions, spatio-temporal overlap of natural enemies with the pests, etc.). In this book, I plan to focus on one phenomenon that adds complexity to the application of biological control. Specifically, what happens when the natural enemies so quickly defined trophically as predator or parasitoid rely nutritionally on things other than prey. In 1996, Jervis et al. listed several reasons why they felt that non-host diets of parasitoids had been largely overlooked: (1) researchers regard adult feeding habits as peripheral to what they consider as the primary ecological function of these organisms, (2) feeding behavior requires a multidisciplinary approach that involves knowledge of plants, and many aspects of insect taxonomy, ecology, and physiology (I would add insect morphology, chemistry, plant physiology, and microbiology among other disciplines to this list), (3) this type of data is difficult to obtain practically, and (4) it isn't always clear what a parasitoid is really doing in a flower. It is my experience that, aside from some conspicuous exceptions (e.g., adults of *Chrysoperla carnea*, *Coleomegilla maculata*, and syrphids), the omnivorous side of predators are equally neglected and the list of Jervis et al. could easily be extrapolated to natural enemies and non-prey foods in a broader sense. At first glance, it seems the goal of a synthesis on non-prey foods for entomophagous arthropods should be a very simple task. I assure you, it was not. Added to the sheer magnitude of omnivory in species popularly thought of exclusively as entomophagous is the diverse nature of the non-prey foods themselves.

1.1 The Non-Prey Foods of Entomophagous Arthropods

Although best appreciated for their propensity for eating insects, entomophagous arthropods routinely consume a wide array of non-prey foods. Facultative phytophagy, frugivory, mycophagy, granivory, pollinivory, and glucophagy abound in most taxonomic groups of natural enemies. Facultative phytophagy in entomophagous species is a book unto itself, and only the surface of this vast topic is breached in the current volume. This leaves fungi, seeds, pollen, and sugar sources on the palette from which to tell a rich story of omnivory as it relates to biological control.

The first point that needs to be established is that non-prey foods vary not only structurally and nutritionally, but also in their affinity for entomophagous organisms. Non-prey foods diverge substantially from prey in terms of their physical characteristics (size, shape, or viscosity), nutrition, and defensive traits, and these foods often necessitate morphological and physiological adaptations in order for entomophagous species to perceive, collect, and digest them. This being said, those species that have evolved to exploit non-prey foods find a rich resource that in many ways is comparable or even superior to prey. Moreover, the raison d'être of some non-prey foods is specifically to attract entomophagous species (extrafloral nectar [EFN], myrmecochorous seeds, etc.) that can benefit the producing organism in various ways (dispersal, protection from more destructive forms of herbivory, etc.). Other non-prey foods are protected tooth and nail from thieves such as predators or parasitoids. Many of the intricacies of these interactions will be discussed systematically and at length from the perspectives of both the non-prey foods and the entomophagous arthropods. Suffice it to say at this point that the individual characteristics that each non-prey food possesses not only constrain which entomophagous species will exploit it, but also a food's nutritional potential for the entomophagous species that do eat it.

1.2 The Functions Served by Non-Prey Foods

A traditional perception is that predatory and parasitoid arthropods feed on non-prey foods to supplement their diets when prey is scarce or of low quality. This is indeed the case, as will be reiterated throughout this book, but a deeper probe reveals that non-prey foods touch many other life processes critical to the success of natural enemies. In other words, non-prey foods touch so many aspects of the lives of natural enemies as to be a truly inseparable component of many species' physiology and behavior. Critical aspects to natural enemy population dynamics that are sustained by omnivory include reproduction, dispersal, diapause, and demographics. Understanding the spatiotemporal occurrence of when non-prey foods are available and when they are actually consumed by natural enemies helps to set the stage for how natural enemies rely on non-prey foods to promote their fitness.

For non-prey foods to hold any function for the consumer and the consumed, the natural enemy and non-prey food must coincide in space and time. Although non-prey foods as a group are ubiquitously on hand in space and time, individual foods vary widely in their predictability and availability. As discussed in the individual sections of this book, pulses of flowers occur throughout the season, EFN protects maturing tissues from herbivores, honeydew accumulates only after sternorrhynchan populations amass, fungus pervades most habitats all season long, and seeds tend to be most available in agricultural systems during the fall. Thus, although these resources may be patchy in their spatio-temporal distribution, they are fairly foreseeable and once a natural enemy has learned when and where to find non-prey foods, they can rely on them being there year after year. This predictability thereby supports a diverse range of functions that vary depending on the natural enemy in question.

The preponderance of reports of spring peaks in the consumption of non-prey foods (Anderson, 1962b; Banks and Macauley, 1967; Benton and Crump, 1981; Culver and Beattie, 1978; Diaz, 1992; Fauvel, 1974; Hemptinne and Desprets, 1986; Skruhavy, 1959; Nalepa et al., 1992; Nystrand and Granstrom, 2000; Sheldon and MacLeod, 1971; Triltsch, 1999; Villaneve et al., 2005) indicate that this is frequently a largely non-predaceous period in the life histories of natural enemies. Two conflicting processes are underway that lead to this conspicuous trend. First, prey is typically at low diversity and abundance during the spring. Concurrently, as natural enemies emerge from hibernation they require a dependable source of nutrition to promote dispersal and reproductive processes key to this time of the year. Prey being scarce, many natural enemies turn to non-prey foods.

Non-prey foods play a different part in the life histories of natural enemies later in the season. Many natural enemies consume non-prey foods during the summer and fall, especially granivorous entomophages (Anderson, 1982; Bebawi and Campbell, 2004; Cardina et al., 1996; Honek et al., 2003; Triltsch, 1999). These resources are devoted to other processes such as maintaining natural enemies during quiescent summer stages, providing fuel for overwintering, and allowing newly enclosed adults to disperse.

1.2.1 Dispersal

Flight in insects requires a ready source of fuel (especially carbohydrates) (Stoffolano, 1995), and non-prey foods fit the bill nicely for many species of natural enemies. Many predators are prey-limited during early spring, and omnivory plays an important role in providing nutrition critical for spring migrations in these species (Ewing, 1913). Coccinellids undergo migrations after diapause, and provision and quality of food often affects the duration of these migratory flights (Nedved et al., 2001; Rankin and Rankin, 1980). Pollen from early spring flowers, such as *Taraxacum officinale*, is critical for the development of flight potential in populations of *Coleomegilla maculata* as they awake from hibernation (Solbreck, 1974). Also, post-eclosion flights of *Coccinella septempunctata* are fueled by pollen and fungal spores in the field (Ricci et al., 2005).

Non-prey foods are rich in carbohydrates, lipids, and proteins, which are all important to the initiation and maintenance of flight in natural enemies. In parasitoids, mono- and oligosaccharides vary in their influence on the total distance flown, number of flights, and longest single flight, and males and females may react differently to these sugars (Wanner et al., 2006). Sugars are also important to flight initiation in parasitoids (Forsse et al., 1992; Hausmann et al., 2005), and sugars vary in their ability to promote flight recovery when parasitoids are exhausted (Hausmann et al., 2005). Parasitoids are quite mobile, and sugar-feeding can fuel flights as far as 0.75 km (Wanner et al., 2006). In fact, *Diadegma semiclausum* routinely forages 80 m to obtain nectar in the field (Lavendero et al., 2005). Lipids, and proteins that can be used for lipogenesis, are another important set of fuels for flight in insects, and many pollens have high levels of these nutrients that are used to generate flight capabilities (Kammer and Heinrich, 1978; Stanley and Linskins, 1974). Proline is one of the most abundant amino acids in pollen and is important in the initiation of flight in insects. Gut dissections of migrating syrphid adults reveal that they consume pollen before initiating their migration, and this food is likely important in sustaining them on long-distance dispersals (Svensson and Janzon, 1984).

1.2.2 Reproduction

A large body of evidence leads to the conclusion that non-prey foods are an important contribution to the reproductive capacity of many natural enemies. On one hand, this may be because non-prey foods extend the lives of parasitoids and predators, allowing them to locate more hosts/prey and lay more eggs. Indeed, in many parasitoids there is a documented trade-off between searching for hosts and non-prey food, the bottom line being that hungry females search for sugar sources to prolong their lives and satiated females search for hosts to maximize reproductive potential (Siekmann et al., 2004; Takasu and Lewis, 1993) (Fig. 1.1). But non-prey foods also contain critical nutrients that facilitate the sexual maturation of females and ovigenesis. Evidence for this assertion is found indirectly in the sex-specific consumption rates of non-prey foods by males and females, and the temporal occurrence and feeding patterns of natural enemies during reproductive periods. Numerous laboratory studies support the notion that non-prey foods improve fecundity over prey/hosts alone, and sometimes can even support reproduction in the absence of prey altogether.

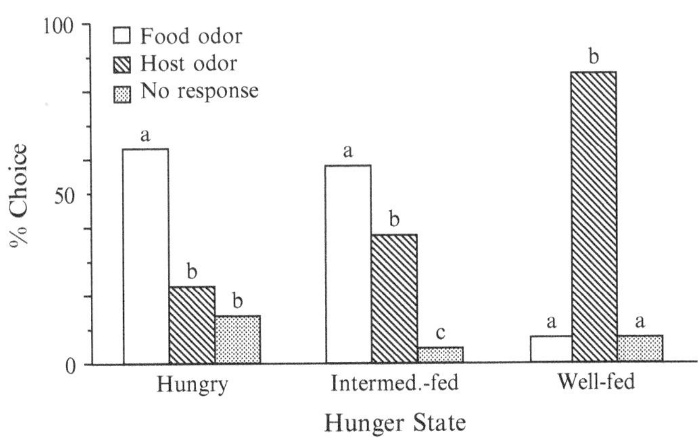

Fig. 1.1 Preference for host- and food-associated odors by *Microplitis croceipes* females without previous ovipositional experience in three different hunger states after preflight experiences with these two different odors. Bars within the same treatment group capped by different letters are significantly different (Reproduced from Takasu and Lewis, 1993. With permission by Elsevier)

1.2.2.1 Sex-Specific Consumption Rates

There is a consistent pattern in entomophagous species that females consume more non-prey foods than male conspecifics. The fact that this statement generally applies most significantly to the more proteinaceous non-prey foods than the sugar sources suggests that these sexual differences may be reflective of the extra or unique nutrition required by females for egg production (Jervis et al., 1996b). The differences in consumption levels is sometimes quite dramatic; females of *C. maculata* consume ten times more pollen than males in the field (Lundgren et al., 2005), and *Chrysoperla* spp. females have an average of 1,100 grains in their stomachs, versus 44 pollen grains in the male guts (Villaneve et al., 2005). In fact, the guts of field-collected male syrphids are sometimes entirely devoid of pollen, even when females of the same species are pollinivorous (Gilbert, 1981; Haslett, 1989). This pattern seems to hold for granivory as well; *Gryllus pennsylvanicus* females consume more seeds and more biomass than males under laboratory conditions (Carmona et al., 1999).

Sex-specific foraging decisions are a first indication of the relative importance of proteinaceous, non-prey foods for females and males. Typically, female predators and parasitoids are more likely to visit flowers than males, and in many cases some level of pollinivory occur during these visits (Fauvel, 1974; Gilbert, 1985a; Jervis et al., 1993). As will be discussed in the glucophagy section, honeydew and EFN are much more accessible sugar sources than floral nectar. Thus not surprisingly, a disproportionate number of females of the tachinid *Senometopia pollinosa* is found on oak trees, apart from their pine-dwelling hosts, presumably feeding on honeydew during their preoviposition period (Herrebout, 1967). Also, female chrysopids (five of six species surveyed) were more attracted to olive orchards sprayed with artificial honeydew than male chrysopids Liber and Niccoli (1988). A conspicuous exception to this pattern is in the spider, *Misumenoides formosipes* (Pollard et al., 1995). In this species, males are the only sex to consistently visit flowers where they drink nectar. But males are also 20 times smaller than females, and are more prone to desiccation; drinking floral nectar in this species reduces this deadly process.

In some species, the stronger reliance on proteinaceous non-prey foods by females manifests itself in sexually dimorphic anatomy and physiology. Some male syrphids have longer proboscises than females, which likely relates to stronger reliance on nectar than pollen (Gilbert, 1985b). Sexual differences in the levels of salivary enzymes used to digest plant-derived compounds (e. g. pectinases) have been noted in omnivorous mirids (Miles, 1972). Also, the females of many chrysopids have expanded tracheation associated with the digestive system than do males; these tracheae help to oxygenate endosymbiotic yeasts that may facilitate a glucophagous and pollinivorous lifestyle (Canard, 2001, Hagen, 1987; see also Chapter 15 in this volume). What is unclear from all of this work is whether the pattern of increased consumption of proteinaceous non-prey food is reflective of the increased energetic needs of the female, or the female-specific requirement for particular nutrients present only in the non-prey food sources.

1.2.2.2 Maintaining Reproductive Potential

Non-prey foods can have important implications for reproduction in natural enemies even when these foods are not nutritionally sufficient to support ovigenesis on their own. Non-prey foods are able to meet many of the metabolic needs of the female, and consequently the females are "primed" for reproduction when foods more suitable for reproduction are encountered. For instance, pollinivory in the spring is important to sexual maturation in many insect species, such as *Chrysoperla carnea* (Sheldon and MacLeod, 1971), anthocorids, and various coccinellids. The coccinellids *Adalia bipunctata* and *Propylaea quatuordecimpunctata* are commonly encountered on pollinating flowers in the spring (Hemptinne and Desprets, 1986; Hemptinne et al., 1988), and although these beetles are seldom able to mature eggs on a pollen-only diet, the majority of spring females feed on Rosaceae pollen. Hemptinne and Desprets (1986) suggest that spring pollinivory allows these females to maintain their reproductive potential in the absence of prey so that these predators can quickly respond to prey availability by immediately devoting prey-based resources to egg production. Burakowski (1967) states that birch seeds are an important source of food for attaining sexual maturity in adult *Amara pulpani,* and that there is

> *a clear phenological correlation of sexual activity of* [A. pulpani] *with the ripening and falling of* [Betula verrucosa] *seeds mentioned in late summer.*

Consumption of seeds by two omnivorous carabids, *Harpalus affinis* and *H. distinguendus*, peaks during their reproductive period, at least indicating that these species consume more food during reproduction and that seeds are a suitable diet for this nutrient acquisition phase (Honek et al., 2006) (Fig. 1.2).

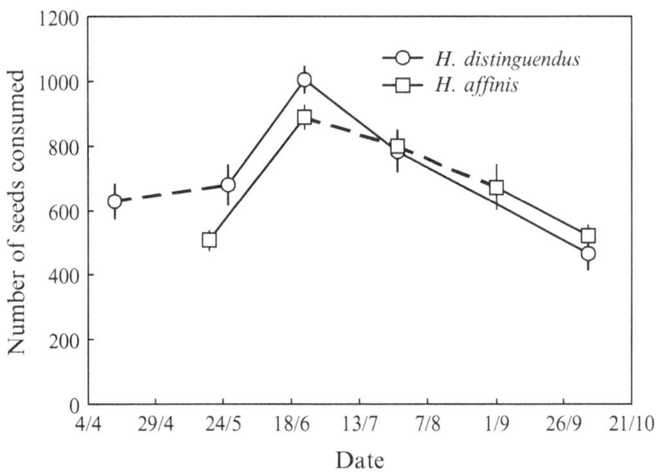

Fig. 1.2 Seasonal variation in the overall seed consumption of *Harpalus distinguendus* (circles) and *Harpalus affinis* (squares). Average number of seeds consumed per replicate (±SE) at different times of a year. Both of these species are spring- and summer-breeders, and this seasonal pattern maximizes seed consumption during peak reproductive periods (Reproduced from Honek et al., 2006. With permission by Blackwell)

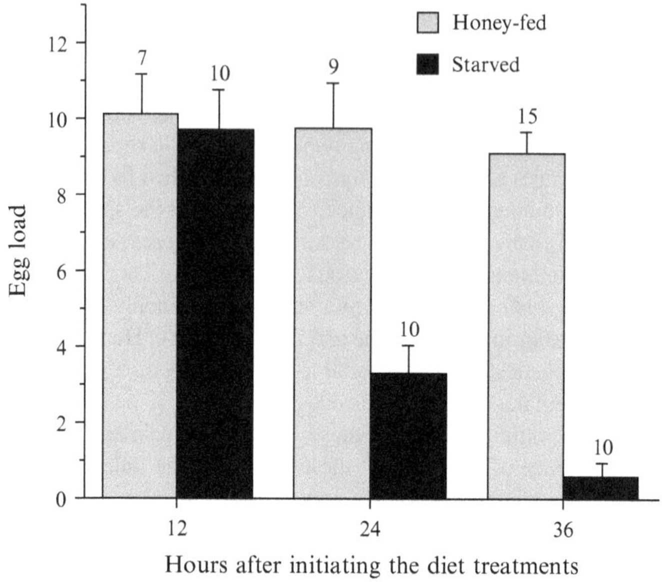

Fig. 1.3 Mean egg loads (S.E.M.) of *Aphytis melinus* females that were either fed honey or starved 12, 24 and 36 h after individuals from the starvation group were deprived of honey (all females were fed honey for the first 2 days post-emergence). Numbers above standard error bars are sample sizes (Reproduced from Heimpel et al., 2007. With permission by Blackwell)

Another instance of non-prey foods maintaining reproductive potential occurs in those predatory mite species which readily feed on nectar, but can't mature eggs fully on sugar-only diets. As soon as proteinaceous foods become available, these nectar-fed mites initiate reproduction more quickly than unfed mites (van Rijn and Tanigoshi, 1999a). Finally, in addition to priming the reproductive system of females, consumption of non-prey foods can reduce or prevent the resorption of viable eggs already present in the female (Quicke, 1997). A good example of this is reported by Heimpel et al. (1997), who show that egg resorption was slowed in honey-fed *Aphytis melinus*, which normally host feeds to achieve maximum fecundity (Fig. 1.3).

1.2.2.3 Reproduction Exclusively on Non-Prey Foods

Some predators can produce eggs on a diet consisting solely of non-prey foods. Sugar sources alone support oviposition in glucophagous adults when nutrients are sequestered by entomophagous larvae. But when prey is entirely absent from their diet, more proteinaceous foods like pollen and seeds are often requisite for reproduction in predaceous species. In many hymenopteran parasitoids and lacewing adults, honey, honeydew, nectar or EFN is sufficient for supporting or improving egg production rates (Burger et al., 2004; England and Evans, 1997; Leatemia et al., 1995; Lee et al., 2004; Rose et al., 2006; Sundby, 1967; Venzon et al., 2006). In part, these glucophagous adults rely on protein and fat reserves accumulated during the larval stage. But when sugarmeals support prolonged reproduction in synovigenic parasitoids, the females begin to produce eggs from their adult nutrition (Olson and Andow, 1998; Schmale et al., 2001; Waage and Ming, 1984). Indeed, feeding on sugar for only 1 day is sometimes sufficient for improving the lifetime fecundity of parasitoids (Hagley and Barber, 1992). Aside from parasitoids and lacewings, most natural enemies require a more nutritious diet to mature eggs.

More proteinaceous foods are necessary for reproduction in other natural enemies. For example, protein is required for reproduction in phytoseiid mites, but they emerge in spring before animal sources of protein are widely available. One of very few protein sources at this time of year is pollen, and *Typhlodromus pyri* matures eggs on a diet of only apple pollen (Chant, 1959). McMurtry and Rodriguez (1987) compiled an entire list of phytoseiids that were able to reproduce better on pollen than on prey. *Orius vicinus* presents another case of an entomophagous species capable of reproducing on pollen-only diets. This species produces eggs in the laboratory when reared on a diet consisting solely of pollen from cherry, pumpkin, or mullein (Fauvel, 1974). In the syrphid, *Episyrphus balteatus*, females are unable to produce eggs when reared on sucrose or honeydew alone, but readily begin oviposition when provided with *Corylus avellana* pollen (Maier, 1978; Schneider, 1948, 1969). Haslett (1989) found that nectar and pollen consumption were inversely proportional in the syrphid, *Rhingia campestris*, and the amount of pollen consumed depended on the stage of reproduction, peaking as the oocytes occupied a greater proportion of the follicle (Fig. 1.4). Some lacewings and parasitoids are able to lay eggs on pollen-only diets in the laboratory as well (Venzon et al., 2006; Zhang et al., 2004). The only coccinellid known to lay eggs when reared from egg to adulthood on pollen is *Coleomegilla maculata* (Lundgren and Wiedenmann, 2004).

Typically, mixtures of non-prey foods support greater reproduction than the individual components alone. Even different foods within the same class (sugar, pollens, seeds, etc.) can support reproduction to varying degrees. Honeydews typically are superior to nectars for lacewings (Hagen and Tassan, 1970), possibly because honeydews contain oviposition stimulants that are not present in other sugar sources. Floral nectars also vary in their suitability for reproduction in hymenopteran parasitoids (Idris and Grafius, 1995) (Fig. 1.5). A commonly reported observation of the benefits of non-prey food mixing is that the combining of pollen and sugarmeals leads to enhanced reproduction over pollen or sugar alone (Geng et al., 2006; Venzon et al., 2006). Sheldon and MacCleod (1971) believe that the starch in pollen is largely inaccessible to non-predaceous lacewing adults, and so they require a carbohydrate source in order to effectively reproduce. Adding yeast-based artificial diets to sugar also sometimes promotes oviposition in lacewings (Sundby, 1967).

Although typically some of the most nutritious of non-prey foods, seeds are not created equally in their ability to support reproduction in natural enemies. Of five seed species tested in the laboratory, *Chenopodium album* allow the

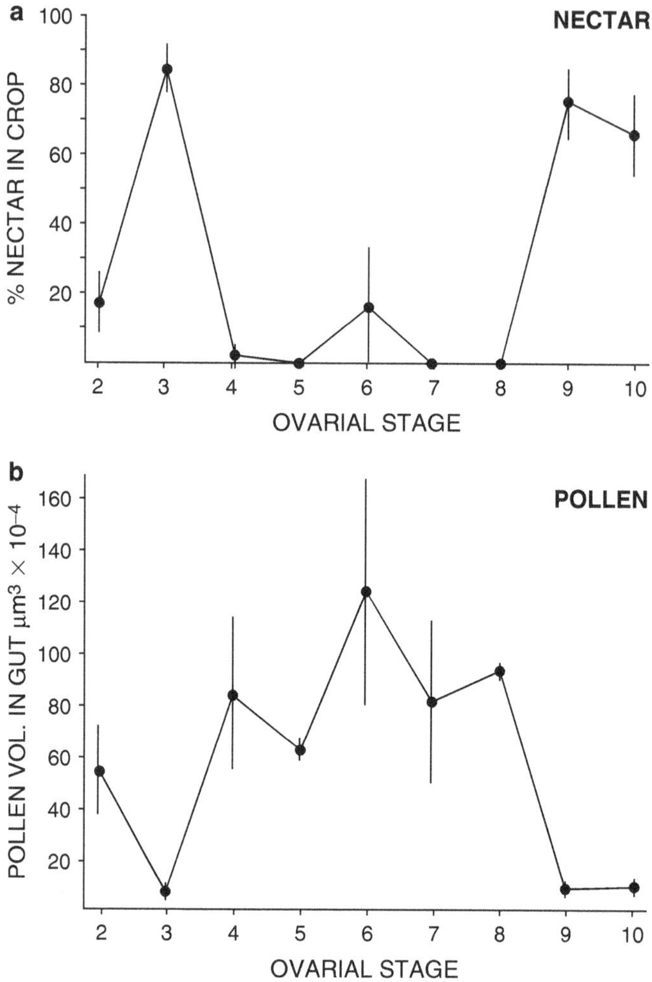

Fig. 1.4 Relationship between ovarial development and diet of adult female *Rhingia campestris*. Points are means from a minimum of four flies. Vertical lines represent one standard error either side of the means. *Stage 1*. The germarium. *Stage 2*. The newly formed follicle, spherical in shape. The oocyte within is not visible. *Stage 3*. The oocyte is distinct from the nurse cells within the slightly ovate follicle, but forms less than 10% of the total follicular volume. *Stage 4*. The oocyte occupies between 10% and 20% of the follicle, which is now distinctly oval in shape. This stage represents the onset of yolk deposition. *Stage 5*. The oocyte (with yolk) occupies 20–30% of the follicle. *Stage 6*. The oocyte occupies 30–50% of the follicle. *Stage 7*. The oocyte occupies 50–75% of the follicle. *Stage 8*. The oocyte occupies up to 90% of the follicle. Yolk deposition is near completion. *Stage 9*. The mature egg, occupying nearly the entire follicle. The egg is more opaque than at earlier stages and is longer and thinner. *Stage 10*. The eggs have been discharged and the ovaries have shrunk and appear rather degenerate (Reproduced from Haslett, 1989. With permission by Springer)

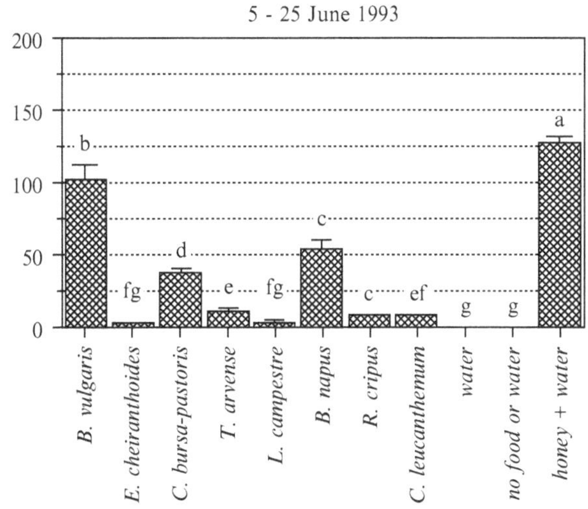

Fig. 1.5 Total fecundity of *Diadegma insulare* females fed on various wildflowers as nectar sources. Columns with different letters are significantly different (Reproduced from Idris and Grafius, 1995. With permission by the Entomological Society of America)

greatest number of eggs to be laid in the laboratory by *Harpalus rufipes* (Briggs, 1965). In, *Amara similata* adults lay more eggs when fed a seed mixture than seed species offered individually (Jorgensen and Toft, 1997b). Of individual seed species, *Poa annua* is less suitable for fecundity than *Taraxacum* sp. and *Tripleurospermum inodorum* seeds.

1.2.2.4 Improving Fecundity over Prey-Only Diets

Although prey is commonly presumed to be an ideal diet for entomophagous species, adding non-prey foods to prey diets often improves fecundity. Eubanks and Styrsky (2005) found that fecundity is improved when predators are fed plant-based foods (especially pollen) in addition to prey more often than not in the literature. This indicates that there are critical nutrients absent or deficient in certain prey items that are necessary for optimal fecundity in natural enemies. Still, many of the studies conducted with predators are done so in the laboratory, and it isn't always clear that the optimal prey is the one tested. Indeed, under more realistic conditions natural enemies probably self-select different prey that optimize their nutrient intake (Greenstone, 1979).

Although glucophagy is at first glance a poor food for reproduction, carbohydrates are a critical nutrient for reproduction and numerous reports indicate that sugar-feeding is crucial for attaining maximum fecundity for a wide range of predators and parasitoids (Brian, 1973; McMurtry and Scriven, 1964b; van Rijn and Tanigoshi, 1999a; Zhimo and McMurtry, 1990). When time limited, honeydew of *Trialeurodes vaporariorum* was equivalent to host feeding for reproduction in *Encarsia formosa* (Burger et al., 2004, but see Burger et al., 2005). Host-feeding without a sugarmeal does not support reproduction in *Aphytis melinus* very well either (Heimpel et al., 1997). Sugar also can improve fecundity of predators provided with low quality prey (a phenomenon likely common in the field). The aphidophagous ladybeetles, *Coccinella septempunctata* and *C. transversoguttata richardsoni*, are unable to produce eggs when reared on alfalfa weevil larvae alone. However, when this diet is supplemented with sucrose, the ladybeetles lay eggs, although at lower rates than aphid-fed beetles (Richards and Evans, 1998). All this is to say that sugar limitation under field conditions can have important implications for the reproductive capacity of natural enemies.

Adding more nutritious non-prey foods like seeds or pollen to a diet consisting solely of prey also frequently improves the fecundity of entomophagous arthropods. Parasitoids often require a source of lipids and proteins to maintain survivorship, and in the case of synovigenic parasitoids, non-prey nutritional resources may be important for maturing eggs (Heimpel and Collier, 1996; Jervis et al., 1996b). For instance, including pollen of *Pinus sylvestris* to the diet of the host-feeding parasitoids, *Scambus buolianae* and *Itoplectis conquisitor*, improves fecundity over those wasps fed only with host-fluids (Leius, 1961a, b). The omnivorous carabids, *Harpalus rufipes* and *Amara similata* produce only a minimum of eggs on a prey only diet, and fecundity is drastically improved when a seed mixture is included (Jorgensen and Toft, 1997a, b). Although pollen on its own is not ideal for reproduction, adding pollen to the diet of prey-fed *Orius* females increases their oviposition by 40% (Cocuzza et al., 1997). *Euseius sojaensis* reproduces poorly when fed only the poor-quality prey, *Tertranychus kansawai*, but oviposition is increased dramatically when *Camellia sinensis* pollen is added to the diets of female mites (Osakabe et al., 1986).

1.2.3 Other Roles of Non-Prey Foods in Natural Enemy Ecology

1.2.3.1 Maternal Diet and Progeny Fitness

A few reports indicate that the relative quality of non-prey foods for adult females has lasting effects on their offspring. When females of the hymenopteran parasitoid *Aptesis basizonia* consume honey, their progeny develop faster and survive better than when mothers are fed raisins (Finlayson and Finlayson, 1957). This is also true for *Cryptus inornatus*, and in this case larvae from females fed raisins are more likely to enter a state of dormancy than the progeny from honey-fed wasps (Simmonds, 1948). Whether the larvae are using maternal nutrient provisions as a guage to assess the availability of resources for their adult stage remains to be tested.

A complication in interpreting the relative effects of non-prey foods on reproduction comes when nutrient deprivation in the mother improves the fitness of her offspring. In some carabids, nutrient limitation in the adults results in lower fecundity, but the eggs that are laid are larger and more fit than those produced by well-fed mothers (Wallin et al., 1992). This phenomenon takes on meaning when examining the case of the carabid, *Amara similata*. The larvae of females who are fed on individual seed species (*Poa annua* or *Tripleurospermum inodorum*) survive better than progeny of females reared on mixed seed species. This is in spite of both larvae and adults surviving better on a diet of mixed seeds. Jorgensen and Toft (1997b) explain these results by suggesting that the nutrient deprived females may lay fewer eggs of higher fitness than females receiving a superior diet. Clearly, maternal inheritance of nutrients is important to some natural enemies, and omnivory in these species is found in the thick of these interactions.

1.2.3.2 Sex Ratio

Mothers fed foods of varying quality produce different sex ratios in their progeny, a consequence of importance to the dynamics of a natural enemy population (Heimpel and Lundgren, 2000). In part, this stems from the fact that many parasitoids lay eggs in a sexual sequence, often laying either more females or males early in reproduction, depending on the species in question. By providing sugar meals to parasitoids, and thus prolonging their lives, a much different sex ratio is realized than in wasps that are starved or fed low quality foods (Olson and Andow, 1998; Simmonds, 1948; Suzuki et al., 1984; Waage and Ming, 1984; Berndt and Wratten, 2005).

Ants also display diet-dependent variability in their sex ratios, a phenomenon well described for myrmechorous species. Resource-stressed ant colonies tend to produce more male offspring, since more nutrition is required to produce workers and queens. In at least two studies, when seed-associated food bodies become abundant, ant colonies produce more queens. In *Aphaenogaster rudis*, larvae destined to be gynes receive at least five times the weight of elaisosomes than larvae destined to be workers or males (Bono and Heithaus, 2002). Thus, the availability of myrmecochorous seeds alters the final sex ratio of progeny. In another study, five times more queens are produced during a pulse of *Sanguinaria canadensis* seeds, and these queens are significantly larger than those from colonies fed other species of seeds (Morales and Heithaus, 1998).

1.2.3.3 Diapause

Overwintering and quiescence are two important processes in the natural history of many insects, and omnivory (especially on pollen) is shown to promote these processes under several circumstances. First, Michaud and Qureshi (2006) show that pollen and *Helianthus annuus* EFN are able to maintain the survival of the coccinellid *Hippodamia convergens* during summer reproductive diapause better than several prey species. They also note that prey scarcity often triggers reproductive diapause of this nature in predatory species, and point out that non-prey foods are likely of widespread importance in sustaining predators during these periods of quiescence.

Food availability is also an important cue in regulating overwintering in many entomophagous insects (Anderson and Hale, 1986). Changes in the abundance of prey and other food resources have been shown to influence how insectivores manage energy budgets and accumulate reserves for diapause and overwintering. Predormancy fat reserves in *Apolinus lividigaster* and *Ileis galbula* are derived from pollens of various plants, especially *Bidens pilosa* (Anderson, 1981, 1982). Fat body reserves in *Coccinella septempunctata* are increased as a result of pollen feeding in the laboratory (Hodek and Honěk, 1996), and *C. septempunctata* relies on pollen most heavily directly before undergoing diapause (Triltsch, 1997). Anderson's (1981) research shows that when *A. lividogaster* feeds on aphids, energy is routed into ovigenesis; fat for dormancy is accumulated when the beetles consume alternative foods, namely pollen. Thus in this case, pollen consumption may actually be a cue for entering dormancy. In the case of *Euseius hibisci,* augmenting pollen densities in California citrus leads to higher overwintering densities relative to untreated plots (Kennett et al., 1979).

Certain nutrients contained in some pollens may also be required for dormancy or diapause in predaceous arthropods. Research has shown that carotenoid pigments are essential for diapause induction in phytoseiid mites, and that different pollens, presumably differing in levels of vitamin A, have differential effects on diapause induction in *Amblyseius potentillae* (Overmeer and van Zon, 1983; Veerman et al., 1983). Feeding mites pollen containing vitamin A, namely from *Mesembryanthemum criniflorum*, is able to restore the diapause capability of predators reared on carotenoid-deficient prey (Dicke et al., 1986). Phytoseiids are eyeless, and therefore it is hypothesized that this group requires extra-retinal photoreception to induce diapause, and carotenoid pigments are thought to play a role here.

1.3 Closing the Introduction; Opening the Rest of the Book

The preceding several pages set the stage for the rest of the book- non-prey foods impinge on nearly every aspect of natural enemy biology. But the literature on the full extent of this topic is often idiosyncratic and without a synthesis on the topic it is difficult to see the bigger picture. When we examine the dietary breadth of entomophagous species as a group, the uniform adaptations to consuming specific non-prey foods that have occurred across broad taxonomic groups is enlightening. Moreover, the influence of non-prey foods on the behavior and physiology of entomophagous species has direct implications to the incorporation of biological control into reliable, biologically intensive IPM systems. Another point that I hope will be driven home are the tremendous contributions, many of which are not widely known, that have been made by our scientific forefathers.

In an unhealthy research climate where only the most recent articles published in the top-ranked journals are cited, it is easy to lose our sense of where we have come from. Peppered throughout the book are quotations from significant contributors to the field of omnivory in biological control from the past several hundred years; these notes serve to illustrate how this field has changed and to indicate where we need to go in order to move forward.

Obviously, each natural enemy is its own case and not all non-prey foods fit the nutritional bill for every entomophagous species. To truly understand the nature of the interactions of natural enemies with non-prey foods, we need to appreciate the extent to which entomophagous taxa consume which non-prey foods, and we need to explore the motivations and natural histories of the non-prey foods themselves.

Section I
Glucophagy

The world is much sweeter than we think it is. Nectar from floral and extrafloral (EFN) sources pervades the plant kingdom, and if nectar seems localized or sheltered, then honeydew is splattered over many of the flat surfaces in the world. To give an idea of the availability of sugar in the environment, it has been estimated that alfalfa fields can yield up to 190 l of nectar per ha per day (but only 1.4 and 2.4 l per day in cantaloupe and cotton fields, respectively) (Butler et al., 1972; McGregor and Todd, 1952). While sugar undeniably constitutes the major nutritional component in nectar and honeydew, other nutrients (e.g., amino acids, water) can serve important functional roles in nectar- or honeydew-mediated interactions. Except for extreme examples, all entomophagous arthropods will accept a sugar meal if encountered, and many have evolved to rely on these sugar sources as an important component of their diet.

> *The suitability of a carbohydrate solution...as an insect food source depends on its detectability, its accessability, and the efficacy with which the sugars can be converted into energy.* (Wäckers, 2000)

Nectar and honeydew sources can vary widely with regard to their carbohydrate composition. As Wäckers pointed out, for a sugar to be a viable food source, it must be detectable and elicit a gustatory response, both processes involving receptor cells. The breadth of sugars a natural enemy can use for its metabolic needs depends on their ability to perceive these different molecules, a relationship best understood at the behavioral level in the parasitoid Hymenoptera (Hausmann et al., 2005; Wäckers, 1994, 1999), and ants (Bluthgen and Fiedler, 2004). First, not all sugars excite the sugar receptor cells of natural enemies. For example, the receptor cells of the syrphid, *Eristalis tenax*, respond most strongly to sucrose and glucose. Fructose, maltose and galactose are less detectable by the flies, and rhamnose, raffinose, and stachyose have no effect on the receptors of this fly (Wacht et al., 2000). In testing the ability of 14 natural sugars to elicit a feeding response in *Cotesia glomerata*, Wäckers, (1999, 2001) found that only eight of the sugars are phagostimulatory, with fructose having the strongest effect on the parasitoids (Fig. I.1). Beach et al. (2003) and Williams and Roane (2007) evaluated the ability of 15 sugars to elicit a gustatory response in *Anaphes iole*. One group of sugars elicited a relatively low frequency of response (around 50% of individuals or less), and another group elicited a response in more than 90% of

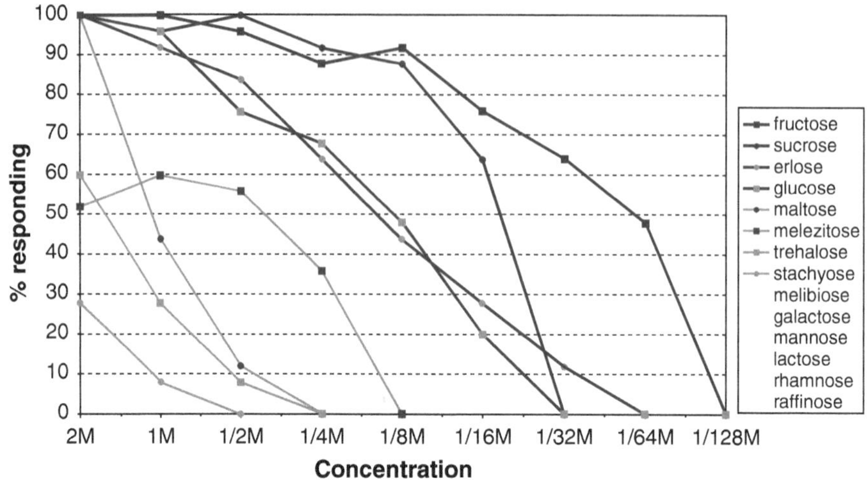

Fig. I.1 Percentage of food deprived *Cotesia glomerata* responding to sugar solutions of various concentrations. Individual cohorts of 25 parasitoids were used for each sugar and each concentration. Two molar solutions of melibiose, galactose, mannose, lactose, rhamnose, and raffinose failed to elicit a feeding response (Reproduced from Wäckers, 1999. With permission by Springer)

individuals tested (this group is comprised of sucrose, fructose, glucose, maltose, melizitose, and trehalulose) (Beach et al., 2003). Glucose invokes the greatest response rate by this wasp, eliciting a gustatory response even at a concentration as low as 4 mM. The relative digestibility of sugars also limits which can be used by natural enemies for their metabolic needs. Oligosaccharides must be enzymatically degraded into monosaccharides, invertase being the enzyme used to hydrolyze sucrose. This enzyme is found in the digestive systems of some natural enemies, but its presence has not been widely explored. Were it absent from a species, one might expect these natural enemies to prefer or specialize on sugars rich in monosaccharides.

The viscosity and concentration of a sugar solution also influences which natural enemies can partake of it (Bartlett, 1962; Brian, 1973; Jervis and Heimpel, 2005; Kingsolver and Daniel, 1995; Stapel et al., 1997). Put quite simply, dried sugar solutions are less accessible to some natural enemies, and necessitate behavioral and physiological adaptations for consumption. Still, many parasitoids and predators are observed to feed on crystallized sugars or dried sugar sources (Bartlett, 1962; Wäckers, 2000). Some dissolve dried sugars in saliva and suck up the solution (Bartlett, 1962; Gilbert, 1981), others scrape the sugars with mandibles and consume them dry (Bartlett, 1962). The bottom line is that desiccated sugar solutions are still appealing food sources. An extreme case in the palatability of aged sugar is with *Chrysoperla carnea*, which is observed to feed in the spring on fall-deposited, dried honeydew on oak leaves (Sheldon and MacLeod, 1971). The concentration of sugar in a solution also influences the response rate of natural enemies (Sutherland et al., 2001), as well as their survival (Azzouz et al., 2004). Ants have repeatedly been shown to prefer more concentrated sugar sources under choice conditions, and may even be able to communicate the concentration of a sugar source to their nestmates (Bluthgen and Fiedler, 2004; Cassill, 2003; Tschinkel, 2006). From the plant's point of view, the sugar complement and floral architecture may actually have evolved to influence the viscosity of nectar to suit the function of the sugar source. Deep flowers help to reduce evaporation of nectar, making it more attractive to a wider array of potentially beneficial insect visitors. But the more complex sugar complement (beyond simply hexoses) of honeydew may encourage drying (Baker, 1975; Baker and Baker, 1983), thereby reducing its attractiveness to insects, many of which are hoping to eat the source of the honeydew. This hypothesis for how sugars may function in evaporation rates and the relative attractiveness to natural enemies merits further attention from ecologists (Wäckers, 2000).

I.1 Interclass Differences in Sugar Sources

Descriptions of the different sugar sources treated in this section are established at the beginning of each chapter, but some preliminary discussion of terms is warranted. First, 'nectary' is more of a functional term than a clearly defined anatomical structure. The location, structure, and physiology of nectaries vary widely, but all are related in that they exude a sugary solution (Pacini et al., 2003; Schmid, 1988). Extrafloral nectaries were likely the first type to evolve, and are distinguished from floral nectaries in that the latter function in the pollination of the plant, a distinction first proposed by Delpino in 1968 (Fahn, 1988, 2000). Another sugar source distinguished from nectaries in the botanical literature is leaf exudate. Leaf veins of cassava (*Manihot utilissima*) and other plants exude sugar-rich solutions, but no cellular differentiation in tissues associated with these exudates is apparent (Bakker and Klein, 1992; Pereira and Splittstoesser, 1987; Sadasivam, 1970). Still, the arguments that these leaf exudates are separate from nectaries are somewhat unsatisfying, and I do not treat them separately in this section of the book.

The abundance, dispersion, volume, and degree of accessibility all affect the quality of sugar sources (Wäckers, 2005; Jervis and Heimpel, 2005), and the relative suitability of nectars and honeydew as resources for specific natural enemies has been explored repeatedly over the years. Generally speaking, floral nectar has the most restricted breadth of sugars available to visitors, with EFNs being intermediate in nutritional breadth and honeydew having the greatest diversity of sugars and other nutrients (Table I.1). Because of its more diverse nutritional profile, Hagen (1962) proposes that honeydew is a superior food source to nectar for natural enemies. Further support for this notion may stem from the fact that natural enemies have had much longer to adapt to feeding on honeydew compared to nectar (likely 100–200 million years); sternorrhynchans evolved during the Permian, whereas flowering plants showed up in the Cretaceous Period (Downes and Dahlem, 1987). In fact, many natural enemies are indeed often more attracted to honeydew:

Table I.1 Sugar contents of floral nectar, extrafloral nectar, and honeydews reported in the literature. ('?' indicates that the sugar reported in one study is specifically questioned in other research)

	Floral nectar	Extrafloral nectar	Honeydew
Arabinose	X[31]	X[4, 5, 7, 18]	–
Bemisiose	–	–	X[6, 19]
Erlose	–	–	X[12, 13, 17, 21, 34]
Fructose	X[8, 9, 14, 15, 23, 24, 27, 28, 29, 30, 32, 33, 35]	X[2, 4, 5, 7, 11, 18, 33]	X[6, 10, 12, 13, 16, 17, 21, 22, 26, 34]
Galactose	X[33]	–	X[21]
Gentiobiose	–	X[3]	–
Glucose	X[8, 9, 14, 23, 25, 27, 28, 29, 31, 32, 33, 35]	X[2, 4, 5, 7, 11, 18, 33]	X[6, 10, 12, 13, 16, 17, 21, 22, 26, 35]
Lactose	X[29]	X[3]	–
Maltose	X[28, 29?, 35]	X[3, 5]	X[21, 22, 13]
Mannitol	X[29]	X[18]	–
Mannose	X[29]	–	X[21]
Melibiose	X[8, 28, 29, 35]	X[3, 24]	X[21]
Melezitose	X[1?]	X[3]	X[6, 10, 12, 13, 20, 21, 26, 34]
Raffinose	X[28, 33, 35]	X[2, 4, 5, 24]	X[6, 21]
Rhamnose	–	X[4]	X[21]
Ribose	X[29]	X[9]	–
Saccharose	X[15]	–	X[21]
Sorbose	X[29]	–	–
Stachyose	X[29]	X[3]	X[6, 21]
Sucrose	X[8, 14, 23, 25, 27, 28, 29, 30, 31, 33, 35]	X[2, 4, 5, 7, 11, 18, 33]	X[6, 10, 12, 13, 16, 17, 20, 22, 26, 34]
Trehalose	X[29]	–	X[13, 21, 34]
Trehalulose	–	–	X[6, 10, 20]
Turanose	–	–	X[20, 21]
Xylose	X[27]	X[4, 5]	–

[1]Baker and Baker (1983)
[2]Baskin and Bliss (1969)
[3]Beattie (1985)
[4]Bentley (1977a)
[5]Bowden (1970)
[6]Byrne et al. (2003)
[7]Caldwell and Gerhardt (1986)
[8]Churchill and Christensen (1970)
[9]Clark and Lukefahr (1956)
[10]Costa et al. (1999)
[11]Elias and Gelband (1975)
[12]Ewart and Metcalf (1956)
[13]Fischer & Shingleton (2001)
[14]Galletto and Bernardello (2004)
[15]Gottsberger et al. (1984)
[16]Gray (1952)
[17]Gray and Fraenkel (1954)
[18]Heil et al. (2000)
[19]Hendrix and Wei (1994)
[20]Hendrix et al. (1992)
[21]Holldobler and Wilson (1990)
[22]Irvin et al. (2007)
[23]Kaczorowski et al. (2005)
[24]Keeler (1977)
[25]Lanza et al. (1995)
[26]Mittler (1958a)
[27]Nicolson and Van Wyck (1998)
[28]Percival (1961)
[29]Petanidou (2005)
[30]Stiles and Freeman (1993)
[31]Torres and Galetto (2002)
[32]Wäckers et al. (2001)
[33]Watt et al. (1974)
[34]Woodring et al. (2006)
[35]Wykes (1952)

> *Whether it be on account of its greater quantity, or because it possesses a more agreeable flavor, I am unable to say, but the honeydew of aphides is far more attractive to all of the ants observed than the nectar from extrafloral glands* (Trelease, 1881).

But aside from its attractiveness, honeydew has repeatedly been shown to be of lower nutritional suitability to at least some natural enemies compared with nectars (Avidov et al., 1970; Lee et al., 2004; Leius, 1961a, b; McMurtry and Scriven, 1965; Nomikou et al., 2003b; Wäckers, 2001). But direct comparisons between these different sugar sources tend to involve single representatives of each food type, and so the real power of these comparisons on a whole is quite low. As will be discussed at length in subsequent chapters, the intraclass nutrition of nectars and honeydews is tremendously heterogeneous, and can measurably influence natural enemy performance (Hagen and Tassan, 1972; Nomikou et al., 2003b). Even nectars from different organs on the same plant can vary nutritionally and attract different suites of visitors (Baker et al., 1978; Elias and Gelband, 1975; Keeler, 1977, 1980, 1981; Koptur, 1994; Percival, 1961). All of these limitations make generalizing, based on nutrition, which is the 'ideal sugar source' for natural enemies difficult.

The origin and function of the different sugar sources has important implications from a plant's perspective. Nectar flow and nutrition is managed by the plant, and while nectar production comes at a cost in terms of production, the plant manipulates its production according to its needs. Honeydew can be considered phloem thieving and is quite another matter. Sternorrhyncha take what phloem they need from the plant, and change it to suit their needs, often at the expense of the plant's fitness. Indeed, Sternorrhyncha-tending ants will actually move these plant parasites to optimal feeding areas on the plant to maximize honeydew production. Still, as will be seen in subsequent chapters, plants can benefit from even these parasitic relationships when herbivore-deterrent defensive chemicals in the honeydew are disseminated onto foliage. Clearly the interactions of sugar sources and all parties involved are much more complex and dynamic than first meets the eye.

Chapter 2
The Sugar Feeders

Because feeding on sugary fluids does not require considerable specialized morphology or physiology, most (if not all) of the arthropods that are commonly designated as entomophagous will readily feed on nectar and honeydew. In addition to taxonomic groups treated in detail in this chapter, several other groups of natural enemies consume sugar, predatory wasps being one of the more conspicuous cases in point (Beggs, 2001; Cuautle and Rico-Gray, 2003; Opler, 1983). Indeed, Krombein (1951) found 93 species of wasps (more than half of those found locally) fed on honeydew of the tulip tree scale, *Toumeyella liriodendri*. Even epigeal predators like Carabidae and Staphylinidae will readily consume sugar sources when offered (Lundgren, personal observations), and when one considers that 35% of native grass plants have root sternorrhynchans in some habitats (Tennant and Porter, 1991), the potential importance of sugar-feeding to this understudied guild of predators may be much greater than is currently appreciated. Although many entomophagous arthropods accept sugar sources as food, these foods are typically insufficient as a sole source of nutrition for all life processes. In short, nectar is great for supporting short-term metabolic requirements and survival. But nectar and honeydew seldom possesses the nutritional wherewithal to support development and reproduction in natural enemies. This notwithstanding, the taxonomic breadth of natural enemies known to consume sugar is testament to the importance of this non-prey food to natural enemy populations.

2.1 Predators

2.1.1 Arachnida: Araneae

> *All spiders are carnivorous. Their prey consists chiefly of insects; but they will feed on other spiders that they can overcome, even on weaker members of their own species.* (J. H. Comstock, 1913)

J.G. Lundgren, *Relationships of Natural Enemies and Non-Prey Foods*,
Progress in Biological Control 7,

Published by Springer Science+Business Media B.V.

This historical view of spider feeding behavior is only recently being challenged, and now it is widely understood that many spiders include nectar and honeydew in their diets, and that sugar promotes longevity in spiders under some conditions (Pollard et al., 1995; Taylor and Foster, 1996; Vogelei and Greissl, 1989). Spiders have infrequently been observed feeding on honeydew; one instance, (the only example of this phenomenon I could find), involves the ant-mimicking salticid, *Myrmarachne foenisex*), which consumes coccid honeydew (as cited in Taylor and Foster, 1996). Some of the best studied examples of glucophagy in spiders relate to foliar- and flower-dwelling species feeding on nectar.

Spiders frequently hunt on flowers, and at least some of these flower-hunting species benefit from floral food sources. Because spiders have short mouthparts, nectar sources buried deep within the flower are difficult to access. But hungry spiders are resourceful, and have devised ways of getting at concealed nectar when in a pinch (Vogelei and Greissl, 1989). 'Wandering spiders' have been observed to visit the flowers of *Terminalia catappa*, *Eupatorium serotinum*, and *Hibiscus tiliaceus*, where they bury their faces in the flowers to ingest nectar meals (Taylor and Foster, 1996). To feed on the nectar of *Daucus carota*, males of *Misumenoides formosipes* pull the flower close to their mouthparts using their first pair of legs. Then they squeeze the nectary with their fangs, presumably to 'milk' the nectar from the flower (Pollard et al., 1995). The spiders will visit several nectaries from the same inflorescence, spending about 40 s at each one.

The size and nectar-status of inflorescences affect spider residence decisions, and while these floral characteristics also influence visitation by prey, spider residence time may be related in part to their use of these nectar resources (Schmalhofer, 2001). Crab spiders are frequently reported to be more attracted to large inflorescences as hunting sites. They also preferred umbels that produced the most nectar, and were likely to leave umbels that did not contain nectar-producing flowers (Morse and Fritz, 1982). Nectar-bearing umbels were more attractive to prey insects, but Morse (1986) was unable to detect differences in predation frequency on umbels of large and medium sizes, even though fewer prey insects arrived at the medium-sized umbels. Whether consuming nectar may have contributed to the satiation of the crab spiders was not determined, but the causation bears investigation.

Extrafloral nectaries (EFNs) are accessible to and frequently consumed by spiders (Edmunds, 1978; Hespenheide, 1985; Lanza, 1988; Taylor and Foster, 1996). Both immature and adult spiders feed on EFN e.g., both life stages of *Hibana tiliaceus* feed on the EFNs of *Ricinus communis* (Taylor and Foster, 1996). Taylor and Pfannenstiel (in press) show that *Cheiracanthium inclusum* hatchlings fed EFN from *Terminalia cattapa* have markedly higher fitness when they are prey limited than those not provided sugar. Spider nymphs offered only low quantities of prey (three *Helicoverpa zea* eggs per week) molt only once. When this same dietary treatment is supplemented with nectar, the nymphs molt up to nine times and survive for 505 days. When the spiders are provided with an intermediate quantity of prey (15 eggs per week), none mature to adulthood; nectar allows this treatment to fully mature and produce eggs in 50% of the emerged adults. Finally, nectar provision allows spiderlings fed 75 eggs per week to grow to a size equivalent to those fed 375 eggs per week in the absence of nectar. Clearly, this sugar source dramatically improves the fitness of developing spiders.

Two studies to date show that glucophagy in spiders benefits plants that produce extrafloral nectaries. Spider abundance is positively associated with plants that produce extrafloral nectar. Jumping spiders, *Eris* sp. and *Metaphiddipus* sp., are six times more attracted to *Chamaecrista nictitans* that have extrafloral nectaries than those without (Rurhen and Handel, 1999). Similarly, *Phryganoporus candidus* colonies are an order of magnitude larger on *Acacia ligulata* plants with extrafloral nectaries than on those without (Whitney, 2004). On *A. ligulata*, seed set is higher on plants with spiders and nectaries. Whitney showed that this was likely because fruit protected by web colonies are less damaged by predispersal seed predators (although it is worth noting that stable isotope analysis could not verify that the spiders were actually consuming the nectar in this case). Excluding the jumping spiders from *C. nictitans* results in lower seed set and fruit production under field conditions (Rurhen and Handel, 1999) (Fig. 2.1).

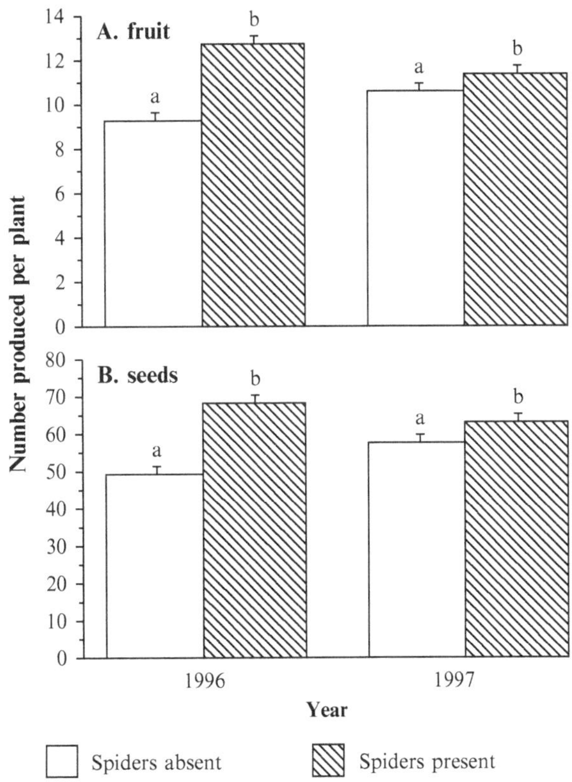

Fig. 2.1 Mean production (plus SE) of fruit and seeds by *Chamaecrista nictitans* as a function of jumping spider (*Eris* sp. and *Metaphidippus* sp.) presence in field patches of varying sizes in 1996 and 1997. Years were analyzed separately. Bars marked with different letters are significantly different (Reproduced from Ruhren and Handel, 1999. With permission by Springer)

2.1.2 *Arachnida: Acari*

Predaceous mites frequently consume sugars, and although the strength of the interactions varies both by the sugar source and the mite species in question, it is generally safe to say that glucophagy is largely used as a fuel for the short-term metabolic needs of the mites. van Rijn and Tanigoshi (1999a) present a list of studies involving sugar feeding in predaceous mites and how the mites have put the sugar to use. Trends from this literature search reveal that sugar feeding in mites drastically improves survival over starved or water-fed individuals, although nectar and honeydew is not sufficient to support reproduction on its own. However, a number of studies show that sugar sources can improve reproduction over a prey-only diet for many species. The two most commonly studied sources of sugar for predaceous mites are sternorrhynchan honeydew and nectar from extrafloral sources.

Honeydews from aphids, coccids, and whiteflies have all been evaluated as food for different predaceous mite species. It appears that mites vary in their ability to use these sugary secretions as a food, and the quality of honeydew from different sources varies in its nutritional suitability (as will be discussed at length in Chapter 5). For instance, nymphs of *Typhlodromips swirskii* could not complete development on the honeydews from three non-whitefly sternorrhynchans (Ragusa and Swirski, 1977), nor could nymphs of *Euseius victoriensis* complete development on honeydew from *Orchamoplatus citri* (James, 1989), although the sugar improved mite survival over a water-only diet. Still, other studies show that some predatory mites can complete development on honeydew-only diets, though survival is invariably less than 31% (Bruce-Oliver et al., 1996; Ferragut et al., 1987; Nomikou et al., 2003b).

The effects of honeydew consumption on reproduction have also been studied extensively for predaceous mites. Similar to nymphal development, some honeydews support reproduction on their own in certain predaceous mites (Nomikou et al., 2003b), but this is far from always being the case (Bruce-Oliver et al., 1996; Ferragut et al., 1987; James, 1989; McMurtry and Scriven, 1964a; Ragusa and Swirski, 1977). More often, it has been reported that honeydews in addition to prey support higher levels of reproduction than when these mites are fed on prey alone (McMurtry and Scriven, 1964b; Ragusa and Swirski, 1977; Zhimo and McMurtry, 1990). Another common observation is that honeydews are able to improve survivorship of adults over unfed individuals (Bruce-Oliver et al., 1996; McMurtry and Scriven, 1964a, 1965; Nomikou et al., 2003b), which may improve lifetime fecundity rates. Honeydew may also serve to shorten the duration of the pre-oviposition period in predaceous mites (Ragusa and Swirski, 1977).

The nutritional suitability of EFN for predatory mites is the focus of several studies, and results are in line with those seen for honeydew. In the field, *Anystis* sp. were seen to feed from the EFNs of *Prunus sargentii* and *Populus tomentiglandulosa*, and *Czenspinkia* sp. was seen feeding from the cup-like EFNs of *Paulownia tomentosa* (Pemberton, 1993). In the latter case, the *Czenspinkia* mites stand on their hind legs to reach the nectar contained in cup-like nectaries on the plant.

In the laboratory, EFN from lima beans improves the survivorship of *Phytoseiulus persimilis* over water alone (Choh et al., 2006), and nectar from various organs on *Ricinus communis* is better able to sustain *Ipheseius degenerans* adults compared to water alone (van Rijn and Tanigoshi, 1999a). Bakker and Klein (1992) found that cassava exudates arrests adult foraging behavior and sustains nymphal development in *Typhlodromalus limonicus*. Thirty-one percent of *Euseius fustis* can complete development on these cassava exudates (Bruce-Oliver et al., 1996). Although *R. communis* nectar supports little reproduction by *I. degenerans* on its own, adding nectar to a pollen diet increases fecundity by 25% over pollen alone (van Rijn and Tanigoshi, 1999a). The importance of EFN to predatory mites can be seen in their numerical response to these nectaries. After 10 days in the laboratory, leaves of *Viburnum tinus* possessing EFNs contain seven times as many phytoseiids and eight times as many total predatory mites than a plant with nectaries excised (Walter and O'Dowd, 1995).

2.1.3 Heteroptera

Many members of predaceous families of Heteroptera are phytophagous to varying degrees, and members of the Anthocoridae, Nabidae, Pentatomidae (Asopinae), Geocoridae, Reduviidae, and Phymatidae all feed on sugar, mainly nectar. For anthocorids, *Orius tristicolor* feeds at EFNs of cotton (Yokoyama, 1978) and at floral nectaries of *Quillaja saponaria* (Bugg, 1987), and *Anthocoris gallarumulmi* feeds on aphid honeydew (Miller, 1971). *Geocoris pallens* also feeds at the EFNs of cotton, and both this species and *O. tristicolor* lay their eggs in close proximity to these nectaries (Yokoyama, 1978). Anthocorids generally have substantially greater longevity when provided with sugar over water alone (Anderson, 1962b; Chu, 1969). Indeed, nectar is essential for development of *G. pallens* raised on different cotton varieties in the absence of prey (De Lima and Leigh, 1984). Up to 17% of this species completes development to the 5th instar on a diet of only nectar and leaf tissue; 55% of insects develop to 4th instars. Adult longevity is four times greater in nectar-fed *G. pallens* than those fed only leaf tissue. Another geocorid that feeds on nectar is *G. uliginosus*, who defends foliar nectaries of *Senna obtusifolia* from visiting *Dorymyrmex* ant (Crocker and Whitcomb, 1980). Finally, observers of EFNs frequently record predatory heteropterans (e.g., Reduviidae, Pentatomidae, and Miridae) as visiting and consuming the nectar (Hespenheide, 1985; Keeler, 1978).

Some indirect evidence of the importance of nectar-feeding in predaceous Heteroptera is summarized by Naranjo and Gibson (1996). In reviewing the literature on population densities of *Orius*, *Geocoris*, and *Nabis* species in nectaried versus nectariless cotton, they found that the vast majority of studies show reduced predator densities in nectariless cotton. Obviously, the presence of nectar is not the only difference between these systems, but the results are at least in line with the hypothesis that nectar-feeding occurs and is important for these heteropterans.

Members of the Phymatidae (ambush bugs) frequently hunt on flowers, where in addition to consuming flower-visiting prey, they appear to also drink nectar. *Phymata pensylvanica* was observed by Balduf (1939b) to insert their mouthparts into the nectar-bearing regions of *Helianthus tuberosus* and *Symphiotrichum ericoides* flowers. He observed this behavior on a windy day when prey were not actively visiting flowers, and he concluded that the ambush bugs were drinking nectar to compensate for poor hunting conditions. Nearly 60 years later, Yong (2003) revisited this system to find that nectar feeding is important to the survival of these normally predaceous bugs. In the laboratory, *P. pensylvanica* adults prefer to drink from sugar water over plain water under choice conditions, indicating their ability to perceive nectar. Although bugs do not attain adulthood on a diet of sugar water alone, *P. pensylvanica* survives for around two months on this non-prey food; on average, sugar-fed individuals survived three times longer than on water alone, although they lose weight over this period. In the field, *P. pensylvanica* moves to inflorescences during and after the 4th stadium. Through the use of cages, the survival of *P. pensylvanica* adults was monitored on inflorescences and leaves in the absence of prey. Nectar provided by the wildflowers is able to sustain the ambush bugs substantially longer than those held on vegetative material. Thus, phymatids provide one more example of predators which hunt on flowers taking advantage of floral resources.

2.1.4 Coleoptera: Coccinellidae

> *Such common aphidophagous coccinellids as* Coccinella *spp.,* Adalia bipunctata *(L.), and* Cycloneda sanguinea *(L.) are often conspicuous as they feed on the leaf nectar of young peach trees in midsummer when aphids are scarce.* (Putnam, 1963)

Floral and extrafloral nectar, as well as honeydew, are a significant food source for many of the Coccinellidae (Hagen, 1962). Indeed, sugar feeding in coccinellids has been going on for a long time; coccinellids found in association with fossilized EFNs of *Populus* date back 35 million years (Pemberton, 1992). From a sensory standpoint, coccinellids are able to distinguish sugar foods from water in choice tests (Koch et al., 2004), indicating their ability to perceive sugar solutions. In addition to prolonging their lives in the absence of prey, glucophagy can promote reproduction when only suboptimal prey is available, and honeydew may be of wide importance in arresting coccinellid foraging behavior.

In addition to frequently feeding on floral nectar sources (Bugg, 1987; Nalepa et al., 1992; Spellman et al., 2006), coccinellid adults are often some of the most frequent visitors to EFNs (Banks, 1957; Keeler, 1978; Ricci et al., 2005; Stephenson, 1982b). In an extensive literature review, Pemberton and Vandenberg (1993) document 41 coccinellid species (representing 19 genera) that are known to consume EFN. These coccinellids do not show strong fidelity to the nectaries of any given plant species or group; in sum 32 plant species, representing 23 genera and 15 families, are visited by the ladybeetles. Springtime is often when ladybeetles are found on EFNs (Ewing, 1913; Rockwood, 1952; Watson and Thompson, 1933), presumably because prey is scarce. One study showed that *Stethorus punctillum*, which normally dies within 4–5 days on water alone, can survive for 43.2 days on a diet of nectar from peach tree secretory glands (Putnam, 1955). Another instance of this is *Exochomus flavipes*, for whom sugar water and honey are able to prolong their lives tenfold in the laboratory (Geyer, 1947). Surprisingly, only two instances of coccinellid larvae consuming nectar sources have been documented, *E. flavipes* larvae feeding on EFNs (Geyer, 1947), and *Harmonia axyridis* larvae feeding at the EFNs of *Prunus, Populus*, and *Sambucus* (Pemberton and Vandenberg, 1993). This is in spite of the importance of pollinivory to this life stage in several coccinellid species (see Chapter 6).

In addition to providing fuel for normal metabolic processes, honeydew can improve reproduction and is used as a cue for locating the sternorrhynchan prey of ladybeetles. Larvae of *Chilocorus bipustulatus* can be maintained for 'a long time' on the honeydew of scale insects in the laboratory (Yinon, 1969). *Stethorus punctillum* feeds the honeydew of *Coloradoa rufomaculata* on chrysanthemum leaves in the laboratory (Putnam, 1955). Evans (2000) reports that *Coccinella transversalis* does not lay eggs on *Helicoverpa armigera* larvae, nor on a diet of sucrose (nutritionally equivalent to honeydew from *Acyrthosiphon pisum* for this coccinellid), when these foods are offered on their own. However, when these two suboptimal foods are offered together, females produce small numbers of eggs (10–15% of fecundity realized on the preferred food of aphids) (Table 2.1). In addition to providing direct nutrition, sternorrhynchan honeydew also functions as a cue for locating sternorrhynchan prey. Carter and Dixon (1984) showed that honeydew of the cereal aphid, *Sitobion avenae*, arrests the movement of *Coccinella septempunctata* larvae. Corn ears coated in honeydew have greater numbers of ladybeetle larvae, and they spent more time searching these sugary ears. Increased residence

Table 2.1 Number of eggs laid per day [mean (s.e.)] by female *Coccinella transversalis* that were maintained on different diets (n = 6 females per treatment). Honeydew was produced by *Acyrthosiphon pisum*. Within a column, values followed by different letters are significantly different (Data is from Evans, 2000. With permission by Blackwell)

Diet	Days 4–10 of experiment
Aphids	20.0 (3.8) A
Helicoverpa armigera (2nd stadium)	0.4 (0.2) C
Sucrose solution only (150 g l^{-1})	0.0 (–) C
Helicoverpa + sucrose	2.7 (1.3) B
Helicoverpa + honeydew	2.0 (1.2) B

time by the larvae is ultimately associated with fewer aphids. Thus in this case, glucophagy by ladybeetles leads to increased pest suppression.

2.1.5 Neuroptera: Chrysopidae

all green lacewings were for a long time considered predaceous, essentially aphidophagous. Further studies showed that most of them have, at least partially, non-live food, mainly honeydew and other sweet juices (Canard, 2001)

Adults of the Chrysopidae, Hemerobiidae (Bugg, 1987), and Mantispidae (Keeler, 1978) feed at sugar sources of various classes under field conditions. Some hemerobiids feed on honeydew in the field (Stelzl, 1990, 1991 as cited in Canard, 2001), and sugar feeding improves their longevity significantly over water alone (Neuenschwander and Hagen, 1980). Still, the most studied taxa within this order are the chrysopids, and most adult green lacewings are glucophagous to some degree. They consume nectar and various plant exudates, as well as honeydew. In addition to maintaining metabolic processes in the adult stage of chrysopids, a growing body of literature indicates that the larval stage may also rely on sugar meals as an important source of non-prey food.

Many chrysopids are well adapted to a non-carnivorous lifestyle during the adult stage (Canard, 2001; Duelli, 1987; Hagen et al., 1976). Only three genera of Chrysopini are considered to be insectivorous as adults (*Anomalochrysa*, *Atlantochrysa*, and *Chrysopa*) (Canard, 2001), and of these half the species of *Chrysopa* are believed to be non-predaceous (Hagen et al., 1976). The mouthparts of many green lacewings are adapted for sugar feeding. They have symmetrical mandibles without any incisor, and spoon-like laciniae that allow them to scrape dried honeydew from substrates (Canard, 2001). Also, patterns in the digestive systems suggest that there are microbial symbioses that support a glucophagous lifestyle. The trunk diameter of the trachea that leads to the diverticulum (and aerobic gut microorganisms), and its ratio to forewing length is a good predictor of diet in various chrysopid species, with greater tracheation to the gut associated with glucophagy and plant-feeding (Canard, 2001). This potential symbiotic relationship will be discussed more at length in Chapter 15, but suffice it to say that adaptations to a glucophagous lifestyle in chrysopid adults are evident.

Sugar feeding promotes longevity, fecundity, and intrinsic growth rates in green lacewing adults, and these functions are reinforced through observations of sugar feeding by these insects in the field. Sugar feeding is particularly prevalent in *Chrysoperla carnea*; which consumes honeydew (Schuster and Calderon, 1986; Sheldon and MacLeod, 1971), and possibly other sugar sources (Bozsik, 1992) in the field. Sheldon and MacCleod (1971) found that 89% of adults (n = 28) consume honeydew on a single sample date during the early spring when other food sources were scarce, and adults scrape the leaf surfaces with their mandibles to gather dried honeydew. In fact, *Chrysoperla plorabunda* adults can be maintained on a sole diet of *Pseudococcus citri* honeydew, which allows full reproduction of 150 eggs daily per female (Finney, 1948). Research has not shown sugar-only diets to be capable of supporting reproduction in other lacewing species, although sugars can improve egg production when combined with other foods (Sundby, 1967; Venzon et al., 2006). Under field conditions, research suggests that more eggs and larvae are typically found in fields where honeydew and nectar is relatively more available compared with fields where sugar availability is reduced (Adjei-Maafo and Wilson, 1983; McEwen et al., 1994).

Sugar feeding by lacewing larvae is best studied in *Chysoperla* spp., which will accept sugar from a variety of sources. *Chrysoperla* larvae visit the EFNs of several plant species in the lab and field (Keeler, 1978; Limburg and Rosenheim, 2001; Schuster and Calderon, 1986). In cotton fields, 21% of *Chrysoperla plorabunda* larvae consume nectar from cotton EFNs, and they are found even more frequently at the EFNs of almond (Limburg and Rosenheim, 2001). Downes (1974) observed *Chysopa* larvae probing the florets of *Cirsium vulgare* searching for and drinking nectar droplets. In the laboratory, these larvae recognize a sucrose solution when it comes in contact with the tips of their palps (Fig. 2.2). Another unspecified chrysopid species does not recognize dried honeydew, but will readily consume a sucrose solution.

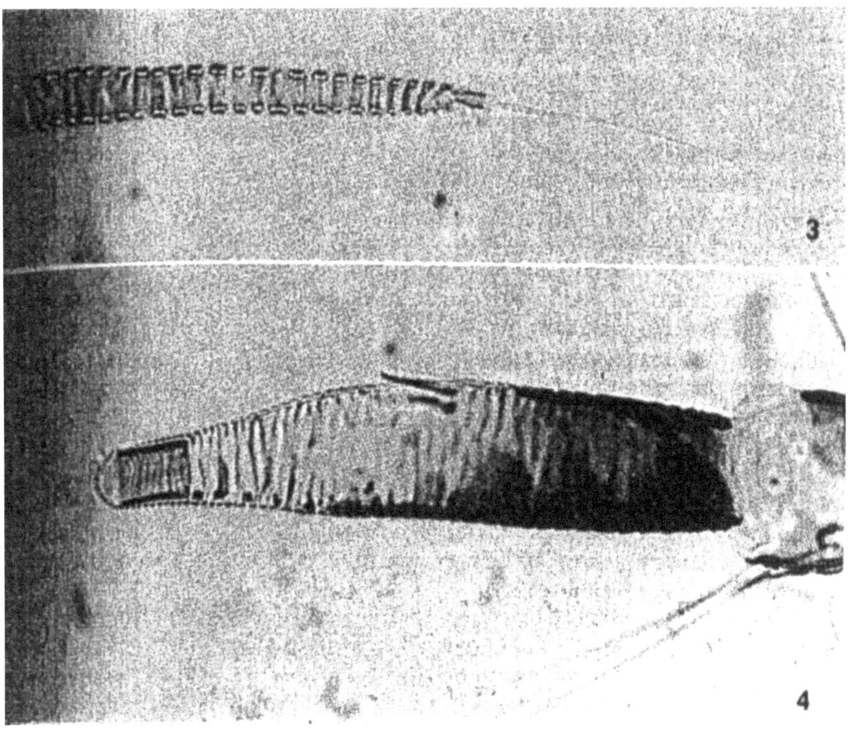

Fig. 2.2 The apex of antenna (top) and labial palp of *Chrysopa* sp. larva to show terminal sensillae, presumably used in contact chemoreception of sugars (Reproduced from Downes, 1974)

The physiological function of carbohydrates appears to be in sustaining the larvae during periods of prey scarcity, and as a result sugar feeding may lead to higher levels of predation by lacewing larvae. Patt et al. (2003) found that *Chrysoperla carnea* larvae are able to metabolize sucrose more efficiently than bee pollen in the laboratory, and suggest that these larvae are better adapted to sugar feeding than pollen consumption. *Chrysoperla plorabunda* larvae fed sucrose do not lose weight as quickly as unfed individuals, suggesting that nectar can be used immediately as fuel that prevents the burning of energy reserves (Limburg and Rosenheim, 2001). Sugar-feeding can reduce prey consumption (McEwen et al., 1993a), and distracts larvae from foraging for prey in the field (Limburg and Rosenheim, 2001). Still, sugar-fed larvae survive substantially longer than those fed water alone (Limburg and Rosenheim, 2001; McEwen et al., 1993b, 1996), and so prey consumption over their extended lifetimes is probably much higher than for those that are sugar-limited. Larvae fed prey in addition to sugarmeals generally perform better than those on sugar alone (Limburg and Rosenheim, 2001; McEwen et al., 1993b), although natural sugar sources seem to provide additional nutrients not found in a prey-only diet (McEwen et al., 1996).

2.1.6 Diptera: Syrphidae

Hoverflies are renowned flower visitors, and the physiological adaptations to sugar feeding are remarkably well described for members of this group. Syrphid adults are fairly fussy over which flower species they prefer, and different fly species specialize on specific flower types (Colley and Luna, 2000). Pollen is one floral reward gleaned from this flower feeding (Gilbert, 1985b), although various adaptations bolster the case that nectar is another important food source. The diet of syrphids is associated with their body size; larger body size is correlated with nectar feeding, whereas smaller-bodied syrphids feed more frequently on pollen (Gilbert, 1985b). Also, longer-winged individuals are more likely to feed on pollen, and less on nectar. Pollinivorous species need to spend more time airborne to collect the pollen grains from anthers, and so the smaller body size facilitates the stationary flight necessary for this task. Honeydew is another important sugar source for syrphids in the field (Hogervorst et al., 2007; Belliure and Michaud, 2001). Adults of some aphidophagous species are attracted to the volatiles of honeydew (Belliure and Michaud, 2001; Budenberg and Powell, 1992), and also show a gustatory response to honeydew that is proportional to the concentration of this substance (Sutherland et al., 2001).

Sensing of sugar appears to be accomplished by the labellar taste hairs, rather than the antennae (Hood Henderson and Wellington, 1982). Interestingly, chemosensilla on the ovipositors of two aphidophagous species (*Eupeodes fumipennis* and *E. volucris*) are sensitive to aphid honeydew, and the honeydew components sucrose, tryptophan (and its oxidation product) and alanine (Hood Henderson, 1982). In the non-predaceous hoverfly, *Eristalis tenax*, the flower-visiting adults have labellar taste hairs with receptor cells that only register sugar (sucrose) solutions (Wacht et al., 1996, 2000). Thus, they have the capabilities to taste sugar on potential food sources.

The process of nectar consumption by syrphids illustrates the complex evolutionary developments that have occurred to allow these flies to consume this important food source. The feeding process was first described by Müller (1883), and was advanced substantially by Gilbert (1981). Once the sugar source is identified by the fly, the proboscis is extended until the labellum comes in contact with the sugary fluid. The folds of the labella then separate so that they lie flush with the nectar droplet. Cibarial pumping motions suck the fluid into the mouth (Gilbert, 1981). The labella does not always have to touch the nectar; it can be wicked up along creases in the corolla through capillary functions of the hydrophilic labella (Gilbert, 1981). Several morphological features of the mouthparts are diagnostic for glucophagous syrphid species. Longer, thinner proboscises are associated with nectar feeding (Gilbert, 1985b), and the pseudotracheae of the labella may have spinose closing apparatuses that filter heterogenous materials out of nectar meals (Zaitsev, 1982). The fidelity to different flower species described above may be reflected by their mouthpart morphology. Gilbert (1985b) found that the length of the tongue is strongly correlated with the corolla length of the flowers that they visit (Fig. 2.3). It is presumed that long-tongued flies feed at deeper flowers because the quality of the nectar is

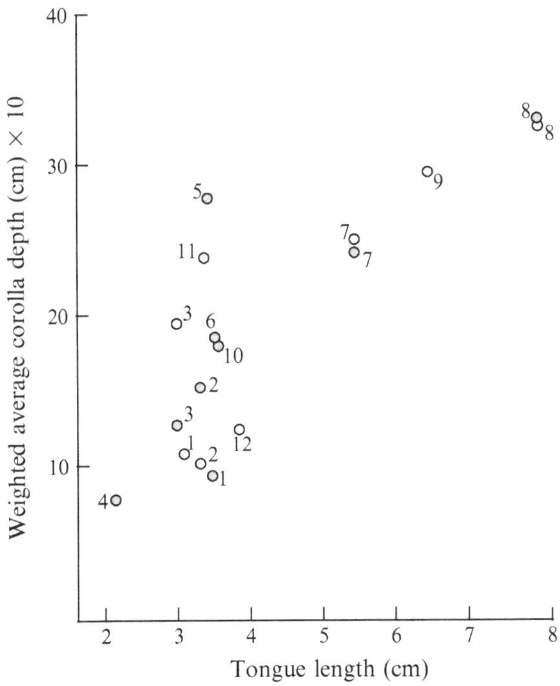

Fig. 2.3 The relationship between the proboscis length and the average corolla depth (CD), weighted by the frequencies of visitation (weighted average = sum of (% x CD/10). Species are: 1. *Syrphus ribesii*, 2. *Eupeodes corollae*, 3. *Episyrphus balteatus*, 4. *Melanostoma*, 5. *Melanostoma mellinum*, 6. *Syritta pipiens*, 7. *Eristalis interrupta*, 8. *Eristalis tenax*, 9. *Eristalis nemorum*, 10–12. *Sphaerophoria* (Reproduced from Gilbert, 1985. With permission by the Royal Society)

superior for these species (see Chapter 3). In this way, a community of syrphids can separate the floral resources to reduce competition much the way that parasitoids do (described below).

2.1.7 Hymenoptera: Formicidae

The vast majority of ants consume sugar of one type or another, and the availability and abundance of sugar strongly shapes the dynamics of ant colonies and communities in which they live. In ants, sugar functions to

> *make water more drinkable, it stimulates worker activity, and it conserves protein* (Brian, 1973)

Within a colony, most sugar is used by the worker castes. As major nutrient classes (carbohydrates, proteins, and lipids) enter the colony, it is clear that workers receive sugars first, while larvae and queens are the first recipients of lipid and proteinaceous foods (the fate of these higher calorie foods will be discussed more at length in Chapter 10) (Beattie, 1985; Brian, 1973; Carroll and Janzen, 1973; Tschinkel, 2006). The workers use sugary fluids as a fuel for foraging, hunting, and hauling their horde back to the nest (Beattie, 1985). The distribution of sugar among the workers is more rapid and widespread when the colony is starved than when the workers are sated, reaffirming that the workers use sugar for maintaining normal activity patterns (Tschinkel, 2006). By examining respiration rates of the different life stages, Brian (1973) presents evidence that larvae and workers are found to burn up sugary foods before digesting the protein sources. In the laboratory, an average worker of *Labidus rubra* consumes 22 µg of sucrose daily (Brian, 1973). Approximately 20–40% of *Solenopsis* workers return to the nest with a sugar meal, often comprising 35% of their body weight (Tennant and Porter, 1991). When fed honey in addition to a standard, animal protein diet, *Solenopsis invicta* colonies survive better and have greater weight gain than those offered only an artificial diet (Williams et al., 1980).

Physiological characteristics of the ants help to understand the intra-colony dynamics of sugar feeding. One reason that nectar and honeydew are such an important source of food is that worker ants are restricted to a liquid diet, and there simply aren't many non-prey sources of nutritious fluids to eat that don't need processing. Solid food particles are prevented from entering the digestive system of ants by the infrabuccal pocket (Eisner and Happ, 1962; Glancey et al., 1981); workers bring foods like seeds and insect fragments to their 4th instars, who are able to masticate and digest the foods and recirculate the nutrients back to the rest of the colony (this process is discussed at length in Section III). The crop of the workers represents an important evolutionary development that increases the sugar feeding capacity of ants. *Solenopsis invicta* ants that take a sugar meal are externally distinguishable because their crop is so full of fluid (Tennant and Porter, 1991). A more extreme example that illustrates the importance of sugar sources to ant colonies is when the colony devotes a portion of their worker caste specifically to sugar storage, these workers are called repletes (Wheeler, 1910; Carroll and Janzen, 1973; Holldobler and Wilson, 1990). The repletes in honeypot ants have grossly distended abdomens filled with sugary solutions that are transferred as needed to their attending sisters. These more extreme examples are best noted in arid environments, where environmental conditions are either unpredictable or predictably harsh. Other ants living in arid environments have simply stopped trying to find rare sugarmeals; *Messor* ants in the desert are not attracted to sugar or water sources (Went et al., 1972). Instead they obtain their nutrients exclusively from seeds, rodent and bird droppings, and dead insects.

One of the best studied forms of glucophagy in ants is honeydew feeding. In 1914, William Morton Wheeler found *Iridomyrmex* preserved with aphids in Baltic amber, dating back to the early Oligocene Epoch (38 mya) (Holldobler and Wilson, 1990), and Beattie (1985) believed ant-sternorrhynchan associations to extend as far back as the Cretaceous (135 mya). Ants are primitively carnivorous, but even the most predatory clades will consume honeydew when it drops to the ground or when they encounter it on foliage (Holldobler and Wilson, 1990; Wheeler, 1910). Carroll and Janzen (1973) postulated that the majority of ants will harvest honeydew from Sternorrhyncha, and this substance can comprise a major portion of the diet of many species. As a case in point, *Oecophylla longinoda longinoda* colonies can thrive when fed solely the honeydew of the scale *Saissetia zanzibarensis* (Way, 1954). Ants often call on multiple species of Sternorrhyncha (Beattie, 1985; Way, 1954), and many sternorrhynchans are visited by several species of ants (Nickerson et al., 1977). On cacao in Africa, at least 50 ant species were found to collect the honeydew from the coccid, *Formicococcus njalensis* (Strickland, 1947). The relative contributions of honeydew and prey to the diet of an ant species varies among species, and may even change based on the physiological status of the colony (Way, 1963). One estimate is that honeydew comprises 62% of the diet of *Formica rufa rufa*, whereas prey constitutes only 33% (Way, 1963, and references therein).

Hundreds of species are know to actually tend sternorrhynchans (although many sternorrhynchans are untended by ants), and the strongest mutualisms result in a fascinating sort of pseudo-domestication of the sternorrhynchans by the ants. The three most phylogenetically advanced subfamilies of Formicidae (Dolichoderinae, Formicinae, and Myrmicinae) are comprised of numerous species that tend sternorrhynchans for their honeydew (Holldobler and Wilson, 1990). The mutualism may have resulted from sternorrhynchans evolving to exploit a trophallactic appeasement behavior in ants (Choe and Rust, 2006). When two ants display aggression toward one another, they offer each other food which serves to curb the aggression. The result of providing ants food is that they are less aggressive to the donor. The importance of the mutualistic association between honeydew-producing sternorrhynchans and ants is evidenced in some of the behavioral and morphological adaptations observed in both participants (Way, 1963). In the absence of ants, sternorrhynchans frequently will kick off or simply drop their honeydew droplets. Ants have been shown to palpate the aphid's abdomen with their antennae (Wheeler, 1910), which then exudes a honeydew droplet (a similar behavioral solicitation occurs with mutualistic lycaenid caterpillars; Way, 1963). In fact, ant-tended sternorrhynchans often possess a ring of small hairs near the anus that actually suspends the honeydew droplet until it can be collected by the ants; untended sternorrhynchan species lack this ring of hairs (Wheeler, 1910) a similar behavior solicits honeydew from lycaenid caterpillars. Morphologically, it is thought that some aphid hind-quarters have evolved to resemble the morphology of a donor ant. Thus, ants use similar behaviors to solicit honeydew from sternorrhynchans as they do when soliciting sugar-meals from their nestmates (Way, 1963). In exchange for providing the ants a meal, the sternorrhynchans receive a number of services. The first is that the ants look after the sternorrhynchans; the ants protect their livestock from natural enemies, move them to prime feeding spots on the plant, and may even provide shelter within the ant's nest or through building special structures that harbor the sternorrhynchans (Way, 1954, 1963). Perhaps as importantly, the ants keep the sternorrhynchan colonies clean from microorganisms that would ordinarily colonize the copious amount of honeydew and cause disease within the sternorrhynchans (Way, 1954, 1963). To keep the ants happy, the tended sternorrhynchans increase their feeding rates to provide more honeydew to their bodyguard/butlers. The end result is greater fitness and higher population growth rates in ant-tended sternorrhynchan colonies versus untended colonies. The ants benefit as well, and some species go to extraordinary lengths to ensure that their relationship with sternorrhynchans continue (Holldobler and Wilson, 1990). The queens of one ant species will actually carry sternorrhynchans in their mandibles during the nuptial flight, to ensure that they have a new food source when they land!

There is no doubt that ants are frequent visitors to EFNs, and many ecologists would go so far as to say that ants are the raison d'êtres of these plant structures. EFNs from a single plant species can attract a wide array of ant species (Barton, 1986; Beattie, 1985; Bentley, 1977b; Keeler, 1978; Stephenson, 1982b), the greatest number that was found in the literature being 22 ants species visiting the EFNs of *Turnera ulmifolia*, eight of which fed only at EFNs (Cuautle et al., 2005). Carroll and Janzen (1973) said it well when they explained that nectar from EFNs "*is eaten by practically any ant that encounters it.*"

In the absence of EFNs, ants are less interested in a plant (Agnew et al., 1982). For example, 75% of nectaried *Cassia fasciculata* plants were tended by ants, compared with <10% of plants lacking nectaries (Barton, 1986). A major benefit provided by ants that visit the EFNs is protection from herbivores. Beattie (1985) laid out several tenets of what makes an ant a good mutualist, some of the most valuable of which are paraphrased as:

1. A dietary need for the nectar
2. Aggressive toward herbivores
3. Are able to defend the plants where and when the plant needs it most

Essentially, the ideal situation for the plant is to develop a strong sense of 'ownership behavior' in their EFN visitors. This term refers to the degree to which an ant protects a plant from all interlopers, and the highest degree of ownership behavior is evident when the ants nest directly on the plant within plant-provided domatia (Way, 1963). Sometimes, this ownership can result in plants being kept free of beneficial insects, such as generalist predators or pollinators that the plant would like to keep around (McLain, 1980).

Compared to the topics of honeydew consumption and EFN visitation, the utilization of floral nectar by ants has received little attention by myrmecologists. Indeed, this has been the case for many years. Wheeler (1910) explains that the dearth of information on flower-visiting ants is

> *probably because* [ants] *treat the flowers very cavalierly, for unlike the bees, they do not concentrate attention on particular plants and make cross-fertilization one of their main avocations.*

Janzen (1977) hypothesizes that floral nectar sources are chemically defended against nectar thieves. It is certainly true that nectars are defended against thieves (see the next chapter), but the bottom line is that floral nectars are not universally toxic to ants, and more work should fill the void on this topic. Given the sexiness of domestication of sternorrhynchans by ants and the use of ants attracted to EFNs as a type of plant immune response against herbivory, it isn't surprising that the comparatively ho-hum consumption of floral nectar has been given short-shrift in the world of ant glucophagy.

2.2 Parasitoids

2.2.1 Parasitoid Diptera

> *The mouthparts of virtually all fly families are specialized fundamentally for fluid feeding* (Gilbert and Jervis, 1998)

With the above statement in mind, it is not surprising to find that parasitoid members of the families Phoridae, Tachinidae, and Bombyliidae consume nectar and honeydew under a variety of circumstances, and that their morphology is well adapted to glucophagy. In fact, honeydew may be a more important food source evolutionarily to calypterate flies than nectar (Downes and Dahlem, 1987). Some of Downes and Dahlem's arguments for the importance of honeydew to the evolutionary history of parasitoid flies (and other non-parasitoid calypterates) are: (1) Flies taste sugars with their feet, so it seems likely that as they evolved, these flies must have walked on sugars. Honeydew is much more likely to be walked upon than nectar. (2) Tachinids skim leaf surfaces while in flight, and regularly touch down at shiny objects on the leaf surface; this is likely related to honeydew foraging. (3) The pseudotracheate labellum of most Diptera is an adaptation to suck up dried honeydews with a minimum loss of water. If nectar were an important source of nutrition, the proportion of long-tongued Diptera would be greater. Moreover, the pseudotracheate labellum is ancestral to the group, which evolved in the Triassic Period. Flowering plants didn't rise to dominance until the Cretaceous, but Sternorrhyncha were widespread in the Permian, long before the radiation of

Diptera. (4) Downes and Dahlem anecdotally point out that there are fewer Diptera found in areas where honeydew is scarce, resulting either as a function of the habitat or resulting from high numbers of ants that protect the honeydew. This final relationship bears further investigation. But certainly there are many instances of honeydew feeding in the literature on parasitoid Diptera. A case in point is with the phorid *Pseudacteon tricuspis*, which survive better on *Aphis gossypii* honeydew than on buckwheat nectar (Fadamiro and Chen, 2005).

Despite the convincing arguments put forth by Downes and Dahlem, parasitoid Diptera frequently are observed feeding on floral and EFN sources (Kost and Heil, 2005; Opler, 1983; Tillman, 2006). In a comprehensive examination, Allen (1929) found many species feeding on floral and extrafloral nectar, and Opler (1983) found that Tachinidae are some of the most frequent nectar-feeding visitors to Costa Rican plants. To some degree, flower feeding parasitoid Diptera can be distinguished from the honeydew feeders based on their mouthpart morphology (Allen, 1929; Gilbert and Jervis, 1998). In Allen's survey, he found that those species that have proboscises that are longer than the height of their head are more inclined to feed on nectar from flowers (only five of 18 feed on non-nectar sources, and never exclusively). Shorter mouthparts are typically associated with feeding at EFN sources (28 of 33 feed on non-nectar sources) (Table 1 of Gilbert and Jervis, 1998, taken from data of Allen, 1929).

> *The habitual flower feeder has a slender, strongly-chitinized, elongate proboscis with a small labella at the tip, well adapted for reaching deeply seated secretions of nectar in flowers. The honeydew and surface-nectary feeder, on the other hand, has a short, stout, more flexible proboscis with a large, fleshy labella, better adapted to feeding from exposed surfaces, but very poorly adapted to sucking nectar from flowers* (Allen, 1929)

Gilbert and Jervis (1998) built upon Allen's early description of the relationship between diet and morphology in the parasitoid Diptera. First, nectar specialists in the parasitoid Diptera and Hymenoptera have developed a concealed nectar extraction apparatus (CNEA; Jervis, 1998), which is basically a longer, thinner labellum with fewer pseudotracheal canals. Bombyliids with this CNEA also have spinose edges that may function as nectar filters (Zaitsev, 1982). Those species that feed on honeydew or dried nectar likely have wider labellar surfaces with more pseudotracheal grooves.

The actual process of nectar feeding is described nicely by Gilbert and Jervis (1998), and is summarized here. First, the pseudotracheal canal transports saliva out of the mouth and onto the food. The fluids containing the nectar are then sucked through the interpseudotracheal folds of the labella and into the labral food canal. The prelabral pump is responsible for generating the pressure necessary for this process. Drinking dried sugars requires dissolving them in saliva, hence the wider labella with more pseudotracheae for salival transport. Another observation that merits further research is that the CNEA is disproportionately found in parasitoid Diptera that live in arid regions. Gilbert and Jervis believe that this form of mouthparts may have been selected for more strongly in arid habitats because these flies use nectar as a water source.

2.2.2 *Parasitoid Hymenoptera*

Several exquisite reviews of glucophagy in parasitoid Hymenoptera are currently available (Jervis, 1998; Jervis et al., 1993, 1996a; Wäckers et al., 2005), and I certainly cannot hope to recreate or pre-empt these sources. Still, all shapes and sizes of parasitoids use sugar, and some of the best instances of the function of this food are known from the parasitoid literature. Therefore, sugar-feeding in parasitoids is a topic worthy of at least a brief discussion here.

A number of gut analyses have been applied to indirectly substantiate sugar-feeding by parasitoids in the field. In crucifer fields, nearly 70% of *Cotesia* (spp.) test positive for the presence of sucrose (Wäckers and Stepphun, 2003). Fructose, a sugar that is not commonly encountered in unfed insects but is commonly found at greater levels in plant tissues and exudates, is detectable in less than 21% of the field populations of *Aphelinus albipodus*, *Aphytis aonidiae*, *Macrocentrus grandii* and *Trichogramma ostriniae* (Heimpel et al., 2004). Around 75% of *Diadegma semiclausum* test positive for fructose in the broccoli fields when *Fagopyrum esculentum* flowers are adjacent (Lavendero et al., 2005).

Members of over 30 families of Hymenoptera consume non-host foods (Jervis et al., 1993). In observing a particular habitat, 33 of 53 flowering plants are visited by parasitoids, roughly 25% of which (of nearly 1,000 specimens) actually feed at the flowers. As if this monumental set of field observations is not enough, Jervis and Kidd go on to report another 330 parasitoid species from the literature also feed at floral nectaries. In another comprehensive examination of flower-visiting parasitoids, Tooker and Hanks (2000) revisit a dataset from 1928 created by Robertson that documents flower visiting insects from 453 plant species found in central Illinois. Parasitoids comprise a large portion of these insects, 15,172 specimens to be exact. Ichneumonids are the most frequent visitors (48 species; almost twice as many as the next most frequent family). Most species feed from a narrow range of flower species; the exceptions to this rule are the largest-bodied families (Tiphiidae, Leucospididae, and Scoliidae) which visit the most flower species per wasp (Tooker and Hanks, 2000). In the Canadian arctic, at least 18 parasitoid species were observed feeding on nectar at flowers (Kevan, 1973). Although floral architecture likely plays a role in which flowers are accessible to a parasitoid (see next chapter), Patt et al. (1997) show that some parasitoids are behaviorally more disposed to feeding from a greater breadth of flowers. Specifically, *Edovum puttleri* feeds on the nectar of a large number of flowers, all except those where the nectaries are completely hidden at the base of a long corolla. *Pediobius foveolatus*, a larger parasitoid, feeds on a very restricted set of these nectar sources, regardless of whether the nectar is accessible. As more data is collected on the feeding patterns of a greater number of wasp species, the factors that ultimately drive floral preferences will likely become more consistently apparent.

Honeydew (Eijs et al., 1998; Fuchsberg et al., 2007; Hagley and Barber, 1992; Irvin et al., 2007; Leius, 1961a) and EFN (Lingren and Lukefahr, 1977; Patt et al., 1997) are other sugar sources important to parasitoids. Some parasitoids (e.g. *Encarsia formosa*) will even take the fluid droplets directly from an aphid's anus (Jervis and Heimpel, 2005)! In the field, 80% of *Cotesia glomerata* and 55% of *Microplitis mediator* consume honeydew (determined using indicator sugars) in crucifer fields (Wäckers and Stepphun, 2003). Most of this feeding occurs in association with field margins where buckwheat is planted, but the majority of these parasitoids also consume honeydew. Sugar profiles found in the guts of field-collected *Diadegma insulare* suggest that this parasitoid will consume both honeydew (from *Aphis glycines*) and nectar (likely from buckwheat) (Heimpel et al., 2004). Parasitoids are also commonly recorded visitors to EFNs; for example, *Cotesia congregata* is one of the most frequent visitors to the EFNs of *Catalpa speciosa* (Stephenson, 1982b).

The first step in glucophagy is that the wasp needs to recognize a sugar source as food; odor is an important cue for recognizing sugar meals (Jervis et al., 1996b). The use of odors by parasitoids to fulfill their physiological needs has been best explored for the wasp, *Microplitis croceipes* (Lewis and Takasu, 1990). Female wasps are eternally torn between feeding themselves and foraging for hosts; feeding prolongs their life, but may come at the expense of laying eggs since food and hosts are not always found together. Work on *M. croceipes* has shown that female wasps learn odors and respond to them in correspondence to their immediate physiological status- starved females are attracted to sugar-associated odors, and satiated females go hunting for hosts (Lewis and Takasu, 1990) (Fig. 2.4). The learned odors appear to be somewhat arbitrary, and wasps can be taught to associate a broad range of chemicals with a potential sugar meal. Research has shown that *M. croceipes* can be trained to use natural odors such as vanilla and chocolate, or entirely synthetic chemicals (cylcohexanone, diisopropylaminoethanol, 3, 4-dinitrotoluene, and 3, 4-octanone & octanol) as a sign to dine (Olson et al., 2003). Inexperienced wasps also show specific innate behavioral responses to odors and colors depending on their physiological state (Siekmann et al., 2004; Wäckers, 1994). One of the few chemicals that is not attractive to many parasitoids is sucrose, one of the prevailing sugars found in most phloem-derived foods (Leius, 1961b; Patt et al., 1997; Rose et al., 2006; Tertuliano et al., 2004; Wäckers et al., 2002). This is likely since sucrose is non-volatile, but upon contact, sucrose is phagostimulatory. Thus it seems that other chemical cues associated with nectar or honeydew beyond the sugars themselves are necessary for parasitoids to locate food (Patt et al., 1997; Rose et al., 2006). Another pattern that is repeatedly shown in the literature is that parasitoids with experience are better at locating specific food-associated odors (Patt et al., 1999). Again, inexperienced parasitoids are also well capable of locating food sources (Siekmann et al., 2004; Stapel et al., 1997; Wäckers, 1994, 2004). As a case in point, Stapel et al. (1997) reveal that naïve wasps locate EFN from cotton faster than sucrose spiked with vanilla. When these same parasitoids are tested a second time with the same food sources, the wasps find the sucrose and nectar equally fast.

Once they have located a sugar meal, wasps have evolved a variety of means for actually eating it. The morphological and physiological adaptations to sugar feeding are easily seen when the feeding process is described. Jervis (1998) adapted the proc-

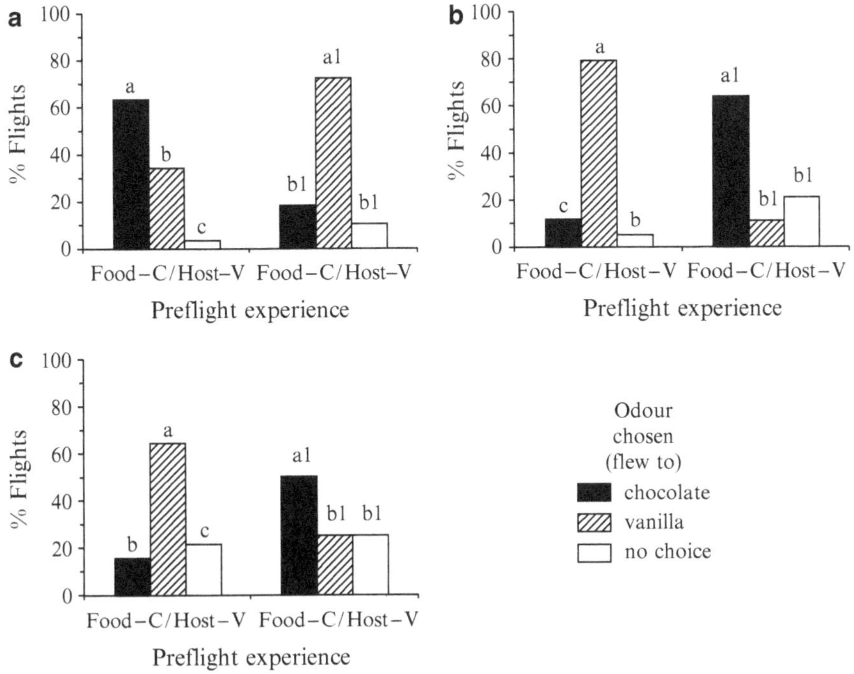

Fig. 2.4 Flight responses to vanilla or chocolate extract by hungry (a), well feed (b), or trained hungry then well fed (c) *Microplitis croceipes* females, with preflight training indicated. Food-C/Host-V: females given training experience of chocolate-to-food and vanilla-to-hosts; Food-V/Host-C: females given training experience of vanilla-to-food and chocolate-to-hosts (the order of food-odour and host-odour training was alternated and the results combined). Bars within same treatment group capped by different letters are significantly different (Reproduced from Lewis and Takasu, 1990. With permission by Nature Publishing Group)

ess of liquid feeding in bees (detailed by Kingsolver and Daniel, 1995) to parasitoid wasps. First, it appears that the mouthparts of the parasitoid need to contact the sugar solution to elicit a gustatory response (Beach et al., 2003; Wäckers, 1999). Acceptable sugars are then loaded onto the glossa, probably because of hydrophilic properties (often involving hairs) present on the tongue of the parasitoid. The glossal hairs likely function to filter heterogenous material out of the nectar. Saliva issues down the glossa and mixes with the sugar meal. The saliva-sugar mixture is then pumped into the oral cavity through movements of the cibarial pump.

The morphology of the mouthparts can be categorized to yield some inference on the feeding ecology of specific parasitoids. Perhaps the morphological structure that has received the most attention in parasitoids is the CNEA. This is a bit surprising, since only 1 in 30 species actually possesses this specialized adaptation to feeding on floral nectar (this is contrast to one-third of dipteran parasitoids) (Jervis, 1998). Jervis groups parasitoids into two major feeding guilds, those with and without the CNEA, which in parasitoids is an elongated labiomaxillary complex (this is reiterated by Olson et al., 2005). He then goes on to subdivide the CNEA species into seven functional groups, based largely on the morphological arrangements of the labiomaxillary process. The Braconidae and Ichneumonidae are particularly noted for having the CNEA which facilitates the removal of nectar from flowers with deep corollas (Jervis et al., 1996b). Beach et al. (2003) raise the idea that gustatory response may reinforce the ecomorphological relationships of the mouthparts. Specifically, wasps with unspecialized mouthparts feed on a broader range of sugar sources in nature, and may then be expected to show gustatory responses to a wider range of sugars. Their observations with *Anaphes iole* support this hypothesis, which merits further examination.

Food must be digested for it to yield any benefits to the wasps, and even oligosaccharides may require a suite of specialized enzymes to digest them. Williams et al. (2005) found that melizitose is not digested by *Anaphes iole*, while other sugars are completely digested within 24 h of feeding. Wäckers (2001) believes that the digestive capabilities of wasps for different sugars are directly affected by the suite of digestive enzymes present in a parasitoid, and this may limit which sugars are fed upon and which can support the life processes of the wasps.

The most conspicuous function of sugar feeding in parasitoids is that sugar improves longevity, even for species that host feed (Heimpel et al., 1997; Leius, 1961b). There is no shortage of laboratory studies that exemplify the importance of sugar sources (floral and extrafloral nectar, and honeydew) in prolonging the lives of parasitoids big and small (Foster and Ruesink, 1984; Heimpel et al., 1997; Lundgren and Heimpel, 2003; Olson et al., 2000; Putnam, 1963; Rose et al., 2006; Takasu and Lewis, 1993; Williams et al., 2005 are just a few). Indeed, as little as one sugar feeding can have dramatic effects on the lifespan of wasps (Wäckers, 2001; Hausmann et al., 2005), although most parasitoids require many feedings to reach maximum longevity. Azzouz et al. (2004) show that *Aphidius ervi* survive best when continuously offered sugar. Olson et al. (2000) demonstrate that sugar-feeding in *Macrocentrus grandii* reduces the catabolism of carbohydrate and lipid reserves over unfed wasps (Fig. 2.5). EFN from cotton promotes longevity of *Campoletis sonorensis* better than floral sources (Lingren and Lukefahr, 1977), and Idris and Grafius (1995) show that different flowers yield nectars of varying suitability in maintaining the lives of *Diadegma insulare*. With these studies in mind, most parasitoids live an abbreviated and likely despondent life if they don't find a sugar meal of one sort or another. The reproductive and dispersal functions of sugar for parasitoids are discussed at length in Chapter 1. Suffice it to say that sugar-feeding may affect the egg viability, rate of egg resorption, diapause rate and primary sex ratio of progeny, and flight initiation and duration in parasitoids.

2.3 Conclusions

The general pattern in the research which has provided the basis for this chapter is that most natural enemies will consume a sugar meal if given a chance. Indeed, many predators and parasitoids have evolved sensory organs specifically for

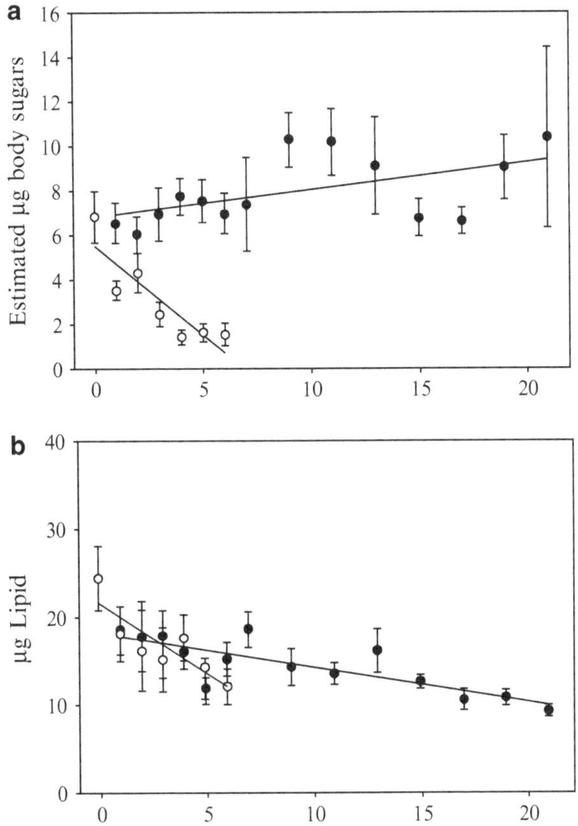

Fig. 2.5 Estimated amounts of body sugars (*top*) and lipids (*bottom*) over the lifespans of female *Macrocentrus grandii* that were provided with sucrose and water (•) and water only (○) (Reproduced from Olson et al., 2000. With permission by Blackwell)

perceiving sugar, and seek this non-prey food source out when in need of food. Another recurring theme in this chapter is that although sugar meals are coveted by a wide range of entomophagous species, very few can persevere through an entire generation without some other nutriment. Thus, simple carbohydrates cannot replace prey in the diets of these arthropods, but can sustain them for long periods of time in the absence of prey, and generally fitness is improved in sugar-fed natural enemies even when prey is available. As we shall see in subsequent chapters, the nutrition and defenses of different sugar sources have led to fascinating ecological interactions among natural enemies and their environments Indeed, their lust for sugar even drives many entomophagous arthropods to kill.

Chapter 3
Floral Nectar

Müller (1883) and Faegri and van der Pijl (1966) present several ways that flowers attract pollinating insects: appeal to the sexual nature of the insects, offer a favorable microclimate or shelter, attract them with the visual (either color or movement) and chemical (odor) aspects of the flower, or offer floral rewards. Obviously, nectar falls into the latter category. Although some flower species can produce thousands of microliters of nectar per nectary, most nectaries generally produce less than 10 µl (Opler, 1983; Pacini et al., 2003; Petanidou et al., 1996; Stiles and Freeman, 1993; Watt et al., 1974). In fact, some stingier plants have no nectar at all; one estimate is that nearly 14% of plants in Mediterranean phrygana are nectarless (Petanidou et al., 1999), but these plants may produce other rewards. A number of factors influence the interactions of floral nectar and insects, most notably when and where floral nectar is supplied and its nutrition and defense.

3.1 Structure of Floral Nectaries

Nectaries typically accomplish a complicated and insecure goal (pollination through transient insect visitation) using a remarkably simple set of adaptations (Fahn, 1988, 2000). In their simplest forms, nectaries occur on plant surfaces as a simple aggregation of cells (Durkee, 1983; Fahn, 1988, 2000) (Fig. 3.1). On a more detailed level, nectaries may appear externally as a protrusion or be deeply embedded in the tissues (Pacini et al., 2003). Invariably, nectaries have three components: (1) an epidermis, (2) a layer of parenchyma, and (3) vascular tissue (usually with a phloem element) that transports water and nutrients to the nectary (Pacini et al., 2003). The nectar is typically secreted through a stomata on the epidermis that has lost its ability to close (Pacini et al., 2003). The size of this stomatal opening has great bearing on how much fluid a nectary secretes (Petanidou et al., 2000).

Phloem is the primary precursor of nectar in plants, but exceptions exist. Xylem is a frequent contributor to nectar, as is the degeneration of associative cells or photosynthate of nearby parenchyma cells (Faegri and van der Pijl, 1966; Pacini

J.G. Lundgren, *Relationships of Natural Enemies and Non-Prey Foods*,
Progress in Biological Control 7,

Published by Springer Science+Business Media B.V.

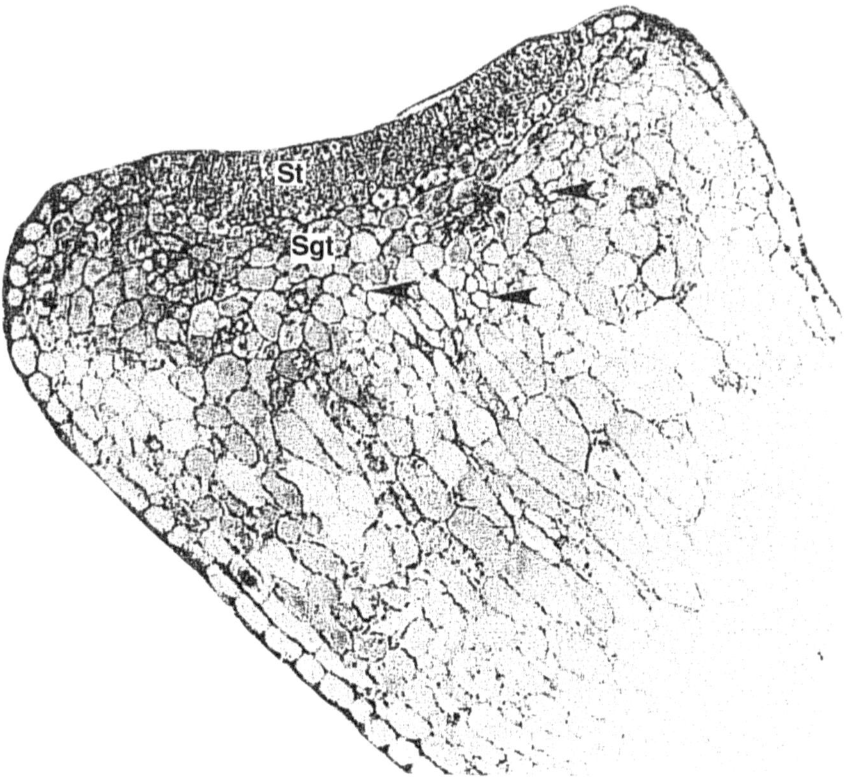

Fig. 3.1 Longitudinal section through a mature petiolar nectary of *Passiflora coerulea*, showing the secretory (St), subglandular (Sgt), and vascular tissues (arrows) (Reproduced from Durkee, 1983. With permission by Columbia University Press)

et al., 2003). Many changes occur in the pre-nectar during the secretion process that alter it nutritionally (Durkee, 1983). First, because it typically stems from phloem, the nutrition of pre-nectar fluctuates as much as phloem sap does. Also, the nectariferous tissue changes the composition of the pre-nectar through enzymatic actions and resorption before the nectar is actually secreted (Fahn, 1988, 2000). Often, the sugar contents of phloem and nectar are similar, but nitrogenous material is filtered out such that lower concentrations are found in nectar than in the phloem sap (Gottsberger et al., 1984).

3.2 Nutrition and Chemistry of Floral Nectar

The primary nutrients in floral nectar are sugars, but amino acids and other phytochemicals are also available at low quantities in most nectars. Generally speaking, the range of sugars found in floral nectar is substantially less than those found per capita in EFN and honeydew, and sucrose, glucose, and fructose are by far the most abundant and pervasive of nectar carbohydrates. Amino acids have received considerable attention from scientists since Herbert and Irene Baker first pointed out their ubiquity in floral nectars in 1973. Small quantities of other chemicals, such as lipids (Baker and Baker, 1983; Keeler, 1977; Koptur, 1994) and micronutrients (e.g., ascorbic acid, vanillin) (Baker and Baker, 1983; Jakubska et al., 2005) are found in some nectars, but a comprehensive investigation of their occurrence and function remains to be conducted.

Sugar concentrations in nectar can be as high as 80%, but values near 20% are more common (Adler, 2000; Baker, 1975; Guerrant and Fiedler, 1981). Generally speaking, sucrose, glucose and fructose dominate nectars (Petanidou et al., 1996). When other sugars are found in floral nectar, they typically comprise 1% or less of the nectar dry weight and are inconsistently present even within a species (Petanidou, 2005; Torres and Galetto, 2002). A notable discrepancy to this pattern is the widespread occurrence of xylose in the nectars of *Protea* and *Faurea* (Nicolson and Van Wyck, 1998; these authors also note that maltose and melezitose recorded by earlier explorations were likely xylose). Starches have also been found in floral nectar (Shuel, 1955). The concentration of sugars within nectar greatly influences the viscosity of the solution and is potentially limiting to certain glucophages. For instance, at 20°C, a 60% sugar solution is 28 times as viscous as a 20% sucrose solution (Baker, 1975).

Nectars are commonly classified based on the ratios of sucrose:hexose quantities, a system first proposed in the middle of the last century (Fahn, 1949; Wykes, 1952). In her analysis of 889 plant species, Percival (1961) categorizes nectars into one of three broad groups: (1) sucrose-dominant, (2) balanced among sucrose, fructose, and glucose, and (3) hexose-dominant. Essentially, while the concentration of sugars can vary substantially intraspecifically, the ratio of sucrose to hexose (glucose and fructose) is highly conserved within a species. Moreover, there appears to be trends in sugar ratios at higher taxonomic levels (Wykes, 1952), suggesting some heritability in this nectar trait.

Amino acids are pervasive in floral nectar, although at low quantities (less than 0.3% of nectar dry weight; Baker and Baker, 1977; Heinrich, 1975). Similar to sugars, the complement of amino acids within a species is often consistent and distinct, but the concentration of nectar amino acids fluctuates greatly (Gardener and Gillman, 2001a). All of the amino acids essential for insect growth and development are present in floral nectars (but not in any particular order), and all told, dozens of amino acids have been isolated from these nectars (Boucher et al., 1982; Keeler, 1977; Petanidou et al., 1996). Between 13 and 15 amino acids are typically found in the majority of nectars (Baker and Baker, 1983; Petanidou et al., 2006). The data from these nectar analyses should be interpreted with care. Generally speaking, proline is considered a pollen amino acid, and nectars with high levels of this amino acid likely have been contaminated with pollen during the collection process (Gottsberger et al., 1984). Only 4 of 32 nectars contain proline when pollen is clearly excluded from the nectar (Gottsberger et al., 1984). Pollen contamination can greatly affect the nutritional constituency of nectar (Erhardt and Baker, 1990), and so any studies that do not specify that precautions were undertaken to prevent pollen contamination (as in Carter et al., 2006) should be viewed with caution.

However, when pollen contamination is common in nature, pollen exclusion will result in artefacts.

The fact that amino acids occur at such low quantities in nectars raises the question of whether they are of nutritional value for insects. In some habitats, specific amino acids have become quite widespread. For example, phenylalanine was the predominant amino acid in the Mediterranean phrygana (Petanidou et al., 2006), but this amino acid occurs in only 55% of plant nectars from a broader sample distribution (Baker and Baker, 1983). Since phenylalanine is attractive to long-tongued bees, the primary pollinators in the phrygana (Petanidou et al., 2006), the phrygana survey lends credence to the theory that the primary function of amino acids is to give nectars a certain taste (Baker and Baker, 1983; Gardener and Gillman, 2002). Based on the taste receptor profiles of flesh flies, Shiraishi and Kuwabara (1970) and Gardener and Gillman (2002) show that floral nectars have a wide range of taste profiles that could help to restrict which nectars that pollinators (and natural enemies) prefer. Moreover, there is very little variation in the taste profile within a plant species, thereby reinforcing the attraction of a consistent set of insects (Baker and Baker, 1977; Gardener and Gillman, 2002).

3.3 Factors That Influence the Production and Nutrition of Nectar

> *Everything that has happened to the plant prior to [flowering] will in greater or lesser degree affect the amount of nectar produced. We should therefore expect to find that secretion is a complex process, and that there is much about it that is variable and unpredictable.* (Shuel, 1955)

The volume and nutrition of nectar is influenced by a number of intrinsic and environmental factors (Cruden et al., 1983; Gottsberger et al., 1984; Shuel, 1955), making it difficult to generalize concerning nectar production, even within a species. A method that has been proposed to classify the myriad factors affecting nectar production is to identify each as pre- and post-secretory in nature. Frankly, our baseline knowledge of the mechanisms that are driving the associations between various factors and nectar nutrition and quantity often makes this impossible to do. Moreover, the factors influencing nectar production are not mutually exclusive. But in spite of all of the factors that operate to change nectar, it is in the plant's best interest to keep their nectar consistent so as to promote pollinator fidelity (Petanidou et al., 1999).

One factor affecting nectar production and nutrition that is clearly pre-secretory in nature is that nectar production is to some degree under genetic control (Baker and Baker, 1976). Therefore it is not entirely surprising to see general phylogenetic patterns to nectar production (Baker and Baker, 1983). First, there is the phylogenetic trend that more derived plant taxa have higher concentrations of amino acids than more basal plant groups (Baker and Baker, 1973; Gottsberger et al., 1984; Stiles and Freeman, 1993). Percival (1961) also found that related plants had similar sugar ratios in their nectar. Still, plants can vary intraspecifically in their nectar production, often more so among populations than within (Lanza et al., 1995; Percival, 1961; Petanidou et al., 1996). Indeed, even male and female flowers of the same species may differ in their nectar concentration and quantity (Pacini et al., 2003). Percival (1961) found that nectar sugar composition varied within a population in only 7% of plants, but when nectar from the same species was compared among populations, 35% of species varied in their sugar ratios. One interpretation of all of these studies is that local pollinator behavior is driving the diversification of plants. These observations do not give a clear idea whether genetics or environment are more important in driving the inter-population differences in nectar nutrition, but certainly a plant's surroundings has a dramatic effect on its nectar phenotype.

The availability of resources to the plant is another factor that affects nectar nutrition. The provision of nutrients in the form of fertilizers has been shown to affect nectar nutrients in several cases, but at this point trends in the literature are inconsistent at best. Shuel (1957) found that N, P, and K are associated with lower nectar production in *Trifolium repens* and *Antirrhinum majus*. One study shows that feeding plants with a general fertilizer under field conditions has no effect on nectar amino acid concentrations (Petanidou et al., 1999). But in *Agrostemma githago*, fertilizing leads to higher levels of glutamine and proline, and reduced concentrations of gamma-amino butyric acid (GABA) (Gardener and Gillman, 2001b) (Fig. 3.2). Gardener and Gillman point out that the fertilizers may affect the amino acid forerunners alpha-ketoglutarate and glutamate, which are common to all three affected amino acids. Interestingly, water availability seems to have more of an effect on nectar nutrition than fertilizers in the Mediterranean phrygana (Petanidou et al., 1999).

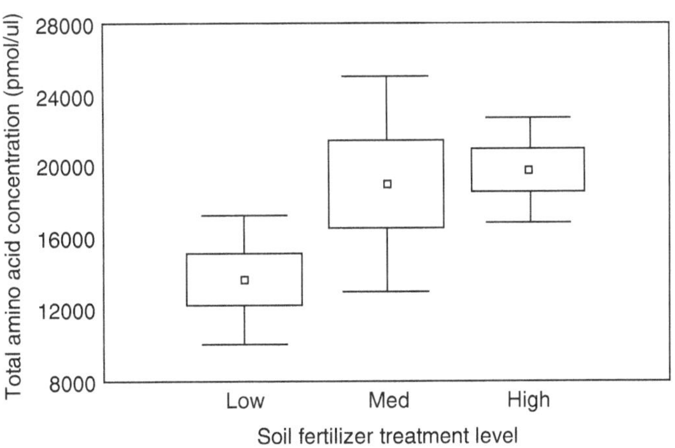

Fig. 3.2 Total concentration of amino acids in nectar of *Agrostemma githago* grown under three soil fertilizer conditions. Shown are: mean, standard error (box) and standard deviation (bar) (Reproduced from Gardener and Gillman, 2001. With permission by Blackwell)

Flower structure has repeatedly been shown to influence nectar nutrition, but the jury is still out on whether nectar traits in different flowers are under genetic control or simply a result of the floral micro-environment. Long-tubed flowers produce nectar with higher sucrose levels, and nectar from shallow flowers tend to be hexose-dominant (Percival, 1961). Baker and Baker (1983) pointed out that hexose-dominant nectars evaporate more slowly than sucrose solutions, and so the correlation between hexose prevalence in open flowers may help to attract a specific group of pollinators. In addition to having sucrose-dominant nectars, long-tubed flowers also produce larger quantities of nectar, although sometimes at lower sugar concentrations (Brink and deWet, 1980; Galletto and Bernardello, 2004; Kaczorowski et al., 2005; Petanidou et al., 2000). In one study, wider flowers produced more nectar than smaller flowers (Petanidou et al., 2000), but Harder and Barrett (1992) found that bigger flowers have lower sugar concentrations (Fig. 3.3). It may therefore be that more nectar does not necessarily mean more nutrition for beneficial insects. In any case, there are some trends in floral structure and nectar production that bear noting when examining insect-plant interactions and the resources available to natural enemies.

Another factor that influences nectar production is time; both the age of the flower and the time of year influence the quality and quantity of nectar resources available to natural enemies. Generally speaking, peak nectar secretion begins with anthesis (Percival, 1961), and tends to be positively correlated with pollinator activity (Cruden et al., 1983; Willmer, 1980). In typical flowers, nectar is secreted at a normal rate until a certain quantity is produced (Cruden et al., 1983). Production

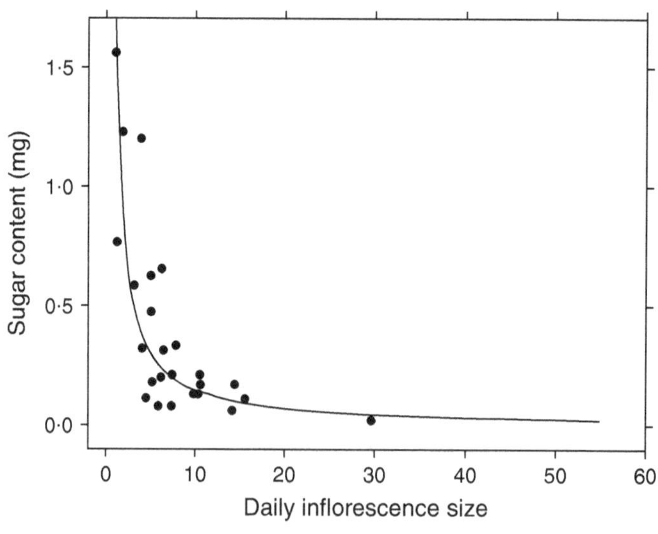

Fig. 3.3 Relation between 12-h sugar production and daily inflorescence size for 27 plant species (Reproduced from Harder and Barrett, 1992. With permission by Blackwell)

then stops, but may resume if nectar is removed below this threshold level (Cruden et al., 1983). As a flower ages, the concentration of nutrients and volume produced is altered. In the case of *Aconitum columbianum*, older flowers secrete more nectar than younger flowers (Brink and deWet, 1980). And in *Capparis spinosa*, sucrose is hydrolyzed into glucose and fructose and amino acids are at higher levels in older flowers (Petanidou et al., 1996; but see Prys-Jones and Willmer, 1992).

The time of day has repeatedly been shown to affect nectar production (Butler et al., 1972; Cruden et al., 1983; Kaczorowski et al., 2005; Mohr and Jay, 1990; Petanidou et al., 1996; Prys-Jones and Willmer, 1992). Willmer (1980) suggests that the nearly tenfold difference in amino acid concentrations over a 24-h period may be the result of insect visitation; insects frequently knock pollen into the nectar or damage surrounding floral tissues, thereby leaking the nutrients into the nectar and changing its nutrition (Fig. 3.4). Nectar production at the species and community levels varies substantially over the season (Opler, 1983; Petanidou et al., 2000). Idris and Grafius (1995) discuss how nectar in a single flower species changes seasonally in its nutritional suitability for parasitoid wasps. Indeed, Petanidou (2005) found that flower nectars in the Mediterranean phrygana community are sucrose dominant during the spring and summer, but shift to hexose-dominant during winter months.

Although temporal patterns in nectar production may be under some genetic control, climatic conditions also play their part in mitigating the quantity and quality of nectar. Obviously, nectar dries out and this affects the final quantity and nutrient

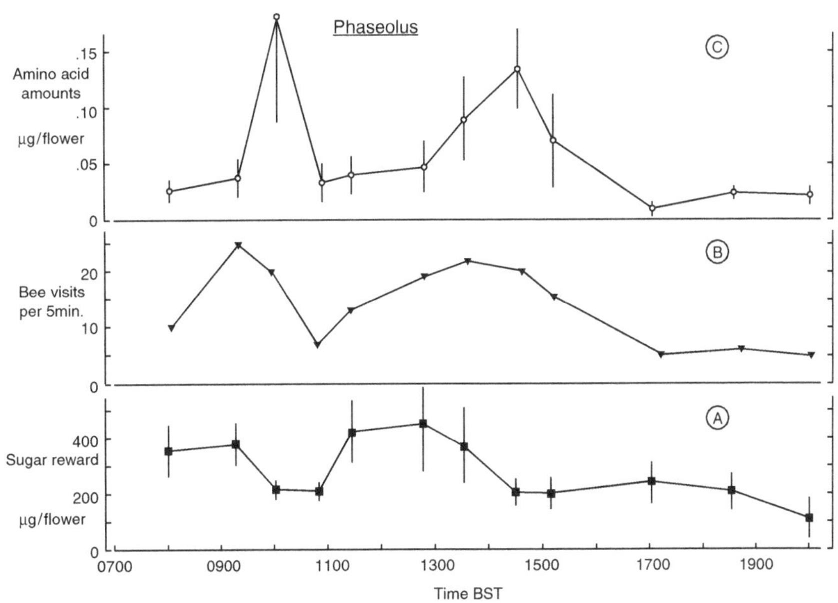

Fig. 3.4 Temporal variation in amino acids, sugar amounts and bee visits to the flowers of *Phaseolus*. Vertical bars represent ±1 SEM (Reproduced from Willmer, 1980. With permission from Springer)

concentration that is encountered by an insect. Therefore, one might expect nectar to have the highest sugar concentrations during the hottest part of the day (Prys-Jones and Willmer, 1992; Willmer, 1980). Another frequent phenomenon is that nectar becomes more concentrated at higher humidities (Butler et al., 1972; Corbet et al., 1979; Shuel, 1955). Finally, elevated levels of atmospheric CO_2 are associated with significant reductions in nectar volume in three of five flower species (Rusterholz and Erhardt, 1998). Although sugar contents were equivalent under different CO_2 levels, amino acids fluctuated for some plant species.

3.4 The Plant-Protective Benefits of Floral Nectar

It is clear that the primary function of floral nectar is to attract pollinators. Nectar secretion is timed to coincide with pollinator activity, and the quantity of nectar is related to the needs of the preferred pollinators (Cruden et al., 1983; Opler, 1983; Pacini et al., 2003). Furthermore, nectar is nutritionally optimized by the plant to attract the preferred guild of pollinators (Baker and Baker, 1973; Kaczorowski et al., 2005; Petanidou et al., 1996; Stiles and Freeman, 1993; but see Galletto and Bernardello, 2004; Torres and Galletto, 2002). Still, evidence suggests that floral nectar may have an additional function of attracting protective arthropods, thereby adding a new level of complexity to the dynamics of floral nectar secretion.

First, plant protectors and pollinators are not exclusive groups, and several taxa fulfill both roles, notably the Syrphidae and Bombyliidae. Certain flowers are clearly adapted to target these flies as their primary pollinators, although it is unsubstantiated whether syrphid pollinated plants are in some way protected. Syrphids prefer hexose-dominant nectars in phrygana habitats, and the plant community produces this nectar type during the winter months when these syrphids are most abundant (Petanidou, 2005). Competitive interactions among these dual-functioned floral visitors can be quite stiff. Reader et al. (2005) show that bees do not visit flowers that have previously been visited by syrphids, but syrphids feed at flowers visited by bees. Presumably, the exclusion of bees from syrphid-visited flowers may reduce pollination of the plant if syrphids are less efficient pollinators. Competition among flower-visiting bombyliids is such that some species are excluded from their preferred nectar sources (Toft, 1983, 1984). The importance of attracting these beneficial insects to flowering plants, beyond just their role in pollination, has yet to be resolved.

Flowers often produce more nectar than is actually needed (Harder and Barrett, 1992). These authors report that pollen is completely removed from flowers by mid-day, and that 45% of nectar had no function in pollination of the plant. On one hand, this excess nectar may simply be a form of bet hedging; by producing more than is needed, the flower somehow has a better chance of getting pollinated. But flowers also change nutritionally as they age (Percival, 1961; Petanidou et al., 1996) – what is the biological significance of this change? Since pollination often occurs soon after nectar is initially produced, perhaps the post-pollination nectar may alter its nutrition to make it more attractive to other visitors, including natural enemies (Prys-Jones and Willmer, 1992).

As mentioned earlier, the amino acid profiles of nectar give floral nectar its taste, and the complement of amino acids likely restricts which natural enemies may come calling. Hagen et al. (1976) is one of the first biological control scientists to suggest that amino acid contents of sugar sources may be important as attractants to natural enemies. In their work ten amino acids (tryptophan in particular) are attractive to *Chrysoperla carnea* in the field, whereas sucrose only arrests the movement of individuals. The potential importance of amino acids as attractants to natural enemies is reiterated in the ant literature (Bluthgen and Fiedler, 2004). Glycine is a gustatory stimulant for the ant *Camponotus japonicus* (serine and methionine have no effect on ant behavior in this study) (Wada et al., 2001). The greater the content of glycine, the more preferred the sugar solution (Fig. 3.5). Adding glycine to sucrose results in the stimulation of a sugar cell on a labial sensillum, a greater stimulation than glycine or glucose alone. Other nectar amino acids that elicit a gustatory response in natural enemies are tryptophan, phenylalanine and GABA (Syrphidae and Coleoptera), and asparagine is repellent to all insects tested (Petanidou et al., 2006). Other nutrients found in nectar that are known to affect natural enemy behavior are fructose (syrphids, wasps, and beetles) (Petanidou et al., 2006), and chemical extracts such as ethanol, eugenol, and methoxyeugenol (syrphids, ants, vespids, coccinellids, cantharids, and ichneumonids) (Jakubska et al., 2005). Lanza and Krauss (1984) show that ant species have a stronger response to artificial nectars when they mimic the amino acid profiles of real floral nectars. In particular, *Monomorium* and *Leptothorax* ants are more attracted to the nectars with alanine, arginine, cysteine, aspartic acid, methionine, and serine. Other work by Lanza (1988) calls this initial study into question, and raises the notion that the physiological status of natural enemies may alter the

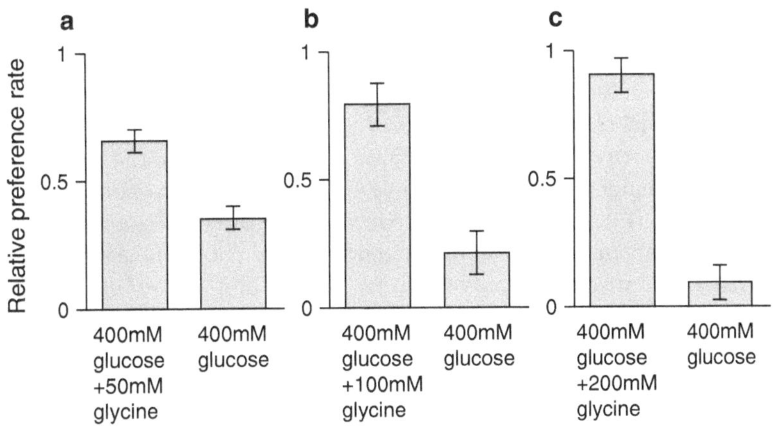

Fig. 3.5 The preference of *Camponotus japonicus* for 400 mM glucose plus (a) 50, (b) 100, or (c) 200 mM glycine is compared with that of 400 mM glucose alone (Reproduced from Wada et al., 2001. With permission by Oxford Journals)

importance of amino acids as foraging cues. In addition to taste, amino acids may be of nutritional importance to insects with low protein diets (such as butterflies) (Jervis and Boggs, 2005; Mevi-Schutz and Erhardt, 2005), a factor that seems less influential for entomophagous species (as suggested by Toft, 1983, 1984, and observed by McDougall and Mills, 1997).

Finally, while floral nectar is often chemically and structurally protected from many natural enemies, at least two cases illustrate that nectar derived from flowers can function protectively for plants in a similar fashion as EFNs. The first example of this is with post-floral nectar in *Mentzelia nuda* (Keeler, 1981). In this case, after the flower senesces the floral nectaries continue to produce nectar (less abundant, but with higher levels of lipids). This post-floral nectar is attractive to ants which cannot access the nectar when the flower is still in place due to the dense stamens. In this case, the floral nectary functions as an EFN, reducing granivory by attracting natural enemies. Another example of the protective benefits of floral nectar is with *Croton suberosus* and the predatory wasp *Polistes instabilis* (Dominguez et al., 1989). This wasp's activity closely coincides with daily nectar production rates in the plant. Excluding the wasps from visiting the nectar leads to increased levels of herbivory by *Hipercombe* caterpillars, largely because the wasps remove more caterpillars from nectaried plants. The problem with this system is that *Croton suberosus* is wind-pollinated, so again the floral nectaries are functioning as EFNs. While neither of these examples provides overwhelming evidence that floral nectar may function protectively, these studies indicate that natural enemies can improve the fitness of the plant if given access to floral sources of nectar.

3.5 Cost of Nectar Production

Given the high nutritional value and copious amount of nectar produced by some plants, it is feasible that nectar production is costly to the plant; the literature offers little consensus on this topic. Using the caloric content of sugars, nectar quantity produced, and total photosynthate produced by plants, the energy devoted to nectar production has been estimated several times. Harder and Barrett (1992) conclude that nectar production comes at a negligible cost to *Pontederia cordata* plants. Only 1.44 J of energy is devoted to nectar production, 3% of the total energy found in each flower. In another study, up to one-third of daily photosynthate produced by *Asclepias syriaca* was transferred into nectar (a range of 4–37% of daily photosynthate), and excess of this costly nectar is resorbed by the plant (Southwick, 1984). Using seed set as a currency, Pyke (1991) found that nectar production reduces the resources that can be devoted to seed production. Specifically, *Blandfordia nobilis* flowers that are artificially pollinated and do not have to replace nectar removed by visiting insects, produce more seeds than those that have access to pollinators. Pyke points out that the efficiency of pollination by insects may overcome the cost of nectar production.

Heinrich (1975) says it well when he points out how difficult it is to make generalizations on this topic. For example, the amount of nectar per flower in *Ochroma lagapus* can vary by as much as 60,000 times (Heinrich, 1975). Still, the costs and benefits of nectar production have only been examined on a per plant basis; the benefits of attracting pollinators and plant protectors on plant communities and populations are likely to outweigh the costs paid by individuals. To summarize, the quantity of nectar that is ultimately produced is a trade-off between mutualistic benefits and metabolic costs of production.

3.6 Defenses of Floral Nectar

Nectar is a valuable source of nutrition for insects, and the act of thievery and robbery (as defined by Inouye, 1980) by non-pollinators has great consequences for the reproductive success of the plant. To deal with this dilemma, Müller (1883) suggests that flowers exclude unwanted nectar thieves through flower color or odor, offering sparse or unappealing resources, and through altering the time and duration of flowering. Feinsinger and Swarm (1978) reaffirm these early thoughts and suggest that flowers defend their nectar either chemically or through floral architecture. Also, Feinsinger and Swarm point out that the metabolic cost of crimes against nectar may not warrant any defense at all since defenses are potentially costly in terms of energy and pollinator efficiency. They suggest that some flowers simply tolerate a certain level of nectar larceny. At first glance, nectar defense seems a paradox: if the whole idea behind nectar is to attract pollinators, then how does one avoid poisoning the well? Adler (2000) suggests that nectar defense actually fosters reproductive success since it promotes pollinator fidelity. Once a pollinator adapts behaviorally or physiologically to collect nectar from a toxic flower species, then it may be more likely to return to this species, thereby increasing the likelihood of cross pollination.

3.6.1 Architectural Defenses Guarding Floral Nectar

Floral architecture is a first, and often the only, line of defense against nectar larceny. The accessibility of nectar tends to increase the more open a flower is, and corollas with wide apertures tend to have a greater diversity of visitors than closed flowers (Galletto and Bernardello, 2004; Koptur and Truong, 1998; Leius, 1967; Opler, 1983; Stang et al., 2006; Tooker and Hanks, 2000; Vattala et al., 2006). Still, this is only true for larger floral visitors, many parasitoid wasps and small predators (e.g., anthocorids) are unhindered by narrow corollas (Baggen et al., 1999; Jervis et al., 1993; Skirvin et al., 2007), and may actually prefer narrower corollas (Fiedler and Landis, 2007a, b). Stamens and petal density may favor larger natural enemies, which can muscle their way through these structures better than smaller, weaker insects (Patt et al., 1997). Long-tubed flowers tend to have higher quantities of nectar and possibly more suitable nutrient concentrations and constituency for certain natural enemies, which may thereby drive the adaptations necessary for feeding on this type of flower (Gilbert and Jervis, 1998; Gilbert, 1981; Idris and Grafius, 1995; Jervis, 1998). Floral area per plant also strongly influences the number of natural enemies found per plant (Fiedler and Landis, 2007b). Beyond the aperture of the corolla and floral area, the factors that influence the suitability of flowers for natural enemies are usually poorly understood. And in certain circumstances, the patterns in floral preferences in natural enemies are not well explained by flower morphology or nutrition (Vattala et al., 2006). In these situations, attention should be given to the chemical defenses of nectar.

3.6.2 *Chemical Defenses of Floral Nectar*

Humans have known that nectar is often poisonous for centuries, this knowledge stemming largely from experience with what is commonly referred as mad honey (Ecker, 1933; Vansell, 1926). One of the earliest records of mad honey comes from Xenophon's writings in the 4th century B.C. (Ott, 1998). Since then, the effects of honey gathered from toxic nectar or honeydew on religious ceremonies and transcendental experiences have popped into human culture repeatedly (Koca and Koca, 2007; Ott, 1998), and it is still a trendy topic in popular bee journals today. Understandably so, with reports such as those of a German naturalist in 1930 Brazil who says that after eating contaminated honey, German soldiers "... *tore off their cloths and butted their heads on nearby objects*" (Krochmal, 1994). While not nearly as amusing as this report, the effects of toxic nectar on insects are varied and widespread.

Toxic nectar is found in the flowers of at least 21 plant families (Adler and Irwin, 2005). Within the plant kingdom, members of the Ericaceae, Orchidaceae, Euphorbiaceae and Solanaceae are particularly notorious for presenting poisonous floral nectar (Koca and Koca, 2007; Krochmal, 1994; Ott, 1998; Wood et al., 1954). A staggering array of defensive chemicals can be isolated from nectars, including phenolics, alkaloids, glycosides, saponins, non-protein amino acids, alcohols, and ammonia (Adler, 2000; Baker and Baker, 1978, 1983; Deinzer et al., 1977; Hagler and Buchmann, 1993; Jakubska et al., 2005; Krochmal, 1994; Prys-Jones and Willmer, 1992; Smith et al., 1990). To give an idea of the preponderance of these chemicals in nectar, Baker and Baker (1983) note that 36% of nectars collected from tropical plant species contain non-protein amino acids, 12% alkaloids, and 40% phenolics. Although certain species are more inclined to producing toxic nectar, it is not always easy to predict when and where this type of nectar will pop up in a plant (Clinch et al., 1972; Macleod-Carey et al., 1959).

The question of why these chemicals are so pervasive in nectars has been discussed by several authors over the years who present the following hypotheses.

First, by poisoning one's nectar, a plant discourages generalist pollinators and nectar robbers from visiting and encourages fidelity by what is hopefully an optimal pollinator (Adler, 2000; Prys-Jones and Willmer, 1992; Stephenson, 1982a). In line with this theory, Rhoades and Bergdahl (1981) predict that (1) species that constitute a small portion of the community are more likely to have toxic nectar (thus areas with low floral diversity are less likely to have species with toxic nectars); (2) flowers that occur sympatrically should have a greater diversity of toxins in the nectar; and (3) closely related species will have closely related nectar toxins. An added benefit to intoxicating floral nectar is that the pollinator becomes inebriated (Adler, 2000). In *Epipactis helleborne*, ethanol is produced by nectar-bound microorganisms and pollinators that drink this spiked nectar move sluggishly (Jakubska et al., 2005). These narcotic effects on pollinators are seen in other nectars as well (Clinch et al., 1972; Palmer-Jones and Line, 1962), and may serve to increase the likelihood of pollination. Drunken pollinators spend more time per flower and fly shorter distances, both beneficial traits from the standpoint of cross pollination on a local scale.

A relatively new idea that is gaining steam is that secondary compounds are not intentionally shunted into nectar nor do they have any deterrent effects on floral visitors (Adler and Irwin, 2005; Landholt and Lenczewski, 1993; Singaravelan et al., 2006). Rather they may be a pleiotropic result of chemical defenses found throughout vegetative portions of the plant (Adler, 2000). Moreover, herbivory on *Nicotiana tabacum* leads to higher levels of alkaloid in floral nectar (Adler et al., 2006) (Fig. 3.6), and fertilization led to higher levels of alkaloids in both leaves and nectar. One interpretation of these results is that secretion of defensive chemicals in vegetative portions of the plant is linked to the levels of defensive chemicals in nectar. But the level of toxicity in nectar needs to be considered in context with other floral resources in a habitat (Gegear et al., 2007). In this study, alkaloids in the nectar of *Gelsemium sempervirens*

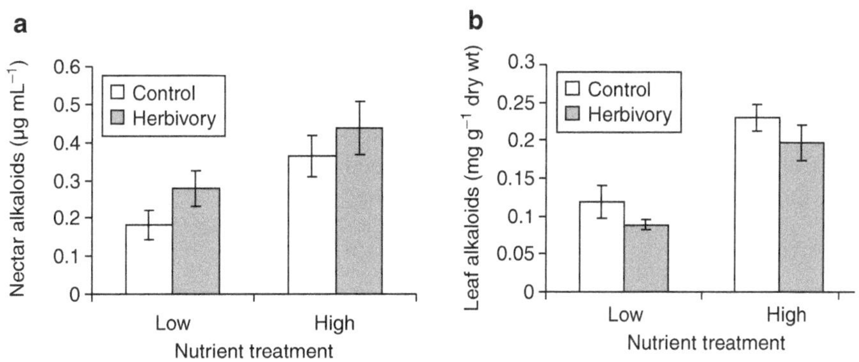

Fig. 3.6 Effect of nutrient treatment and herbivory on alkaloid concentration in (a) nectar and (b) leaves. Note different units for nectar and leaf concentrations. Error bars represent standard error. (With permission by Adler et al 2006).

deter bumblebees, but the magnitude of this deterrence depends on how appealing alternative nectar sources were. When flowers without alkaloids are offered in addition to the *G. sempervirens* flowers, the alkaloid-rich nectar is avoided almost entirely. When the sucrose content of the nectar is artificially increased, the nutritional benefit over-rides the toxicity of the nectar. In other words, we are only beginning to understand the ecological functions of secondary chemicals in nectar, and their effects on higher trophic levels seem like one productive avenue for future research.

Surprisingly, the role of toxic nectar on natural enemy behavior has been poorly explored, the exception being ants. In a 132-word paragraph published in Biotropica, Janzen (1977) instigated a long series of papers aimed at proving or disproving the notion that toxic floral nectar discourages flower visitation by ants. The first response to this hypothesis was to document that ants visit floral nectar more than was originally believed; not as often as they visit honeydew or EFN, but still often enough (Baker and Baker, 1978; Churchill and Christensen, 1970; Haber et al., 1981). In a survey of tropical flowers, 27 species (even those that contained high levels of phenolics and alkaloids) were visited by ants in the field (Baker and Baker, 1978). Koptur and Truong (1998) tested the nectars of 29 flowers, and found that at least five ant species would feed on at least some of the nectars. One ant species, *Paratrechina longicornis*, consumed the nectar from 17 of the flower species, while other ants consumed only a few of the tested nectars. Nevertheless, some floral nectar is known to be repellent or toxic to ants (Feinsinger and Swarm, 1978). A great example of this was the iridoid glycosides found in *Catalpa speciosa* (Warder ex Barney) floral nectar (Stephenson, 1982a). Ants preferred to feed on pure sucrose solutions over those that contained the glycosides, and those ants that did feed on the spiked sucrose solution became disoriented and often fell from the tree. Research has repeatedly shown that floral architecture often excludes walking thieves before they ever come into contact with the nectar (Feinsinger and Swarm, 1978; Schubart and Anderson, 1978). Guerrant and Fiedler (1981) suggested that in some cases the floral tissue was actually more toxic than the nectar itself. Placing macerated petals from 11 of 17 species into nectar deterred ant feeding, and nearly all floral tissue had secondary phytochemicals known to repel insects. Interestingly, although multiple nectars have been tested against individual ant species, I was not able to find any work that tested the range of toxicity of a single nectar against multiple ant species. So the breadth of effectiveness of toxic nectar to ants remains to be evaluated. To conclude, the consensus from Janzen's hypothesis that ants are repelled by toxic nectar is in, and the answer is: sometimes, but not as often as proposed.

3.7 Conclusions

Floral nectar is abundant and nutritious, and is indisputably the best guarded of the sugar sources presented in this book. Because of its importance to pollination and the reproductive success of the plant, the nutrition of floral nectars is well described for numerous plant species. While pollinator-flower interactions have been well explored, much more research is needed before we begin to understand how natural enemies respond to the nutrition and defenses of floral nectars. Several gaps in our knowledge of these interactions are ripe for exploration. First, the attractiveness of sugars to natural enemies has been nicely established, but how entomophagous arthropods respond to the amino acids that give nectar its taste remains unclear. Second, how do natural enemies respond to toxic nectars? In reviewing the data I hypothesize that the presence and nutrition of floral nectar attracts natural enemies, and thereby indirectly provides protection to the plant from herbivory. Several trends suggest that this may be the case.

1. Natural enemies are frequent flower visitors, sometimes even functioning as pollinators.
2. Natural enemies are known to use certain amino acids found in nectar (that give nectar its 'taste') as gustatory cues.
3. Flowers produce more nectar than they need, and nectar changes nutritionally over the life of the flower. What is more, older nectar often seems to become more hexose-rich, which has been repeatedly been shown to be more attractive to natural enemies than sucrose solutions.
4. Although there have been no studies clearly demonstrating that a natural enemy who visits floral nectar helps to protect the plant from herbivory, there are several cases when floral nectaries function as EFNs to reduce herbivory, and biological control of crop pests is generally improved when sources of floral nectar are available.

Clearly, biological control scientists need to gain a better understanding of how entomophagous species are using floral nectar in order to improve the reliability of biological control in farmland.

Chapter 4
Extrafloral Nectar

Extrafloral nectaries are sugar-secreting organs that can be found on numerous tissues throughout numerous plant species worldwide. They differ from their floral counterparts in that their primary function appears to be the attraction of protective arthropods. The main nutrients found in EFN are the oligosaccharides fructose, glucose, and sucrose, although many amino acids, other sugars, and micronutrients have been isolated from EFN. The protective benefits of EFNs stem from their attractiveness to beneficial arthropods, ants being the best studied. In some plant species, the flow and composition of EFN changes in response to herbivory in similar ways to other induced plant defenses, underscoring the importance of natural enemies to EFN-bearing plants.

4.1 Phylogenetic and Geographical Trends

Extrafloral nectaries are widespread phylogenetically, and these organs likely evolved multiple times within plants. At least 93 families in 39 orders of plants bear EFNs (Pemberton, 1992). Dicotyledons most frequently bear EFNs, but grasses and ferns also secrete EFN (Elias, 1983). Beattie (1985) mentions that ferns are some of the first plants to show signs of EFNs, dating back to the Cretaceous, but specific details of these plants are not presented. One early fossil record of plant tissues bearing EFNs is from a 35 million year old leaf of *Populus crassa*, from the Florrisant Formation of the Oligocene (Pemberton, 1992). It is notable that ants were one of the most abundant insects found in this same fossil formation, which provides at least circumstantial evidence that contemporary ant-EFN mutualisms may have been in place millions of years ago.

Several geographic trends in EFN-bearing plants are pointed out in the literature. First, similar to many insect genera, there is a strikingly clear positive correlation between the prevalence of EFN-bearing species and their proximity to the equator (Pemberton, 1998). This pattern also holds for the percent cover occupied by EFN-bearing species within given habitats (Pemberton, 1998) (Fig. 4.1). Bentley (1977a)

J.G. Lundgren, *Relationships of Natural Enemies and Non-Prey Foods*, Progress in Biological Control 7,

Published by Springer Science+Business Media B.V.

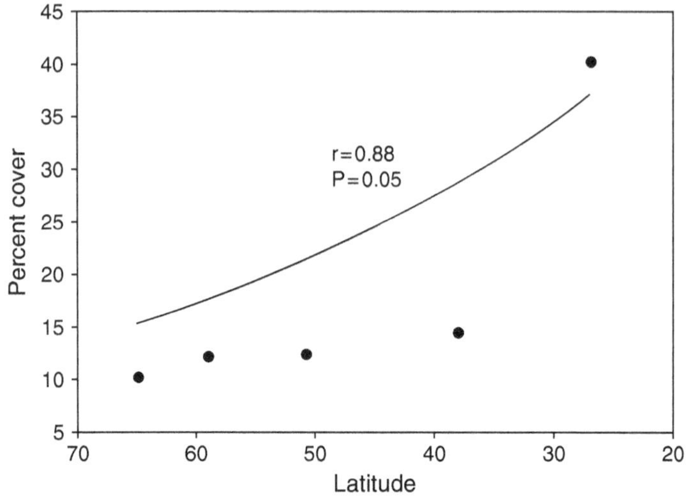

Fig. 4.1 Percent cover (log transformed) of plants with EFNs as a function of latitude (Reproduced from Pemberton, 1998. With permission from Blackwell)

suggests that ant species follow this same trend and may partially explain the incidence pattern (it is notable that other beneficial insects that feed on EFN are more diverse at lower latitudes as well). On a more local scale, EFN-bearing plants constitute 2.5–8% of the flora of various habitats around the world, and can occupy 0–55% of the plant cover in these habitats (Pemberton, 1992, 1998; Pemberton and Vandenberg, 1993).

4.2 Physiology and Nutritional Composition of EFN

The EFNs of many plants are localized on tissues that most require protection, often the developing leaves. But EFNs also occur on fruits, flowers (outside the corolla), bracts or pedicels bearing the flowers, petioles, and stipules (Baker et al., 1978; Beattie, 1985; Bentley, 1977a; Elias, 1983; Oliveira and Leito-Filho, 1987). In fact a single plant species may have EFNs on several organs; a case in point is *Campsis radicans* which has four sets of independently operating EFN systems (Elias and Gelband, 1975). Their location on the plant (and possibly relative production rates of the different nectaries) determines which insects frequent them. For instance, 92% of *Solenopsis invicta* visit the EFNs located near cotton blooms, while only 9% visit the foliar EFNs (Agnew et al., 1982). Extrafloral nectaries vary in their morphology and physiology, and range in structure from only a few cells to being well-developed, cup-like organs. Elias (1983) propose that there a re seven structural categories of EFNs, six of which were first coined by Zimmerman in 1932. These EFN types are formless, flattened, pit, hollow, scalelike, elevated, and embedded nectaries. EFNs are not necessarily vascularized (Elias, 1983), but always have photosynthesizing parenchyma cells in close approximation (Pacini et al., 2003). Sometimes, the only cellular differentiation of EFNs from other plant tissues is a slight gap between the epidermis and the subtending layer of cells, which stems from the accumulation of nectar at these sites (Elias, 1983). Because EFN often originates from photosynthate of nearby cells, sugars are only produced during the day and are stored as starch, which is hydrolyzed if nectar is required at night (Pacini et al., 2003). Vascularized EFNs are comparatively uniform in structure, and usually are larger than the unvascularized variety. Usually both phloem and xylem are associated with vascularized EFNs.

The primary nutritional components of EFN are mono- and di-saccharides, with glucose, fructose and sucrose dominating the sugars of most species (Bowden, 1970; Caldwell and Gerhardt, 1986; Elias and Gelband, 1975; Heil et al., 2000; Wäckers, 2001; Yokoyama, 1978). I only note one species, *Ipomoea leptophylla*, in which one of these three sugars was missing from the EFN (sucrose) (Keeler, 1980). A dozen other sugars are present in EFN, but these often occur at lower concentrations than the 'big three'. The actual sugar profiles of EFN harvested from EFNs on different organs of a plant may be quite different (Elias and Gelband, 1975; Keeler, 1977), and may be a reflection of attempts to target specific taxa to offer protection to the different tissues.

Amino acids and micronutrients are also frequently found in EFN. Serine is consistently one of the most abundant amino acids in EFN (Baker et al., 1978; Caldwell and Gerhardt, 1986; Heil et al., 2000; Keeler, 1977, 1980; Pickett and Clark, 1979), and more than 22 others (including all of those essential to insects) are present in the EFN of plants (Baker et al., 1978). In fact, EFN of some species can possess nearly 20 amino acids (Caldwell and Gerhardt, 1986; Hanney and Elmore, 1974; Smith et al., 1990). Baker et al. (1978) survey 21 plants that produced both floral and EFN, and report that universally the amino acids differed between these two sources. EFN generally was richer in cysteine, lysine, isoleucine, tryptophan, methionine and valine than floral nectar. Also, Keeler (1977) notes that at least in *Ipomoea carnea*, the breadth of amino acids is greater in EFN than in floral nectar. The community of amino acids is often dominated by 4–5 individuals that comprise the bulk of the amino acid concentration. For instance, asparagine/glutamine, serine, valine, proline, and threonine comprise nearly 80% of the 280 mM amino acid concentration in the EFN of *Opuntia acanthocarpa* (Pickett and Clark, 1979). This may be related to the fact that their protectors rely on these specific amino acids for nutrition or taste recognition of the nectar.

Other nutrients within EFN are seldom reported. For instance, reports of lipids within EFN were particularly hard to come by, although anecdotally the EFN of at least 15 of 83 plant species were found to contain lipids (Beattie, 1985). A more empirical report is that of Caldwell and Gerhardt (1986), who found lauric, myristic, palmitic, palmitoleic, stearic, oleic, and linoleic acids in the EFN of peach (*Prunus persicae*). Another species, *Ipomoea carnea*, had no lipids in its EFN (Keeler, 1977). Caldwell and Gerhardt also reported inositol, an important nutrient for insect nervous system function. The complement of nutrients and minerals in EFN tended to be very consistent within a plant species, but the concentrations of individual nutrients changed dramatically in response to the environment or physiological needs of the plant (Baker et al., 1978; Heil et al., 2000; Keeler, 1977; O'Dowd, 1979; Pickett and Clark, 1979; Smith et al., 1990).

Generally speaking, plants are only too happy to have visitors drink their EFN, and it is much less chemically defended than floral nectar. Still, secondary chemicals are present in some EFN, including phenolics (Keeler, 1977) and tannins (Knapheisowna 1927; as cited in Putnam, 1955). Indeed, one survey reports that non-protein amino acids are more abundant in EFN than in their floral counterparts (75% of those species examined had EFN with non-protein amino acids) (Baker et al., 1978). This notwithstanding, the chemical defenses of floral nectar have received much more attention than secondary compounds in EFN; it isn't clear whether this is a case of evidence of absence, or absence of evidence.

The concentration of nutrients within EFN is dependent on the production level of the nectaries, and this varies substantially within the plant kingdom. Nectar flow rate of *Opuntia acanthocarpa* is 0.3 μl per gland per day (Pickett and Clark, 1979). On a per leaf basis, 0.2 μl are produced by multiple EFNs per day in *Ipomoea carnea* (Keeler, 1977), but as much as 125 μl are produced per leaf per day in *Ochroma pyramidale* (O'Dowd, 1979). The amount of EFN (and subsequent concentration of the nutrients) even varies among the nectaries on a plant species. Nectaries on the petioles of *Ochroma pyramidale* produce 15 times more nectar than those on the leaf veins. Not surprisingly, the amino acid contents of the petiole EFN are much more dilute than those on the leaf veins (1.5–2 vs. 2.5 units on the histidine scale) (O'Dowd, 1979). Ultimately, the amount of nectar and nutrients therein dictate which visitors will attend the plant. This simple fact has had broad implications for the evolutionary function of EFNs.

4.3 Protective Benefits of EFNs

> *Thus it appears that the secretion of nectar by extrafloral glands on poplars attracts to the plants many insects, of which at least three kinds- ants, ichneumonids, and lady-birds- are of benefit to them* (Trelease, 1881)

While the main function of EFNs is currently believed to be as an attractant to plant-protecting natural enemies, this has not always been the popular stance. Indeed, the ecological significance of EFNs was hotly contested for many years (Bentley, 1977a). Wheeler (1910) presented a nice early discussion of the debate over the ecological function of EFNs, and detailed that a large group of scientists believed that EFNs were strictly excretory and that any attractive benefits to predacious insects were secondary. Baker et al. (1978) suggested that indeed the original function of EFNs may have been strictly excretory, but as natural enemies adapted to this abundant and easily obtained food source, the mutualism evolved.

Without question, ants are the best studied group of natural enemies to visit EFNs, and their protective behavior is well documented.

> *the ants as a rule show a disposition to fight, rather than give up their places by the glands, over which they sometimes remain for hours, and some species are so pugnacious that the slightest jar to the branch upon which they are is sufficient to cause them to assume the offensive, and, with mandibles open, they rush about in search of the cause of the disturbance.* (Trelease, 1881)

This early observation isn't entirely true; ants vary substantially in their protective capabilities, as pointed out by Elias and Gelband (1975). These authors report that *Formica* sp. are excellent guards of EFNs, while *Crematogaster lineolata* are quick to yield to other visitors. One explanation for their predominance as EFN visitors is that the ants simply are able to outcompete other visitors through numbers and organization. Kost and Heil (2005) determine that ants visit lima bean tendrils more frequently when they know that the tendrils have produced nectar in the past. They even go so far as to prefer the tried and tested source of nectar over new artificial nectaries.

With all of the attention given to ants, other entomophagous visitors to EFNs are often given short shrift. Research finds that these alternative natural enemies (particularly predatory wasps and flies) can be equally as important as ants in protecting the plant from herbivory (Beattie, 1985; Choh and Takabayashi, 2006; Kost and Heil, 2005). Moreover, non-ant protectors may explain the production by plants of EFNs in habitats or circumstances where the ants are not effective protectors, such as on Hawaii where no ants are native (Pemberton and Vandenberg, 1993). However, a number of plants have lost the EFN trait in habitats lacking ants such as *Gossypium tomentosum* in Hawaii (Wäckers and Bonifay, 2004). This is evidence that ants are a major selective force. Ants predominate the community of visitors to *Ipomoea carnea* EFNs, but the only observed event of predation by an EFN visitor on a caterpillar is by *Polybia occidentalis*, a predaceous wasp (Keeler, 1977). Peach EFNs are attractive to ants and parasitoid wasps, both of which contribute to the nearly 90% reduction in damage caused by the oriental fruit moth (*Grapholita molesta*) (Matthews et al., 2007) (Fig. 4.2). Another example comes from Central America, where micropezid flies make up 17% of the visitors of the EFNs of *Macaranga tanarius* (Heil et al., 2004). These flies do not protect the plants from herbivory as well as ant visitors, but they will actively defend the EFNs from interlopers. Heil et al. report them chasing off several arthropods

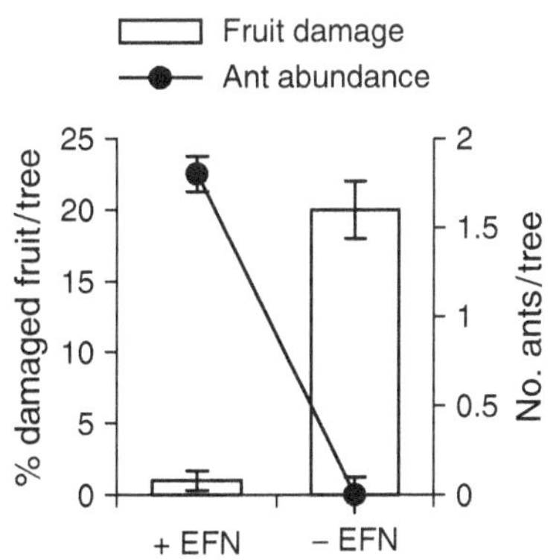

Fig. 4.2 Effect of leaf EFNs (±EFN) on percentage of fruit (15/tree) infested with *Grapholita molesta* larvae (Y1 axis) and number of ants foraging in tree canopy (Y2 axis). Back-transformed means are plotted with 95% CI. (Reproduced from Matthews et al., 2007. With permission by the Entomological Society of America)

visiting the nectaries, including some ants that are larger than the micropezids! With the diverse community of potential protectors, it becomes a challenge for the plant to produce the appropriate nectar for the protector du jour. For example, the physiological needs of an ant colony are likely to change over the season (Beattie, 1985).

The end result of all of these natural enemies visiting the EFNs of plants is that EFN-producing plants suffer less herbivory. In the literature, there are numerous examples of improved plant fitness associated with ant attendance on a plant. Ants reduce herbivory (Barton, 1986; Heil et al., 2001; Pickett and Clark, 1979), decrease damage to flowers (Keeler, 1980), reduce the destruction of fruits (Barton, 1986), and improve overall seed set in plants (Keeler, 1980; Schemske, 1980). These protective insects may also chase off insects that seek to rob floral nectar (Keeler, 1977). Removal of the EFNs customarily reduces the benefits provided by the ants.

A question that remains is why sugar-loving ants are apt to kill insects on the plants that have increased levels of EFN (Heil et al., 2004)? On one hand, the ants may just be foraging for prey, and the provision of EFN allows them to forage more efficiently. But also the ants may be attempting to preserve the health and vigor of the EFN-bearing plant. Heil et al. (2004) suggest that this latter hypothesis may be more relevant to those species with obligate mutualists. Evidence for this plant preservation hypothesis is when the ants don't consume herbivores on EFN-protected plant species, simply tossing invaders off the plant. In this case, the only benefit received by the ants for their intervention is the continuance of their EFN-bearing host.

4.4 Temporal Occurrence of EFN

Therefore, during periods of growth a potentially large number of maturing structures [on plants] *are exposed to short periods of acute vulnerability.* (Beattie, 1985)

EFN production tends to peak in reproductive and pre-reproductive plants; young vegetative tissues require protection from herbivores, as do flowers and young fruit (Bugg, 1987; Elias and Gelband, 1975; Heil et al., 2000; Keeler, 1977, 1980; Pickett and Clark, 1979; Rose et al., 2006; Ruffner and Clark, 1986; Tilman, 1978; Trelease, 1881; Wäckers and Bonifay, 2004; Yokoyama, 1978). Older plant tissues tend to be tougher, have accumulated higher levels of secondary compounds, and developed physical defenses such as thorns and spines, thus alleviating the need for insect-coerced protection from EFN (Beattie, 1985; Heil et al., 2000; O'Dowd, 1979). Some plants also increase the number of their EFNs to attract more natural enemies at key times (Mondor and Addicott, 2003) (Fig. 4.3), or to those structures that are particularly valuable in terms of plant fitness. Nevertheless, even though EFN production is localized temporally, it is typically available for substantially longer periods of time than floral nectar sources (Rose et al., 2006). Another consideration is that for EFNs to have any protective benefit to the plant, they must coincide with their bodyguards (as in Stephenson, 1982b) (Fig. 4.4). Attracting ants can be a bit dicey, they are central place foragers after all and cannot be found in all places at all times. Plants with EFNs have adapted one of two strategies for surmounting this problem:

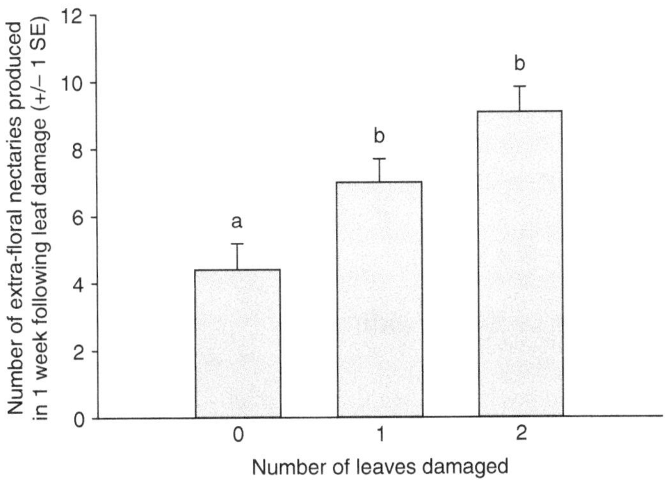

Fig. 4.3 The number of EFNs produced by *Vicia faba* 1 week following leaf damage. Columns with different letters are significantly different. (Reproduced from Mondor and Addicott, 2003. With the permission of Blackwell)

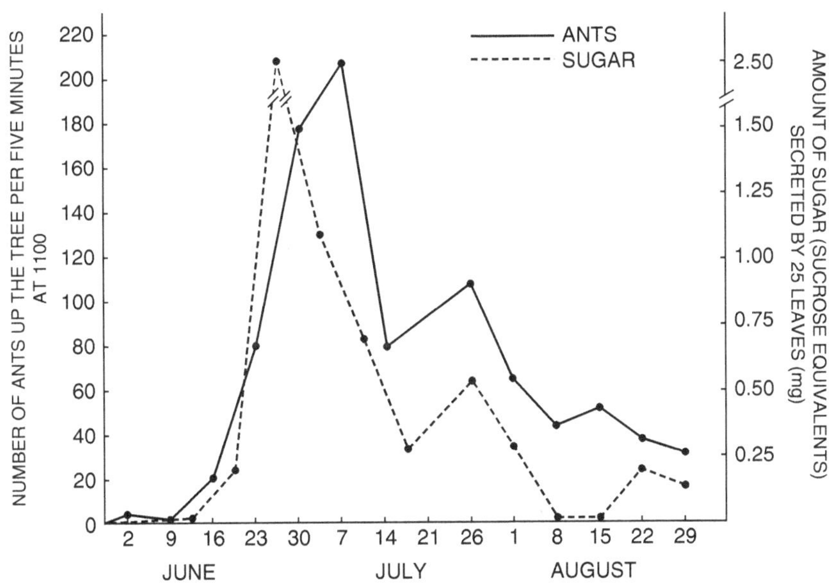

Fig. 4.4 Season pattern of EFN secretion and the seasonal foraging activity of ants on *Catalpa speciosa*. (Reproduced from Stephenson, 1982. With permission by the Ecological Society of America)

(1) intensify the relationship with key ant species known to occur when the plant needs them, and (2) attract as broad a range of natural enemies as one can in the hopes that some will always be on hand when the plant needs them. Ruffner and Clark (1986) provide one example of a plant that alters its EFN production specifically to attract ants. The barrel cactus *Ferocactus acanthodes* var. *lecontei* produces four

times more EFN during its reproductive stages, and this nectar has both more glucose (50% higher concentration) and 2.5 times as much water produced per day than during other growth stages. Activity of the ants, *Crematogaster depilis* and *Forelius pruinosus*, closely track the nectar levels. Other studies reaffirm these findings of ants being more attracted to plants at critical times in the plant's life history thanks to alterations in EFN production (Agrawal and Rutter, 1998; Bentley, 1977b; Pickett and Clark, 1979; Stephenson, 1982b; Trelease, 1881).

4.5 Regulation of EFN Production

A series of recent studies is shedding light on exactly how plants initiate increased EFN production in response to herbivory, and the cascade of events traces the same pathway as other induced plant defenses. At least in some plants, small quantities of EFN is produced over extended periods of time, and the plant is able to kick up production very quickly, thus EFN can be considered both a constitutive and induced defense in some plants (Wäckers and Bonifay, 2004; Wäckers et al., 2001). In *Gossypium*, the bracteal EFNs of cotton produce high levels of nectar, irrespective of herbivory during the flowering and fruiting, while foliar EFN is secreted largely in response to herbivory. Herbivory-induced production of EFN is very localized (both temporally and spatially) in some plant species, arising only near where the plant is actually being damaged and only for a short duration following an herbivory event (Wäckers et al., 2001).

Jasmonic acid (JA) is an important elicitor that initiates the induced production of EFN. JA, a component of the octadecanoid signal pathway that induces a range of plant defenses, occurs at higher levels in plants experiencing herbivory, and simply placing aqueous JA solution upon a plant tissue increases the neighboring EFN production 30–75 times in lima beans (Heil, 2004), and 2–5 times in *Macaranga tanarius* (Heil et al., 2001). Heil et al. (2001) also show that mechanically removing the EFNs and placing JA on plant tissue has the same effect on ant visitation as having no JA, ruling out the idea that the ants may be attracted to the JA itself (a fact reiterated experimentally by Kost and Heil, 2005). *Macaranga tanarius* treated with JA experience an increase of ant visitors within 3 h of treatment, and incur less overall herbivory than those where the EFNs have not been up-regulated (Heil et al., 2001).

Another remarkable phenomenon is when a plant produces more EFN in response to herbivory on neighboring plants. Specifically, lima beans produce herbivore-induced plant volatiles (HIPVs) in response to herbivory from phytophagous mites and other herbivores, and neighboring lima beans that are not experiencing herbivory respond as though they are, thanks to these HIPVs (Choh et al., 2006) (Fig. 4.5). The active compound appears to be (3Z)-Hex-3-enyl acetate, which is a common HIPV in many plant species. When a plant is being consumed, it emits this chemical, and the EFN production in its neighbors is substantially increased (Kost and Heil, 2006). When the herbivory stops, the EFN production also ceases. But both the neighbors and damaged

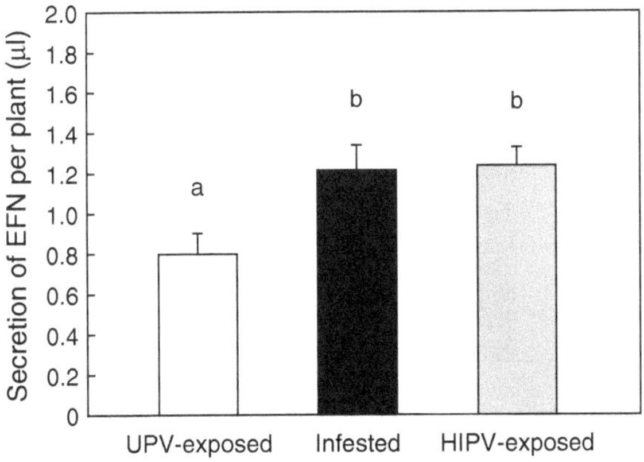

Fig. 4.5 The amounts of EFN secreted by *Phaseolus lunatus* plants exposed to volatiles from an uninfested plant (UPV), *Tetranychus urticae*-infested plants, and plants exposed to herbivore-induced plant volatiles (HIPV) (mean ± SE). The letters above each bar indicate significant differences among treatments (Reproduced from Choh et al., 2006. With permission by Springer)

plants are 'primed', and begin producing EFN even faster when herbivory resumes (Choh and Takabayashi, 2006). Predators search both the damaged and undamaged plants more intensively, and attack any herbivores that are found (Choh et al., 2006; Kost and Heil, 2006). In this amazing case of plant group defense, the uninfested plants cry for help in the anticipation of herbivory. But not having any prey to keep their predators around, they offer their saviors a little sugar instead (Fig. 4.6).

Extrafloral nectar appears to be a fairly cheap defense for plants, and has been shown at least twice to be easily accounted for in the reduction of herbivory experienced by EFN-producing plants. One explanation for why EFNs are such an efficient defense is that their very presence may be a simple and benevolent byproduct of the herbivory event itself. Damaged tissues require less carbon than intact tissues, and one theory is that the extra carbon produced by damaged plants is shunted out the nearest EFN in the form of sugar (Agrawal and Rutter, 1998). O'Dowd (1979) calculates that 1.51 cal per leaf per day is required to produce EFN; essentially 1% of the energy contained within the leaf. He also shows that 13.2 cm^2 of leaf tissue is spared from herbivory in EFN-bearing leaves over leaves where the predators are excluded (this is roughly equivalent to 2–5% of the area of each leaf). Wäckers et al. (2001) come to a similar conclusion when they calculate that EFN production amounts to only 0.1% of a leaf's daily photosynthate production. Of course, other considerations such as indirect costs of herbivory (exposing the plant to secondary pathogens, reducing the photosynthetic efficiency of the leaf) and the mutualism (any damage the natural enemy may inflict on the plant, or complex interactions involving other herbivores such as honeydew-producing sternorrhynchans) may also affect the cost of EFN production. Also, the fact that plants lose this trait in the absence of bodyguards suggests that the cost of EFN production is not negligible. A comprehensive energy

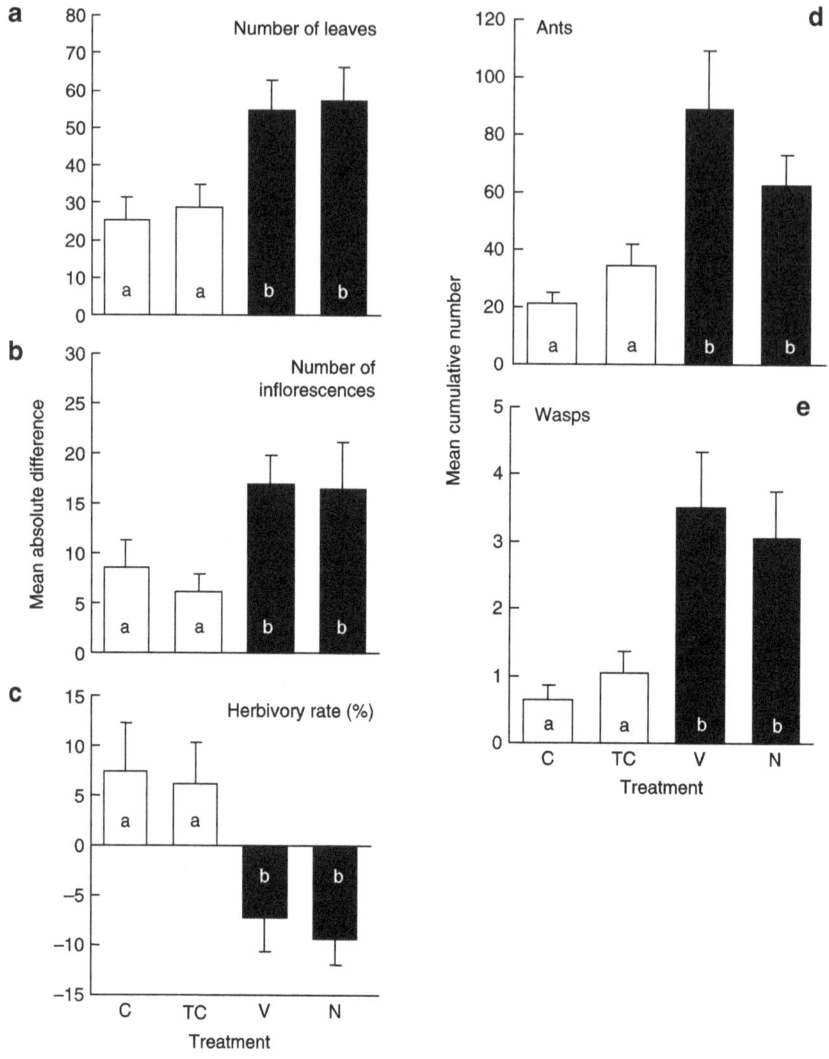

Fig. 4.6 Effects of volatile and nectar treatment on fitness-relevant plant parameters (**a–c**) and beneficial insects (**d** and **e**). Differences between fitness measurements at time 0 and 25 days later, or the number of insects present (mean ± SEM) are presented. Groups of *Phaseolus lunatus* tendrils were left untreated (C), or were treated at regular intervals (3 days) with lanolin paste (TC), artificial volatile blend dissolved in lanolin paste (V), or an artificial extrafloral nectar (N). Different letters indicate significant differences among treatments (Reproduced from Kost and Heil, 2006. With permission from Blackwell)

budget for EFN producing plants would greatly improve our understanding of the efficiency of this defensive system.

4.6 Conclusion

EFN is much more widely used by natural enemies than its floral counterpart, and it is clear that the benefits accrued through the activities of entomophagous species far outweigh the cost to the plant of producing this sugary substance. EFNs can function both as a constitutive or induced defense, and plants communicate their distress to neighbors such that entire groups of plants can respond to herbivory, using entomophagous arthropods as part of their immune system. To the human eye, EFNs are easily overlooked; clearly predators and parasitoids do not suffer from our myopia, to the benefit of the plants.

Chapter 5
Honeydew

A source of nutriment at once so rich and so inexhaustible, could hardly remain unnoticed and unexploited by the ants in their interminable search for food.

Wheeler, 1910

Numerous herbivores, taxonomically aggregated in the Sternorrhyncha and Lepidoptera, produce sugary secretions that are consumed by entomophagous species. Within the Sternorrhyncha, Aphididae, Cercopidae, Cicadellidae, Coccidae, Fulgoridae, Membracidae, Pseudococcidae, and Psyllidae all have honeydew-producing members (Beattie, 1985; Nickerson et al., 1977). Within the Lepidoptera, the Lycaenidae and Riodinidae (it should be noted that some regard the Riodininae as a subfamily of the Lycaenidae) have a Newcomer's gland and associated organs on or near the dorsum of the seventh abdominal segment that produces sugary secretions when solicited by ants (Beattie, 1985; Holldobler and Wilson, 1990). These lycaenid-ant interactions are well reviewed by Holldobler and Wilson (1990). Honeydew provides the basis for a fascinating series of ecological interactions with implications for plants, the honeydew-producing herbivores, and the natural enemies of these herbivores.

5.1 Nutritional Value of Honeydew

Carbohydrates typically comprise more than 80% of honeydew by dry weight (Ewart and Metcalf, 1956; Lamb, 1959). This should not be entirely surprising, since honeydew is largely derived from phloem sap, which is comprised primarily of sucrose. Sugar concentrations in the honeydew are often very similar to those seen in the phloem (Mittler, 1958a), although the types of sugars found in the phloem sap changes as it passes through the sternorrhynchan digestive system. Most studies report fewer than six sugars in the honeydew of sternorrhynchans. Upon ingestion, the enzyme sucrase metabolizes the sucrose-laden phloem sap into fructose and glucose molecules (Karley et al., 2005). Ashford et al. (2000) showed

J.G. Lundgren, *Relationships of Natural Enemies and Non-Prey Foods*, Progress in Biological Control 7,

Published by Springer Science+Business Media B.V.

that *Acyrthosiphon pisum* only uses the fructose moiety of sucrose for its metabolic needs, and excretes the glucose molecule in the form of di- and oligo-saccharides. These oligosaccharides have important implications for the physiology of the sternorrhynchans, and the interactions of other species with honeydew.

In addition to sucrose, fructose, and glucose, there are many insect-synthesized di- and oligo-saccharides found in honeydew. While some of these sugars may be produced de novo by the insects, microbial endosymbionts also synthesize tri-saccharides for their host (Davidson et al., 1994). These sugars function in part to reduce the osmotic differential between the sternorrhynchan and the phloem it consumes (Byrne et al., 2003; Karley et al., 2005 and references therein). Most phloem-feeding sternorrhynchans reduce osmotic pressure in their haemocoel by altering the sugar content of the phloem sap; essentially, they create fewer large sugar molecules from many small ones. As a case in point, the manufacture of melezitose from glucose and sucrose reduces the osmotic pressure experienced by the aphid *Metopeurum fuscoviride* by 25–35% (Woodring et al., 2006). The more honeydew that is produced, the greater the osmotic pressure experienced by the insects, and so one would expect the production of trisaccharides to increase in sternorrhynchans that exude more honeydew (Woodring et al., 2006), but not all data support this hypothesis (Costa et al., 1999). Many of these sternorrhynchan-made sugars are found few other places in nature, and their presence in the guts of natural enemies may be used as an indicator of honeydew consumption in the field. Turanose, erlose, trehalulose, raffinose, stachyose, and bemisiose are all possible indicator sugars (Hendrix and Wei, 1994; Wäckers, 2000; Wäckers et al., 2006 and the references therein) (see Table I.1). Wäckers points out that for a sugar to function as an indicator of honeydew feeding, it needs to be synthesized by phloem-feeding insects, can't occur in other sugar sources, and can't be synthesized by the natural enemies being studied. Recent studies show that a number of natural enemies are capable of synthesizing some of the oligosaccharides that are used as indicators of honeydew feeding (Wäckers et al., 2006; Hogervorst et al., 2007). In those instances, other indicators of honeydew feeding can be used such as the ratios of sugars in the insect guts (Hogervorst et al., 2007; Lee et al., 2006), or other nutrients such as specific amino acids that may be unique to honeydew.

It should be noted that not all honeydews are sugar-rich. Xylem-feeding sternorrhynchans, such as *Homalodisca vitripennis*, may produce honeydews completely devoid of sugar (Irvin et al., 2007). Also, the secretions from lycaenid caterpillars are sometimes low in sugars. DeVries and Baker (1989) show that honeydew of the lycaenid, *Thisbe irenea*, when reared on *Croton billbergianus* is extremely low in sugars, but contained substantial quantities of amino acids.

Although sugars predominate the nutritional landscape of honeydew, other nutrients are also frequently encountered, especially amino acids (DeVries and Baker, 1989; Ewart and Metcalf, 1956; Gray, 1952; Lamb, 1959; Maltais and Auclair, 1952; Wada et al., 2001; Woodring et al., 2006; Yao and Akimoto, 2002). Holldobler and Wilson (1990) report that 0.2–1.8% of honeydew is nitrogenous, mostly consisting of amino acids. Other work reports that amino acids can constitute up to 13.5% of dry weight in honeydew (Way, 1963). The amino acid profile of honeydew can be quite speciose, represented by more than 20 amino acids in a single honeydew (Maltais and Auclair, 1952). Unlike pollens, single amino acids are not ubiquitously predominant in most honeydews. Ultimately, the amino acid content of the honeydew is contingent on the amino acids present in the phloem; honeydew usually mimics the amino acid profile of the phloem, but at lesser concentrations (Mittler, 1958a; Woodring et al., 2006, but see Gray, 1952). Mittler (1958a) shows that *Tuberolachnus salignus* incorporates approximately 55% of the nitrogenous material found in phloem and diverts the remainder into its honeydew. The physiological status of both the plant tissues consumed and the sternorrhynchan species influences the nitrogen content of the honeydew (Gray, 1952; Mittler, 1958a, b).

Aside from sugar and amino acids, honeydews also may contain phytosterols and other micronutrients essential to natural enemy physiology. *Myzus persicae* produces honeydew containing phytosterols (more than 90% of which were cholesterol and β-sitosterol) when reared on *Raphanus sativus* (Forrest and Knights, 1972). It appears that the aphids synthesize the cholesterol from phloem-derived brassicasterol and campesterol. They found sterols in the honeydew from *M. persicae* reared on other plants as well. Aphids are also able to generate their own sterols, thanks to endosymbionts. Citric acid is found in the honeydew of the pineapple mealybug (Gray, 1952), and *Brevicoryne brassicae* (Lamb, 1959). Way (1963) also reports that honeydew can be a significant source of B vitamins. And finally, honeydew is often the substrate for fungi and other microorganisms that can change the nutrition of the honeydew dramatically.

5.2 Factors That Influence Honeydew Production

> *Most persons, including many entomologists, are unaware of the abundance and ubiquity of honeydew.* (Downes and Dahlem, 1987)

Honeydew is extremely abundant in most ecosystems, and is widely available to natural enemies both spatially and temporally. One extreme example is the 'honeydew beech forests' of New Zealand, where the scale insect, *Ultracoelostoma assimile*, produces copious amounts of honeydew (estimated up to 1,200 l per ha, annually) which then forms the basis for a complex set of community interactions (Beggs, 2001; Hughes, 1976). Zoebelein (cited by Way, 1963), estimated that a single colony of *Formica rufa rufa* collect 500 kg of honeydew in 1 year. Still, the production of honeydew is a dynamic process, influenced both by the inherent status of the sternorrhynchan and the environment.

The age and physiological condition of the sternorrhynchan has a large effect on the amount of honeydew that is produced (Auclair, 1963). Mittler (1957) shows that the youngest nymphs of the aphid *Tuberolachnus salignus* are fairly slow feeders, and feed passively on the phloem sieve tubes. The more aggressively feeding adults are shown to actually suck the fluids out of the plant. A similar pattern is observed in *Bemisia tabaci*, where the younger nymphs feed in shorter bouts, and the fourth instars feed nearly continuously (Costa et al., 1999). Feeding rates are linked to how much honeydew is eventually produced, so the age-structure of a sternorrhynchan colony will affect its production of honeydew.

Factors extrinsic to the sternorrhynchan also influence the production rate of honeydew (Auclair, 1963). For instance, the quality of the host plant can affect the production rate of honeydew under some circumstances. In nutritionally poor plants, honeydew production is increased (Mittler, 1958a); the aphids need to imbibe much more of the phloem in order to make a living. Ant attendance also influences the rate of honeydew production, sometimes at the expense of the sternorrhynchans' fitness. As was discussed in a previous chapter, ants solicit honeydew droplets from sternorrhynchans, and increasing ant abundance can result in greater honeydew production in the sternorrhynchans that they attend (Del-Claro and Oliveira, 1993). Yao and Akimoto (2002) found that ant (*Formica yessensis*) attendance actually reduces the fitness of *Tuberculatus quercicola*. In this case, the ant-tended aphids devote nutrients into making attractive honeydew (and more of it), and so grow less than untended aphids. Moreover, ant-tended aphids maintain consistent levels of amino acids, even though the nitrogen content of the phloem fluctuates substantially over the season. Thus, keeping your body guards happy can be a costly game for a sternorrhynchan.

5.3 Honeydew in Mutualistic Interactions

Once phloem-feeding in sternorrhynchans evolved to produce copious amounts of excess nutritious fluids, these secretions began to shape the interactions of aphids and their natural enemies. As we have seen repeatedly in this and previous chapters, it is not long before sweet substances attract ants, and several aspects of honeydew have evolved to maintain ant-sternorrhynchan mutualisms. The creation of insect-synthesized sugars in honeydew may be reinforced through the relatively poor suitability of these sugars on the fitness of sternorrhynchan parasitoids. In sum, natural enemies can be either behaviorally or physiologically distracted from consuming sternorrhynchans thanks to the production of honeydew.

5.3.1 Honeydew-Guarding Ants

A critical development in the ant-sternorrhynchan mutualism is that the ants have learned to distinguish sternorrhynchans from prey. Ants recognize the cuticular chemicals of honeydew-producing insects (Choe and Rust, 2006). The relationship between *Linepithema humile* and the coccid *Coccus hesperidum* is strengthened as the ant learns that this insect secretes sugar. The honeydew itself contains sugars that are attractive to ants, melezitose specifically (Fischer and Shingleton, 2001; Wäckers, 2000; Woodring et al., 2006). In addition to the direct attractiveness of melezitose to ants in choice assays, the fact that ant-tended and untended sternorrhynchans produce different sugars in their honeydew reinforces the idea that sugar profiles of honeydews are cues for mutualistic partners. In two systems, Woodring et al. (2006) and Fischer and Shingleton (2001) document that ant-tended aphids produce honeydew containing melezitose, while congeners that produce alternative sugars were untended by the ants. Fischer and Shingleton (2001) go on to show that when the ant-tended species are not being attended, they reduce the melezitose fraction of their honeydew.

As discussed in the previous chapter, sternorrhynchans receive several benefits from their ant mutualists. Perhaps most important is that the sternorrhynchans suffer less predation from the ants. Additionally, the ants clean excess honeydew away from the sternorrhynchan colonies, create shelters for them, and protect the sternorrhynchans from other natural enemies (Beattie, 1985). Honeydew is important enough to the ants that it is worth fighting for. Majerus et al. (2007) present a nice discussion of the competitive interactions between ants and coccinellids that hope to eat from the ant's lunchbox (or the lunchbox itself). Generally, ants come out on top in these confrontations, chasing off adult coccinellids and carrying the coccinellid larvae away from the sternorrhynchans. The level of ownership behavior displayed by the ants dictates the degree of benefits that sternorrhynchans receive from their mutualists.

There is some debate over the degree to which ant-tended sternorrhynchan infestations actually benefit the plant, and whether plants may actually try to disrupt the ant-sternorrhynchan mutualism. On one hand, sternorrhynchans damage the plant, sometimes extensively. One estimate is that if the aphid, *Tuberolachnus salignus*, is allowed to feed continuously all day on the phloem of *Salix triandra*, each individual will consume the photosynthetic product created by 2–10 cm^2 of leaf material (Mittler, 1958a). Indeed, there is some evidence that supports the premise that plants may actually produce EFN to distract ants from sternorrhynchans (Becerra and Venable, 1991), although more research is needed on this topic. Still, not all ants are so easily distracted from their sternorrhynchan cattle. Ants are not dissuaded from the membracid, *Guayaquila xiphius*, when other sugar sources become available; perhaps because this membracid produces more honeydew when the ant activity on the plant is stimulated by alternative sugar sources (Del-Claro and Oliveira, 1993) (Fig. 5.1). In fact, ants attracted to honeydew remove other herbivores from the plant, and several studies show that the presence of sternorrhynchans indirectly improves plant fitness. For example, Compton and Robertson (1988) found that ants which tend a honeydew-producing tettigometrid on *Ficus sur* also protect the developing fruits from parasitoids of the primary pollinator and from seed predators. These trees produce significantly more fruits than trees lacking the sternorrhynchans. Another interesting consideration on the relative contributions of EFN and honeydew to plant defense comes from the fern, *Pteridium aquilinum*. Rashbrook et al. (1992) found that the ants attracted to the EFNs of *P. aquilinum* are not sufficient to protect the plant from herbivory. However, ant activity is increased on plants where coccids are present, and herbivory is diminished. Thus, because the ants prefer honeydew, the plant should favor infestation by the coccids. The within-plant distribution of

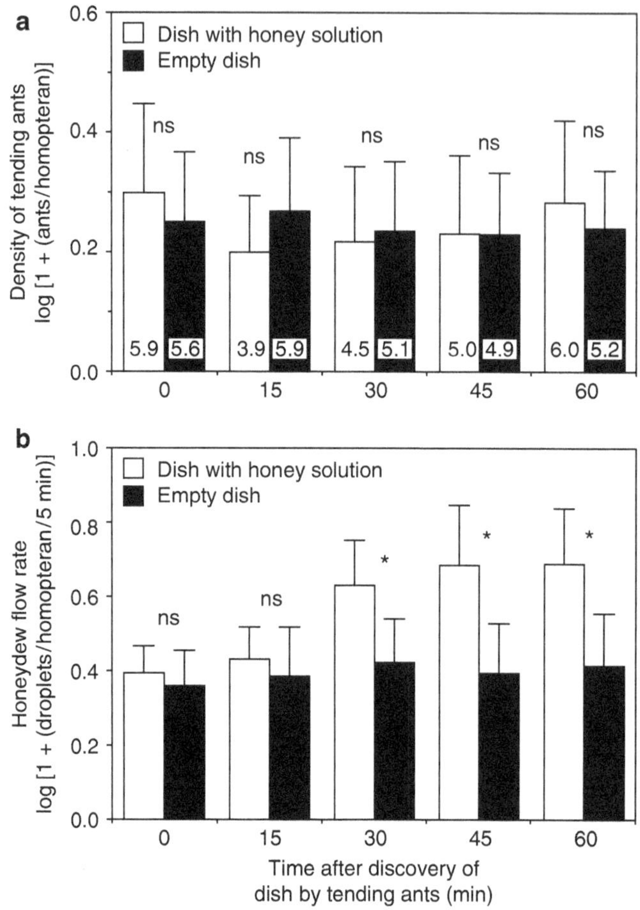

Fig. 5.1 Ant tending (*top*) and honeydew production (*bottom*) in aggregations of *Guayaquila xiphias* on treated (white bars) and control (black bars) shrubs of *Didymopanax vinosum*. Values within bars indicate the number of attending ants per sternorrhynchan aggregation, and asterisks above the bars indicate significant differences (Reproduced from Del-Claro and Oliveira, 1993. With permission by Blackwell)

Aphis nerii on oleander plants is another instance where aphid infestations are encouraged and manipulated by the plants (Bristow, 1991). Populations of *Aphis nerii* excel on the floral tips of this plant species, but languish on the leaf tips. The secondary compounds are expressed more strongly in the leaves, and ants avoid the honeydew produced by aphids on the leaves, presumably because it contains these toxins. Thus, by selectively directing their secondary compounds to allow aphids to colonize their floral tips, the oleander plant is aggregating protective ants to its developing flowers. A simple conclusion to this discussion is that sternorrhynchan-ant mutualisms are more likely to be fostered by plants if pressure from herbivory exceeds the costs of the sternorrhynchan infestation.

5.3.2 Antinutrient Properties of Honeydew

In addition to alleviating osmotic pressure differentials and maintaining ant mutualisms, the chemical composition of honeydew may also render it less suitable or even toxic as a food source for natural enemies of sternorrhynchans (Avidov et al., 1970; Leius, 1961a; Way, 1963). Sucrose sustains greater longevity in *Aphidius ervi* than many of the honeydews produced by five aphids reared on various plants (Hogervorst et al., 2007). This same pattern is observed in *Trichogramma platneri* fed *Dysaphis plantaginea* honeydew; honeydew-fed wasps survive better than unfed wasps, but the sucrose is more suitable for prolonging wasp lives (McDougall and Mills, 1997). Honeydew-derived oligosaccharides are relatively unsuitable for parasitoid wasps (Wäckers, 2001; Williams et al., 2005). Wäckers (2000) points out that trisaccharides frequently are less likely to elicit a gustatory response in parasitoids than monosaccharides. Also, some parasitoids are unable to use insect-synthesized sugars as kairomones, and trisaccharides tend to crystallize more quickly than sucrose, fructose, and glucose, thereby limiting their accessibility to some natural enemies. Still, honeydews from different sternorrhynchans vary widely in their nutritional value, and one should be careful not to paint with too broad of a brush. For example, the honeydew of *Homalodisca vitripennis* (a xylem-feeder) does not prolong the life of *Gonatocerus* egg parasitoids any better than water, whereas honeydew from *Coccus hesperidum* improves their longevity substantially (Irvin et al., 2007).

A number of more conspicuous, plant-derived defensive chemicals occur in honeydew, but whether these secondary compounds hamper natural enemies is largely unknown. For instance, alkaloids and cardenolides are in honeydews from several species (Molyneux et al., 1990; Wink and Romer, 1986). The honeydew of *Aphis nerii*, which feeds on milkweed plants (Asclepiadaceae), contains bitter cardenolides (up to 19 of them) (Malcolm, 1990; Rothschild et al., 1970). Malcolm (1990) suggests that by coating the leaves in this bitter honeydew, the aphids and the plant both benefit by deterring large herbivores. It shouldn't be assumed that the toxins found in chemically defended plants will show up in aphid honeydew, but the likelihood of this increases if the chemicals are transported in the phloem (Molyneux et al., 1990). A final deterring factor associated with honeydew is that it is sticky, and small predators have frequently been reported to become mired in it. Predatory mites (McMurtry and Scriven, 1964a; Nomikou et al., 2003b), nymphs of predatory heteropterans (De Lima and Leigh, 1984), parasitoids (Hulspas-Jordaan and van Lenteren, 1978; Lundgren and Heimpel, 2003) and coccinellid larvae (Putnam, 1955) all suffer this fate occasionally. Thus, let it be said that eating honeydew isn't all fun and games for natural enemies.

5.4 The Downside of Honeydew: Its Kairomone Effects

For a sternorrhynchan, attracting natural enemies to oneself is a dangerous game; one never knows when the natural enemy will tire of its sugary snack and want something meatier. It really is not surprising that numerous natural enemies use the presence of honeydew as a kairomone for locating potential food sources. Specifically, honeydew has been observed to act as a volatile cue for locating a host or prey from a distance, as a contact oviposition stimulant, and as an arrestant for foraging natural enemies.

Parasitoids and predators are attracted to volatile properties of honeydew. Numerous aphidophagous parasitoids respond to honeydew in olfactometer studies (Hagvar and Hofsvang, 1989; Wickremasinghe and van Emden, 1992). It is not surprising that polyphagous parasitoids (as many sternorrhynchan-specific parasitoids are), should use host-derived chemicals during foraging since plant-based cues are likely to vary widely for different specific hosts (Hagvar and Hofsvang, 1989). The fact that only females of *Aphidius nigripes* are attracted to the volatiles of aphid honeydew suggests that this parasitoid uses the kairomones for host finding, rather than appetitive foraging (Bouchard and Cloutier, 1985). The chemical cues from honeydew that attract *Chrysoperla carnea* have been well studied over the past half century. Both larvae (Kawecki, as cited by Szentkiralyi, 2001) and adults are attracted by honeydew (Duelli, 1987). At first, tryptophan was implicated as the attractive agent (Dean and Satasook, 1983). But tryptophan is not volatile, and so attention fell on some of its breakdown products, such as tryptamine and indole acetaldehyde (van Emden and Hagen, 1976). A consensus cue remains to be discovered in this system; Harrison and McEwen (1998) question whether even the acid hydrolyzed tryptophan is the causative agent. They suggest that this chemical may be damaging the plant, and that the lacewings could be attracted to an induced plant defensive volatile. This is the case for *Diaretiella rapae* which responds to feeding-induced chemicals rather than the honeydew of its host, *Myzus persicae* (Girling et al., 2006). Other honeydew-derived volatiles are less elusive; indole-3-acetaldehyde is the attractive agent from honeydew that is used by *Aphidius rhopalosiphi* (Wickremasinghe and van Emden, 1992).

More commonly, honeydew functions as a contact kairomone for sternorrhynchan natural enemies, and often these entomophages respond to honeydew by laying eggs (Bargen et al., 1998; Belliure and Michaud, 2001; Evans and Dixon, 1986; Scholz and Poehling, 2000; Sutherland et al., 2001) (Fig. 5.2). Hood Henderson (1982) demonstrates that the ovipositors of two syrphid species, *Eupeodes fumipennis* and *E. volucris*, have chemosensilla that are sensitive to aphid honeydew, and the components of honeydew, sucrose, tryptophan (and its oxidation product), and alanine. Honeydew from *Sitobion avenae* is found to be as strong an oviposition stimulant for *Episyrphus balteatus* as the aphids themselves (Budenberg and Powell, 1992) (Fig. 5.3). But not all honeydews are attractive to this syrphid; of the honeydews from four aphid species tested, *Microlophium carnosum* honeydew does not elicit oviposition by *E. balteatus*.

Upon contact with honeydew, many natural enemies alter their foraging behavior to explore areas with honeydew more intensively. Natural enemies slow their walking speed, turn more frequently, stop more, and in general increase their residency time near honeydew. The arresting effect of honeydew on natural enemy foraging is known for coccinellid adults and larvae (van den Meiracker et al., 1990; Banks, 1957; Carter

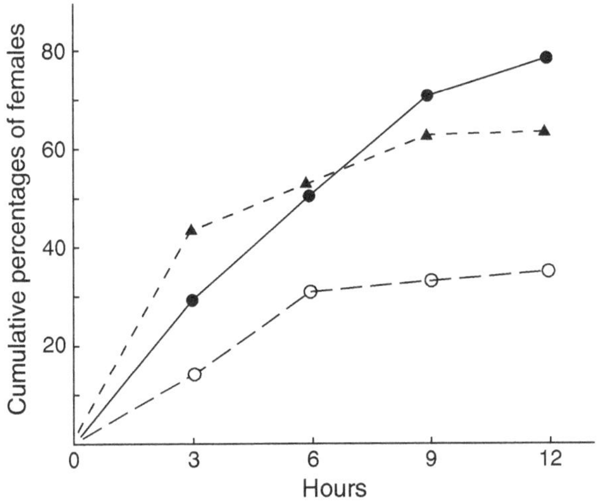

Fig. 5.2 The cumulative percentage of female *Coccinella septempunctata* laying their first batch of more than five eggs within a given number of hours after being placed in vials with *Acyrthosiphon pisum* (•), previously with *A. pisum* (▲), and without aphids (◦) (Reproduced from Evans and Dixon, 1986. With permission by Blackwell)

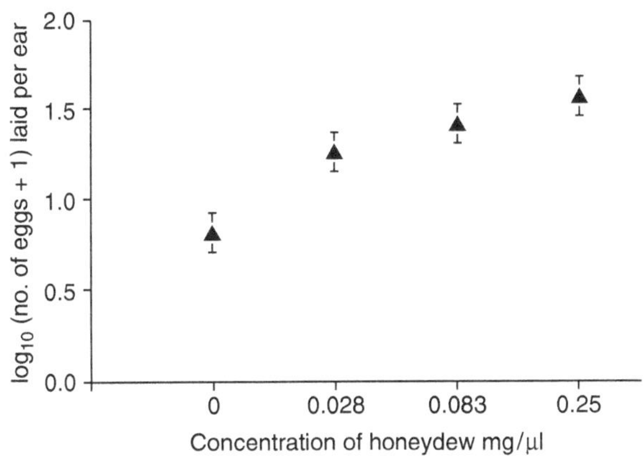

Fig. 5.3 The mean number of eggs laid by *Episyrphus balteatus* on ears of wheat treated with different concentrations of *Metopolophium dirhodum* honeydew and on untreated ears (Reproduced from Budenberg and Powell, 1992. With permission by Blackwell)

and Dixon, 1984; Han and Chen, 2002) (Fig. 5.4), predatory mites (Fleschner, 1950), parasitoid Hymenoptera (Bouchard and Cloutier, 1984; Budenberg, 1990; Carter and Dixon, 1984; Han and Chen, 2002; Romeis and Zebitz, 1997; Vinson et al., 1978), chrysopid adults and larvae (Downes, 1974; Han and Chen, 2002), and syrphid larvae (Bargen et al., 1998). Often times, when the natural enemy comes in contact with the

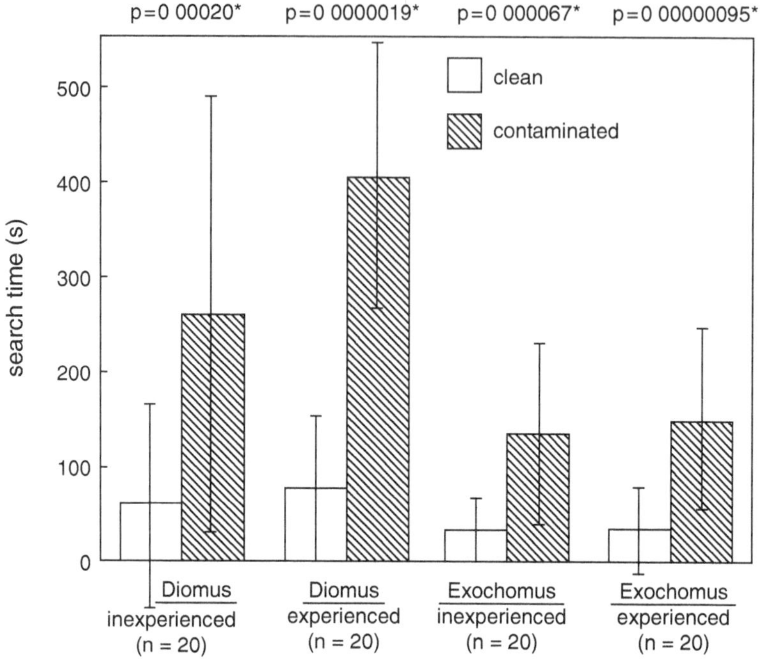

Fig. 5.4 Search time by two coccinellid predators on clean cassava leaves and on leaves contaminated with *Planococcus citri* wax and/or honeydew. All differences within column pairs are significantly different (Reproduced from van den Meiracker et al., 1990. With permission by Blackwell)

honeydew, it assesses it with its antennae and mouthparts (Bouchard and Cloutier, 1984; Budenberg, 1990; Han and Chen, 2002), suggesting that some contact kairomones are at play. Sugars have repeatedly been shown to arrest the foraging of natural enemies (Romeis and Zebitz, 1997; Vinson et al., 1978). Also, proteins and amino acids may operate in arresting the movement of natural enemies (Vinson et al., 1978). Ken Hagen and colleagues suggested that tryptophan functioned as a contact kairomone for *Chrysoperla carnea* adults (Hagen et al., 1976; van Emden and Hagen, 1976), a concept supported by data of McEwen et al. (1993a). Sucrose on its own is not attractive to lacewings from a distance, but does arrest their foraging in the field (Carlson and Chiang, 1973; Hagen et al., 1976). Indeed, contact with prey or prey products appears to be a primary mode of detection for many predators; Banks (1957) found that *Propylaea quatuordecimpunctata* was unable to detect aphid prey from as little as 3 mm away! Downes (1974) observed *Chrysoperla* larvae following along an artificial trail of sucrose (eating all the while) to find aphid prey. Interestingly, aphids that were not associated with this trail were not of interest to the lacewing larvae. Carter and Dixon (1984) found that increasing the foraging time of the predator, *Coccinella septempunctata*, on corn ears with honeydew from *Sitobion avenae* led to reduced numbers of these aphids compared to honeydew-free corn ears.

The age and quantity of honeydew can also be assessed by natural enemies, and affects their foraging decisions. These two characteristics of honeydew are correlated with the presence and size of aphid colonies. For instance, a patch of old honeydew may have been produced by an aphid colony that has moved on (Shaltiel and Ayal, 1998). And more honeydew found in an area likely means more sternorrhynchan prey or hosts are present to produce it (Bouchard and Cloutier, 1984; Budenberg, 1990; Romeis and Zebitz, 1997). It makes sense that a natural enemy would evolve to respond to a short-lived kairomone signal, which is more closely tied to the prey/host's density (Shaltiel and Ayal, 1998). But if the kairomone is too short-lived, then the natural enemy is not likely to find it before it degrades. Romeis and Zebitz (1997) are surprised to find that the age of *Trialeurodes vaporariorum* honeydew does not affect its attractiveness to the parasitoid *Encarsia formosa*, and suggest that it may be an artifact of the laboratory conditions.

Finally, in spite of these numerous examples of honeydew being used against its producer by natural enemies, Jervis et al. (1996b and references therein) point out that many natural enemies do not use honeydew to locate their hosts. Certainly, not all syrphids respond to honeydew as an oviposition cue (Chandler, 1968b). As a case in point, the presence of honeydew is used by *Episyrphus balteatus* as an oviposition cue, but not by *Melanostoma mellinum* (Budenberg and Powell, 1992).

5.5 Conclusion

Given the current data, it seems likely that honeydew is much more pervasive than humans realize. Its high nutrition is a double-edged sword for the sternorrhynchans that produce it. On one hand, the sternorrhynchans can hope that their enemies prefer to feed on sugar, and this is often the case with ants. But this is not always so, and honeydew has become a red flag that attracts entomophagous arthropods to an easy meal (with an accompanying dessert!). But since phloem feeding leaves no other solution than to produce honeydew, sternorrhynchans have devised a plethora of ways to reduce the palatability and nutritional suitability of their excrement. This honeydew-mediated arms race between sternorrhynchans and their natural enemies has over eons led honeydew to be one of the most abundant and nutritionally diverse, but least agreeable, sugar sources for many natural enemies.

Section II
Pollinivory

Pollen is likely one of the first substances to attract insects to flowers (Faegri and van der Pijl, 1966), and it pervades nearly every terrestrial habitat of importance to insects and man. Up to 10,000 kg of pollen km^{-1} are produced annually in the vegetated regions of the Earth, and some plant species can produce more than 10^7 grains per plant (O'Neal & Waller, 1984; Stanley & Linskins, 1974). During peak anthesis of trees in Japan, more than 350 pollen grains cm^{-2} settle onto the ground (Iwanami et al., 1988). One plant species alone, *Zea mays*, produced an estimated 4.84 trillion kg of pollen in the United States during 2005 (J.G. Lundgren, unpublished data, 2007). Honeybees consume between 62–228 × 10^6 kg of pollen annually (O'Neal & Waller, 1984). Clearly this is a dependable and important food source for numerous organisms. Although the importance of pollen to insects is best understood in honeybees and other pollinators, many species of arthropods unimportant as pollinators, including many natural enemies, take advantage of this widespread source of food.

The abundance and nutrition of pollen and the commonplace occurrence of pollinivory within entomophagous taxa confirm the importance of this food to natural enemies. Pollen is unique from the sugar solutions discussed in the last section in that it has a much more diverse nutritional profile and presents unique challenges to natural enemies wishing to extract nutrients from its tiny grains. The simple objective of this section is to document the importance of pollen feeding to entomophagous arthropods, but in depth examination of pollen and natural enemies reveal a fascinating ecology of interactions that science is only beginning to uncover.

In reviewing and summarizing the literature on the topic of pollinivory, several caveats should be mentioned. First, one must distinguish between observations of pollinivory in the laboratory versus in the field. Only a minority of studies report the gut contents of arthropods collected from the field (immature stages are particularly overlooked), and gut dissections are critical to understanding the dietary breadth of entomophagous arthropods under natural conditions. One limitation of gut dissections is that it is difficult to distinguish low levels of true pollinivory from incidental pollinivory that occurs when pollen is consumed with other food (i.e. prey or nectar). Examples of ancillary pollen consumption during prey meals have been reported in carabids, where small amounts of pollen in the guts of field-collected individuals are likely present because the pollen is on or in the prey when it is consumed (Dawson, 1965). Although laboratory experiments involving pollen feeding may be criticized for over-inflating pollinivory rates under artificial conditions, laboratory studies are still useful in identifying cases of potential pollinivory within the expanse of entomophagous species, and consequently both laboratory and field studies are included in this review.

Chapter 6
The Pollen Feeders

Most, if not all, major classes of arthropod predators and parasitoids contain members that consume pollen in some fashion. Given the tremendous diversity of entomophagous arthropods, it is necessary to curtail the list of species treated in this chapter. To this end, the focus of this chapter is primarily on taxa for which the ecological function of the group is fairly well understood or where importance to biological control is demonstrated. While the scope of this review is intended to be comprehensive, the state of the literature on this topic prevents an entirely exhaustive treatment. Thus please view this as an attempt to point out the diversity of pollen feeding taxa, and the range of pollens that they are known to consume.

6.1 Predators

In addition to the families treated in detail below, other predatory taxa routinely consume pollen as part of their diet. As examples, Scoliodea: Myzininae, Pompilidae, Nyssonidae, Sphecidae, and Vespidae (subfamilies Eumeninae, Masarinae, Polistinae, Vespinae) (Hymenoptera) contain predaceous members that will accept pollen as food (Hunt et al., 1991). Also, some adults of Cleridae (Coleoptera) (*Enoclerus* spp. specifically) are pollinivorous (Balduf, 1935), and *Frankliniella occidentalis* (Thysanoptera: Thripidae) is a natural enemy of spider mites that actually experiences higher survival and reproduction on a diet of *Gossypium barbadense* pollen than on certain prey in the laboratory (Trichilo and Leigh, 1988). Many phalangiids also consume pollen under field conditions (Acosta and Machado, 2007). Suffice it to say that the instances reported here are intended to point out some of the specific relationships between entomophagous arthropods and pollen species, and to illustrate how groups of predators approach pollinivory.

6.1.1 Arachnida: Araneae

Spiders are traditionally regarded as strict carnivores, although investigations have revealed that pollen is of dietary importance for at least a handful of species.

J.G. Lundgren, *Relationships of Natural Enemies and Non-Prey Foods*,
Progress in Biological Control 7,

Published by Springer Science+Business Media B.V.

Exceptions to strict carnivory are accumulating in web-building and flower-dwelling spiders (Smith and Mommsen, 1984; Vogelei and Greissl, 1989) (Table 6.1). Immature spiders collect their webs periodically regardless of whether the webs contain prey (Lubin, 1978). A number of microorganisms, pollen, and spores are often adhered to the strands of silk, which can augment the nutrition of spiders, especially when prey is scarce (Agarwal, 1976; Bera et al., 2002; Linskins et al., 1993; Smith and Mommsen, 1984). Linskins et al. (1993) found that horizontal webs have more pollen than those arranged vertically, and that individual webs can contain thousands of pollen grains. This is confirmed by Ludy and Lang (2006b), who report 1,044 ± 1,193 pollen grains per web of *Araneus diadematus* near cornfields. The farther the webs are from the cornfield, the less corn pollen is found in the webs. Bera et al. (2002) report dozens of species of pollen in six spider webs.

There are some anatomical hurdles that spiders need to overcome in order to ingest pollen. Cuticular platelets covering the oral cavity prohibit spiders from ingesting materials larger than 1 μm. Yet, the fact that pollen is larger than 1 μm does not apparently prevent spiders from ingesting it, although it isn't exactly clear how they accomplish this (Smith and Mommsen, 1984; Vogelei and Greissl, 1989). Linskins et al. (1993) suggest that when the pollen grains germinate in the sticky fluids on the web, their nutrients become accessible to spiders, which digest foods extra-orally.

Pollen feeding improves the fitness of spiders, especially with regard to adult longevity and survivorship of immatures. Ingesting webs that contain pollen from *Betula papyrifera* results in greater longevity, survivorship, and web-regeneration capacities in second and third instars of the orb-weaver, *Araneus diadematus*, compared to spiders that ingest webs containing no pollen (Smith and Mommsen, 1984). In another case, young *Thomisus onustus* subsist on pollen for more than 40 days in the laboratory, and this may be an important spring food for flower-inhabiting immatures that have not yet accumulated substantial fat reserves (Vogelei and Greissl, 1989). Thus, feeding on non-prey foods such as pollen may be important for the survival of some spider immatures when prey is unavailable. It should be noted that not all web-building spiders ingest pollen; offering the lyniphiid, *Frontinella communis* pollen from *Pinus elliottii* var. *densa* does not improve spider fitness over unfed individuals, nor are the spiders actually observed to feed on the pollen (Carrel et al., 2000).

6.1.2 Arachnida: Acari

Predatory mites in several families feed on pollen to varying degrees. McMurtry and Croft (1997) classify the feeding behavior of predatory Phytoseiidae into four groups, three of which feed on pollen. McMurtry & Croft's Group IV is comprised of the genus *Euseius*, members of which can subsist on pollen in the absence of prey with minimal reductions in fitness. Indeed, pollen is as suitable for larval development and reproduction as prey for a number of entomophagous

Table 6.1 Predaceous families in Arachnida known to feed on plant pollen

Family	Species	Life stage	Pollen species	Reference
Class: Araneae				
Aranae	*Araneus diadematus*	Immature	*Betula papyrifera*; *Zea mays*	(Ludy and Lang, 2006b; Smith and Mommsen, 1984)
Argiopidae	*Meta segmentata*	Unknown	Pollen of numerous species found in webs (unknown whether this was ingested)	(Linskins et al., 1993)
Argiopidae	*Zygiella x-notata*	Unknown	Pollen of numerous species found in webs (unknown whether this was ingested)	(Linskins et al., 1993)
Thomisidae	*Misumenoides formosipes*	Adult males	*Daucus carota* (seen brushing grains into mouth)	(Vogelei and Greissl, 1989)
Thomisidae	*Thomisus onustus*	Immature	*Erigeron annuus*, *Bellis perennis*	(Pollard et al., 1995)
Class: Acari				
Cheyletidae	*Cheletogenes ornatus*	Immature/adult	*Phoenix dactylifera*, *Z. mays*	(Zaher et al., 1981)
Erythraeidae	*Balaustium florale*		*B. perennis*, *Ranunculus*, *Acer campestre*	(Childers and Rock, 1980; Grandjean, 1959)
Erythraeidae	*Balaustium putnami*	Immature	Unidentified yellow pollen, *Malus*	(Cadogan and Laing, 1977; Childers and Rock, 1980)
Phytoseiidae	*Amblyseius andersoni*	Immature/adult	*Mesembryanthemum criniflorum*	(Duso and Camporese, 1991)
Phytoseiidae	*Amblyseius largoensis*	Immature/adult	*Lysichiton camtschatcensis*, *Salix udensis*, *Betula platyphylla*, *Acer negundo*, *Plantago lanceolata*, *Sambucus sieboldiana* var. *miquelii*, *Pinus nigra*, *Heracleum dulce*, *Carpobrotus acinaciformis* (anthers), *Z. mays*, *Citrus limon* anthers, [completed development on a diet consisting solely of *Pyrus communis*, *Diospyros kaki*, *Camellia*, *Pinus*, *Tulipa*, *Castanea* (chestnut), *Portulaca grandiflora*, *Podocarpus macrophylla*, *Salix lucida lasiandra*, *Quercus argifolia*, an unidentified palm, *Ricinus communis*, *Quercus virginiana*, *Typha domingensis*, *Phoenix roebelenii*	(Kamburov, 1971; Kennett and Hamai, 1980; Mori, 1977; Yue and Tsai, 1996)

(continued)

Table 6.1 (continued)

Family	Species	Life stage	Pollen species	Reference
Phytoseiidae	*Amblyseius potentillae*	Immature/adult	*Vicia faba*, *M. criniflorum*	(Overmeer and van Zon, 1983)
Phytoseiidae	*Amblyseius similoides*	Adult	*Persea americana*	(Flechtmann and McMurtry, 1992b)
Phytoseiidae	*Euseius*	Adult/immature	Appear to be quite pollinivorous	(Kennett et al., 1979; McMurtry, 1985)
Phytoseiidae	*Eusius addoensis addoensis*	Immature	*Carpobrotus muirii*, *Casuarina cunninghamiana*	(Grout and Richards, 1992)
Phytoseiidae	*Euseius aleyrodis*	Immature/ adult	*Gossypium*, *Antirrhinum majus*	(Elbadry, 1968)
Phytoseiidae	*Euseius finlandicus*	Immature/ Adult	*Malus sylvestris*, *Prunus avium*, *Betula*, *Prunus persica*, *Prunus armeniaca*, *Juglans regia*, *Papaver rhoeas*, *P. communis*	(Broufas and Koveos, 2000; Schausberger, 1992)
Phytoseiidae	*Euseius gossipi*	Immature/adult	*Z. mays*, *P. dactylifera*, *Gossypium barbadense*	(Elbadry and Elbenhawy, 1968a, b)
Phytoseiidae	*Euseius mesembrinus*	Immature/adult	*Malephora crocea*	(Abou-Setta and Childers, 1989)
Phytoseiidae	*Euseius ovalis*	Immature	*Z. mays*	(Shih et al., 1993)
Phytoseiidae	*Euseius scutalis*	Adult	*R. communis*, *Z. mays*	(Allawi, 1991)
Phytoseiidae	*Euseius sojaensis*	Immature/adult	*Camellia sinensis*	(Osakabe, 1988; Osakabe et al., 1986)
Phytoseiidae	*Eusieus tipulatus*	Adult	*M. crocea*	(Flechtmann and McMurtry, 1992b)

Phytoseiidae	*Euseius tularensis*; = *Euseius hibisci*	Immature/adult	*Salix*, *S. l. lasiandra*, *Nicotiana glauca*, *Quercus agrifolia*, *Brassica nigra*, *Aesculus californica*, *Rubrus*, *Typha latifolia*, *Z. mays*, *Ligustrum japonicum*, *Ananas comosus*, *M. crocea*, *M. sylvestris*, *P. communis*, *Prunus dulcis*, *Olea europaea*, *Lolium multiflorum*, *Secale cereale* var. 'Merced', *Avena sativa* var. 'California red', *Hordeum vulgare* var. 'UC476', *Bromus mollis* var. 'Blando', *Bromus carinatus*, *Hordeum brachyantherum*, *Festuca rubra*, *Triticum aestivum*, *P. persica*, *Carpobrotus edulis*, *P. americana*, *R. communis*, *Capsicum annuum*, *Mesembryanthemum*, *Eucalyptus*, *J. regia*, *Citrus*, *G. barbadense*, *V. faba*, *Pisum sativum* var. '*arvense*', *Trifolium repens*, *Vicia sativa*, *Vicia villosa* ssp. *dasycarpa* var. 'Lana', *Trifolium incarnatum* var. 'Flame', *Trifolium hirtum*, *Trifolium pratense*	(Grafton-Cardwell et al., 1999; Kennett et al., 1979; Kennett and Hamai, 1980; McMurtry and Scriven, 1964b; Ouyang et al., 1992; Swirski et al., 1970)
Phytoseiidae	*Euseius victoriensis*	Immature	*Typha orientalis*	(James, 1993)
Phytoseiidae	*Galendromus occidentalis*	Adult	*Prunus* (plum), *Z. mays*, *Carya* (pecan), *Pistacia* (pistachio), *Malus* (apple), *A. chinensis*, *P. dulcis*, *P. communis*	(Afifi et al., 1988)
Phytoseiidae	*Iphiseiodes quadripilis*	Adult/ immature	*M. crocea*, *Quercus*	(Villanueva and Childers, 2007)
Phytoseiidae	*Iphiseius degenerans*	Adult/immature	>25 species of pollen	(Eveleigh and Chant, 1982; van Rijn and Tanigoshi, 1999b)
Phytoseiidae	*Kambimodromus aberrans*	Immature/adult	*Malus*, *Prunus* (cherry), *Betula*, *Corylus avellana*	(Chant, 1959; Schausberger, 1992)
Phytoseiidae	*Metaseiulus arboreus*	Immature/adult	*Q. argifolia*, *S. l. lasiandra*, unidentified palm	(Kennett and Hamai, 1980)
Phytoseiidae	*Neoseiulus chilenensis*	Immature/adult	*R. communis*, *Z. mays*, *C. edulis*, *Amygdalus communis*, *P. americana*	(Swirski et al., 1970)

(continued)

Table 6.1 (continued)

Family	Species	Life stage	Pollen species	Reference
Phytoseiidae	*Neoseiulus cucumeris*	Adult/immature	>25 species of pollen; *R. communis, Quercus*	(Skirvin et al., 2007; van Rijn and Tanigoshi, 1999b; Marisa and Sauro, 1990)
Phytoseiidae	*Neoseiulus fallacis*	Immature/adult	*Prunus, Z. mays, Carya, Pistacia, Malus, Actinidia chinensis, P. dulcis, P. communis, Malus, Campsis grandiflora, P. persica*	(Afifi et al., 1988; Zaher et al., 1969)
Phytoseiidae	*Neoseiulus idaeus*	Adult	*Manihot esculenta, Elaeis guineensis, R. communis*	(Tanigoshi et al., 1993)
Phytoseiidae	*Neoseiulus longispinosus*	Immature/adult	*S. udensis, A. negundo, Rosa rugosa, Z. mays*	(Mori, 1977)
Phytoseiidae	*Neoseiulus paraki*	Immature/adult	*L. camtschatcensis, S. udensis, A. negundo, P. lanceolata, Z. mays*	(Mori, 1977)
Phytoseiidae	*Neioseiulus umbraticus*	Not specified	*Rubus, C. avellana*	(Chant, 1959)
Phytoseiidae	*Phytoseiulus persimilis*	Adult	*Prunus* (plum), *Z. mays, Carya* (pecan), *Pistacia* (pistachio), *Malus* (apple), *A. chinensis, P. dulcis, P. communis*	(Zaher et al., 1969)
Phytoseiidae	*Phytoseius plumifer*	Immature/adult	*P. dactylifera, G. barbadense, Althaea rosea*	(Afifi et al., 1988)
Phytoseiidae	*Proprioseiopsis athiasae*	Immature/adult	*Gossypium* (can complete development and reproduce on)	(Swirski et al., 1967)
Phytoseiidae	*Proprioseiopsis rotundus*	Immature/adult	*M. crocea, Q. virginiana, T. latifolia*	(Abou-Setta et al., 1997)

Phytoseiidae	*Typhlodromus caudiglans*	Immature/adult	*Chenopodium album*, *Ambrosia artemisiifolia*, *Setaria pumila pumila*	(Putnam, 1962)
Phytoseiidae	*Typhlodromips deleoni*	Immature/adult	*R. communis*	(Shou-Jian and Fen-Wei, 1982)
Phytoseiidae	*Typhlodromus doreenae*	Immature	*T. orientalis*	(James, 1993)
Phytoseiidae	*Typhlodromus kerkirae*	Immature/adult	*V. faba*	(Koveos and Broufas, 1999)
Phytoseiidae	*Typhlodromalus limonicus*	Immature/adult	*Persea* (avocado), *C. annuum*, *R. communis*, *Mesembryanthemum*, *Citrus*	(McMurtry and Scriven, 1965)
Stigmaeidae	*Zetzellia mali*	Immature/adult	*Malus* (apple)	(White and Laing, 1977)
Phytoseiidae	*Typhlodromus pyri*	Immature/adult	*V. faba* (reared in culture for 6 months), *P. dactylifera*, *G. barbadense*, *A. rosea*, *Malus* (apple), *M. criniflorum*, *Alnus glutinosa*, *Salix*, *Citrus* (orange), *P. americana*	(Chant, 1959; Chant and Fleschner, 1960; Dicke, 1988b; Dosse, 1961; Duso and Camporese, 1991; Zaher and Shehata, 1971)
Phytoseiidae	*Typhlodromus rickeri*	Immature/adult	*P. americana* pollen and anthers	(Overmeer, 1981; Overmeer, 1985; McMurtry and Scriven, 1964a)
Phytoseiidae	*Typhlodromips swirskii*	Immature/adult	*Gossypium* (can complete development and reproduce), *R. communis*	(Abou-Awad and Elsawi, 1992; Swirski et al., 1967)
Phytoseiidae	*Typhlodromalus peregrinus*	Immature/adult	*M. crocea*, *Q. virginiana*, *T. latifolia*	(Fouly et al., 1995)

acarids (Elbadry and Elbenhawy, 1968a; Ouyang et al., 1992; Swirksi et al., 1970; Zaher and Shehata, 1971). Perhaps the best-studied member of this genus with regard to pollinivory is *Euseius tularensis*, which can develop and reproduce on dozens of species of pollen (Grafton-Cardwell et al., 1999; Kennett et al., 1979; McMurtry and Scriven, 1964b; Ouyang et al., 1992; Swirski et al., 1970). Natural peaks in pollen densities in California orchards are consistently followed by increases in the densities of *Euseius tularensis*, although their attractiveness and ability to induce reproduction varies among pollen species (Kennett et al., 1979). The ability of these highly pollinivorous predators to regulate pest populations has historically been questioned, but some examples suggest that pollen feeding may allow *Euseius* spp. and other polyphagous feeders to control pest mite populations before outbreaks occur (Elbadry, 1968; Wiedenmann and Smith, 1997) (see Chapter 16).

Pollen plays a different role in the life histories of entomophagous acarids that are relatively less reliant on pollen. In these taxa, pollen functions as a dietary supplement to prey of low quality or when prey is absent. The presence of pollen during spring before prey become abundant is important for several species of mites, which aggregate to spring flowering plants like the catkins of *Corylus avellana* and *Salix*, and flowers of *Malus* (Chant, 1959). For example, apple pollen was sufficient to sustain spring-collected *Typhlodromus pyri* and prompt egg production, whereas those fed only water soon died (Chant, 1959). But even for these more entomophagous mites, pollens can be equivalent to prey in terms of nutrition; two pollen species are equivalent to citrus rust mites for supporting development of *Iphiseiodes quadripilis* (Villanueva and Childers, 2007). Different pollens vary in their suitability for mites, and the importance of factors such as pollen nutrition and pollen grain structure are relatively well studied for acarid carnivores compared with other entomophages (see Chapter 7).

6.1.3 Coleoptera: Carabidae

The feeding behavior of carabids as a group is difficult to characterize, and while pollen is acceptable to a handful of species, the scope of pollinivory within this family requires more attention. Carabids range from nearly complete carnivory (as in most Carabini, Cicindelini), to nearly complete herbivory (as in some Harpalini, Zabrini). In actuality, most species are best described as omnivorous, feeding on fungal spores, sugar, seeds, plant tissue, and dead and living prey (Allen, 1979; Larochelle, 1990). Quantifying the dietary breadth of omnivorous carabids is difficult at best. Of the approximately 40,000 described species of Carabidae, feeding habits are only described for 1,054 (2.6%) of species (Larochelle, 1990). Of this minority, only 36 are recorded as feeding on pollen (Table 6.2). This proportion is undoubtedly a serious underestimate of pollinivory in carabids. For instance, in the laboratory, 14 carabid species never previously observed to consume pollen readily eat corn pollen under no-choice conditions (Ahmad et al., 2006; Mullin et al., 2005). Indeed, *Scarites quadriceps* (a predominantly entomophagous species) and

Table 6.2 Species in selected families of Insecta that are predaceous at some point during their lives (excluding parasitoids)

Family	Species	Life stage	Species of pollen consumed	Reference
Mantodea				
Mantidae	*Tenodera aridifolia sinensis*	Adult	Mixed species, bee-collected	(Beckman and Hurd, 2003)
Mantidae	*Tenodera aridifolia sinensis*	Nymph	Mixed species, bee-collected	(Beckman and Hurd, 2003)
Coleoptera				
Carabidae	*Agonum cupripenne*	Adult	*Z. mays*	(Mullin et al., 2005)
Carabidae	*Agonum decorum*	Adult	Leguminosae, Cruciferae, Caryophyllaceae, Umbelliferae, and Compositae	(Dawson, 1965)
Carabidae	*Agonum fuliginosum*	Adult	Leguminosae, Cruciferae, Caryophyllaceae, Umbelliferae, and Compositae	(Dawson, 1965)
Carabidae	*Agonum muelleri*	Adult	*Z. mays*	(Mullin et al., 2005)
Carabidae	*Agonum placidum*	Adult	*Z. mays*	(Mullin et al., 2005)
Carabidae	*Amara aulica*	Adult	pollen	(Forsythe, 1982a)
Carabidae	*Amara eurynota*	Adult	*Scabiosa*	(Jänner, 1905; Larochelle, 1990)
Carabidae	*Amara pennsylvanica*	Adult	*Z. mays*	(Mullin et al., 2005)
Carabidae	*Amphasia sericea*	Adult	Grass pollen	(Forbes, 1881)
Carabidae	*Anisodactylus sanctaecrucis*	Adult	*Z. mays*	(Mullin et al., 2005)
Carabidae	*Bembidion biguttatum*	Adult	Pollen	(Davies, 1953)
Carabidae	*Bembidion lampros*	Adult	Pollen	(Davies, 1953; Mitchell, 1963)
Carabidae	*Bembidion obtusum*	Adult	Pollen	(Davies, 1953)
Carabidae	*Bembidion quadrimaculatum oppositum*	Adult	*Z. mays*	(Mullin et al., 2005)
Carabidae	*Calathus gregarius*	Adult	Grass pollen	(Forbes, 1883)
Carabidae	*Carterus*	Adult	Anthers of Graminae and Umbelliferae	(Bonadona, 1971; Jeannel, 1941; Larochelle, 1990)
Carabidae	*Chlaenius tricolor tricolor*	Adult	*Z. mays*	(Mullin et al., 2005)
Carabidae	*Ditomus capito*	Adult	Graminaceous anthers	(Auber, 1965; Larochelle, 1990)
Carabidae	*Ditomus tricuspidatus*	Adult	*Ammi majus*	(Burmeister, 1939; Larochelle, 1990)

(continued)

Table 6.2 (continued)

Family	Species	Life stage	Species of pollen consumed	Reference
Carabidae	*Harpalus affinis*	Adult	*Z. mays*	(Mullin et al., 2005)
Carabidae	*Harpalus caliginosus*	Adult	Pollen (Compositae), *A. artemisiifolia*, *Z. mays* (mixed with dog food)	(Ahmad et al., 2006; Forbes, 1883; Mullin et al., 2005; Webster, 1881)
Carabidae	*Harpalus herbivagus*	Adult	*Z. mays*	(Mullin et al., 2005)
Carabidae	*Harpalus pensylvanicus*	Adult	*A. artemisiifolia*, *Koeleria macrantha*, *Z. mays* (mixed with dog food)	(Ahmad et al., 2006; Forbes, 1883)
Carabidae	*Harpalus rufipes*	Adult	Pollen	(Cornic, 1973; Larochelle, 1990)
Carabidae	*Lebia atriventris*	Adult	Pollen and anthers of grass, probably *Poa* (bluegrass)	(Forbes, 1883)
Carabidae	*Loricera pilicornis pilicornis*	Adult	Pollen	(Davies, 1953)
Carabidae	*Microlestes maurus*	Adult	Pollen	(Davies, 1953)
Carabidae	*Notiophilus biguttatus*	Adult	Pollen	(Davies, 1953)
Carabidae	*Notiophilus rufipes*	Adult	Pollen	(Davies, 1953)
Carabidae	*Patrobus longicornis*	Adult	*Z. mays*	(Mullin et al., 2005)
Carabidae	*Poecilus chalcites*	Adult	*Z. mays*	(Mullin et al., 2005)
Carabidae	*Poecilus lucublandus*	Adult	*Z. mays*	(Mullin et al., 2005)
Carabidae	*Pterostichus diligens*	Adult	Leguminosae, Cruciferae, Caryophyllaceae, Umbelliferae, and Compositae	(Dawson, 1965)
Carabidae	*Pterostichus melanarius*	Adult	*Z. mays*	(Mullin et al., 2005)
Carabidae	*Scarites quadriceps*	Adult	*Z. mays*	(Mullin et al., 2005)
Carabidae	*Stenolophus mixtus*	Adult	Pollen	(Gersdorf, 1937; Larochelle, 1990)
Coccinellidae	*Adalia bipunctata*	Larva	Bee pollen	(De Clercq et al., 2006)
Coccinellidae	*Adalia bipunctata*	Adult	Aceraceae, Amaryllidaceae, Berberidaceae, *Betula*, *Corylus*, Brassicaceae, Fagaceae, Grossulariaceae, Liliaceae, Pinaceae, Rosaceae, Salicaceae, bee pollen	(Hemptinne and Desprets, 1986; Putnam, 1964; De Clercq et al., 2006)

(continued)

Table 6.2 (continued)

Family	Species	Life stage	Species of pollen consumed	Reference
Coccinellidae	*Anatis ocellata*	Adult	*Pinus banksiana*	(Allen et al., 1970)
Coccinellidae	*Anisosticta novemdecimpunctata*	Adult	Pollen	(Goidanich, 1947; Hodek and Honěk, 1996)
Coccinellidae	*Apolinus lividigaster*	Adult	*Acacia, Bidens pilosa*	(Anderson, 1982)
Coccinellidae	*Bulaea*	Adults & larvae	Chenopodaceae	(Capra, 1947; Hodek and Honěk., 1996)
Coccinellidae	*Bulaea lichatschovi*	Adults & larvae	*Euphorbia, Artemisia, Eurotia, Atriplex, Nitraria, Tamarix, Clematis*	(Hodek and Honěk., 1996; Savoiskaya, 1983)
Coccinellidae	*Chilocorus kuwanae*	Adult	*Euonymus*	(Nalepa et al., 1992)
Coccinellidae	*Coccinella*	Adult	Compositae	(Forbes, 1883)
Coccinellidae	*Coccinella reitteri*	Adult	*Leontopodium alpinum*; exclusively pollinivorous	(Hodek and Honěk., 1996; Savoiskaya, 1970)
Coccinellidae	*Coccinella repanda*	Adult	pollen	(Smith, 1961)
Coccinellidae	*Coccinella septempunctata*	Larva	pollen	(Triltsch, 1999)
Coccinellidae	*Coccinella septempunctata*	Adult	*Ribes, Stellaria, Pinus, Solidago, Tanacetum vulgare, T. aestivum*, Graminae, *Helianthus anuus*, mixed pollen, Umbelliferae, Compositae, *Laserpitium garganicum, Centaurea rupestris, Cachrys ferulacea*, also *Leucanthemum vulgare, Centaurea cyanus, Gentiana lutea, Cirsium eriophorum, Achillea collina, Heracleum sphondylium*	(Bozsik, 2006; Hodek and Honěk., 1996; Nedved et al., 2001; Ricci et al., 2005; Savoiskaya, 1970; Savoiskaya, 1983; Triltsch, 1997, 1999)
Coccinellidae	*Coccinella transversoguttata richardsoni*	Adult	Pollen	(Hodek and Honěk., 1996; Ibrahim, 1955)
Coccinellidae	*Coccinella trifaciata*	Larva	*Z. mays, Betula populifolia, T. latifolia, Carpinus caroliniana, Cannibas sativa, Tsuga canadensis, Carya ovata, Quercus rubra, Juglans cinerea*	(Anderson and Hale, 1986)

(continued)

Table 6.2 (continued)

Family	Species	Life stage	Species of pollen consumed	Reference
Coccinellidae	*Coccinella undecimpunctata aegyptiaca*		Pollen	(Hodek and Honěk., 1996; Ibrahim, 1955)
Coccinellidae	*Coccinula crotchi*	Adult	*Taraxacum officinale, Rumex acetosella, P. lanceolata*	(Hoshikawa, 1995)
Coccinellidae	*Coleomegilla maculata*	Larva	*Z. mays, B. populifolia, T. latifolia, C. caroliniana, C. sativa, T. canadensis, C. ovata, Q. rubra, J. cinerea, Pinus resinosa, A. artemisiifolia*, flowers of grass and Compositae, *Polygonum, M. crocea, Gossypium*, wildflower pollen, *Helianthus annus, Sorghum bicolor*, bee pollen	(Cottrell and Yeargan, 1998; Giroux et al., 1994; Harris, 1969; Lundgren et al., 2005; Lundgren et al., 2004; Lundgren and Wiedenmann, 2004; Michaud, 2000; Smith, 1960, 1961)
Coccinellidae	*Coleomegilla maculata*	Adult	*Z. mays, T. officinale, Populus deltoides, Caltha palustris*, wildflower pollen, *H. annus, S. bicolor*, bee pollen	(Benton and Crump, 1981; Forbes, 1881, 1883; Giroux et al., 1994; Harmon et al., 2000; Lundgren et al., 2005; Lundgren et al., 2004; Michaud and Grant, 2005; Putnam, 1964; Rondon et al., 2006; Solbreck, 1974; Webster, 1881)
Coccinellidae	*Cycloneda munda*	Adult	Pollen	(Putnam, 1964)
Coccinellidae	*Cycloneda sanguinea*	Larva	*Z. mays, B. populifolia, T. latifolia, C. caroliniana, C. sativa, T. canadensis, C. ovata, Q. rubra, J. cinerea*, other Compositae	(Forbes, 1883; Smith, 1961)
Coccinellidae	*Exochomus childreni childreni*	Adult	*A. artemisiifolia*	(Balduf, 1935)
Coccinellidae	*Exochomus flavipes*	Larva	Pollen	(Geyer, 1947)
Coccinellidae	*Exochomus flavipes*	Adult	Pollen	(Geyer, 1947)
Coccinellidae	*Harmonia axyridis*	Larva	*Z. mays, Spiraea douglasii*	(LaMana and Miller, 1996; Lundgren et al., 2004)

(continued)

Table 6.2 (continued)

Family	Species	Life stage	Species of pollen consumed	Reference
Coccinellidae	*Harmonia axyridis*	Adult	*Z. mays* (only 3% of adults had fed on corn pollen), *M. crocea*	(Lundgren et al., 2004; Michaud, 2000)
Coccinellidae	*Harmonia conformis*	Adult	*Serenoa repens*, *Crotalaria striata* (flowers and blossoms), *Erechtites hieracifolia* (blossoms, including pistils)	(Watson and Thompson, 1933)
Coccinellidae	*Hippodamia convergens*	Adult	*T. officinale*, Compositae, grass, bee pollen	(Forbes, 1883; Michaud and Qureshi, 2006)
Coccinellidae	*Hippodamia glacialis*	Adult	Compositae	(Forbes, 1883)
Coccinellidae	*Hippodamia notata*	Larvae	Umbelliferae, Graminaceae, Chenopodiaceae, other pollens	(Ricci and Ponti, 2005)
Coccinellidae	*Hippodamia notata*	Adult	*Carduus*, Graminaceae, *Z. mays*, Compositae, Umbelliferae, *Heracleum sphondylium*	(Ricci and Ponti, 2005)
Coccinellidae	*Hippodamia parenthesis*	Adult	Pollen	(Putnam, 1964)
Coccinellidae	*Hippodamia tredecimpunctata*	Adult	Pollen	(Goidanich, 1947; Hodek and Honěk., 1996)
Coccinellidae	*Hyperaspis notata*	Larva	*M. esculenta*; could not complete 1st instar	(Dreyer et al., 1997)
Coccinellidae	*Hyperaspis notata*	Adult	*M. esculenta*; doubtful whether it consumed pollen in the laboratory	(Dreyer et al., 1997)
Coccinellidae	*Ileis galbula*	Adult	*Ligustrum*, *Acacia*	(Anderson, 1982)
Coccinellidae	*Micraspis discolor*	Larva	*Z. mays*	(Omkar, 2006)
Coccinellidae	*Micraspis discolor*	Adult	*Z. mays*	(Omkar, 2006)
Coccinellidae	*Micraspis frenata*	Adult	*Chloris gayana*, *Chloris truncata*, *Paspalum urvillei*, *Setaria gracilis*, *S. bicolor* ssp. *bicolor*, *Themeda australis*	(Hawkeswood and Turner, 2002)
Coccinellidae	*Propylaea japonica*	Larva	*Oryza sativa*	(Bai et al., 2005)
Coccinellidae	*Propylaea japonica*	Adult	*O. sativa*	(Bai et al., 2005)
Coccinellidae	*Propylaea quatuordecimpunctata*	Adult	*Lamium album*, *Endymion nonscripta*	(Hemptinne et al., 1988; Hodek and Honěk., 1996)

(continued)

Table 6.2 (continued)

Family	Species	Life stage	Species of pollen consumed	Reference
Coccinellidae	*Rhyzobius litura*	Adult	Graminae, Asteraceae, Labiatae, Boraginaceae, Cruciphereae, Euphorbiaceae, Malvaceae, Ranunculaceae, Rubiaceae, Graminae, *Mercurialis annua*	(Ricci, 1986b)
Coccinellidae	*Spiladelpha barovskii kiritschenkoi*	Adult	*L. alpinum*; exclusively pollinivorous	(Hodek and Honěk., 1996; Savoiskaya, 1970)
Coccinellidae	*Tytthaspis sedec-impunctata*	Larvae	*Lolium perenne*, *Lolium multiflorum*	(Hodek and Honěk., 1996; Ricci, 1982)
Coccinellidae	*Tytthaspis sedec-impunctata*	Adult	*L. perenne, Alopecurus pratensis,* Gramineae, Compositae	(Hodek and Honěk., 1996; Ricci, 1986a; Ricci et al., 1983)
Coccinellidae	*Tytthaspis trilineata*	Larvae	*L. perenne*, *L. multiflorum*	(Hodek and Honěk., 1996; Ricci, 1982)
Coccinellidae	*Verania*	Larvae and adults	Exclusively pollinivorous	(Hodek and Honěk., 1996)
Neuroptera				
Chrysopidae	*Brinckochrysa scelestes*	Adult	*R. communis*	(Krishnamoorthy, 1984)
Chrysopidae	*Ceraeochrysa*	Adult	Pollen	(Albuquerque et al., 2001; Brooks and Barnard, 1990)
Chrysopidae	*Ceraeochrysa cubana*	Adult	Reared on pollen in aboratory	(Venzon and Carvalho, 1992)
Chrysopidae	*Chrysopa formosa*	Adult	Pollen	(Bozsik, 1992)
Chrysopidae	*Chrysopa nigricornis*	Adult	Pollen	(Sheldon and MacLeod, 1971)
Chrysopidae	*Chrysopa oculata*	Adult	Pollen	(Sheldon and MacLeod, 1971)
Chrysopidae	*Chrysopa pallens*	Adult	Pollen	(Bozsik, 1992)
Chrysopidae	*Chrysopa perla*	Adult	Pollen	(Bozsik, 1992)
Chrysopidae	*Chrysopa viridana*	Adult	Pollen	(Bozsik, 1992)
Chrysopidae	*Chrysoperla affinis*	Adult	Pollen from trees and herbaceous plants	(Villaneve et al., 2005)
Chrysopidae	*Chrysoperla carnea*	Larva	*Z. mays,* commercial bee pollen	(Patt et al., 2003; Pilcher et al., 1997)

(continued)

Table 6.2 (continued)

Family	Species	Life stage	Species of pollen consumed	Reference
Chrysopidae	*Chrysoperla carnea*	Adult	*Catalpa bignoniodes, Acer saccharum, Ulmus, Carya, Celtis occidentalis, Z. mays,* other grass pollen, *Phleum pratense,* up to 40 species of pollen found in guts	(Bozsik, 1992; Sheldon and MacLeod, 1971; Sundby, 1967; Villaneve et al., 2005)
Chrysopidae	*Chrysoperla externa externa*	Adult	Reared on pollen in laboratory; *Cajanus cajan, Crotalaria juncea, R. communis*	(Canedo and Lizarraga, 1988) (Venzon et al., 2006)
Chrysopidae	*Chrysoperla lucasina*	Adult	Ten species of pollen found in guts	(Villaneve et al., 2005)
Chrysopidae	*Chrysopodes nigripilosus*	Adult	Pollen	(Brooks and Barnard, 1990)
Chrysopidae	*Dichochrysa prasina*	Adult	Pollen	(Bozsik, 1992)
Chrysopidae	*Eremochrysa*	Adult	Exclusively pollinivorous	(Adams and Garland, 1981; Brooks and Barnard, 1990; Sheldon and MacLeod, 1971)
Chrysopidae	*Eremochrysa brevisetosa*	Adult	*Atriplex* (Chenopodiaceae)	(Adams and Garland, 1981)
Chrysopidae	*Eremochrysa fraterna*	Adult	Exclusively pollinivorous	(Sheldon and MacLeod, 1971)
Chrysopidae	*Eremochrysa sabulosa*	Adult	Pollen	(Brooks and Barnard, 1990)
Chrysopidae	*Hypochrysa*	Adult	Exclusively pollinivorous	(Brooks and Barnard, 1990; Canard, 2001; Tjeder, 1966)
Chrysopidae	*Hypochrysa elegans*	Adult	Pollen	(Canard, 2001)
Chrysopidae	*Kimochrysa*	Adult	Exclusively pollinivorous	(Brooks and Barnard, 1990; Canard, 2001; Tjeder, 1966)
Chrysopidae	*Pamochrysa stellata*	Adult	Exclusively pollinivorous; Dipsacacae, Compositae found in gut contents	(Brooks and Barnard, 1990; Canard, 2001; Tjeder, 1966)
Chrysopidae	*Parachrysopiella*	Adult	Pollen	(Brooks and Barnard, 1990)
Chrysopidae	*Pimachrysa*	Adult	Exclusively pollinivorous	(Brooks and Barnard, 1990; Canard, 2001)
Hemerobiidae	*Drepanepteryx phalaenoides*	Adult	Regularly feeds on pollen	(Canard, 2001; Stelzl, 1990, 1991)

(continued)

Table 6.2 (continued)

Family	Species	Life stage	Species of pollen consumed	Reference
Hemerobiidae	*Hemerobius lutescens*	Adult	Regularly feeds on pollen	(Canard, 2001; Stelzl, 1990, 1991)
Hemerobiidae	*Hemerobius nitidulus*	Adult	Regularly feeds on pollen	(Canard, 2001; Stelzl, 1990, 1991)
Hemerobiidae	*Micromus angulatus*	Adult	Reared on pollen in laboratory	(Stelzl and Hassan, 1992)
Hemerobiidae	*Micromus tasmaniae*	Adult	Buckwheat (*Fagopyrum esculentum*)	(Robinson et al., 2008)
Hemerobiidae	*Micromus lanosus*	Adult	Regularly feeds on pollen	(Canard, 2001; Stelzl, 1990, 1991)
Heteroptera				
Anthocoridae	*Anthocoris confusus*	Adult	Yellowish granules in the rectum of overwintering individuals	(Anderson, 1962b)
Anthocoridae	*Anthocoris nemoralis*	Adult	Aggregates to male catkins of *Salix*, and leaves when flowers senesce	(Anderson, 1962b)
Anthocoridae	*Anthocoris nemorum*	Adult	Yellowish Granules in the rectum of overwintering individuals	(Anderson, 1962b)
Anthocoridae	*Orius albidipennis*	Nymph	Bee-collected pollen, *C. annuum*	(Vacante et al., 1997)
Anthocoridae	*Orius albidipennis*	Adult	Bee-collected pollen, *C. annuum*	(Cocuzza et al., 1997)
Anthocoridae	*Orius insidiosus*	Nymph	*Glycine max*, *Z. mays*, *Acer*, *Verbascum thapsus*	(Corey et al., 1998; Elden and McCaslin, 1997; Kiman and Yeargan, 1985; McCaffrey and Horsburgh, 1986; Pilcher et al., 1997)
Anthocoridae	*Orius insidiosus*	Adult	*Z. mays*, *Acer*, *Abutilon theophrasti*, cotton	(Barber, 1936; Corey et al., 1998; Kiman and Yeargan, 1985; Richards and Schmidt, 1996) (Iglinsky, 1950, as cited by Chu, 1969)
Anthocoridae	*Orius laevigatus*	Nymph	Bee-collected pollen, *C. annuum*	(Vacante et al., 1997) J.G. Lundgren, unpublished data, 2007

(continued)

Table 6.2 (continued)

Family	Species	Life stage	Species of pollen consumed	Reference
Anthocoridae	*Orius laevigatus*	Adult	Bee-collected pollen, *C. annuum*, *R. communis*	(Cocuzza et al., 1997; Hulshof and Jurchenko, 2000; Skirvin et al., 2007)
Anthocoridae	*Orius majusculus*	Nymph	*Z. mays*	(Obrist et al., 2006a)
Anthocoridae	*Orius majusculus*	Adult	*Z. mays*	(Obrist et al., 2006a)
Anthocoridae	*Orius minutus*	Nymph	*C. annuum*, *Corylus americana*, *Cucumis sativa*	(Carayon and Steffan, 1959; Fauvel, 1974; Lattin et al., 1989)
Anthocoridae	*Orius niger*	Nymph	*Z. mays*	(Baniameri et al., 2005)
Anthocoridae	*Orius pallidicornis*	Adult	*Ecballium elaterium*, exclusively; flower pollen	(Carayon et al., 1959, cited in Chu, 1969)
Anthocoridae	*Orius sauteri*	Nymph	*Momordica charantia*, *Luffa cylindrica*, *Dolichos lablab*, *Rosa chinensis*, *C. sativus*, *Z. mays*	(Funau and Yoshiyasu, 1995; Vacante et al., 1997; Zhou and Wang, 1989)
Anthocoridae	*Orius tristicolor*	Nymph	Bee-collected pollen	(Salas-Aguilar and Ehler, 1977)
Anthocoridae	*Orius tristicolor*	Adult	Bee-collected pollen	(Salas-Aguilar and Ehler, 1977)
Anthocoridae	*Orius vicinus*	Nymph	*V. fabae*, Rosaceae, bee-collected pollen, Cucurbita pepo, *V. thapsus*, *M. annua*, *Datura stramonium* (stamens), *P. persica*, *Prunus cerasus*, *Prunus domestica*/*P. cerasus* (cross), *Malus pumila* (all of these Rosaceae included stamens); *Z. mays*	(Fauvel, 1974; Heitmans et al., 1986)
Anthocoridae	*Orius vicinus*	Adult	*V. fabae*, Rosaceae, *V. thapsus*, *C. pepo*, *Cucurbita melo*	(Fauvel, 1974; Heitmans et al., 1986)
Anthocoridae	*Paratriphleps laeviusculus*	Adult	*Manilkara zapotilla*	(Bacheler and Baranowski, 1975; Lattin et al., 1989)
Geocoridae	*Geocoris punctipes*	Nymphs and adults	*T. officinale* (bee-collected)	(Stoner, 1970)
Nabidae	*Nabis alternatus*	Nymphs (first instars)	*T. officinale* (bee-collected)	(Stoner, 1972)

(continued)

Table 6.2 (continued)

Family	Species	Life stage	Species of pollen consumed	Reference
Nabidae	*Nabis americoferus*	Nymphs (first instars)	*T. officinale* (bee-collected)	(Stoner, 1972)
Nabidae	*Nabis capsiformis*	Nymphs (first instars)	*T. officinale* (bee-collected)	(Stoner, 1972)
Pentatomidae: Asopinae	*Tylospilus acutissimus*	Nymphs	*T. officinale* (bee-collected)	(Stoner et al., 1974)
Reduviidae	*Sinea complexa*	Nymphs	*T. officinale* (bee-collected)	(Stoner et al., 1975)
Reduviidae	*Sinea confusa*	Nymphs	*T.officinale* (bee-collected)	(Stoner et al., 1975)
Reduviidae	*Zelus renardii*	Nymphs	*T. officinale* (bee-collected)	(Stoner et al., 1975)
Reduviidae	*Zelus tetracanthus*	Nymphs	*T. officinale* (bee-collected	(Stoner et al., 1975)
Diptera				
Syrphidae	*Allograpta ropalus*	Adult	Compositae, *Taraxacum*, *Hebe*	(Holloway, 1976)
Syrphidae	*Cheilosia albitarsis*	Adult	*Ranunculus repens*	(Haslett, 1983)
Syrphidae	*Cheilosia splendida*	Adult	*Saxifraga hirculus, Galium uliginosum*	(Olesen and Warncke, 1989)
Syrphidae	*Chrysogaster hirtella*	Adult	*S. hirculus, Lychnus flos-cuculi, G. uliginosum, Ranunculus acris, Lotus uliginosus*	(Olesen and Warncke, 1989)
Syrphidae	*Chrysotoxum bicinctum*	Adult	*S. hirculus, L. flos-cuculi*	(Olesen and Warncke, 1989)
Syrphidae	*Episyrphus balteatus*	Adult	Almost exclusively pollinivorous, *Pulicaria dysenterica, Eupatorium cannabinum, Centaurea nigra, Echium vulgare, Crepis capillaris, Epilobium hirsutum, Rumex crispus, Achillea millefolium, Lythrum salicaria, Dipsacus fullonum, P. lanceolata, C. avellana*	(Gilbert, 1981; Goulson and Wright, 1998; Schneider, 1948, 1969)
Syrphidae	*Eriozona syrphoides*	Adult	*Bidens, Epilobium angustifolium, Succisa pratensis, Calluna vulgaris, Erica cinerea, Erica tetralix, Liguliflora, Serratula*	(Haslett and Entwistle, 1980b)
Syrphidae	*Eristalis abusivus*	Adult	*S. hirculus, L. flos-cuculi, G. uliginosum, R. acris, Cirsium palustre, Valeriana sambucifolia*	(Olesen and Warncke, 1989)

(continued)

Table 6.2 (continued)

Family	Species	Life stage	Species of pollen consumed	Reference
Syrphidae	*Eristalis tenax* NOT PREDACEOUS AS A LARVA	Adult	*Taraxacum*, *Matricaria*, *Metrosideros*, Rosaceae, Cruciferae, Leguminosae, *Achillea*, *Quintinia*, *Leptospermum*, *Ranunculus sardous*, *Raphanus maritimus*, *Raphanus sativus*	(Holloway, 1976)
Syrphidae	*Eupeodes corollae*	Adult	*Senecio jacobaea*, *Rubus*, *Artemisia*, *Hypericum*, *Epilobium*, *Scabiosa*, *Achillea*, *C. avellana*	(Barlow, 1961; Svensson and Janzon, 1984)
Syrphidae	*Helophilus campbellicus*	Adult	*Hebe elliptica*	(Holloway, 1976)
Syrphidae	*Helophilus hochstetteri*	Adult	*Taraxacum*, *Matricaria*, Cruciferae, Rosaceae, Umbelliferae, Ranunculaceae, Graminae,	(Holloway, 1976)
Syrphidae	*Helophilus hybridus*	Adult	*S. hirculus*, *L. flos-cuculi*, *G. uliginosum*, *R. acris*, *C. palustre*	(Olesen and Warncke, 1989)
Syrphidae	*Helophilus montanus*	Adult	Compositae, *Ranunculus*	(Holloway, 1976)
Syrphidae	*Helophilus trilineatus*	Adult	*Leptospermum*, Compositae, *Taraxacum*, *Hebe*	(Holloway, 1976)
Syrphidae	*Lejogaster metallina*	Adult	*S. hirculus*, *L. flos-cuculi*, *G. uliginosum*, *R. acris*, *C. palustre*, *L. uliginosus*, *V. sambucifolia*	(Olesen and Warncke, 1989)
Syrphidae	*Lejops contracta*	Adult	*S. hirculus*, *L. flos-cuculi*, *G. uliginosum*, *R. acris*, *C. palustre*, *L. uliginosus*	(Olesen and Warncke, 1989)
Syrphidae	*Lejops lineatus*	Adult	*S. hirculus*, *L. flos-cuculi*, *G. uliginosum*, *R. acris*, *Epilobium palustre*, *C. palustre*, *L. uliginosus*	(Olesen and Warncke, 1989)
Syrphidae	*Melangyna novaezelandiae*	Adult	*Phacelia tanacetifolia tanacetifolia*, *Bulbinella*, *Ranunculus*, *Stellaria*, *Hebe*, Compositae	(Holloway, 1976; White et al., 1995)
Syrphidae	*Melanostoma fasciatum*	Adult	*P. t. tanacetifolia*, *P. lanceolata*, Graminae, *R. maritimus*, *R. sardous*, *Eschscholtzia californica*, *D. carota*, Malvaceae, Compositae: Liguliflorae	(Holloway, 1976; Leereveld, 1982; White et al., 1995)

(continued)

Table 6.2 (continued)

Family	Species	Life stage	Species of pollen consumed	Reference
Syrphidae	*Melanostoma mellinum*	Adult	*S. hirculus, L. flos-cuculi, C. palustre,* Graminae, *P. lanceolata,* Cyperaceae, Cruciferae, Compositae: Liguloflorae, Rosaceae, Ericaceae, Tiliaceae, Pteridophyta, Caryophyllaceae, *Pinus, Picea, Ranunculus*-type, Umbelliferae, *Convolvulus arvensis, Solanum nigrum, P. lanceolata, Anemone*-type, *Aster*-type, *Veronica*-type, *Stellaria*-type, *T. latifolia*	(Leereveld, 1982; Olesen and Warncke, 1989; van der Goot and Grabandt, 1970)
Syrphidae	*Melanostoma scalare*	Adult	Almost exclusively pollinivorous, Graminae, *P. lanceolata, Veronica*-type, Umbelliferae, Ranunculaceae, Roscaceae, Compositae: Liguloflorae	(Gilbert, 1981; Leereveld, 1982; van der Goot and Grabandt, 1970)
Syrphidae	*Neoascia meticulosa*	Adult	*S. hirculus, R. acris*	(Olesen and Warncke, 1989)
Syrphidae	*Neoascia tenur*	Adult	*S. hirculus, L. flos-cuculi, G. uliginosum, R. acris, C. palustre, L. uliginosus*	(Olesen and Warncke, 1989)
Syrphidae	*Paragus*	Adult	Compositae, *Forstera, Sebea, P. lanceolata, Hebe,* Cruciferae	(Holloway, 1976)
Syrphidae	*Parthelophilus frutetorum*	Adult	*S. hirculus, R. acris, E. palustre*	(Olesen and Warncke, 1989)
Syrphidae	*Platycheirus*	Adult	*Ranunculus gracilipes, Ranunculus*	(Holloway, 1976)
Syrphidae	*Platycheirus angustatus*	Adult	Graminae, *P. lanceolata, Pinus* Chenopodiaceae, Cruciferae	(Leereveld, 1982; van der Goot and Grabandt, 1970)
Syrphidae	*Platycheirus clypeatus*	Adult	Graminae, *P. lanceolata, Rumex*-like, *Melampyrum*	(Leereveld, 1982; van der Goot and Grabandt, 1970)
Syrphidae	*Platycheirus fulviventris*	Adult	Graminae, *P. lanceolata,* Caryophyllaceae	(Leereveld, 1982; van der Goot and Grabandt, 1970)

(continued)

Table 6.2 (continued)

Family	Species	Life stage	Species of pollen consumed	Reference
Syrphidae	*Platycheirus granditarsus*	Adult	Graminae, *P. lanceolata*, Cruciferae, *Aster*-type, Compositae: Liguliflorae, Rosaceae, *Stellaria*-type, *Ranunculus*-type, *Achillea*-type, Umbelliferae, *Hypericum*-type	(Leereveld, 1982; van der Goot and Grabandt, 1970)
Syrphidae	*Platycheirus immarginatus*	Adult	Grass, Cyperaceae	(Leereveld, 1982)
Syrphidae	*Platycheirus manicatus*	Adult	Chenopodiaceae, Cruciferae, *Aster*-type, Compositae: Liguliflorae, *Ranunculus*-type, *Cerastium*-type	(van der Goot and Grabandt, 1970)
Syrphidae	*Platycheirus peltatus*	Adult	*P. lanceolata*, Cyperaceae, Compositae: Liguliflorae, Liliaceae, Rosaceae, Caryophyllaceae, Rosaceae, *Ranunculus*-type, *Achillea*-type, *Cerastium*-type, *Allium*-type, *Rumex*-like, *Gladiolus*-type, *Urtica*, *T. latifolia*	(Leereveld, 1982; van der Goot and Grabandt, 1970)
Syrphidae	*Platycheirus scambus*	Adult	*P. lanceolata*, grass, Cyperaceae, *Rumex*, *Secale*, *Tilia*, *Pinus*, Pteridophyta	(Leereveld, 1982; van der Goot and Grabandt, 1970)
Syrphidae	*Platycheirus scutatus*	Adult	Chenopodiaceae, Cruciferae, *Aster*-type, Compositae: Liguliflorae, Rosaceae, Rosaceae-type, *Veronica*-type, *Stellaria*-type, *Ranunculus*-type, *Achillea*-type, *Allium*-type, Umbelliferae, *Cirsium*-type, *Melandrium*-type, *Polygonum*-type	(van der Goot and Grabandt, 1970)
Syrphidae	*Rhingia campestris*	Adult	*Stachys*, *Prunella*	(Haslett, 1983)
Syrphidae	*Sphaerophoria*	Adult	*S. hirculus*, *L. flos-cuculi*, *G. uliginosum*	(Olesen and Warncke, 1989)
Syrphidae	*Sphaerophoria abbreviata*	Adult	*S. hirculus*, *L. flos-cuculi*, *G. uliginosum*, *R. acris*, *C. palustre*	(Olesen and Warncke, 1989)

(continued)

Table 6.2 (continued)

Family	Species	Life stage	Species of pollen consumed	Reference
Syrphidae	*Sphaerophoria menthastri*	Adult	*S. hirculus, L. flos-cuculi, G. uliginosum, R. acris, C. palustre*	(Olesen and Warncke, 1989)
Syrphidae	*Syrphus ribesii*	Adult	Almost exclusively pollinivorous, *P. dysenterica, E. cannabinum, C. nigra, C. capillaris, E. hirsutum, D. fullonum, S. jacobaea, Mentha aquatica, Lapsana communis, Lathyrus latifolius, R. repens*	(Gilbert, 1981; Goulson and Wright, 1998)

Harpalus pensylvanicus survive for up to 1 year on a diet of only corn pollen (Mullin et al., 2005). Few of the records presented in Larochelle (1990) are actual gut dissections of field-collected beetles, and many of the feeding reports result from experimental manipulations of feeding behavior such as providing factitious prey for rearing programs. Thus, the true dietary range of most carabids is still a mystery. Additional research is required to understand the role of pollinivory by carabids under field conditions, and the reader is diverted to Section III for a deeper discussion on the feeding ecology of this group.

6.1.4 Coleoptera: Coccinellidae

Coccinellidae as a family is primarily insectivorous, although many studies on the dietary breadth of ladybeetles under field conditions have revealed pollinivory (Ewing, 1913; Forbes, 1881, 1883; Hoogendoorn and Heimpel, 2004; Lundgren et al., 2004, 2005; Putnam, 1964; Ricci and Ponti, 2005; Ricci et al., 2005; Triltsch, 1999). Some species of coccinellids, i.e. *Coccinella reitteri* and *Spiladelpha barovskii kiritschenkoi*, are thought to be exclusively pollinivorous. Aphids do not occur within the high-altitude habitats of these ladybeetles, and they cope with the dearth of prey by feeding on pollen, especially that of *Leontopodium alpinum* (Hodek and Honěk, 1996; Savoiskaya, 1970). Comparatively few coccinellid species are exclusively phytophagous or mycophagous, and most species in this family are best described as omnivorous.

Of the entomophagous species of ladybeetles, beetles reared on a diet consisting solely of pollen seldom develop entirely or mature eggs without supplemental nutrition; thus pollen represents an "alternative food" (as defined by Hodek, 1967) for many species of ladybeetles. An important exception is *Coleomegilla maculata*, which can complete development on pollen from a number of plants (Lundgren and Wiedenmann, 2004; Michaud and Grant, 2005; Smith, 1961), and can produce

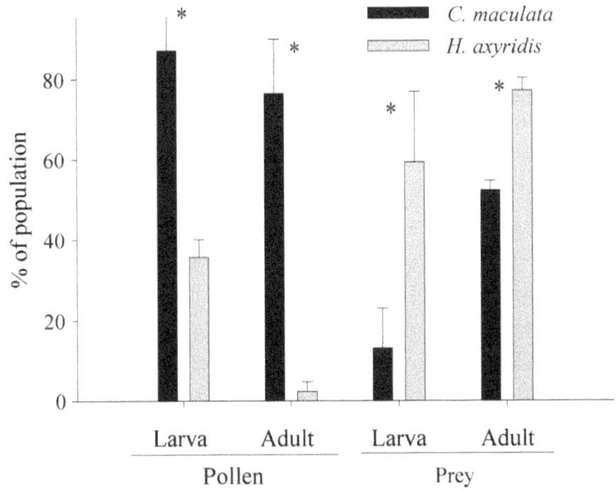

Fig. 6.1 Frequency that *Harmonia axyridis* and *Coleomegilla maculata* collected during anthesis from field corn had pollen and prey remains in their guts (Reproduced from Lundgren et al., 2004. With permission from the Entomological Society of America)

viable eggs on a diet consisting solely of *Zea mays* (Lundgren and Wiedenmann, 2004). Although pollen is an alternative food for most ladybeetles, it still serve a critical role in sustaining these natural enemies during periods of low prey availability (Ricci et al., 2005), or when prey is of low quality. Furthermore, pollen is important for building fat reserves used in diapause and migratory flights in some ladybeetles, and pollen engages sexual maturation of overwintered beetles. Most research has focused on pollinivory in adult ladybeetles, although some experiments suggest that larvae of ladybeetles can be quite pollinivorous (Hoogendoorn and Heimpel, 2004; Lundgren et al., 2004). For example, Lundgren et al. (2004) show that larvae of *C. maculata* and *H. axyridis* feed on *Zea mays* pollen more frequently than do adults, and that *C. maculata* larvae are more frequently pollinivorous than insectivorous during *Zea mays* anthesis (Fig. 6.1).

6.1.5 Neuroptera: Chrysopidae

The dietary breadth of lacewing adults is more extensively explored than for the larval stage. Most chrysopid adults are phytophagous to some degree, feeding on plant tissue (leaf-scraping; described by Sheldon and MacLeod, 1971), nectar, and pollen (Bozsik, 1992, 2000; Canard, 2001). The adults of some genera within Chrysopidae appear to be predominately pollinivorous (*Eremochrysa*, *Hypochrysa*, *Kimochrysa*, *Pamochrysa stellata*, and *Pimachrysa*). There is some evidence that adults of *Chrysoperla*, an important genus to biological control, are also predominately phytophagous. Up to 40 species of pollen (especially from the Brassicaceae,

Chenopodiaceae, Rocaceae, and Graminae) are reported from the guts of field-collected *Chysoperla* spp. (Villaneve et al., 2005), and adults have been shown to aggregate to *Z. mays* tassels during anthesis (Sheldon and MacLeod, 1971), presumably to feed on pollen, although aphids are often found at high densities on *Z. mays* tassels. *Phleum pratense* pollen sustains low levels of oviposition in *C. carnea*, but egg production is lower on a pollen-only diet versus a more diversified one (Sundby, 1967). Exploring alternative pollen species may reveal pollens that could better promote reproduction in *Chrysoperla carnea* and other lacewing taxa.

Omnivory in lacewing larvae has gained attention due to several recent reports. In the past, lacewing larvae were thought to digest prey extra-orally and consume liquid foods exclusively. A series of experiments shows that larvae of *C. carnea* do not digest food extra-orally, nor do they possess the digestive physiology for extra-oral digestion (Yazlovetsky, 2001, and references therein). Because lacewing larvae were thought to be largely carnivorous, literature on pollinivory and glucophagy (see Chapter 2) by this life stage is scarce. Nevertheless, research is beginning to reveal phytophagy in *Chrysoperla* larvae. Larvae of *C. carnea* feed on *Z. mays* pollen in the laboratory (Pilcher et al., 1997), and even though this food on its own is suboptimal for larval survival, high quality non-prey foods such as *Z. mays* pollen may help to improve larval fitness when prey quality is poor (Patt et al., 2003).

6.1.6 Heteroptera

Many families of Heteroptera are described as omnivorous, and feeding behavior is best understood in the Anthocoridae, Nabidae, Reduviidae, Geocoridae, and Asopinae (Pentatomidae). Of these more carnivorous heteropterans, pollinivory research has focused almost exclusively on anthocorids. Many species of anthocorids (especially in the Anthocorinae) feed on pollen, although gut content analyses are scarce (one example is Corey et al., 1998). Anthocorids can often be captured in flowers, presumably feeding on nectar and pollen when prey is absent. Some species are quite herbivorous, preferring to feed on flowers or pollen even in the presence of prey (Bacheler and Baranowski, 1975; Carayon and Steffan, 1959; Corey et al., 1998; Dicke and Jarvis, 1962). Aggregations of *Orius insidiosus* shift between *Glycine max* (soybean) and *Z. mays* fields depending on the respective flowering periods of these crops (Dicke and Jarvis, 1962; Isenhour and Marston, 1981), and these bugs change their seasonal within-plant distribution in corn to best exploit available pollen during anthesis (Coll and Bottrell, 1991). Several species of *Anthocoris* aggregate on male *Salix* catkins during anthesis, and leave as the catkins senesce (Anderson, 1962a). Similarly, spring populations of *Orius vicinus* aggregate on flowers of Rosaceae, feeding on the flowers even in the presence of prey elsewhere on the plant (Fauvel, 1974). Some species of anthocorids can complete development on a diet consisting solely of *Z. mays* pollen (Kiman and Yeargan, 1985; Pilcher et al., 1997; Salas-Aguilar and Ehler, 1977), and nymphs have been observed to feed on corn pollen in the field (Dicke and Jarvis, 1962).

In the laboratory, *O. vicinus* completes development more quickly on the pollen of several rosaceous species than on some prey species, although resulting adults are smaller (by 10%), have more malformations, and lay fewer eggs than prey-fed individuals (Fauvel, 1974). But even "unsuitable" pollens (which do not permit pupation) such as that of *Mercurialis annua* are able to sustain *O. vicinus* nymphs for up to 40 days. Another laboratory experiment shows that nymphs and adults of *Orius majusculus* gain the most weight when fed on a diet of corn pollen and water; adults do even better on the pollen diet than when fed on prey (spider mites) alone (Obrist et al., 2006a).

Although nabids, pentatomids, reduviids and geocorids feed on plant tissues other than pollen, the only research on pollinivory by these insects was conducted in the 1970s by Stoner and colleagues (Stoner, 1970, 1972; Stoner et al., 1974, 1975). Pollen was invariably found to be the most suitable plant substance for survival in the absence of prey for these predatory heteropterans, and allowed nymphs to survive for up to 20 days with no other food.

6.1.7 Diptera: Syrphidae

All adults of Syrphidae are believed to use nectar and pollen as principle foods, but the degree to which they rely on pollen varies from species to species (Gilbert, 1981) (see also Chapter 2). For example, of eight common species of syrphids in the U. K., Gilbert found three that consume almost exclusively pollen, and the other five divide their diets fairly evenly between nectar and pollen. Care must be taken when describing a syrphid species as exclusively pollinivorous based on limited observations, because reproductive and physiological status predetermines what foods the flies will forage for (Haslett, 1989). The locale where pollen is consumed differs for syrphids; some consume pollen on the flowers, whereas others (*Xylota* spp.) feed on pollen from the phylloplane (Gilbert, 1986a; Holloway, 1976). Diurnal flight activity is strongly correlated with pollen availability for at least some species (Maier and Waldbauer, 1979), and syrphids sometimes display flower constancy, selectively visiting flowers of a single species (Goulson and Wright, 1998; Olesen and Warncke, 1989; van der Goot and Grabandt, 1970). Not all species have such fidelity to specific plants; *Scaeva pyrastri* visits flowers somewhat indiscriminately (Schneider, 1948; van der Goot and Grabandt, 1970).

Larvae of Syrphidae display a wide range of feeding habits, and are categorized as insectivorous, phytophagous, scavengers, or saprophytic (Gilbert, 1986b; Maier, 1978; Parmenter, 1953a; Schneider, 1969). All entomophagous syrphid larvae occur in the subfamily Syrphinae, and can be classified as strictly or factultatively insectivorous (Gilbert, 1986b). As an example of the latter, the genus *Platycheirus* feeds on both aphids and rotting plant material (Gilbert, 1986b). Although pollen-feeding has not been observed in entomophagous syrphid larvae, it is conceivable that some of the more polyphagous taxa will accept pollen if it were offered; more research on the dietary breadth of syrphid larvae is required.

6.1.8 Hymenoptera: Formicidae

> *If there was ever a scoundrel in the pollination drama, that role has been assigned to ants.*
> (Faegri and van der Pijl, 1966)

Ants are key components of insect communities, and while they are known to consume pollen, it is surprising that this component of the ant diet does not receive more attention. Ants are frequent visitors to flowers and are commonly observed to be transporting pollen on their bodies (Peakall et al., 1991). Because adult ants consume primarily fluids, filtering all but the smallest particles from liquid meals, they must transport pollen grains to their larvae for digestion (see Chapter 10 for more discussion). Indeed pollen feeding by larvae is widespread in certain groups of ants, and may be a universal component in the diets of polyphagous species. For example, nearly all of the larvae of Pseudomyrminae ants feed on different pollens to various degrees (Wheeler and Bailey, 1920), but the dietary range of most species of ant larvae remains a mystery.

Given that ants are numerically abundant in nearly every terrestrial habitat worldwide, and that ants frequently visit flowers, it is surprising that plants have not come to rely on these insects as pollinators (Beattie and Hughes, 2002; Holldobler and Wilson, 1990). Peakall et al. (1991) present a compelling case that the reason that plants avoid ant pollinators is because the ants produce antibiotic substances that kill the pollen before it can be transferred to another flower. Nevertheless, some plant species that produce copious amounts of pollen and rely on general flying insects as pollinators may occasionally be pollinated by ants (Beattie and Hughes, 2002), and other plants actually specialize on ants through several adaptations outlined by Hickman (1974). How changes in the pollen grains alters their nutritional qualities for the ants is worthy of additional study.

6.1.9 Mantodea: Mantidae

Tenodera aridifolia sinensis can complete the first stadium on a diet consisting only of pollen collected by honeybees (Beckman and Hurd, 2003). Although mantids fed only pollen do not gain body mass as well as prey-fed individuals, adding pollen to the diet of prey-fed individuals results in significantly higher body mass accumulation (Beckman and Hurd, 2003). In the field, mantises that dwell on flowers produce more eggs and have a greater body mass relative to individuals collected from non-flowering plants (Hurd, 1989). Hurd hypothesizes that this may be a reflection of pollen feeding under field conditions.

6.2 Parasitoids

Flower-visiting is commonly recorded in parasitoids (Allen, 1929; Jervis et al., 1993; Leius, 1967), though it is not always clear whether parasitoids visit flowers for nectar, pollen, or other reasons. Direct feeding on pollen is apparently restricted

Table 6.3 The occurrence of pollinivory within selected parasitoid families of Insecta. Pollen feeding is observed only for the adult stages of parasitoid species

Family	Species	Pollen species consumed	Reference
Hymenoptera			
Braconidae	*Asobara*	*A. glutinosa*	(Eijs et al., 1998)
Eulophidae	*Edovum puttleri*	*Anethum graveolens*	(Jervis, 1998)
Ichneumonidae	Parasitoids of spruce web-worm sawfly (*Cephalcia abietis*)	Grass pollen	(Kanecka, 1993)
Ichneumonidae	*Itoplectis conquisitor*	*Pinus sylvestris*	(Leius, 1961a)
Ichneumonidae	*Liotryphon strobilellae*	*C. avellana*	(Györfi, 1945; Jervis et al., 1993)
Ichneumonidae	*Rhyssa persuasoria*	*Pinus*	(Hocking, 1967; Jervis et al., 1993)
Ichneumonidae	*Scambus buolinae*	*P. sylvestris*, *Papaver orientale*, *D. carota*, *Pastinaca sativa*, *A. artemisifolia*, *Solidago canadensis*, *A. majus*, *B. populifolia*, *P. resinosa*, *Tulipa*, *Z.mays*, *Malus*	(Leius, 1961b, 1963)
Ichneumonidae	*Tryphon signator*	Grass pollen	(Hassan, 1967; Jervis et al., 1993)
Mutillidae		Morphological adaptations to pollinivory	(Jervis, 1998)
Scoliidae		Morphological adaptations to pollinivory	(Jervis, 1998)
Tiphiidae		Pollen	(Quicke, 1997)
Trichogrammatidae	*Trichogramma bourarachae*	*Lycopersicon* (tomato)	(Rohi et al., 2002)
Trichogrammatidae	*Trichogramma brassicae*	*Z. mays*	(Zhang et al., 2004)
Trichogrammatidae	*Trichogramma chilonis*	*Gossypium hirsutum*	(Geng et al., 2006)

to medium-sized or large parasitoids, and is highly dependent on mouthpart structure (Jervis et al., 1996b). Pollen is one of the most nutritious non-prey food sources for parasitoids based on its protein levels (Jervis et al., 1996b); still, pollen-feeding by parasitoids has been studied much less frequently than sugar-feeding.

6.2.1 Diptera

Although tachinid adults visit flowers and some species are considered pollinators (Allen, 1929), actual pollinivory is not frequently documented (Herting, 1960; as reported by Jervis et al., 1996b). Sources of protein do not appear to be critical for many Nearctic species in this group, and adults survive for weeks in cages provided with only a sugar source (Wood, 1992). Mouthparts of tachinid adults are long and siphoniphorous or small and sponging (Allen, 1929; Wood, 1992). Thus, it does not appear that Tachinidae are capable of feeding on pollen grains directly because of their mouthpart morphology, although they may consume pollen indirectly if its contents are solubilized in fluid.

Pollinivory seems to be more important to parasitoid bombyliids than the Tachinidae. Initially believed to be exclusively glucophagous, direct observations proved that many (if not all) species of bee flies consume pollen in the field (Deyrup, 1988). In fact, these flies were observed to feed at plants in the Commelinaceae, which produce no nectar. Deyrup also describes several morphological adaptations that facilitate pollen collection and consumption in bombyliids.

6.2.2 Hymenoptera

Much of the literature suggests that pollen is either avoided by parasitoid Hymenoptera or is consumed only indirectly when it contaminates nectar, honeydew, and water sources (Jervis, 1998). This being said, few studies on pollinivory in parasitoid Hymenoptera have been published, and so making definitive conclusions regarding the importance of pollinivory to parasitic wasps is premature. Parasitoid wasps frequently visit flowers, and a number of studies support the notion that these insects are predominately feeding on nectar. Jervis et al. (1993) did not find pollen in the dissected guts of 42 ichneumonoid wasps (representing seven species) that were collected on flowers. Leius (1963) showed that *Scambus buolianae* did not ingest dry pollen in the laboratory. Microscopic examination has revealed that mouthpart structure in some families of parasitoid Hymenoptera does not allow them to feed on pollen, and some taxa actually have mechanisms for filtering pollen out of nectar. Instances of pollen-filtering were reported from Perilampidae and Eucharitidae, whose digitate labrum is used to filter pollen grains from liquid foods (Darling, 1988). Still, *Asobara* was able to increase fat reserves when reared on pollen, suggesting that some parasitoids may obtain nutrients directly from pollen (Eijs et al., 1998). Although Jervis et al. (1993) did not find pollen in the guts of parasitoids, they did observe what may amount to pollen feeding in the field, when parasitoid wasps would spend considerable time probing the sides of the corollas of flowers with open mouthparts.

Although indirect pollen feeding appears at first glance to be trivial to the nutrition of insects, there is evidence that this form of consumption can have important implications for the fitness of wasps and other natural enemies. In fact, pollen is naturally found contaminating nectar sources of many plant species (Todd and Vansell, 1942). Pollen loses amino acids and proteins within minutes of being placed into sucrose solution (Linskins and Schrauwen, 1969; Stanley and Linskins, 1965). Nectar contaminated with pollen absorbs proteins and amino acids from the grains, such that amino acid constituency and abundance of contaminated nectar are altered relative to clean nectar (Gilbert, 1972), and amino acid content of pollen-contaminated nectar increases over time (Erhardt and Baker, 1990) (these relationships are also discussed in Chapter 3). The quantity of pollen grains, and the species of pollen ultimately affect how many nutrients make their way into a nectar solution (Erhardt and Baker, 1990).

As non-prey foods go pollen is quite nutritious, and even as a nectar contaminant, pollen improves fecundity and longevity of other insects (Gilbert, 1972; Todd and Vansell, 1942) and possibly parasitoids. Leius (1961a, b, 1963, 1967) reports that sucrose solutions containing pollen significantly improve the longevity and fecundity of *Scambus buolianae* and *Itoplectis conquisitor* in the laboratory, but feeding on pollen without host-feeding is insufficient to produce eggs. *Trichogramma bourarachae* fed *Lycopersicon* sp. pollen had double the lifespan of unfed females, but the results in these treatments were not statistically different (Rohi et al., 2002). Fitness of other *Trichogramma* (*T. chilonis* and *T. brassicae*) species is promoted through the provision of pollen mixed in solution, but its suitability depends on what the pollen is mixed with. In one instance, mixing pollen with honey prolonged longevity of *T. chilonis* over honey alone, especially when host eggs accompany these non-prey foods (Geng et al., 2006). Mixing pollen with water improves longevity of *T. brassicae* over water alone, but for this species adding pollen to honey yields no measurable benefits (Zhang et al., 2004).

Given that most parasitic Hymenoptera appear to have mouthpart morphology that is suited to pollinivory, that closely related groups of carnivorous Hymenoptera are known to feed on pollen (Hunt et al., 1991), and that flower-visiting commonly exposes these wasps to pollen, it is surprising that pollinivory is not more prevalent within this group of entomophages. Two possible explanations for this phenomenon are (1) parasitoid wasps as a guild are not able to digest pollen, and therefore avoid feeding on pollen grains, or (2) parasitoid wasps are particularly susceptible to the defenses of pollen grains. The influences of pollen defenses and the structure and function of arthropods to pollinivory are discussed in subsequent chapters. Finally, it is possible that pollinivory is more prevalent in parasitoids than currently recognized, but glucocentric (to coin a term…) research on parasitoid nutrition has dominated the literature (Jervis and Kidd, 1996) and the importance of pollen feeding has been overlooked. Gut dissections of field-collected wasps are desperately needed, but are complicated by the fact that dissected specimens are difficult to identify (Jervis et al., 1993). Or if a pollen-specific molecular marker can be developed, then other forms of gut analysis (e.g., PCR, ELISA) may be employed to identify the importance of pollen feeding to this group.

6.3 Conclusions

An array of economically important natural enemies consumes pollen from a diversity of plants, including that of crop species. Furthermore, even fluid-feeding species not generally thought of as pollinivorous can benefit from pollen nutrition, which is readily released into sugar solutions. Recognizing the fact that many biological control agents consume pollen as part of their diet is the first step in designing effective biological control programs to conserve these natural enemies in our cropland. But pollen differs in its structure and nutrition from arthropod prey, and these natural enemies possess specialized adaptations that allow them to fully exploit pollen as food, a topic that will be discussed more in the next chapter.

Chapter 7
Adaptations to Pollen feeding

Behavior is often manifested in the physiological structure and function of an organism. Pollen has nutritional and structural features that are unique from prey, and consequently predators and parasitoids that are suited to consuming insects may not necessarily possess the means to efficiently consume and digest pollen. If pollen feeding is important to the fitness of predators and parasitoids, one should expect physiological and morphological adaptations that facilitate pollen feeding. This is indeed the case for many natural enemies, and the pollen-feeding specializations in arthropod structure and function are characterized as being sensory, morphological, or digestive in nature. Clearly the specific modifications observed in certain species are not ubiquitous in all entomophages, but documenting the adaptations allows us to have a search image for identifying pollinivory in other species.

7.1 The Pollen-Feeding Process in Entomophages

The process of pollen-feeding by entomophagous arthropods is detailed in at least two cases, with syrphid adults and phytoseiid mites. These thorough studies of how pollen is consumed are useful for identifying exact sensory and morphological features used specifically for manipulating and extracting nutrients from pollen grains. Furthermore, studying the process of pollen feeding can elucidate how individual adaptations interact with others to extract the nutrients from pollen.

The feeding process has been recorded in the adults of several syrphids, and species-specific behaviors reveal different tactics employed during pollen consumption. Müller (1883) was the first to provide a detailed description of pollen consumption in syrphids, focusing on adults of *Eristalis*, *Episyrphus*, and *Rhingia*. Modern optical technology has greatly improved our understanding of this process. Gilbert (1981, and references therein) gave a good account of the different roles of the mouthparts in pollinivory. First, the flies needed to dislodge the pollen grains from the anthers, which was accomplished either with the laciniae or with prostomal teeth. How the nutrients were then extracted from within the pollen exine

J.G. Lundgren, *Relationships of Natural Enemies and Non-Prey Foods*,
Progress in Biological Control 7,

Published by Springer Science+Business Media B.V.

differed among species. In some species, the labella grasped the pollen grains and lacerated them to relieve their contents. Other species consumed the grains whole and crushed them with a bristle like structure in the crop (Haslett, 1983, and references therein). However, in the majority of species it appeared that the entire pollen grain was consumed through the labrum and was left intact until it reached the midgut, where either enzymes worked to degrade the exine, or the pollen became solubilized in the fluids of the midgut (Gilbert, 1981; Haslett, 1983). The pollen exine thus remained visible throughout the digestion process. In summary, a number of morphological adaptations of the mouthparts and digestive tracts in syrphids assist them in extracting nutrients from pollen, and the heterogeneity in gastrointestinal structure has potentially resulted in multiple pollen-feeding tactics in this family.

Flechtmann and McMurtry (1992b) give a detailed account of how omnivorous phytoseiids (*Euseius stipulatus* and *Amblyseius similoides*) consume pollen. First, pollen grains are grasped singly by the chelicerae, and are brought intact toward the mouth. The pollen grain is then lacerated with several back and forth movements of the chelicerae. The fluid that escapes from the collapsing grain is ingested. This entire process is predicated on the morphology of the chelicerae, which are shaped to give them a 'side bite' so that they can pick up and hold the pollen grain without lacerating it. In another study involving electron microscopy, Flechtmann and McMurtry (1992a) note that the most pollinivorous of phytoseiids have a spoonlike surface on the inner side of the fixed digit of the chelicerae, presumably to aid in holding the pollen grains. Furthermore, these species have a relatively enlarged deutosternal groove relative to strictly predaceous phytoseiids, which may allow them to funnel the liquid contents of the lacerated pollen grains into the oral cavity (Fig. 7.1).

7.2 Sensory Adaptations for Detecting Pollen

One of the first events that must occur in pollinivory is that the entomophage must recognize pollen as food. Exact chemical and visual stimulants for pollen-feeding are not well known for most natural enemies. Nevertheless, some chemical and visual cues for feeding responses in natural enemies have been isolated from pollen, laying the groundwork for future research in this area.

7.2.1 Vision

Vision is important for identifying habitats where pollen can be found, and for eliciting a feeding response to pollen-meals. Color preferences are reported for a wide range of predators and parasitoids, and may be used to identify plants with floral resources that are suitable for natural enemies. For example, the attractiveness of specific floral color may play a role in plant fidelity, as demonstrated with syrphids like *Eriozona syrphoides*, which prefers to feed at mauve flowers (Haslett and Entwistle, 1980a).

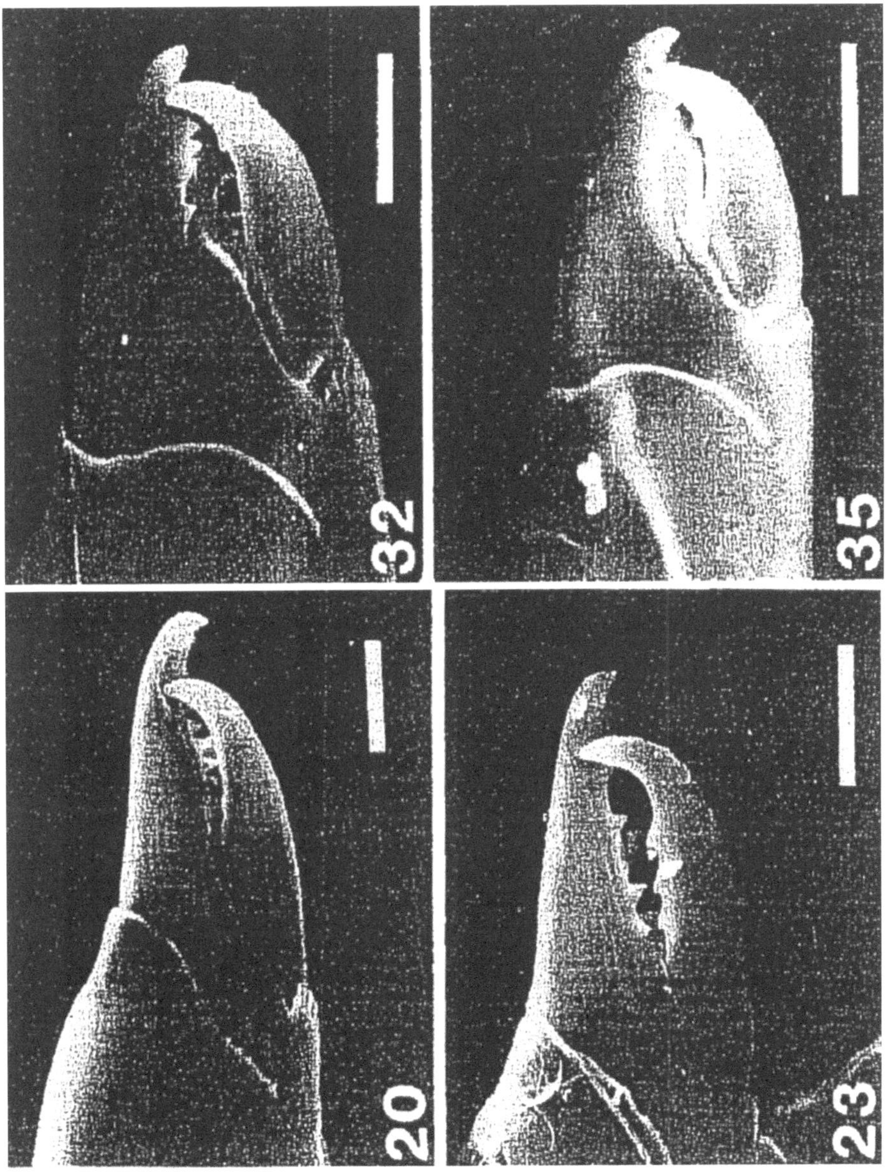

Fig. 7.1 Female cheliceral paraxial aspect for predatory (*left*) and pollinivorous (*right*) Phytoseiidae. Note the spoon-like surface of the top chelicerae in the pollinivorous taxa, presumably to aid in manipulating pollen grains (Reproduced from Flechtmann and McMurtry, 1992b. With permission from Indira Publishing House)

The physiology of color reception and subsequent feeding responses from hoverflies has revealed that specific wavelengths must be received from a food source before many syrphids will begin to feed, and that these wavelengths correspond to the color of some pollens (Lunau and Wacht, 1994; Wacht et al., 1996). The spectral wavelengths that most frequently elicited an extension of the proboscis range from 520–600 nm (green-yellow wavebands), and this same proboscis elongation behavior was observed when the flies fed at yellow anthers or floral guides (Lunau and Wacht, 1994).

7.2.2 Chemical Cues

Much of the research on chemoreception and pollinivory in natural enemies centers on the mouthparts. As an example of sensory adaptations to pollen feeding in phytophagous insects, lepidopterans that feed on pollen possess long mechanoreceptor sensillae proximally on their mouthparts, whereas these sensillae are smaller or absent in lepidopterans that do not consume pollen (Gilbert, 1972). Likewise, the mouthparts of carnivorous arthropods also possess sensory hairs that may be important in detecting non-prey foods. Cohen (1990) describes chemoreceptors in the labial tips of predatory heteropterans, and carnivorous lacewing larvae possibly detect sugarmeals with chemoreceptors in their maxillary palps (Downes, 1974). The antennae may also be important in distinguishing pollen-meals; mantids touch the pollen with their antennae before accepting it as food (Beckman and Hurd, 2003).

So what chemical cues from pollen do entomophages use to elicit feeding? It appears that the chemical elicitors used to initiate feeding are located in the pollen exine, and these chemicals are soluble in methanol or water (McMurtry et al., 1991; Wacht et al., 1996). *Euseius tularensis* aggregates to and lays eggs on leaf discs treated with extracted chemicals of *Malephora crocea* pollen; this is in contrast to untreated leaf discs, on which no eggs are laid (McMurtry et al., 1991). In *Eristalis tenax*, pollen feeding is initiated by triggering a salt receptor cell in the labellar sensory hairs (Wacht et al., 1996). In fact, proline (the predominant amino acid found in most pollens) is the only amino acid that excites this receptor, which allows the flies to identify at least five species of pollen in the laboratory (Wacht et al., 2000). Tryptophan and phenylalanine actually inhibit the salt-receptor cell. It should be noted that while most of this receptor cell work focuses on *E. tenax*, a non-predaceous syrphid, at least the results involving proline are also observed in *Episyrphus balteatus*, which has a predaceous larval stage (Wacht et al., 2000). *Orius insidiosus* is attracted to a lipophilic substance present in corn silks, and Reid and Lampman (1989) infer that these chemical cues are how this predator distinguished between silks, pollen, and leaf tissue in order to find *Helicoverpa* eggs. Thus, even in instances when the insects are thought to be primarily predaceous (e.g. lacewing larvae and mantids), there are sensory adaptations that allow them to identify non-prey foods such as pollen.

7.3 Morphological Adaptations

Structures that allow predators to collect, manipulate, and ingest pollen are pervasive throughout entomophagous arthropods. Often, these structural adaptations involve specialized morphology that is used to collect pollen from entomophilous plants, to separate pollen from liquid food sources, and to crush the hard walls of the pollen grains.

7.3.1 Setae

Pollen-collecting hairs are commonly observed on the bodies of pollinivorous natural enemies, a conspicuous example being syrphid flies. Many syrphids possess hairs that allow them to collect pollen from plants that they consume later while cleaning their bodies (Holloway, 1976; Olesen and Warncke, 1989). These hairs may be locally present in the form of a cleaning-comb (Holloway, 1976). Holloway (1976) hypothesizes that the abundance of these specialized setae may give an indication of whether specific syrphids feed on entomophilous versus anemophilous plants, since entomophilous plants produce pollen that is adapted to sticking to insect pollinators. But it should also be recognized that these hairs may have alternative functions; J. R. Haslett notes that only males of *Platycheirus* sp. have combs of dense hairs on their front femorae, but the females consume considerably more pollen than the males, which suggests that in this case the combs may serve purposes other than pollen cleaning (Gilbert, 1981). Combs are also important in mycophagy, and this topic will be revisited in Chapter 14.

7.3.2 Mouthparts

Mouthparts possess a diverse set of appendages that can be used for manipulating pollen grains or extracting their nutrients. The pollinivorous genus *Euseius* has chelicerae specialized for pollen-feeding; these chelicerae have small, convex inner surfaces with minute denticles only at the distal end of the fixed digit (McMurtry and Croft, 1997). The mandibles of pollinivorous adult lacewings are smaller than those of entomophagous species (Canard et al., 1990). Pollen rakes have been identified on the mandibles of the coccinellids, *Tytthaspis sedecimpunctata* and *T. trilineata* larvae (Hodek and Honěk, 1996; Ricci, 1982, 1986a). Pollen-feeding is correlated with larger labellums in syrphid adults, and this enlarged labellum may aid in removing pollen grains from the anthers of flowers. Dry pollen grains are grabbed and twisted with the inner labellar surface, and moved into what are termed 'food-furrows' (Gilbert and Jervis, 1998; Schuhmacher and Hoffmann, 1982). In these furrows, saliva is mixed with the grain, and the mixture is then sucked into the mouth (Schuhmacher and Hoffmann, 1982). Prestomal teeth and brush-like features on the tips of the labellum facilitate dislodging of pollen grains from the anthers in parasitoid Diptera, although these structures are not widespread in this group (Gilbert and Jervis, 1998). Bombiliids employ their tarsi to collect the grains and place them in the oral cavity (Gilbert and Jervis, 1998). Finally, the mouthparts of Mutellidae and Scoliidae have elongated maxillary palps that are covered in dense hairs, presumably to assist in manipulating pollen (Jervis, 1998). O'Toole observed that species in these families regurgitate liquid onto anthers, and then use their palps to collect the pollen grains that adhere to the fluid (Jervis, 1998).

Once the pollen is obtained, extracting the nutrients from within the pollen wall is accomplished using a variety of morphological adaptations. Many entomophagous arthropods prefer not to swallow the entire pollen grain, and ingest only its liquid contents. Members of Phytoseiidae and Anthocoridae pierce the grains with their stylets and extract the liquid contents (Chant, 1959; Elbadry, 1968; Elbadry and Elbenhawy, 1968a; Fauvel, 1974). In calypterate Diptera, the pollen grains are probably dissolved in saliva and then are transported in solution up the proboscis via the pseudotracheal canals of the labella (Gilbert and Jervis, 1998). A large number of pseudotracheal canals and a short labella is associated with pollinivory in flies, and the width of these canals dictates the size of pollen grain that can ultimately be consumed. Those species with a CNEA are likely entirely prohibited from directly consuming pollen (Gilbert and Jervis, 1998).

7.3.3 Internal Anatomy

For arthropods that consume the entire grain of pollen, internal anatomy may be specially adapted to dismantle the pollen wall and to ensure that maximum energy is extracted from this food. Some syrphids are believed to simply crush the grains in order to access the nutrients (Parmenter, 1953b). One way that this is accomplished is with bristle-like structures of the crop, which reportedly lacerate the grains and drain their contents into the stomach (Zimina, 1957). Other internal features may aid in swallowing pollen grains and protecting the gastrointestinal tract from damage from the tough grains. Along the oesophagus of the foregut in some syrphids there are backward-facing spines that presumably prevent pollen grains from moving anteriorly (Nayar, 1965). Also, some syrphids lack a peritrophic membrane, and possess a compensatory thickening of the midgut epithelium that reduces the risk of damage from solid pollen grains (Nayar, 1965). Finally, the length of the digestive tract is longer in phytophagous Coccinellidae than in predaceous species (Sakurai, 1968), and elongation in the gastrointestinal tracts of these and other omnivores may be a necessary adaptation that has evolved to extract nutrients from these less digestible foods.

Pollen is more difficult to digest than nectar, and so it is necessary for some insects to separate the two foods. As an example, *Apis mellifera* possesses a structure in the proventriculus that is used to separate pollen from nectar (Bailey, 1952), although in this case the insect can digest the pollen. Although the internal anatomy of the proventriculus of granivorous carabids is known to have similar functional structures (see Chapter 10), this topic has not been well studied for pollinivory in natural enemies.

7.4 Adaptations to Digesting the Nutrients of Pollen

In order to derive energy from pollen, insects must possess a means of digesting nutrients and overcoming defensive hurdles that are unique to plants. Digesting plant nutrients poses some unique challenges for arthropods that are specialized to feeding on insects. For example, arthropods that cannot mechanically destroy the pollen wall require specialized physiological conditions that either degrade or disrupt the pollen exine (Roulston and Cane, 2000). Also, plants are composed of starches, pectin, and other polymers that are not present in animal tissues, and the most entomophagous of natural enemies may not be privy to the appropriate digestive enzymes that degrade these phytochemicals. Entomophagous arthropods use several non-mechanical strategies to access and digest pollen nutrients. First, many natural enemies produce enzymes that allow them to digest the polysaccharides in the pollen wall. Arthropods that do not produce digestive enzymes capable of breaking down plant-specific polymers may acquire the appropriate digestive peptides from the pollen itself, or from endosymbiotic relationships with microbes. Also, it may be possible for zoophagous arthropods to avoid digesting plant-specific polymers entirely, and they may access the internal nutrients within the pollen grain by producing a favorable environment within the gut that allows the pollen to germinate and exude its nutrients.

7.4.1 Enzymatic Adaptations to Digesting Pollen

Examples of nutrient polymers that do not occur in insect prey are certain polysaccharides (starch, pectin), and amylases and pectinases are necessary for the digestion of these polysaccharides (Boyd and Cohen, 2002; Cohen, 1996; Zeng and Cohen, 2000). Many entomophagous arthropods possess these enzymes, and their presence in predatory insects is an indication of phytophagy or pollinivory. Amylases are important in breaking down starch and glycogen (Amylase Research Society of Japan, 1995), and amylase is commonly reported in natural enemies. In addition to producing these enzymes in their salivary secretions, some arthropods have these enzymes in their guts. For instance, α-amylase and α-glucosidase are present in the gut of the spider, *Tegenaria atrica* (Mommsen, 1978). Although glycogen is used as a substrate in Mommsen's research, these groups of enzymes also function to cleave the amylopectin portion of starch molecules (Lehninger, 1982), and their presence in the spider gut may assist in digesting plant starches consumed with the web. Still, it should not be forgotten that amylases are used in metabolizing glycogen from animals, and do not unequivocally indicate the digestion of starch-containing tissues, as is assumed in some literature. The coccinellid, *Exochamus flavipes*, has amylases in the guts of both larvae and adults, but larvae are supposedly strictly carnivorous (Geyer, 1947). The predatory midge, *Aphidoletes aphidimyza*, has amylase in its salivary gland secretion, and although starch is used as a substrate for this enzyme, the authors hypothesize that the amylase metabolizes glycogen in its prey (Laurema et al., 1986). Amylases are also detectable in a range of predatory Heteroptera, including *Orius insidiosus* (Zeng and Cohen, 2000) and *Podisus maculiventris*, and *Zelus renardii* tests positive for amylase in their salivas and guts (Chapman, 1998; Cohen, 1990). *Geocoris punctipes* and *Sinea confusa* have amylase only in their guts and *Nabis alternatus* only has amylase in its saliva (Cohen, 1990, 1996).

Pectins are polysaccharides that lend rigidity to the pollen wall (Roulston and Cane, 2000), and pectinases are necessary to degrade these molecules. Because pectin is only found in plants, possessing pectinases is a better indication of phytophagy than amylases. Indisputably, amylases are more commonly reported than pectinases. In part, this may be because pectins do not occur in high enough concentrations to make it nutritionally worthwhile for entomophages to harbor enzymes necessary for their degradation. Nevertheless, some predatory mirids (like *Helopeltis clavifer* and *Deraeocoris nebulosus*) possess pectinases in their saliva and midguts (Boyd and Cohen, 2002; Miles, 1972, 1987). The Miridae aside, pectinases are infrequently reported from entomophagous species.

In addition to producing enzymes specific for plant tissue, some entomophages may be able to acquire necessary enzymes from the pollen itself. Honeybees utilize pollen-derived enzymes to digest specific nutrients within the pollen, and these enzymes may also be used by entomophages to assist in pollinivory. A number of proteases can be found in pollen depending on the species, and pollen-derived chymotrypsin, trypsin, and carboxypeptidases may contribute to a substantial proportion of the protease activity in the gut of *Apis mellifera* during pollen digestion (Grogan and Hunt, 1979). Gut proteases in entomophagous arthropods are generally not limiting, but other pollen-incorporated enzymes, such as amylases and pectinases that are abundant in pollen (Stanley and Linskins, 1974) may play a role in pollen digestion. Empirical research is necessary to substantiate this hypothesis.

7.4.2 *Other Strategies for Digesting Pollen*

It is possible that no digestive processes are necessary to access the nutrients of pollen. As mentioned earlier, pollen readily germinates in sucrose solutions and upon germinating the grains release their nutritional contents (Linskins and Schrauwen, 1969; Stanley and Linskins, 1965) (see Chapter 3). Some insects prompt this germination externally, by disturbing the pollen grains in nectar, and drinking the pollen nutrients with the nectar (Gilbert, 1972; Nicolson, 1994). In 1883, Müller noted that the syrphid *Rhingia rostrata* deliberately moved pollen grains into the nectar of *Lythrum salicaria* during the feeding process in the field, likely resulting in the leaching of pollen nutrients into the nectar. Additionally, it may be possible for entomophagous arthropods to germinate the pollen grains in their stomachs, thereby accessing the nutrients. In two species of syrphids, intact grains simply rupture in the insect midgut without mechanical disruption (Haslett, 1983) (Fig. 7.2). Haslett hypothesizes that the grains ingested by the flies are yielding their contents to the sugar solutions contained within the insect gut, but does not discount that digestive enzymes may also play a role here.

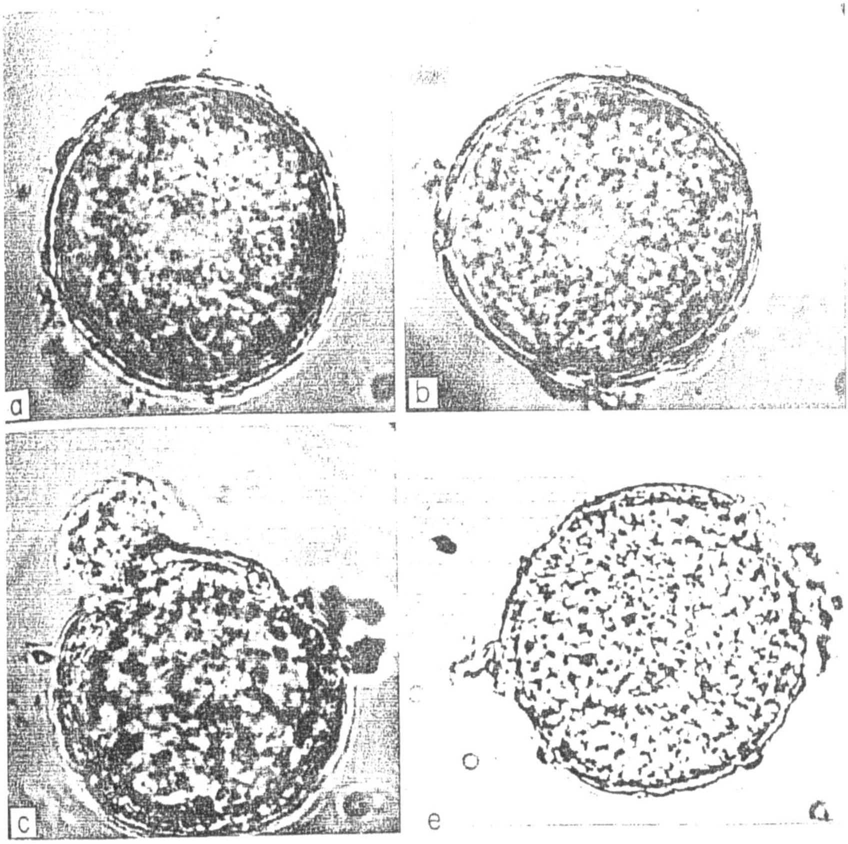

Fig. 7.2 *Ranunculus repens* pollen grains (unstained) taken from difference regions of the gut of *Cheilosia albitarsis*. (a) Pollen from the esophagus, (b) pollen from the crop, (c) pollen from the mid-gut, (d) pollen from the rectum (Reproduced from Haslett, 1983. With permission from Blackwell)

7.5 Conclusion

In summation, a range of adaptations to collecting, consuming, and digesting pollen abound within species ordinarily considered as primarily entomophagous. Indeed, the large carabid beetle, *Harpalus caliginosus*, can consume 50 mg of corn pollen per day in the laboratory (Mullin et al., 2005). Although this is a phylogenetically diverse group of organisms, analogous inventions have been put to use to collect the pollen (e.g., rakes, combs, and hairs), violate the pollen exine and gain access to the nutrients inside (e.g., crushing mandibles, internal bristles and the proventriculus, or solubilizing the grains in fluid), and catabolize the nutrients within (various digestive enzymes, some of which may be pollen-derived). The topic to be explored in the next chapter is why pollen is worth all of this fuss for natural enemies. As it turns out, this widespread plant material is a uniquely rich source of nutrients.

Chapter 8
Pollen Nutrition and Defense

The chemical and structural features of pollen influence its relative attractiveness to natural enemies. Superficially, pollen grains are very simple structures. They consist of a largely indigestible exine and intine, and most nutrients are found within the protoplast (Faegri and van der Pijl, 1966). A deeper examination reveals that these tiny grains have a complex nutritional profile and ecology. Pollens possess high protein and oil contents relative to most other plant tissues, making them nutritionally suitable for many of the life processes of arthropods that ordinarily consume insect prey. Not all pollinivorous arthropods have the plant's best interests in mind, and so pollens contain a variety of morphological and physiological mechanisms of defense. Ultimately the nutritional and defensive properties of pollen mitigate the interactions among pollinivorous entomophages and the degree to which these arthropods rely on pollen in their diet.

8.1 Nutrition

There are numerous essential nutrients that arthropods must obtain from their diet. Pollen is unique among plant-based foods in that it contains many of the essential nutrients required for insect development. It is rich in proteins, lipids, carbohydrates, and minerals, and consequently it is one of the most nutritious non-prey food sources (Jervis et al., 1996a) (see Chapter 19). One nutrient that is largely lacking from pollen is water, and natural enemies typically require a separate source of water for maximum fitness when pollen is their sole nutriment (De Clercq et al., 2006; Michaud and Grant, 2005).

Pollen nutritional profiles vary considerably among species, and these differences may alter the nutritional suitability of pollen for natural enemies and dictate their foraging behaviors in the field. In one of the few studies on the caloric content of pollens, Petanidou and Vokou (1990) calculate the overall mean of 40 species of pollen as 5.61 kcal g^{-1}, with a range from 4.41 to 6.30. Anemophilous species have consistently lower caloric contents than entomophilous species. As a consequence

J.G. Lundgren, *Relationships of Natural Enemies and Non-Prey Foods*,
Progress in Biological Control 7,

Published by Springer Science+Business Media B.V.

of this variability, predators and parasitoids vary significantly in their survival, developmental and reproductive rates when reared on different pollens (Hulshof and Jurchenko, 2000; Kennett et al., 1979; Leius, 1961a, b; Smith, 1960, 1961; van Rijn and Tanigoshi, 1999b).

Pollen nutrition can also vary intraspecifically, and these differences have important implications for predator and parasitoid life histories. The nutritional profile of a pollen changes depending on a number of environmental factors, including temperature, moisture, and season (Baker and Baker, 1979; Roulston and Buchmann, 2000). Also, phenotypic variation in pollen nutrition exists that can impact predator fitness. Ten different *Z. mays* hybrids have substantial differences in micronutrient and phytochemical concentrations, and variations in one or more of these constituents are significantly correlated with mortality rates in pollen-reared *Coleomegilla maculata* (Lundgren and Wiedenmann, 2004). Similarly, in Table 8.1 and Fig. 8.1 we see that 12 different hybrids of corn pollen vary substantially in their ability to support development in the omnivorous bug, *Orius insidiosus*, in the absence of prey. Survival analysis indicates that corn hybrids have a significant effect on the survival of developing nymphs (Mantel $\chi^2_{14} = 136.03$, $P < 0.001$). Some pollens do not support development of this predator at all, while others allow a moderate number of *O. insidiosus* to complete development in the absence of prey.

Once it is shed, the nutritional make up of pollen changes rapidly. For instance, water content of pollen decreases rapidly in pollen grains during anthesis, which

Table 8.1 Corn pollens whose nutritional suitability for *Orius insidiosus* was evaluated in the laboratory. *Orius insidiosus* were reared to adulthood with pollen from one of the designated hybrids along with a green bean segment as a water source (J. G. Lundgren, unpublished data)

Corn hybrid	Proportion that survived to adulthood (number of insects tested)
NK 4640[a]	0.20 (35)
DK 493[b]	0.20 (35)
Pioneer 3563[c]	0.20 (35)
SGI 912 × SGI 785[d]	0.17 (36)
NK 4242[a]	0.16 (37)
LH 330 × LH 273[d]	0.14 (37)
NK N45-T5[a]	0.14 (37)
Pioneer 3730[c]	0.11 (35)
DK 440[b]	0.11 (37)
DKC 60–12[b,e]	0.03 (39)
NB 4703	0 (35)
DKC 57–30[b]	0 (36)
Green bean only	0 (34)
Unfed	0 (38)

[a] Northrup King Company, Golden Valley, MN.
[b] DeKalb Seeds, Monsanto Company, St. Louis, MO.
[c] Pioneer Hi-Bred International, Johnston, IA.
[d] Illinois Foundation Seeds, Champaign, IL.
[e] This hybrid expresses Cry3Bb1 (MON863 event).

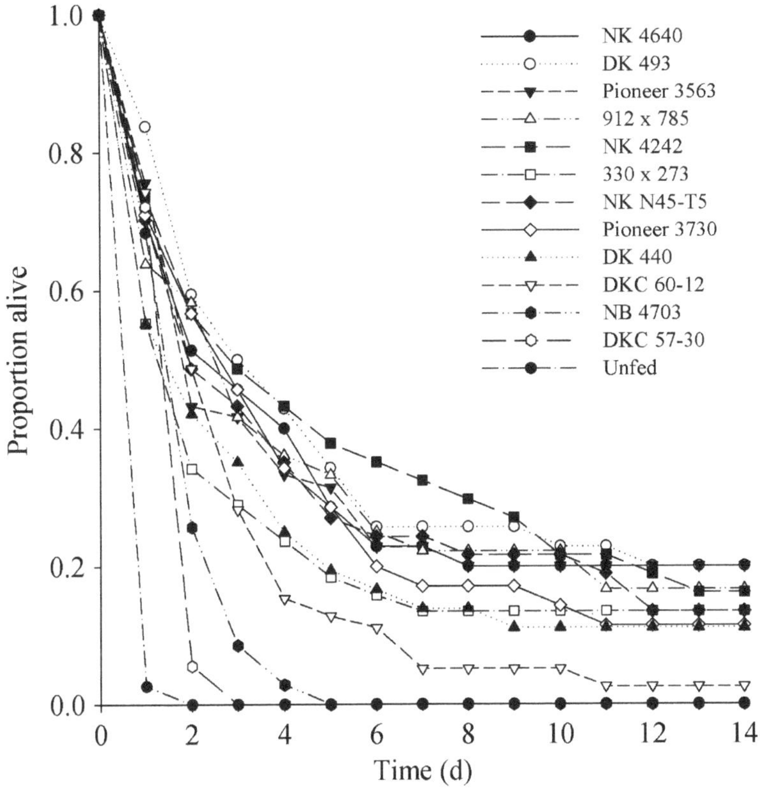

Fig. 8.1 Survivorship curves over time of *Orius insidiosus* nymphs raised on different corn pollens (J. G. Lundgren, unpublished data)

explains why the ultimate water content is quite low in most pollens (Iwanami et al., 1988). Amino acid concentrations also change substantially within a few days of being shed, even at low temperatures (2°C) (Linskins and Pfahler, 1973). Thus, the duration that pollen remains in the field (or the laboratory) can affect its suitability as food for insects.

For most insects, including most natural enemies, the physiological functions of certain nutrients and minerals are poorly understood (Chapman, 1998; Hagen, 1987). Also, it is generally unclear whether entomophagous insects are limited in certain nutrients (but see Bracken, 1966). Certainly, self-selection of dietary items by predators is governed by the nutritional attributes of different foods (Dicke et al., 1986; Greenstone, 1979), though which nutrients that pollen contributes to improve predator fitness has not been explicitly studied; an exception being carotenoid pigments (Dicke et al., 1986; Overmeer and van Zon, 1983). The nutrition of pollen is primarily studied within the context of pollinator feeding and foraging behavior (especially with regard to *Apis mellifera*). As a result, major nutritional constituents are known and quantified from an array of pollens, and some of the most conspicuous nutritional classes found in pollens are discussed below within the context of insect nutrition.

8.1.1 Carbohydrates

Pollens often contain both complex and simple carbohydrates; the complex carbohydrates are often stored as starch. Pollens store energy to support pollen tube growth, and starches are a primary group of biochemicals used in energy storage (Baker and Baker, 1979, 1982). In general, 65% and 53% of dicot and monocot families, respectively, contain 0–2.5% starch (Grayum, 1985), with starchier pollens ranging up to 15% of dry weight (Roulston and Buchmann, 2000). The two starchiest pollens of 89 angiosperm species tested are *Z. mays* (16.6%) and *Typha latifolia* (30.6%) (Roulston and Buchmann, 2000). It deserves noting that the suitability of these two pollen species as food is surprisingly well studied for a wide range of natural enemies (see Chapter 6). In general, small-sized pollens contain less starch (Baker and Baker, 1979, 1982; Grayum, 1985). Another polysaccharide, pectin, is very abundant in the pollen of some species. Mizuno (1958; Stanley and Linskins, 1974) found that pectin is the principle polysaccharide in the pollen of *Cryptomeria japonica*. Finally, there is an inverse relationship between starches and the simple sugars commonly encountered in pollen (Roulston and Buchmann, 2000, and references therein).

Mono- and oligosaccharides are frequently found in many pollens. The most common simple sugars are fructose, glucose, and sucrose (Stanley and Linskins, 1974). Many other simple sugars are found in different pollens, and the sugar profiles are highly species specific. Melezitose has been isolated from at least one species of pollen, *Cydonia vulgaris* (Stanley and Linskins, 1974). Melezitose is a sugar often regarded as being produced exclusively in the honeydew of sternorrhynchan phloem-feeders, and it is typically believed to be absent from plant sap and most nectars (Burgin and Hunter, 1997; Hunter and Ossowski, 1999; Wäckers, 2001) (but see Table I.1). As pollen nutrients are readily leeched into nectar, it may be necessary to consider potential contamination of nectars with pollen-derived melezitose before using this sugar as an indicator of honeydew feeding, as is done in field and laboratory experiments pertaining to sugar feeding by insectivorous arthropods and saprophagous flies.

Insects rely on carbohydrates for fuels and as building blocks for fats and amino acids (Chapman, 1998). Some carbohydrates are not digestible by certain insects; starch, cellulose, and lactose are good examples of these molecules (Cohen, 2004). The ability to use certain carbohydrates may be stage specific in some insects; an example of this given by Chapman (1998) is the mosquito, *Aedes*, in which the larva can use starch and glycogen, while the adult cannot. Stark contrasts in the frequencies of pollinivory by different life stages of a predator are observed in the coccinellid *Harmonia axyridis* (Lundgren et al., 2004) (Fig. 6.1), but it is not clear what is driving this pattern.

8.1.2 Proteins

Although pollen protein concentrations vary considerably across plant orders, it appears that closely related taxa contain similar levels of this nutrient, especially among congeners (Roulston et al., 2000). For angiosperms, protein ranges from

12–61% of pollen dry weight (Roulston et al., 2000; Standifer, 1967). Also, a range of enzymes is reported in pollen, although the concentration of these enzymes varies interspecifically (Grogan and Hunt, 1979; Stanley and Linskins, 1974).

All insects require the 'rat essential' amino acids: methionine, threonine, tryptophan, valine, isoleucine, leucine, phenylalanine, lysine, arginine, and histidine. Others are also required by insects, but many of them can be created metabolically from this suite of essential amino acids, which are abundant in pollen (Barbier, 1970). Proline is often the most abundant amino acid, comprising up to 3% of pollen dry weight (Erhardt and Baker, 1990; Lundgren and Wiedenmann, 2004; Stanley and Linskins, 1974). In fact, proline is not commonly found in any abundance in nectar, and its discovery in nectar is probably an indication of pollen contamination (Carter et al., 2006). Proline is used in flight initiation in many insects, basically providing the initial kick start to begin insect flight (sustained flight is accomplished using sugars as fuel) (Carter et al., 2006; Kammer and Heinrich, 1978). Gilbert (1985a) hypothesizes that selectively feeding on pollens with high levels of proline may allow syrphids to fly at fairly low temperatures. Proline is also important in egg maturation in bees (Carter et al., 2006). Amino acid deficiencies can be fatal to arthropods, and the inability of pollen-reared spiders to complete molting was hypothesized to have resulted from deficiencies in tyrosine, an amino acid important in ecdysis that is at low levels in many pollens (Smith and Mommsen, 1984; Stanley and Linskins, 1974). *Coleomegilla maculata* requires a dietary source of phenylalanine, isoleucine, and valine, but many other amino acids can be synthesized by this omnivore (Atallah and Killebrew, 1967).

8.1.3 Lipids

Lipid content of pollen ranges from 1.5–18.9% of dry weight (Barbier, 1970; Standifer, 1966; Stanley and Linskins, 1974). In addition to starches, oils are an energy reserve put to use during germination and fertilization (Baker and Baker, 1979). Oils contain more energy per unit than starches, and consequently smaller grained pollens that are size-limited tend to rely more on oils than starches for energy reserves (Baker and Baker, 1979), a relationship also seen in seeds (Chapter 11).

Fatty acids are commonly encountered in pollen, and unsaturated fatty acids are major components of this lipid fraction (Stanley and Linskins, 1974). Fatty acids are required in the diets of some insects, but the importance of these acids to entomophagous arthropods is not always clear (Chapman, 1998; Cohen, 2004). Nevertheless, all pollens tested contain linoleic acid, and stearic, palmitic, palmitoleic, oleic, and linolenic acids are also found in some pollens (Stanley and Linskins, 1974). Phosphoinositol and phosphocholine are required in the diets of many insect species, functioning in the nervous system (inositol and choline) and in spermatogenesis and oogenesis (choline) (Chapman, 1998; Hagen, 1962). Phosphoinositol and phosphocholine are some of the most abundant lipids in some pollens, such as in *Typha latifolia* and *Zea mays* where they comprise approximately 37% of all lipids (Stanley and Linskins, 1974).

Insects cannot manufacture sterols, and so are dependent upon dietary sources of sterols to create hormones that are critical to development and behavior (Chapman, 1998; Cohen, 2004; Svoboda et al., 1978). Carnivorous insects must use diet-derived cholesterol, and at least some cannot use phytosterols to synthesize cholesterol (Chapman, 1998; Dadd, 1985; Svoboda et al., 1978). This being said, a number of predators can complete development on a diet consisting solely of pollen. This means that either cholesterol is present at low levels in some pollens, they derive some cholesterol maternally, or that these omnivorous predators can metabolize phytosterols. Sterol-like material ranges from 0.36–3.4% of pollen dry weight (Standifer, 1966), and cholesterol is found in some pollens (Barbier, 1970). Other classes of sterols frequently isolated from pollen are fucosterol, β-sitosterol, 24-methylene cholesterol (or campesterol), and stigmasterol (Standifer et al., 1968; Stanley and Linskins, 1974). Sterol deficiencies in the pollens of some corn hybrids are hypothesized to have led to poor eclosion rates in pollen-reared *Coleomegilla maculata* (Lundgren and Wiedenmann, 2004).

8.1.4 Vitamins

Pollens often have a full complement of water-soluble vitamins that are critical to insect development, though these nutrients are at low levels (less than 100 $\mu g\ g^{-1}$ of dry weight; (Stanley and Linskins, 1974). Specifically, insects require B vitamins, ascorbic acid, and choline (Chapman, 1998; Cohen, 2004), which are present in pollens at varying levels. In insects and pollen, the B vitamins, such as thiamine, riboflavin, niacin, biotin, and folic acid, function in wide variety of metabolic pathways (reviewed in Cohen, 2004 and Stanley and Linskins, 1974). Ascorbic acid is relatively less important for carnivorous insects than in herbivores, for whom it functions as a phagostimulant and in metabolic functions (Cohen, 2004).

Lipid-soluble vitamins are also encountered in pollen. Carotenoid pigments are present in virtually all pollens, although at lower levels in some pine and grass species (Lundgren and Wiedenmann, 2004; Stanley and Linskins, 1974). These pigments are in part responsible for pollen coloration and other physiological functions. Pollen coloration resulting from high levels of carotenoids is attractive to some insects (Stanley and Linskins, 1974), although this is not well studied in natural enemies. Similar to vertebrates, carotenoids are incorporated into the eye pigments of insects and extra-retinal photoreceptors of mites, and are also involved in growth processes and cuticle pigmentation (Chapman, 1998; Cohen, 2004; Dicke et al., 1986). Vitamin A, or its precursors, is critical for diapause in the eyeless predatory mite, *Amblyseius potentillae* (Overmeer and van Zon, 1983; Veerman et al., 1983), and the attractiveness of prey and suitability of pollens for immatures is affected by the vitamin A content of the food and the nutrient status of the mite (Dicke, 1988b;

Dicke et al., 1986). Vitamin E (tocopherol) plays a role in insect reproduction, and its ingestion improves fecundity and spermatogenesis in some species (Chapman, 1998). Vitamin E is present in some pollens at low levels, ranging from 21–170 μg g^{-1} dry weight (Stanley and Linskins, 1974), but research on this topic is far from exhaustive and the function of Vitamin E within pollen grains is not well understood.

8.1.5 Inorganic Minerals

Centuries of examining pollen mineral content have revealed an array of different inorganic elements in pollen (Stanley and Linskins, 1974). Most pollen contains between 2–7% inorganic ash (Stanley and Linskins, 1974). As with other nutritional constituents of pollen, mineral content is species specific; pollen from *Atriplex patula*, the salt bush, contains 26.8% ash! Most pollens are a good source of K, S, P, B, Ca, and Mg (Stanley and Linskins, 1974). Common trace elements include Al, Cu, Fe, Mn, and Ti. Minerals have different roles in pollen physiology, and many of these functions are yet to be discovered. Boron is important in pollen germination, and other elements, such as iron or magnesium, are present as part of cellular structures such as enzymes or ribosomes.

Mineral requirements for insects are poorly understood (Cohen, 2004). Nevertheless, Na, K, Ca, Mg, Cl, Zn, Mn and phosphate are essential for insect development (Chapman, 1998; Cohen, 2004). It does not seem likely that predators ingest pollen to compensate for the poor mineral content of prey, although more research in the bioavailability of minerals to entomophagous arthropods is required. Minerals are provided in many meridic and holidic diets of predatory species as salt mixtures (Atallah and Newsom, 1966; Cohen, 2004; Sahayaraj, 2002; Yazlovetsky, 2001), and these minerals function in many cellular and metabolic processes critical to insect survival (see Cohen, 2004 for review). Within the scope of this chapter, it is important to note that the digestion of starches (which are prevalent in many pollens) by amylases is chloride dependent (Cohen, 2004).

8.2 Defense

Given that pollen is a rich source of nutrition, it simply won't do for a plant to lose all of its pollen to arthropods looking for an easy meal, especially when these insects don't actually provide the service of pollination. Entomophagous arthropods often fall into this category of larcenous pollinivores, and so are the likely targets of a range of defensive properties present in pollen.

8.2.1 Floral Morphology

Perhaps the first line of pollen defense is to develop flower morphology and coloration that inhibits thieves from accessing the pollen (Allen, 1929; Stang et al., 2006), a defense that also pertains to guarding floral nectar (see Chapter 3). Mouthparts may restrict which flowers can be exploited for their pollen (Holloway, 1976). Yellow coloration elicits feeding responses in syrphids that are not necessarily good pollinators (Lunau and Wacht, 1994). White flowers of *Parnassia palustris* display yellow false stamens that are more attractive to non-pollinator syrphid flies (such as *Eristalis tenax*) than the white, pollen-containing stamens (Lunau and Wacht, 1994). Thus the plant is apparently using adaptive coloration to reduce pollen thieving. Once pollinivores gain access to the pollen, plants are able to defend pollen with inherent qualities that are categorized here as 'structural', 'antinutritive', and 'antifeedant'.

8.2.2 Structural Defenses

Grain size, thickness of the pollen wall (exine & intine), and the size and abundance of external appendages have been shown to influence the ability of natural enemies to consume pollen (Fig. 8.2). It is likely that these structural features have several functions in aiding in pollen dissemination, protection, and germination, and are not solely intended to reduce pollinivory by insects. For example, spines on pollen grains in insect-pollinated plants likely aid in adhering to the body of pollinators, but may also impede consumption by small arthropods. Indeed, it should be noted that structural defenses are most deleterious to small arthropods such as anthocorids and phytoseiids.

Some entomophages have restrictions on the size of pollen grains that can be used as food. Pollen ranges in size from 10 to 200 μm, with the average diameter being 30 μm (Iwanami et al., 1988). Cotton and pine pollen grains (75–100 μm in diameter) may be too large for *Euseius tularensis* to effectively manipulate and ingest (Kennett et al., 1979). However, the phytoseiids *Iphiseius degenerans* and *Neoseiulus cucumeris* are apparently undeterred by larger pollen grains (van Rijn and Tanigoshi, 1999b). Fauvel (1974) reports that pollen of *Mercurialis annua* is unsuitable for the development of the anthocorid, *Orius vicinus*, presumably because the grains were too small for this 'lacerate and flush' feeder to insert its rostral tip into the grains and remove the contents. *Orius insidiosus* practices this technique on corn pollen, of which they suck out the pollen contents rather than ingest the entire grain.

The thickness of the pollen wall, which includes the exine and intine, is hypothesized to impede the digestion of pollen by some animals (Haslett, 1983; Roulston and Cane, 2000), particularly for some 'lacerate and flush' predators. Indeed, the pollen wall contains sporopollenin and is extremely resistant to

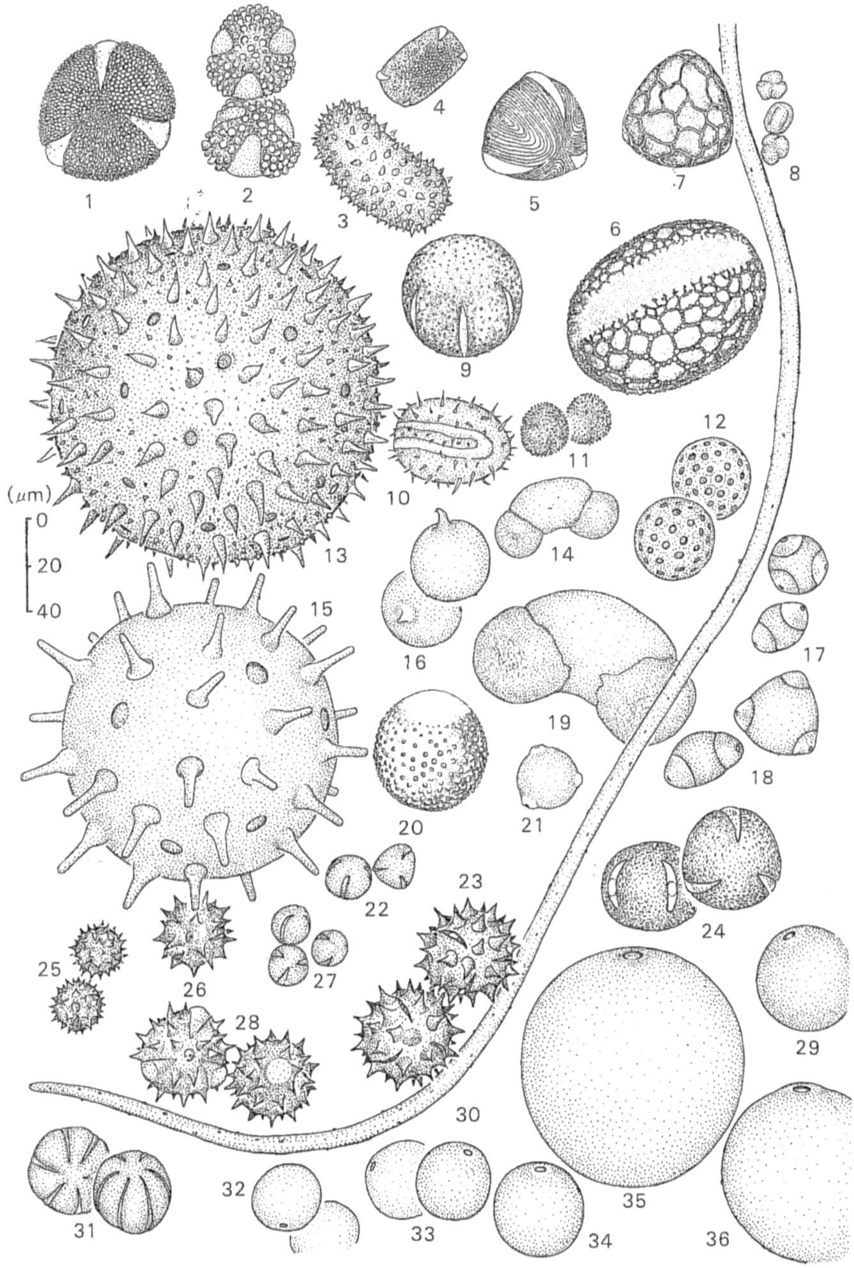

Fig. 8.2 A diversity of pollen grains (Reproduced from Iwanami et al., 1988. With permission by Springer)

degradation, sometimes persisting in the soil for thousands of years (Iwanami et al., 1988). Kennett et al. (1979) and Ouyang et al. (1992) hypothesize that thicker pollen exines may impede pollinivory by the phytoseiid, *Euseius tularensis*. This species develops well on *Malus sylvestris* pollen, which possesses a thin exine, and poorly on *Avena sativa* pollen, which has a relatively thick pollen exine.

Pollen surface structure and appendages reduce the ability of some predators to access the interior of the grain and these traits may reduce the foraging efficiency of the arthropod. Pollen grain surface structure is believed to influence feeding by *Neoseiulus fallacis* (Zhang and Li, 1989; as cited in Ouyang et al., 1992). Conversely, the sculpture of a number of pollen grains does not impede feeding by the phytoseiids, *Iphiseius degenerans* and *Neoseiulus cucumeris* (van Rijn and Tanigoshi, 1999b). Minute spines of *Gossypium* pollen attach to the legs of predatory mites (Elbadry, 1968), and Zaher and Shehata (1971) hypothesize that these spiny grains (and those of *Althaea rosea*) encumber movement of *Typhlodromus pyri*. Thus, the foraging activity of these mites is reduced which results in slower developmental rates than those fed prey or smooth-coated pollen from *Phoenix dactylifera*. The spines on the exterior of *Helianthus anuus* pollen are lethal to larvae and adults of *Coleomegilla maculata* in the laboratory, because the grains stick to the integument of the insects, which subsequently can't move or function correctly (Michaud and Grant, 2005).

8.2.3 Antinutritive Qualities

> *Poor nutritive status may also help to protect the pollen...from being eaten by pollen thieves.* (Baker and Baker, 1979)

Early thoughts on this matter suggest that wind pollinated plants (anemophilous) should have starchier pollens than insect-pollinated species (zoophilous). Starchy pollen does not grant as many calories as pollen that possesses oil energy reserves, and anemophilous pollens are consequently hypothesized to be less attractive to insects (Baker and Baker, 1979). In spite of their objections to the hypothesis that anemophilous plants have starchier pollen, Roulston and Buchmann (2000, data from Fig. 2) published a frequency of 27.4% and 51.3% starchy pollen in zoophilous and anemophilous species, respectively. But they advise that appropriate statistical considerations must be applied when relating the starch and protein contents of pollens to their pollinator types in order to eliminate statistical biases in these analyses (Roulston and Buchmann, 2000; Roulston et al., 2000). Protein content is another nutritional quality that could be altered in plants to attract or deter arthropods. Indeed, anemophilous plants have a greater frequency of low protein content relative to other groups (means of 25.8% and 39.3% protein in anemophilous and zoophilous species, respectively), but these numbers need to be interpreted with care (Roulston and Buchmann, 2000). Evolutionary shifts in pollination mode do not characteristically result in changes to protein content, which would be expected if the protein contents of pollens was driven primarily by pollination source

(Roulston et al., 2000). Furthermore, Roulston et al. (2000) point out that the relationship between low protein contents in anemophilous species may simply be a reflection of a sampling bias for the prolific number of species within certain anemophilous clades. Also, other mechanisms may explain the protein contents of pollen, including grain size and floral morphology (Roulston et al. 2000). Finally, protein content is highly conserved at the family level, and so an overabundance of protein-poor species within a single family may affect statistical analyses that treat each species as an independent entity (Roulston et al. 2000).

To summarize, although there are correlations between pollination mechanisms and nutrient content of pollen, it cannot be stated that plants clearly alter their nutrient content to attract or detract arthropods. Nevertheless, both the nutritional aspects of pollen and the suitability of a pollen species as food for predaceous arthropods are dependent on the phylogenetic location of the plant (Roulston et al., 2000; van Rijn and Tanigoshi, 1999b). For instance, gymnosperms tend to have poor pollen for mite development, whereas pollens from Rosaceae tend to consistently favor the development of predatory mites (van Rijn and Tanigoshi, 1999b). So, while pollinivorous entomophages may not be driving the evolution of pollen nutrition, there are certainly some phylogenetic patterns in the nutrient content of pollens from different plants that could alter their relative suitability as foods for arthropods. Future research will do well to focus on whether pollen provision improves plant fitness as nectars do, and whether plants may foster these relationships through changes in pollen nutritional quality.

8.2.4 Toxic Pollens

The idea that floral rewards may contain chemical defenses to deter feeding by insects that do not function in pollination is supported by the selective toxicity of some pollens (Hitchcock, 1959). Reports on pollen toxic to entomophages are scarce. Some pollens may have a detrimental effect on parasitoid fecundity, as indicated when *Scambus buolianae* females are fed wild parsnip pollen (Leius, 1963). Much more commonly, pollen toxicity is reported in honeybees, and many of these reports have been compiled into Table 8.2.

The toxic properties of pollen are seldom identified. In *Aesculus* and *Tilia*, saponin content is charged with killing honeybees (Maurizio, 1945; as reported in Stanley and Linskins, 1974). In *Hyoscyamus* sp. and *Zigadenus venenosus*, the toxic phytochemical is an alkaloid (Shaginyan, 1956; as reported in Stanley and Linskins, 1974), and pollen from *Asclepias* spp. contains galitoxins (Pryce-Jones, 1942; as reported in Stanley and Linskins, 1974). Also, Lundgren and Wiedenmann (2004) found high concentrations of the antifeedant quercetin in corn pollen, and they reiterate previous hypotheses that this substance may be involved in pollen defense (Stanley and Linskins, 1974). Pollens commonly contain phytotoxins such as alkaloids, phenolics, and tannins, but often at lower levels than those found in leaf tissue; actual toxicity assays of pollen defensive chemicals with insects are rare (Roulston and Cane, 2000).

Table 8.2 Species of pollen that have been identified as toxic to honeybees (*Apis mellifera* L.)

Family	Species	Common name	Reference
Asclepiadaceae	*Asclepias*	Milkweed	(Crane, 1990; Stanley and Linskins, 1974)
Corynocarpaceae	*Corynocarpus*		(Stanley and Linskins, 1974; Crane, 1990)
Ericaceae	*Andromeda*		(Stanley and Linskins, 1974; Crane, 1990)
Ericaceae	*Rhododendron*		(Crane, 1990; Stanley and Linskins, 1974)
Fabaceae	*Astragalus lentiginosus*	Spotted loco	(Vansell and Watkins, 1934)
Fabaceae	*Stryphnodendron polyphyllum*		(Pimentel de Carvalho and Message, 2004)
Hippocastanaceae	*Aesculus* spp., (at least *A. californica*)	Buckeye	(Vansell, 1926; Maurizio, 1945; O'Neal and Waller, 1984)
Liliaceae	*Veratrum californicum*	Western false hellebore	(Vansell and Watkins, 1933)
Liliaceae	*Zigadenus venenosus*	Death camas	(Goolsbey, 1998; Hitchcock, 1959)
Polygonaceae	*Fagopyrum* (after drying only)	Buckwheat	(Crane, 1990; Stanley and Linskins, 1974)
Polygonaceae	*Polygonum bisorta*		(Stanley and Linskins, 1974; Crane, 1990)
Ranunculaceae	*Aconitum* spp.	Monkshood	(Koptev, 1948; Poltev, 1956; O'Neal and Waller, 1984)
Ranunculaceae	*Ranunculus* spp.	Buttercup	(Maurizio, 1941, 1945; Muller, 1948; O'Neal and Waller, 1984; Pryce-Jones, 1944)
Scrofulariaceae	*Digitalis purpurea*	Foxglove	(Muck, 1939; O'Neal and Waller, 1984)
Solanaceae	*Hyoscyamus*	Henbane	(Stanley and Linskins, 1974; Crane, 1990)
Tiliaceae	*Tilia* spp.		(Maurizio, 1945; Stanley and Linskins, 1974)

8.3 Conclusions

Pollen is an extremely rich source of many of the nutrients essential to the development and reproduction in entomophagous arthropods. In fact, the protein and oil contents of pollens are superior to that of most vegetative tissues and even many prey items. Given its nutritional status, it is not surprising that many insects that ordinarily do not consume plant tissues eat pollen, and that some can even complete development upon it in the absence of prey.

Although pollen grains are superficially quite simple, they are not proverbial sitting ducks and defend themselves through a variety of means. First, the structure and appendages of the pollen grains deter some of the smaller entomophages from consuming them. Also, even though predation on pollen may not have driven the evolution of pollen nutrition, the pollens of zoophilous plants are nutritionally better suited for insect development than anemophilous species. This suggests that anemophilous pollens may be partially defended against pollinivory because of reduced nutrition. Finally, pollens contain a range of phytochemicals that are actually toxic to insects. Understanding the nutrition and defensive capabilities of pollen gives us the means to explore how this food source can be incorporated into integrated pest management programs that seek to conserve and promote the consumption of insect pests by facultatively pollinivorous natural enemies.

Section III Granivory

It isn't surprising that entomophagous insects consume seeds, given the sheer abundance of seeds in even the most inhospitable places on Earth (Brown et al., 1979), and the rich nutritional content that seeds offer to these ordinarily carnivorous insects. Pulliam and Brand (1975) estimate that 358 billion seeds per hectare are produced in grassland plains of the southwestern U.S., and Tevis (1958) reckons nearly 3.6 billion seeds per hectare in Coachella Valley desert (California). And neither of these habitats is particularly productive in terms of plant biomass or diversity! Furthermore, unlike prey, seeds don't spoil and can be stored for long periods of time. Thus, these nutritious and abundant foods can be stored and support the persistence of predatory insects even during periods of scarce resources (Carroll and Janzen, 1973).

The number of seeds consumed by entomophagous insects and the diversity of species that accept seeds as food are staggering, and clearly indicate the importance of this dietary item to their life histories. For thousands of years, ants consuming seeds have been the symbols of industriousness in historical writings by ancient human cultures ("Go to the ant, thou sluggard; consider her ways," the Book of Solomon). Indeed, hundreds of species of ants harvest seeds, and omnivorous carabids and crickets are frequently some of the most abundant granivores within agricultural habitats. One astounding estimate of seed consumption is by the desert-dwelling harvesting ant *Messor pergandei,* colonies of whom consume an estimated 37 million seeds per hectare pre year (Tevis, 1958). In the Czech Republic, Honek et al. (2005) calculate that carabids consume 643 seeds of *Taraxacum officinale* per m^2 daily. Thus, granivorous entomophages are diverse and abundant, and are adapted to a life consuming seeds in addition to arthropod prey.

Seeds come in a variety of shapes and sizes, in part as a result of selection pressure by predation. Defensive capabilities stem from the chemistry and morphology of the diaspore, and have important implications for seed preferences in granivorous entomophages. Also, mutualisms between seeds and microorganisms (endophytic symbioses) and the granivores themselves (insect-dispersed seeds) also play an important role in determining which seeds are collected and consumed by predators. Seeds are unique from other forms of non-prey foods in that they sometimes give rise to a pest, and so seeds are the target of biological control programs in agricultural systems. The taxonomic diversity and feeding adaptations of granivores, as well as the nutrition and defense of the seeds and their external appendages all influence the outcome of weed management programs that involve seed predation.

Finally, this section of the book largely focuses on post-dispersal granivory, which differs from pre-dispersal granivory in a number of ways. Pre-dispersal granivory influences the number of seeds falling into the seedbank, while post-dispersal granivory changes the dispersion pattern of the seed shadow of a plant (Harper, 1977). This occurs because post-dispersal predators feed on seeds selectively, consuming largely those that occur within their foraging areas (Harper, 1977). Pre-dispersal predation reduces the seed production and regeneration capacity of plants (Andersen, 1988; Griffeths and Swanton, 1999; Louda et al., 1990). However, most pre-dispersal granivory is generally inflicted by insects that specialize on developing seeds (Crawley, 2000; Hulme, 1998), whereas generalist feeders predominate the community of post-dispersal granivores. Thus, pre-dispersal granivory is often not applicable to entomophagous species. Also, I think that it is defensible that post-dispersal granivores have been better studied than pre-dispersal granivores with regard to seed establishment for agricultural weeds, the seeds of which are the target of biological control programs involving granivorous entomophages.

Chapter 9
The Seed Feeders

Seed predation by entomophagous species occurs almost exclusively after the seeds have fallen to the ground. Here, species of epigeal omnivores come into contact with the seeds, and the seeds are consumed or cached in burrows or nests. Geographically, ground beetles and crickets are probably most important as post-dispersal granivores in temperate ecosystems (Honek et al., 2003; Lundgren et al., 2006; O'Rourke et al., 2006), and ants dominate the guild of post-dispersal insect granivores in tropical and desert habitats. Most work on post-dispersal granivory by entomophages has focused on Carabidae and Formicidae, with a growing interest in crickets (Gryllidae) as granivores and opportunistic predators. The relative importance of seeds as a food source to natural enemies is reflected by the diversity of species within these groups that consume seeds and their relative fitness when provided seeds as a food source compared with insect prey.

9.1 Carabidae

Carabidae are a behaviorally and physiologically heterogeneous group. Feeding behavior ranges from complete carnivory to complete phytophagy, with most granivorous species falling somewhere in the middle of this continuum (Table 9.1). Very few ground beetles have a narrow food range, and many are both predaceous, herbivorous, and fungivorous (Davies, 1953; Dawson, 1965; Sunderland, 1975). This notwithstanding, food specialization does exist for some carabids. A relevant example of this occurs with *Carterus* spp., which specialize on seeds of umbellifers, and *Ditomus* spp. specialize on seeds of *Plantago* (Larochelle, 1990). *Notiobia* species in the Amazonian rainforests also show some degree of specialization on certain fruit falls (Paarmann et al., 2001, 2002a, b, 2003). Within a habitat *Notiobia pseudolimbipennis* and *N. flavicinctus* consume seeds of fig species, while *N. glabrata*, *N. incerta*, and *N. nebrioides* specialize on seeds in the Melastomataceae, and their foraging patterns reinforce these specializations; each group aggregates to the appropriate fruit falls. Because of the diversity of food relationships within Carabidae, it is a dangerous game to make sweeping generalizations on the feeding behavior of carabids at the family level, as is too often done.

J.G. Lundgren, *Relationships of Natural Enemies and Non-Prey Foods*,
Progress in Biological Control 7,

Published by Springer Science+Business Media B.V.

Table 9.1 Granivorous species of carabidae. Included are the species and stages (A = adult; L = larva) of granivorous ground beetles, the seed species consumed, the location of the study (L = Laboratory; F = Field), and the citation

Species	Life stage	Seed species consumed	Study location	Citation
Abax parallelepipedus	A	Germinating seeds	F	(Balachowsky, 1963) as reported in Larochelle (1990)
Acinopus picipes	A	Seeds		(Bonadona, 1971; Larochelle, 1990)
Acupalpus meridianus	A	A seed mixture	L	(Honek et al., 2005)
Agonum	A	*Chenopodium album, Amaranthus etroflexus, Digitaria sanguinalis, Panicum dichotomiflorum*	L	(Brust, 1994)
Agonum cupripenne	A	*Z mays*	L	(Larochelle, 1990)
Agonum extensicolle	A	*Z. mays*	L	(Larochelle, 1990)
Agonum lutulentum	A	Grass (probably *Poa annua*, but not clear from paper)	L	(Johnson and Cameron, 1969)
Agonum muelleri	A	*Capsella bursa-pastoris, P. annua*	L	(Saska et al., 2008)
Agonum placidum	A	*Z. mays*	L	(Larochelle, 1990)
Agonum punctiforme	A	Seed mixture		(Brust and House, 1988b; Tooley and Brust, 2002)
Agonum thoreyi	A	Grass (probably *P. annua*, but not clear rom paper)	L	(Johnson and Cameron, 1969)
Agonum variolatum	A	Weed seeds	F	(Forbes, 1883)
Amara	A	Seed mixture	F	(Brust and House, 1988b)
Amara aenea	L	*Capsella bursa-pastoris, Stellaria media, Tussilago farfara, Plantago major, Urtica dioica Potentilla argentea*	L	(Saska and Jarosik, 2001; Hurka and Jarosik, 2003)
Amara aenea	A	*Triticum* and Graminae, *Z. mays, Glycine max, Cirsium arvense, C. bursa-pastoris*; a seed mixture, *P. annua, S. media*	F and L	(Burakowski, 1967; Honek et al., 2003; Honek et al., 2005; Larochelle, 1990; Menalled et al., 2001; Saska et al., 2008)
Amara angustata	A	*Poa pratensis*, on the heads in the field	F	(Webster, 1881, 1903)
Amara anthobia	A	Seed mixture	L	(Honek et al., 2005)

(continued)

Table 9.1 (continued)

Species	Life stage	Seed species consumed	Study location	Citation
Amara apricaria	A	Grass (probably *P. annua*, but not clear from paper), a seed mixture	L and F	(Allen, 1979; Honek et al., 2005; Johnson and Cameron, 1969; Larochelle, 1990; Zhavoronkova, 1969)
Amara aulica	L	*Artemisia vulgaris, C. arvense, Tripleurospermum inodorum, U. dioica*	L	(Saska, 2005)
Amara aulica	A	*Achillea millefolium, Cirsium oleraceum, C. palustre, Centaurea nigra, Leontodon danubionis*, grass, *C. bursa-pastoris*, thistle seeds, *Daucus* and *Pastinaca, Lactuca, C. arvense*; a seed mixture	F and L	(Burakowski, 1967; Forsythe, 1982a; Honek and Martinkova, 2001; Honek et al., 2003, 2005; Larochelle, 1990)
Amara bifrons	A	*Gutierrezia serothrae*; a seed mixture	F and L	(Honek et al., 2005; Lavigne, 1976)
Amara carinata	A	Seeds	F	(Forbes, 1883)
Amara communis	A	Principally granivorous, *Z. mays*	?	(Brandmayr, 1972; Burakowski, 1967; Larochelle, 1990)
Amara consularis	A	*C. arvense*; a seed mixture	L	(Honek et al., 2003, 2005)
Amara convexior	A	Principally granivorous	?	(Brandmayr, 1972; Larochelle, 1990)
Amara convexiuscula	L	*A. vulgaris, T. inodorum*	L	(Saska, 2005)
Amara convexiuscula	A	Ripening seeds of *Hordeum* and *Fragaria*; seed mixture	F and L	(Blaszyk, 1963; Honek et al., 2005; Larochelle, 1990; Zhavoronkova, 1969)
Amara cupreolata	A	*Agrostis, P. annua, Poa trivialis, P. pratensis, Festuca rubra* var. *commutata, Lolium perenne, Digitaria, S. media*	L	(Barney and Pass, 1986; Brust, 1994; Johnson and Cameron, 1969)
Amara equestris	A	Principally granivorous	?	(Brandmayr, 1972; Larochelle, 1990)
Amara erythrocnema	A	*Agropyron, A. campestre*	F	(Larochelle, 1990; Therond, 1975)
Amara eurynota	L	*A. vulgaris, T. indorum*, and *U. dioica*	L	(Saska, 2004)

(continued)

Table 9.1 (continued)

Species	Life stage	Seed species consumed	Study location	Citation
Amara eurynota	A	*C. bursa-pastoris*, principally granivorous, a seed mixture	F	(Blunck, 1925; Brandmayr, 1972; Honek et al., 2005; Larochelle, 1990; Zhavoronkova, 1969)
Amara familiaris	L	*S. media*	L	(Saska and Jarosik, 2001)
Amara familiaris	A	Unripe seeds of hickweed (*Cerastium*), grass seed, *Z. mays, C. arvense, C. bursa-pastoris*; a seed mixture, *P. annua, S. media*	F and L	(Aubrook, 1949; Burakowski, 1967; Honek et al., 2003, 2005; Johnson and Cameron, 1969; Saska et al., 2008)
Amara fodinae	A	Seed material	F	(Zhavoronkova, 1969)
Amara fulva	A	Grass seed	F	(Burmeister, 1939; Larochelle, 1990)
Amara fulvipes	A	Graminae	F	(Larochelle, 1990; Therond, 1975)
Amara idahoana	A	*G. serothrae*	F	(Lavigne, 1976)
Amara impuncti-collis	A	*Z. mays, Digitaria*	L	(Barney and Pass, 1986; Larochelle, 1990)
Amara ingenua	A	Seed mixture	F and L	(Honek et al., 2005; Zhavoronkova, 1969)
Amara littorea	A	*C. arvense, C. bursa-pastoris*; a seed mixture	L	(Honek et al., 2003, 2005)
Amara montivaga	A	Several species, but a specialist on *Taraxacum officinale*	F and L	(Honek et al., 2005)
Amara nitida	L	*C. bursa-pastoris*	L	(Saska and Jarosik, 2001)
Amara nitida	A	Principally granivorous	?	(Brandmayr, 1972; Larochelle, 1990)
Amara ovata	A	Seeds or cereals and crucifers (young siliques), *Reseda luteola, C. arvense C. bursa-pastoris*; a seed mixture	F and L	(Blunck, 1925; Burmeister, 1939; Honek and Martinkova, 2001; Honek et al., 2003, 2005; Kittel, 1873/1874; Larochelle, 1990)
Amara pallipes	A	Grass seed, *P. pratensis*	F and L	(Johnson and Cameron, 1969)
Amara plebeja	A	*C. bursa-pastoris, P. annua, Lamium amplexicaule, S. media*	L	(Saska et al., 2008)
Amara pulpani	A	Primarily seed-feeding as an adult; *Betula verrucosa, Alnus incana, Triticum aestivum*	F and L	(Burakowski, 1967)

(continued)

Table 9.1 (continued)

Species	Life stage	Seed species consumed	Study location	Citation
Amara quenseli	A	*Wahlbergella affinis*, *Polygonum aviculare*, *Trisetum spicatum*, *Luzula campestris multiflora*	F	(Burmeister, 1939; Larochelle, 1990)
Amara sabulosa	A	A seed mixture	L	(Honek et al., 2005)
Amara similata	L	*C. bursa-pastoris*, *T. inodorum*, *Taraxacum*, *P. rhoeas*, *Trifolium repens*, *P. annua*, *Brassica napus*, *C. album*, *Cerastium triviale*; a seed mixture;	L	(Honek et al., 2005; Jorgensen and Toft, 1997b; Saska and Jarosik, 2001; Fawki and Toft, 2005)
Amara similata	A	Grass, *Triticum*, *Nasturtium*, *R. luteola*, *Deschampsia flexuosa*, *Cynosurus cristatus*, *C. arvense*, *C. bursa-pastoris*, *P. annua*, *Taraxacum*, *T. inodorum*, *P. rhoeas*., *T. repens*, *B. napus*, *C. album*, *C. triviale*	F and L	(Burmeister, 1939; Honek et al., 2003; Jorgensen and Toft, 1997b; Larochelle, 1990; Therond, 1975; Fawki and Toft, 2005)
Amara spreta	A	*C. bursa-pastoris*, *P. annua*, *L. amplexicaule*, *S. media*	L	(Saska et al., 2008)
Amara strenua	A	Grasses	F	(Heike, 1970; Larochelle, 1990)
Amara tricuspidata	A	*Secale*, *Z. mays*	F ?	(Burakowski, 1967; Fassati, 1957; Larochelle, 1990)
Amphasia sericea	A	*P. pratensis*, *Agrostis vulgaris*, seeds	F	(Webster, 1900, 1903; Forbes, 1881)
Anchomenus dorsalis	A	*C. arvense*	L	(Honek et al., 2003)
Anisodactylus	A	Seed mixture, *Phleum*	F	(Brust and House, 1988b; Forbes, 1883)
Anisodactylus binotatus	A	*Fragaria*	F	(Kirchner, 1939; Larochelle, 1977)
Anisodactylus caenus	A	'Granivorous tendencies', *Pastinaca sativa*	F	(Brandmayr, 1972; Briel, 1964; Larochelle, 1990)
Anisodactylus discoideus	A	Grass (probably *P. annua*, but not clear from paper)	L and F	(Johnson and Cameron, 1969)
Anisodactylus harrisii	A	Grasses and other plants	F	(Forbes, 1883)

(continued)

Table 9.1 (continued)

Species	Life stage	Seed species consumed	Study location	Citation
Anisodactylys merula	A	*C. album, A. retroflexus, D. sanguinales, P. dichotorlorum*	L	(Brust, 1994)
Anisodactylus opaculus	A	Seeds of grasses	F	(Forbes, 1883)
Anisodactylus rusticus	A	Grass (probably *P. annua*, but not clear from paper), *Z. mays, C. album, A. retro-flexus, D. sanguinales, P. dichotorlorum*	F and L	(Brust, 1994; Forbes, 1881; Johnson and Cameron, 1969; Larochelle, 1977)
Anisodactylus sanctaecrucis	A	*Medicago sativa, Abutilon theophrasti, Brassica oleracea, F. rubra, Ipomoea hederacea, C. album, D. sanguinalis, Koeleria,*	F and L	(Forbes, 1881; Hagley et al., 1982; Johnson and Cameron, 1969; Larochelle, 1990; Lundgren, 2005; Forbes, 1883)
Anisodactylus signatus	A	*C. arvense*; a seed mixture	L	(Honek et al., 2003, 2005)
Anisodactylus similis	A and L	'Granivorous tendencies', *D. carota,* umbellifers, *Daucus, Foeniculum*	F and L	(Brandmayr, 1972; Larochelle, 1990; Therond, 1975; Brandmayr et al., 1980)
Anisodactylus similis	A	*Foeniculum vulgare, Anethum graveolens*	F	(Burmeister, 1939; Larochelle, 1990; Therond, 1975)
Anisodactylus verticalis	A	*Z. mays*	L	(Larochelle, 1977)
Bembidion	A	Seed mixture	F	(Brust and House, 1988b)
Bembidion lampros	A	*Matthiola* and *Pisum*	F	(Larochelle, 1990; Lundblad, 1927)
Bradycellus badipennis	A	*Z. mays*	L	(Larochelle, 1990)
Bradycellus ganglbaueri	L	*Satureja subspicata*	L	(Brandmayr and Brandmayr, 1978)
Bradycellus harpalinus	A	Seed heads of Graminae	F	(Bonadona, 1971; Larochelle, 1990)
Bradycellus semipubescens	A	*Z. mays*	L	(Larochelle, 1990)
Calathus ambiguus	A	*Fragaria, C. arvense, C. bursa-pastoris*; a seed mixture	F and L	(Honek et al., 2003, 2005; Karl and Hartleb, 1963; Larochelle, 1990)
Calathus fuscipes	A	*Picea, Triticum,* 'seed consumers', *C. bursa-pastoris*	F and L	(Blunck, 1925; Goldschmidt and Toft, 1997; Honek and Martinkova, 2001; Larochelle, 1990; Pierce, 1917)
Calathus gregarius	A	*Phleum*	F	(Webster, 1903)

(continued)

Table 9.1 (continued)

Species	Life stage	Seed species consumed	Study location	Citation
Calathus melanocephalus	A	*C. bursa-pastoris*	L	(Goldschmidt and Toft, 1997)
Calathus micropterus	A	*Pinus sylvestris*	F	(Nystrand and Granstrom, 2000)
Carterus	A	Specialist on seeds of umbellifers		(Larochelle, 1990)
Carterus fulvipes	A	*Plantago* and *Daucus*	F	(Larochelle, 1990; Therond, 1975)
Chlaenius	A	Seed mixture	F	(Brust and House, 1988b)
Chlaenius lithophilus	A	*Z. mays*	L	(Larochelle, 1974)
Chlaenius pennsylvanicus pennsylvanicus	A	*Z. mays*	L	(Larochelle, 1974)
Clivina australasiae	A	Grass and dicotyledonous seeds *Z. mays*	F	(Sunderland et al., 1995; Muggeridge, 1939)
Clivina fossor	A	*Z. mays*, *Beta*	F	(Anglade, 1971; Burger et al., 1984; Larochelle, 1990; Sunderland et al., 1995)
Clivina impressefrons	A	*Z. mays*	F	(Bigger and Blanchard, 1959; Larochelle, 1990; Lindroth, 1961–1969; Pausch and Pausch, 1980)
Diplocheila obtusa	A	*Z. mays*	L	(Larochelle, 1990)
Ditomus	A	Specialist on seeds of *Plantago;* specifi cally *P. major* and *P. maritima*	L	(Theile, 1977)
Ditomus calydonius	A and L	*Daucus;* larvae feed on seeds of *Daucus gingidium polygamus*	F	(Brandmayr, 1973/1974; Therond, 1975; Zetto Brandmayr, 1990; Brandmayr and Brandmayr, 1987)
Ditomus tricuspidatus	A	*Ammi majus*	F	(Burmeister, 1939; Larochelle, 1990)
Dixus capito	A	Graminae and umbellifers	F	(Auber, 1965; Larochelle, 1990)
Dixus clypeatus	L	*P. lanceolata*	F	(Brandmayr and Brandmayr, 1987)
Dixus clypeatus	A	Grass, *P. lanceolata*	F	(Webster, 1903; Larochelle, 1990; Schremmer, 1960)
Dixus sphaerocephalus	A	*Plantago cornuti* and *P. crassifolia*	F	(Larochelle, 1990; Therond, 1975)
Dolichus halensis	A	Seed material	F	(Zhavoronkova, 1969)
Dyschirius	A	*Z. mays*	L	(Larochelle, 1990)
Euryderus grossus	A	Seeds; seed heads of *Bromus tectorum*	F	(Ball, 1960; Lavigne, 1977)

(continued)

Table 9.1 (continued)

Species	Life stage	Seed species consumed	Study location	Citation
Galerita janus	A	Grasses	F	(Forbes, 1883)
Harpalus spp.	L	Can complete development on seeds	L	(Brandmayr et al., 1980)
Harpalus affinis	A	Grass, dicotyledonous seeds, *Picea*, *Fragria*, trees, *S. media*, *T. officinale*, *D. sanguinalis*, *P. aviculare*, *Trifolium pratense*, *C. arvense*, *C. bursa-pastoris*; a seed mixture, *P. annua*, *L. amplexicaule*, *L. perenne*, *C. album*, *C. arvense*	F and L	(Hagley et al., 1982; Honek et al., 2003, 2005, 2006; Jeannel, 1941; Johnson and Cameron, 1969; Larochelle, 1990; Martinkova et al., 2006; Sunderland et al., 1995; Saska et al., 2008; Briggs, 1965; Pierce, 1917)
Harpalus amputatus	A	*Triticum*	F	(Bugbee, 1942)
Harpalus atratus	L	*D. carota*; a seed mixture	L	(Bertrandi and Zetto Brandmayr, 1991; Honek et al., 2005)
Harpalus atratus	A	Shrubby trees, *C. arvense*	F and L	(Burmeister, 1939; Honek et al., 2003; Larochelle, 1990)
Harpalus brevis	A	*Amaranthus*	F	(Zhavoronkova, 1969)
Harpalus calceatus	A	Millet and *Linum; Setaria viridis*, seed material	F	(Burmeister, 1939; Larochelle, 1990; Znoiko, 1935; Zhavoronkova, 1969)
Harpalus caliginosus	A	Grass (probably *P. annua*, but not clear from paper); *Trifolium*, *Fragria*, *Phleum* seeds from the head, *A. artemisiifolia*, *Arabis canadensis*, *Triticum*, *Datura stramonium*, *A. retroflexus*	F and L	(Blatchley, 1910; Brust and House, 1988b; Forbes, 1881; Johnson and Cameron, 1969; Tooley and Brust, 2002; Webster, 1881, 1900, 1903)
Harpalus cautus	A	*Pseudostsuga menziesii*	L and F	(Dick and Johnson, 1958; Johnson et al., 1966)
Harpalus compar	A	Grass; but only 'under stress' in the lab	L	(Johnson and Cameron, 1969)
Harpalus dimidiatus	L	*Daucus*	L	(Bertrandi and Zetto Brandmayr, 1991)

(continued)

Table 9.1 (continued)

Species	Life stage	Seed species consumed	Study location	Citation
Harpalus dimidiatus	A	'Granivorous tendencies'	F	(Brandmayr, 1972; Larochelle, 1990)
Harpalus distingeundus	L	*Daucus*	L	(Bertrandi and Zetto Brandmayr, 1991)
Harpalus distinguendus	A	*Fragria*, cereals, *C. arvense*, *C. bursa-pastoris*; a seed mixture, *P. annua*, *L. amplexicaule*, *S. media*	F and L	(Balachowsky and Mesnil, 1935; Burmeister, 1939; Honek and Martinkova, 2001; Honek et al., 2003, 2006; Larochelle, 1990; Saska et al., 2008)
Harpalus eraticus	L	Grass seeds in burrows	F	(Kirk, 1972)
Harpalus griseus	A	*Fragria*	F	(Kirchner, 1939; Larochelle, 1990)
Harpalus herbivagus	A	Grass, *F. rubra*, seeds	F and L	(Forbes, 1881, 1883; J. G. Lundgren, unpublished data, 2004)
Harpalus honestus	L	*Daucus*, *Taraxacum*, *Peucedanum*	L	(Bertrandi and Zetto Brandmayr, 1991; Zetto Brandmayr, 1990)
Harpalus luteicornis	A	A seed mixture	L	(Honek et al., 2005)
Harpalus marginellus	A	'Granivorous tendencies'	F	(Brandmayr, 1972; Larochelle, 1990)
Harpalus pensylvanicus	L	Grass seeds in burrows	F	(Kirk, 1972)
Harpalus pensylvanicus	A	*A. artemisiifolia*, *P. pratense* from the upright stem, *Panicum*; *Triticum*, *S. viridis* var. *major, Setaria pumila pumila*, *Amaranthus hybridus*, *Setaria pumila*, *Rumex altissimus*, *Echinochloa crus-galli*, *S. media*, *A. retroflexus*, *P. pratensis*, *P. dichotomiflorum*, *D. sanguinalis*, *B. tectuorum*, *D. stramonium*, *Sorghum halepense*, *C. album*	F and L	(Barney and Pass, 1986; Best and Beegle, 1977; Brust, 1994; Brust and House, 1988b; Harrison et al., 2003; Kirk, 1973; Lund, 1975; Lund and Turpin, 1977; Menalled et al., 2001; Webster, 1881, 1900, 1903)
Harpalus plenalis	A	*Z. mays*	L	(Larochelle, 1990)
Harpalus rubripes	A	'Granivorous tendencies'	F	(Brandmayr, 1972; Larochelle, 1990)

(continued)

Table 9.1 (continued)

Species	Life stage	Seed species consumed	Study location	Citation
Harpalus rufipes	L	*C. album, A. retroflexus, T. officinale, Brassica rapa, Galeopsis tetrahit, Trifolium alexandrinum, L. perenne*, also *Elytrigia repens* and *E. crusgalli*, but only to a very small degree in choice tests; *Senecio vulgaris, Agrostis tenuis, Festuca ovina, P. pratense,* cereals and ryegrass, *T. repens, B. napus, P. rhoeas, T. inodorum, P. annua*	L	(Briggs, 1965; Hartke et al., 1998; Jorgensen and Toft, 1997a)
Harpalus rufipes	A	Forest trees, *Z. mays, Linum, Capsella, Polygonum, Rumex,* cereals, pines, *Larix*, deciduous trees, *Fragria, C. album, S. media, P. aviculare, Triticum, E. crusgalli, S. arvensis, C. bursa-pastoris, P. annua, Taraxacum, T. inodorom, P. rhoeas, B. napus, T. repens, Viola arvensis*; 'seed consumers', *C. arvense*; a seed mixture, *L. amplexicaule, Picea*, conifer	F and L	(Briggs, 1965; Burmeister, 1939; Goldschmidt and Toft, 1997; Hartke et al., 1998; Honek and Martinkova, 2001; Honek et al., 2003, 2005; Jorgensen and Toft, 1997a; Larochelle, 1990; Martinkova et al., 2006; Zhang et al., 1997; Zhavoronkova, 1969; Saska et al., 2008; Pierce, 1917; Dick and Johnson, 1958; Nusslin and Rhumbler, 1922)
Harpalus servus	A	*Secale, Hordeum*, forest trees, *Camelina sativa*	F	(Blunck, 1925; Burmeister, 1939; Larochelle, 1990)
Harpalus signaticornis	A	*C. arvense, C. bursa-pastoris*; a seed mixture	F and L	(Honek and Martinkova, 2001; Honek et al., 2003, 2005)
Harpalus solitaris	A	*Carex pilulifera*	F	(Kjellsson, 1985)
Harpalus tardus	L	*Daucus carota*	L	(Bertrandi and Zetto Brandmayr, 1991)
Harpalus tardus	A	'Seed consumers', *C. arvense, C. bursa-pastoris*	F and L	(Honek and Martinkova, 2001; Honek et al., 2003)

(continued)

Table 9.1 (continued)

Species	Life stage	Seed species consumed	Study location	Citation
Harpalus tenebrosus	A	*Daucus, Foeniculum*	L	(Brandmayr et al., 1980)
Lebia	A	Seed mixture	F	(Brust and House, 1988b)
Microlestes linearis	A	*Z. mays*	L	(Larochelle, 1990)
Neoaulacoryssus speciosus	L	Millet	F	(de Costa Lima, 1952; Larochelle, 1990)
Notiobia	L	*P. rhoeas*	L	(Paarmann et al., 2006)
Notiobia acuminata	A	*Goupia glabra*	L	(Arndt and Kirmse, 2002)
Notiobia aulica	L	*Hypericum perforatum*	L	(Paarmann, 2002)
Notiobia flavicinctus	L	*Papaver rhoeas, H. perforatum, Ficus americana guianensis, Ficus donell-smithii, Ficus americana subapiculata*	L	(Paarmann, 2002)
Notiobia glabrata	L	*Bellucia dichtoma, H. perforatum, P. rhoeas*		(Paarmann, 2002)
Notiobia glabrata	A	*G. glabra*	L	(Arndt and Kirmse, 2002)
Notiobia incerta	L	*B. dichtoma, H. perforatum, P. rhoeas*	L	(Paarmann, 2002)
Notiobia nebrioides	L	*B. dichtoma, H. perforatum, P. rhoeas*		(Paarmann, 2002)
Notiobia nebrioides	A	*G. glabra*	L	(Arndt and Kirmse, 2002)
Notiobia pseudolimbipennis	L	*P. rhoeas, H. perforatum, F. a. guianensis, F. donell-smithii, F. a. subapiculata, Ficus hebetifolia*	L	(Paarmann, 2002)
Notiobia terminata	A	grass (probably *P. annua*, but not clear from paper) *Erechtites hieracifolia*	L and F	(Johnson and Cameron, 1969; Blatchley, 1910)
Notiobia variabilis	A	*G. glabra*	L	(Arndt and Kirmse, 2002)
Ophonus	L and A	Primarily granivorous	L	(Brandmayr et al., 1980)
Ophonus ardosiacus	L	*Foeniculum, Bupleurum junceum, Seseli, D. carota, Sium sisarum*	L	(Brandmayr et al., 1980; Zetto Brandmayr, 1990; Brandmayr, 1976b)
Ophonus ardosiacus	A	*D. carota* and *Pastinaca*	F	(Derenne, 1957; Larochelle, 1990; Therond, 1975)

(continued)

Table 9.1 (continued)

Species	Life stage	Seed species consumed	Study location	Citation
Ophonus azureus	A and L	*C. arvense, C. bursa-pastoris, Foeniculum*, specialists on seed-heads of *Daucus* and umbellifers	L	(Brandmayr et al., 1980; Honek et al., 2003)
Ophonus cordatus	A	'Granivorous tendencies'	F	(Brandmayr, 1972; Larochelle, 1990)
Ophonus diffinis	L	Umbelliferae	L	(Brandmayr, 1976a)
Ophonus diffinis	A	*D. carota*	F	(Larochelle, 1990; Therond, 1975)
Ophonus melletii		Specialists on seed-heads of *Daucus*: umbellifers		(Larochelle, 1990)
Ophonus punticeps	A and L	*Daucus, Foeniculum*	L	(Brandmayr et al., 1980; Briel, 1964; Larochelle, 1990; Zetto Brandmayr, 1990)
Ophonus rupicola		Specialists on seed-heads of *Daucus*; umbellifers		(Larochelle, 1990)
Ophonus sabulicola	A	Seeds of umbellifers	F	(Jeannel, 1942; Larochelle, 1990)
Ophonus stictus		Specialists on seed-heads of *Daucus*: umbellifers		(Larochelle, 1990)
Osimus ammophilus	A	*Medicago minima*	F	(Burmeister, 1939; Larochelle, 1990; Honek et al., 2005)
Parophonus maculicornis	A	A seed mixture	L	
Platynus decentis	A	*Z. mays*	L	(Larochelle, 1990)
Poecilus chalcites	A	*Z. mays, S. media*	L	(Lund and Turpin, 1977) but see O'Rourke et al. (2006)
Poecilus cupreus	A	*C. arvense, C. bursa-pastoris, P. annua, Taraxacum*	F and L	(Goldschmidt and Toft, 1997; Honek et al., 2003)
Poecilus lepidus	A	Seeds, *Picea*, cereals	F	(Burmeister, 1939; Larochelle, 1990; Pierce, 1917)
Poecilus lucublandus	A	*S. media, A. retroflexus, Z. mays*	L	(Lund and Turpin, 1977) but see O'Rourke et al. (2006)
Pterostichus algidus	A	*P. menziesii*	L	(Johnson et al., 1966)
Pterostichus amethystinus	A	*P. menziesii*	L	(Johnson et al., 1966)
Pterostichus coracinus	A	*Z. mays*	L	(Larochelle, 1990)

(continued)

Table 9.1 (continued)

Species	Life stage	Seed species consumed	Study location	Citation
Pterostichus herculaneus	A	*P. menziesii*	L	(Johnson et al., 1966)
Pterostichus lama	A	*P. menziesii*	L	(Johnson et al., 1966)
Pterostichus luctuosus	A	*Z. mays*	L	(Larochelle, 1990)
Pterostichus madidus	A	Seeds only occasionally	F	(Luff, 1974)
Pterostichus melanarius	A	Grass (probably *P. annua*, but not clear from paper), *Triticum*, cereals, *C. arvense* (but only a few; wouldn't feed on *C. bursa-pastoris*), *Taraxacum*, *P. menziesii*	L and F	(Brust and House, 1988b; Chiverton and Sotherton, 1991; Geiler, 1956/1957; Goldschmidt and Toft, 1997; Hagley et al., 1982; Hansen and Larsson, 1968; Honek et al., 2003; Johnson and Cameron, 1969; Larochelle, 1990; Dick and Johnson, 1958; Johnson et al., 1966)
Pterostichus niger	A	*C. bursa-pastoris*	L	(Goldschmidt and Toft, 1997)
Pterostichus oblongopunctatus	A	*P. sylvestris*	F	(Nystrand and Granstrom, 2000)
Pterostichus pumilus	A	*P. menziesii*	F	(Johnson et al., 1966)
Pterostichus versicolor	A	*C. bursa-pastoris*, *P. annua*, *Taraxacum*	L	(Goldschmidt and Toft, 1997)
Selenophorus	A	*C. album*, *A. retroflexus*, *D. sanguinales*, *P. dichotorlorum*	L	(Brust, 1994; Brust and House, 1988b)
Stenolophus	A	*C. album*, *A. retroflexus*, *D. sanguinales*, *P. dichotorlorum*	L	(Brust, 1994)
Stenolophus comma	A and L	*Z. mays*; food of larvae in captivity consists of *Avena*, *Hordeum*, *Triticum*. Johnson says that the larvae are strictly carnivorous	F and L	(Hagley et al., 1982; Johnson, 1949; Larochelle, 1990; Pausch, 1979; Starks and Lilly, 1955)
Stenolophus lecontei	A	*Z. mays*	F	(Bigger and Blanchard, 1959; Johnson and Cameron, 1969; Larochelle, 1990; Pausch, 1979)
Stenolophus lineola	A	*Z. mays*	F	(Johnson, 1949)
Stenolophus ochropezus	A	*Z. mays*	L	(Larochelle, 1990)

(continued)

Table 9.1 (continued)

Species	Life stage	Seed species consumed	Study location	Citation
Stenolophus pallipes	A	Grass (probably *P. annua*, but not clear from paper)	L	(Johnson and Cameron, 1969)
Synuchus impunctatus	A	*Melampyrum lineare*	L	(Manley, 1971)
Trechus quadristriatus	A	'Seed consumers', *C. bursa-pastoris*	F and L	(Honek and Martinkova, 2001; Honek et al., 2003)
Zabrus tenebrioides	A and L	*Triticum*, *Avena*, *Hordeum*, and *Secale*	F	(Pierce, 1917)
Zabrus tenebrioides	L	*Triticum*; a seed mixture	F and L	(Bassett, 1978; Honek et al., 2005; Zhavoronkova, 1969)

Considerable effort over the past 130 years has been given to characterizing patterns in the feeding mechanisms of and the major foods consumed by ground beetles, and these analyses can be useful in dividing the Carabidae into feeding guilds.

9.1.1 Adult Feeding Behavior

The first, and still one of the most useful, efforts to characterize the feeding of carabids was made by Forbes (1883). He describes three general feeding guilds that can be grouped by morphological differences in the mouthparts. These feeding guilds are presented as

> *first, those which seem usually to seize their prey and suck its juices, and take vegetation rarely, if at all; second, those which take a much larger ratio of animal food than of vegetable, but masticate and swallow it, as a rule, including indigestible fragments; and third, those whose habit is essentially vegetarian, but which still take solid animal food in diminished ratios*

In addition to developing these feeding guilds, Forbes (1881) is one of the first to recognize the granivorous nature of many carabid species.

Subsequent efforts have often expanded upon, with small changes, the feeding guilds proposed by Forbes, or have found additional evidence to support the guilds developed by him. Zhavoronkova (1969) divided adult carabids into three feeding guilds based on gut dissections, and the structures of the mandibles and proventriculus. One group consisted of strict predators (obligate zoophages), another was more predaceous than phytophagous (predominant zoophages), with a second group of predaceous omnivores that consume more plant material than animal tissue (predominant phytophages). Zhavoronkova pointed out that the plant food consumed by predominant zoophages and predominant phytophages consists mainly of seeds. Hengeveld (1980b) classified ground beetles as specialists or generalists.

The specialists had a narrower food range, and tended to fall within the Carabinae. Nevertheless, it wasn't clear how distinct these categories were, since 'specialists' still could consume up to eight types of food! In part, the feeding guilds were distinguishable based on size; the specialists tended to be smaller species, which are more limited in the types of prey items they can consume compared with larger species.

The phytophagous ground beetles have been further partitioned into guilds by Brandmayr (1990). She distinguished two general groups of phytophagous ground beetles: (1) a general opportunistic form of phytophagy on a range of fleshy plant tissues, and (2) strictly "spermophagous" (or granivorous) species. This may have simplified the phytophagous Carabidae to too great a degree, one of the dangers in creating too fine a distinction among feeding guilds of the Carabidae. For example, with the proposed groupings, there were only a few known strict granivores. Therefore the rest of the phytophagous beetle species would fall into Group 1, which included other plant tissues that are not always consumed by granivorous ground beetles. Recognizing additional degrees of phytophagy and granivory may be helpful when considering the evolutionary progression of feeding behavior in ground beetles.

The different feeding guilds that have ensued have prompted attempts to describe the evolution of feeding behaviors within the Carabidae. In general, the primitive state of feeding within the Carabidae appears to have been liquid feeding, as in the Carabinae. The next step was the acceptance of solid materials, primarily animal tissue and high protein plant material (i.e. seeds). Finally, the physiological and morphological adaptations that allowed these taxa to consume seeds then facilitated the consumption of additional plant material. The evolutionary progression toward granivory may have come about in association with the diversification of carabids into new habitats (Brandmayr, 1990). In her theory, Brandmayr postulated that the first granivores existed along waterside habitats. Then populations moved into the larger steppe and prairie biomes where herbaceous habitats persist. Finally, the provisioning of larvae with seed stores, as in the Ditomini, was a step toward sociogenesis in presocial carabids.

Caveats need to be recognized when evaluating evolutionary patterns in feeding among the Carabidae. First, many of the feeding records are based on gut dissections. One problem of relying on gut dissections to define the feeding behavior of ground beetles is the presumption that beetles with liquid diets consume only animal tissue. For example, *Carabus* and *Calosoma* feed extra-orally, and so never contain seed fragments in their guts. This may erroneously lead to the interpretation of these species being exclusively predatory, although *Carabus* will feed on fir seeds (Johnson and Cameron, 1969). Another caveat pertains to whether the feeding record was generated in the field or in the laboratory. The unnatural, often no-choice, conditions of laboratory analyses likely distort the importance of seed feeding, and should be interpreted carefully. Finally, the feeding behavior of ground beetle larvae remains largely unexplored, and this life stage should be considered when assigning species to feeding guilds of ground beetles.

9.1.2 *Granivory by Larvae*

Many larvae of Carabidae are granivorous, and their feeding behavior can differ ontogenetically between immature and adult stages. Seed feeding occurs both in free-ranging larvae, and in those that reside in burrows. The burrowing ground beetles sometimes create caches of seeds within their burrows. While the behavior of this life stage is still drastically understudied, seed feeding by immatures appears to be very important in the life history of some species of carabids. To date, granivory by larvae is described in large for *Harpalus*, *Ophonus*, *Amara*, and the seed specialists in Ditomini.

9.1.2.1 Feeding Preferences of Larvae

Some carabid larvae feed exclusively on seeds, and prey are unsuitable for larval development. *Ophonus puncticeps* can only be reared in the laboratory on seeds of *Daucus, Foeniculum,* and some other umbellifers (Zetto-Brandmayr, 1990). Furthermore, the larvae of the congener *O. ardosiacus* cannot complete development in the absence of seeds and displays a clear correlation between seasonal larval abundance and the peak seed shed of *Daucus carota* (Brandmayr, 1976b, 1983). *Amara similata* larvae reared on the aphids *Sitobion avenae, Rhopalosiphum padi*, or *Metopolophium dirhodum* do not survive through pupation, and those fed *S. avenae* do not even complete the first stadium. But larvae complete development on a diet consisting solely of seeds (Jorgensen and Toft, 1997b). It should also be noted that maximum eclosion rates in these experiments are around 40%, and come from diets with mixed seeds or solely *Capsella bursa-pastoris*. A final example of obligate granivory comes with seed specialists, like *Carterus, Dixus,* and *Ditomus* species in the Mediterranean region (Brandmayr and Brandmayr, 1987).

Omnivory is more frequently encountered in carabid larvae than obligate granivory, and mixing prey with seeds often supports larval development better than prey or seeds alone. Still, seeds may have higher food value than insect prey for some species of carabid larvae, and thus are a critical component of these mixed diets (Fawki and Toft, 2005; Hurka and Jaroski, 2003). An example of this is with *Harpalus honestus*, which had faster development and comparable pupation rates when fed solely seeds of *Taraxacum* versus animal-based diets (Zetto-Brandmayr, 1990). Another case comes with *Amara aulica* and *A. eurynota*, whose larvae require seeds to complete development (and can complete development on a seed-only diet), but develop fastest when fed a diet of seeds and prey (Saska, 2004, 2005).

Intuitively, not all seeds are created equally, and the relative food value of different seed species is manifested in the foraging behavior of carabid larvae. In one experiment, larvae of *Harpalus rufipes* show distinct preferences for the seeds of *Chenopodium album* and *Sinapis arvensis* over the seeds of grasses *Elytrigia repens* and *Echinochloa crus-galli*, and these preferences remained constant throughout the larval stage (Hartke et al., 1998). *Daucus* seeds are one of the worst foods for developmental rates of *H. honestus*, although the larvae survive well on these seeds (Brandmayr, 1990). But, somewhat contradictorily, Bertrandi and Zetto Brandmayr (1991) later show that *Daucus* seeds are one of the best foods for the development of *Harpalus honestus*, *H. distinguendus*, and *H. dimidiatus* larvae. The larvae of some carabids forage based on food availability and suitability. *Harpalus rufipes* larvae are most commonly encountered near aggregations of *C. album*, whose seeds are of particular nutritive value for the larvae (Briggs, 1965). Larvae of *Harpalus affinis* are most often found in association with annual meadow grass, a food found to be acceptable in the laboratory (Briggs, 1965).

9.1.2.2 Seed Caching

One fascinating behavior of some granivorous carabid larvae is the creation of seed caches by burrowing species. Seed-caching behavior has only been studied for the larvae of *Harpalus* and *Ophonus* species (Brandmayr, 1983; Hartke et al., 1998; Kirk, 1972; Luff, 1980) (Fig. 9.1). The burrows of ground beetle larvae vary in their size according to larval age and species. The diameter of the burrow is around 3–5 mm (Kirk, 1972; Luff, 1980). First and second instars burrow to an average depth of 10–13 cm (Alcock, 1976; Hartke et al., 1998; Luff, 1980). The burrows of third instars extend up to 17 cm into the soil (Luff, 1980). *Harpalus rufipes* will burrow up to 45 cm over winter (Briggs, 1965), and Kirk (1972) found the maximum burrowing depths of other *Harpalus* species to be 70 cm, but depths were more common ly 8–20 cm. The burrows are invariably straight, and the third instar of some species create a cell at the terminal end of the burrow (Alcock, 1976; Luff, 1980). This terminal cell is absent from the burrows of *H. pensylvanicus* and *H. eraticus* (Kirk, 1972). First and second instars forage outside the burrow for seeds at night, though not every night (Alcock, 1976). The foraging bouts of an unidentified harpaline larva last an average of 30 min, when 13 seeds can be transported back to the burrow (Alcock, 1976). Third instars of *H. rufipes* are not as likely to forage as earlier stadia, and Brandmayr (1983) found that overwintered third instar *Ophonus ardiosiacus* do not consume seeds at all. The seeds are pressed into the sides of the burrow (Alcock, 1976); in *H. rufipes* these seeds occur directly above the terminal cell (Luff, 1980), and in *H. pensylvanicus* and *H. eraticus* these seeds are pressed into the walls of the burrow approximately midway down the burrow (Kirk, 1972).

Burrowing larvae exhibit preferences for seed species, and these seed preferences are manifested in the seed caches in their burrows (Alcock, 1976). For instance, third instar *H. rufipes* prefer to consume *Amaranthus retroflexus* and *Chenopodium album* seed over *Trifolium alexandrinum* and *Sinapis arvensis*. First instars do not show preferences in the seeds that are cached, but second instars do, preferring to collect those of *C. album* (Hartke et al., 1998). Larvae of *H. pensylvanicus* and *H. eraticus* only cache foxtail seeds, even when other seed species are available (Kirk, 1972). It also noteworthy that larvae will also cache insect remains (Jorgensen and Toft, 1997a).

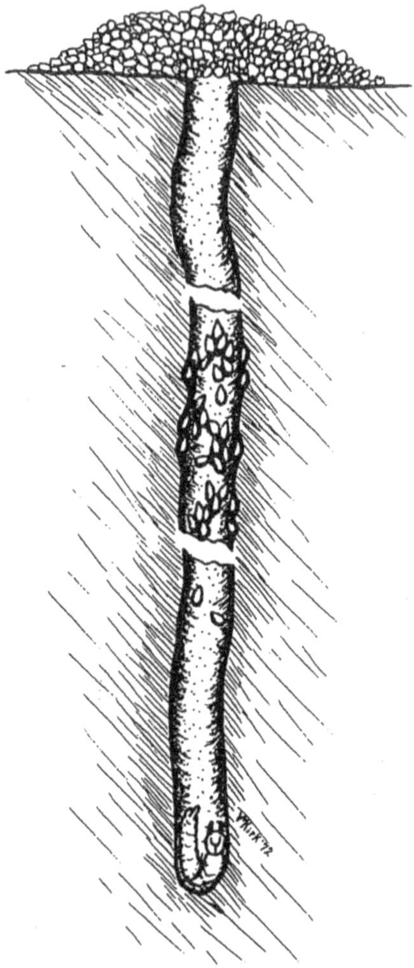

Fig. 9.1 Sketch of a burrow of *Harpalus eraticus* showing tumulus, cached *Setaria* seeds 8–20 cm deep, and the larva as typically found at the bottom of its burrow (Reproduced from Kirk, 1972. With permission by the Entomological Society of America)

The function of the seed caches is not entirely clear at this point. Some species seem to consume the seeds in the their burrows (i.e. *H. rufipes*) (Alcock, 1976; Hartke et al., 1998; Luff, 1980), while others do not. Kirk (1972) records that the seeds are not consumed in the burrows of *Harpalus eraticus* and *H. pensylvanicus*. One possibility is that these seeds may serve as a substrate for microorganisms that are consumed by the larvae.

Seed caching is also observed in the adults of some species. Seeds are stored by the adult stage of seed specialists within the Ditomini (Zetto-Brandmayr, 1990). Tallamy and Wood (1986) state that the underground brood chamber of the presocial carabid, *Ditomus calydonius*, is provisioned with enough seeds to sustain the larva through pupation. Another adult beetle that caches seeds is *Synuchus impunctatus*, but they do not store these seeds for later consumption or to provision their young. Rather, *S. impunctatus* forages for seeds of *Melampyrum lineare*, and upon finding a seed, will bring it beneath a log or branch to feed. There, the beetle consumes the caruncle, a portion of the seed near the hilum and micropyle, leaving the rest of the seed intact (Manley, 1971). The beetles have high fidelity to a given foraging area, and a seed cache is the result of this behavior.

9.1.2.3 Ontogenic Changes in Carabid Diets

A single seed species can be of different suitability for larvae and adults of a carabid species, likely because of the relative accessibility of a seed's nutrition, chemistry, and morphology for the different insect life stages. For example, *Amara similata* adults produce eggs on the nutrition obtained from seeds of *Tripleurospermum inodorum* or *Taraxacum* at a reduced rate compared to females fed *Poa annua* seeds or a seed mixture. However, larvae reared on *Tr. inodorum* or *Taraxacum* seeds survive poorly compared to a diet of *Capsella bursa-pastoris* seeds or a seed mixture. Jorgensen and Toft (1997b) describe how the larvae can't handle the larger seed size of *Tr. inodorum* and *Taraxacum*, even if these seeds had an acceptable level of nutrition for larval survival and development.

9.2 Formicidae

Carnivory is the ancestral state of ants, but seed consumption has developed numerous times in different capacities over the evolutionary history of this group (Wheeler and Bailey, 1920). The development of herbivory in ants may have followed a path outlined by Beattie (1985). The first step toward herbivory in his theory involved ants that consumed sugar-rich prey products such as honeydew. Next, the carnivorous species may have made the "small jump" to consuming plant products from herbivore damaged plant tissues. Once plants were recognized as a food source, other adaptations developed to exploit and specialize on other plant tissues. Beattie (1985) went on to suggest that the rise of ant-angiosperm associations may have occurred largely in semi-arid habitats, based in part on strong selective pressures placed on the angiosperms, and the diversity and dominance of ants in these types of habitats that can be observed even in the present day. Brown et al. (1979) also note that granivory by ants developed in desert environments, largely because insect prey were often scarce in these ecosystems.

Two major types of seed consumption are represented in ants; ants that consume the seeds themselves, and those that consume the food bodies associated with the seeds (Beattie, 1985). Of course, these two types of seed consumption are not exclusive of each other, and a number of species consume both the seed and the food body that accompanies it. Here, I define granivory as the process where the entire seed is consumed, and myrmecochory as when the food body, a specialized nutrient-laden structure attractive to animals, is consumed but the embryo is left intact. Ant species that consume myrmecochorus seeds are numerous and this behavior is phylogenetically extensive in ants. True granivory is more restricted, largely to species dubbed 'harvesting ants'.

9.2.1 Harvester Ants

Harvesting ants are a polyphyletic group with representatives primarily within the subfamilies Ponerinae, Myrmicinae, and Formicinae (Briese and Macauley, 1981; Holldobler and Wilson, 1990). Although harvester ants comprise more than 150 species, in more than 18 genera, half of these species occur within the New World genus *Pogonomyrmex* (MacMahon et al., 2000). Most harvesting ants are polyphagous on both seeds and insect prey (Holldobler and Wilson, 1990). It is suggested that harvester ants evolved toward granivory from carnivorous ancestors (Chreighton, 1952; Wheeler and Bailey, 1920). Evidence that *Pogonomyrmex* ants have become more granivorous over time comes from the primitive *P. huachucanus*, which doesn't store seeds. Thus this trait appears to be a more derived behavior within *Pogonomyrmex*.

9.2.1.1 The Diet of Harvesters

While all harvester ants are granivorous, the degree to which they rely on seeds and insects varies considerably by a number of physiological and environmental conditions. Seeds constitute a variable proportion of the diet in different species of harvesting ants. As examples, seeds comprise 16%, 33%, and 40% of the diets of *Melophorus*, *Tetramorium*, and *Meranoplus* species, all of which are predaceous on insects as well (Briese and Macauley, 1981). However, in other harvesting species, more than 90% of the diet is comprised of plant seeds (Azcarate et al., 2005; Briese and Macauley, 1981). *Pogonomyrmex mayri* consumes living prey, but only reluctantly and only those prey that don't fight back (Kugler and Hincapie, 1983). Nevertheless, insect prey can comprise up to 73% of their diet. Tennant and Porter (1991) estimate that seeds make up 4.8% and 29.6% of the solid food in the diets of *Solenopsis invicta* and *S. geminata geminata*, respectively, under field conditions. This study demonstrates quantitatively the relative importance of seeds in the diets of two congeners, and shows that seeds clearly play a more important role in the diet of *S. g. geminata* than in *S. invicta*. Where harvesting ants abound (deserts and shrub-steppe habitats), insect prey is scarce or sporadic. Thus the degree of granivory in many harvester ants is partly a reflection of the absence of other high-quality foods. When insect prey becomes available, most harvesting ants will accept it readily (Brown et al., 1979).

Harvesting ants often take advantage of a tremendous diversity of seeds within a habitat. For instance, *Messor* species harvests the seeds of at least 18 families of plants (Moggridge, 1873). Another species, *Messor pergandei*, consumes the seeds of 17 species in 14 plant genera in Coachella Valley (Tevis, 1958), and 29 species from 24 genera in Death Valley, California (Rissing and Wheeler, 1976). Thirty-two species of seeds are taken by a *Pheidole* species (Briese and Macauley, 1981), and as many as 40 seed species are collected by *Messor barbarus* (Azcarate et al., 2005). In spite of collecting a tremendous diversity of seeds in nature, harvesting ants clearly prefer certain seeds over others for a range of complex reasons (see Chapter 13).

9.2.1.2 Harvesters Within Plant and Animal Communities

Upon collecting the seeds, harvesters return the seeds to their nests. Harvester ants often reside in xeric habitats, and their nests can be quite extensive in order to reach subterranean moisture sources. The granaries of some species, where many seeds are stored, can be buried more than 3.5 m beneath the soil surface (Tevis, 1958), although most granaries reside within the top 15 cm of the nest (Briese and Macauley, 1981). Against tradition, MacKay (1990) believes that these granaries are not a consistent food source for winter persistence of the colony. Seeds that are stored in these granaries are consumed during the fall and spring, and the use of these stored seeds appears to be a method to avoid predation for the ants. Also, it should be noted that being collected by a harvester ant is not a death sentence for the seed, as will be discussed more in Chapters 12 and 18. Not all seeds make it back to the nest, and the seeds stored in the granary are not always consumed. Additionally, unpreferred seeds may be discarded to external middens; thrown out with the trash. But it turns out that the trash heap has unique characteristics that may actually foster germination of the plant.

Harvester ants are often critical and dominant species within the habitats where they occur. An example is in the Namib Desert, where harvester ants comprise 95% of the total forager biomass (Holldobler and Wilson, 1990). Harvesting ants are also particularly abundant in Australia, where vertebrate granivores (rodents and birds) are less important than in other regions. Another example of the importance of harvesting ants to a habitat is with *Pogonomyrmex* spp., which are the dominant granivores in certain grasslands (compared with birds and rodents), and could consume more than 6.5 million seeds per hectare over 1 year (Pulliam and Brand, 1975). Similarly, in Michigan old fields, ants are the dominant granivores of a number of plant species, particularly of *Centaurea* and *Oenothera* seeds (Mittelbach and Gross, 1984). Within ant communities, harvesting ants may not be the most speciose guild, but are frequently one of the most numerically dominant groups. An example of this is in saltbush and grassland systems, where only 14% of ant species are harvesters, but these species account for 53% of the ant colonies in the habitat (Briese and Macauley, 1981).

In order to sustain themselves, species within harvesting ant communities fulfill distinct niches. Within seed-feeding ant communities, there appears to be clear niche separation based on frequency of foraging. Interspecific differences in diel, seasonal, and yearly patterns of ant activity allow for resources to be exploited without exclusion of individual species (Whitford and Ettershank, 1975). Also, the harvester ants can be grouped into two broad feeding guilds, one in which colony members forage as groups and exploit aggregated seed sources, and another guild that forages individually and specializes on dispersed seed sources (MacMahon et al., 2000). Thus, by specializing on different seed densities, multiple ant species can persist in a single habitat.

9.3 Gryllidae

Although our understanding of the feeding behavior of crickets is still poor, we have advanced to the point that it is understood that these insects have several beneficial traits. This is in contrast to the limited view of crickets as beneficial insects held early last century.

> *The only valuable trait to which crickets can be credited is their habit of digging up and devouring grasshopper eggs* (Criddle, 1925)

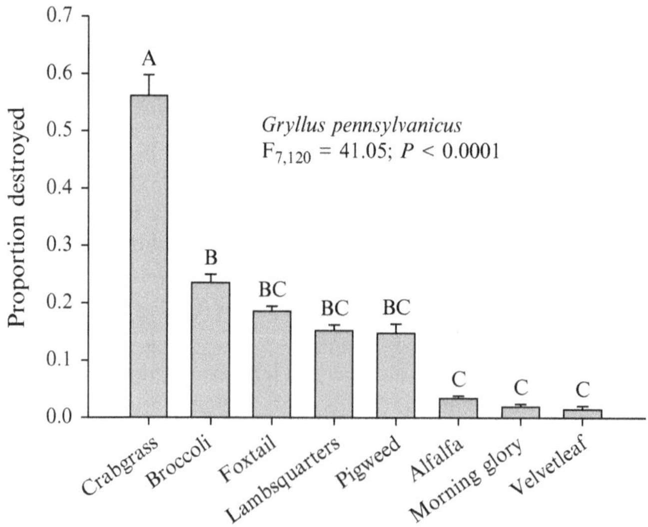

Fig. 9.2 The proportion of seeds destroyed by *Gryllus pennsylvanicus* adults under choice conditions (Reproduced from Lundgren and Rosentrater, 2007. With permission by Springer)

However, the beneficial omnivorous tendencies of crickets is being promoted as more recent research shows this group to be a major consumer of weed seeds in temperate agricultural systems.

Within cropland in temperate North America, crickets often are some of the most numerous granivorous entomophages captured (Lundgren et al., 2006; O'Rourke et al., 2006), especially *Gryllus* and *Allonemobius* species. According to one recent study, in several North American field crops examined, predation on *Setaria faberi* seeds is well correlated with activity of *Gryllus pennsylvanicus* and *Allonemobius allardi* (O'Rourke et al., 2006). Studies show that like most epigeal insects, *Gryllus* favors more stable habitats like no-till cropland (Brust and House, 1988b). Within this cropland, seed preferences by crickets are only rarely explored. In one study, *Gryllus pennsylvanicus* prefer *Chenopodium album* seeds to *Amaranthus retroflexus* seeds (Brust, 1994). In choice studies, *Gryllus pennsylvanicus* again displayed a distinctive preference for certain seeds, namely those of crab grass which are consumed at more than twice the rate of the second most preferred seed under choice conditions (Lundgren and Rosentrater, 2007) (Fig. 9.2). The characteristics that drive this preference are poorly understood; it does not seem that seed strength or structural characteristics play a role in crickets' zeal for grass seeds (Lundgren and Rosentrater, 2007). Given their tremendous abundance within cropland, it is almost criminal that we do not have a better handle on the feeding ecology of these insects.

9.4 Conclusions

Granivorous entomophages are a diverse guild of predominantly ground dwelling insects, many of which are also biological control agents of insect pests. Primitively, these insects were strict carnivores and have evolved to find and accept seeds in their diet to varying degrees. The more specialized species can completely replace prey with seeds without reductions in health, vigor, and fitness. But seeds are unique from prey in many respects, and those entomophagous species that have evolved toward granivory develop morphological and physiological adaptations that allow them to exploit seeds as food. These adaptations will be discussed at length in the next chapter.

Chapter 10
Adaptations to Granivory

Seed feeding is a specialized process and entomophagous arthropods that consume seeds possess a series of adaptations that allow them to exploit this unique food resource. On one level, arthropod populations coincide spatially and temporally with their seed food source. For example, ants are well adapted to living with booms and busts in the abundance of seed resources (Brown et al., 1979). These population characteristics are important in allowing the seed predator to encounter the seeds of choice, and will be discussed in more detail in subsequent chapters. But inevitably the insect that comes in contact with the food needs to know what to do with it, and this is where the present chapter will focus.

A host of morphological and physiological adaptations in the insect are necessary for them to identify, collect, and consume a seed. Anatomically, mandibular shape and other external morphological variations, as well as the structure of the alimentary canal, influence the dietary range of an arthropod. Moreover, the size and morphology of the arthropod will ultimately influence the techniques that are used to consume seeds. Analyses of seed feeding techniques grant an understanding of how the different anatomical adaptations function together to attain the intended outcome: a digested seed.

10.1 Morphological Adaptations to Seed Feeding

The size of an arthropod is one of the most apparent restrictions to its diet. An insect can only consume foods that it can reasonably manipulate and consume. Frequently, the size of a seed is one of the factors that restricts its attractiveness to a seed predator (see Chapters 11 and 13). In order to maximize their caloric intake per unit of foraging 'work', insects are generally expected to consume the largest food item they can manage. Of course, a number of intrinsic and extrinsic factors also play a role in the selection of seeds, but often it is the case that the larger the insect, the larger the seed it consumes.

J.G. Lundgren, *Relationships of Natural Enemies and Non-Prey Foods*,
Progress in Biological Control 7,

Published by Springer Science+Business Media B.V.

Were it always that larger insects manipulated seeds more easily, then one might expect social insects, (i.e., ants) to produce castes of varying sizes that can exploit a range of seed sizes within their habitat. Indeed, ants often do have these differentially sized castes, and these castes carry particles proportional to their size (Wilson, 1978). But the regulation of body size in ant workers is not necessarily driven by the food sizes available in the environment. At first, the observation of consistent seasonal variations in the size of *Messor pergandei* was thought to be a reflection of this species' response to the increasing availability of seed resources; the colony produces larger workers to exploit the seeds when they were available. To the contrary, Rissing (1987) found that worker size is simply a matter of low food quality during certain times of the season, rather than a prediction by the ants of better food to come. *Messor pergandei* worker sizes aside, caste specializations tailored for seed consumption do occur within the ants. One case in point is *Solenopsis geminata geminata*, *Pogonomyrmex badius*, and some species of *Acanthomyrmex* and *Pheidole*, in which a specialized caste occurs specifically for milling seeds (Holldobler and Wilson, 1990). Major workers break apart the endosperm and crack the seed coat with their large mandibles. This task is simply not possible for the minor workers. This milling caste becomes more numerous in a colony as the species' dependence on granivory increases. For example, workers in the seed specialists *Messor* all possess strong mandibles used for milling seeds.

10.1.1 Adaptations in Adult Granivores

10.1.1.1 Seed Collection

Other external morphological adaptations besides physical size are involved in seed collection and consumption by granivorous entomophages. One example of this is the psamnophore, a beard-like structure on the ventral region of the head, in harvester ants. Brown et al. (1979) report that some North American species use this beard to carry a second seed on harvesting expeditions. Also, Ettershank (1966) believes that the clypeal teeth observed in the granivorous ant genus *Monomorium* are modified to aid in seed-collecting. This is a phenomenon further explored by Andersen (1991), who presents the idea that alterations in the ant clypeus is related to the degree of seed specialization; species of *Solenopsis*, *Mayriella*, *Monomorium*, *Melophorus*, and *Meranoplus* are noted specifically for the size of their toothed clypeus.

10.1.1.2 Mouthparts

In carabids, the process of crushing food is divided between the mandibles and the proventriculus, and understandably the morphology of these structures often is suggestive of dietary habits. One early analysis of carabid mandibular morphology with relation to feeding behavior is by Stephen Forbes (1883). Describing the mandibles of the Carabidae, Forbes states *if they are short and quadrate, blunt at the tips, and provided either with strong basal processes or broad opposed surfaces, vegetable food is found to predominate.*

This description forms the basis for numerous additional interpretations of the mandibles.

After detailed examinations on the evolution of mouthparts in adult Carabidae, Zhavoronkova (1969), Forsythe (1983) and Acorn and Ball (1991) point out the following features found in the mandibles of herbivorous/granivorous ground beetles, particularly evident in the mandibular shapes in *Amara*, *Zabrus*, and the Harpalini. First, mandibles of granivorous species tend to be more stout (triangular in shape) in granivorous species, and the tips are more rounded. Second, the mandibles tend to be asymmetrical, the left longer than the right, which aids in manipulating cylindrical and round seeds. Next, the left terebral ridge, in occlusal aspect, is sinuate in granivorous species. Also, the incisor region of the mandible is chisel-shaped for increased vertical shear. Finally, the retinacular, or food grinding, region of the mandibles tends to be better developed in granivorous species, comprising up to 67% of the internal surface of the mandible. The ridges of the retinaculum form a compact basin for producing a bolus of chewed food. The observations by these scientists basically take the initial theories postulated by Forbes to the next level (Fig. 10.1).

Other structural features of the head and mouthparts that appear to be associated with granivory and herbivory in adult carabids is a reduced gular width to head ratio relative to more carnivorous species (Forsythe, 1982a). More developed gular regions of the head accompany extensive musculature to the maxillae. These maxillae are used to hold struggling prey while the mandibles crush them. In *Amara aulica* and *Harpalus rufipes*, two facultative granivores, the gular region is reduced. The mandibular musculature is better developed in these species, to facilitate crushing seeds.

10.1.1.3 Internal Structures

The proventriculus is another structure that varies in its anatomy depending on the diet of the arthropod species. The function of the proventriculus in carabids is to triturate hard food particles and protect the midgut epithelium (Evans, 1965; Forsythe, 1982a), which is demonstrated nicely by Cheeseman and Pritchard (1984). Much of the digestion of food particles occurs in or anterior to the proventriculus, and nutrient absorption occurs downstream in the mid- and hindgut. The observed grinding adaptations are present on the internal topography of the proventricular lining. In strict fluid-feeding entomophagous species (e.g., *Calosoma* and *Carabus*), the interior cuticle of the proventriculus is coated with hairs (Balfour-Browne, 1944; Hengeveld, 1980a). In more omnivorous species, the anterior of the proventriculus is covered with sharp raduli and sclerotized plates, presumably to assist grinding food (Forsythe, 1982a; Zhavoronkova, 1969). The posterior lining is covered in hairs (but see Balfour-Browne [1944] and Judd [1947] for a description of the *Chlaenius* proventriculus, in which the spines are located in the posterior portion of the intima). Zhavoronkova categorizes the raduli into teeth, spines, and combs. Teeth predominate in omnivores,

Fig. 10.1 Mandibles of primarily entomophagous (*top*) and granivorous (*bottom*) carabid beetles. Ventral aspect (Photos by J. G. Lundgren)

and spines and needle-like raduli are commonly found in the more granivorous species (Fig. 10.2). Evans (1965) believes that the spine-like appendages keep the solid food from entering the midgut, from whence digestive juices flow into the crop.

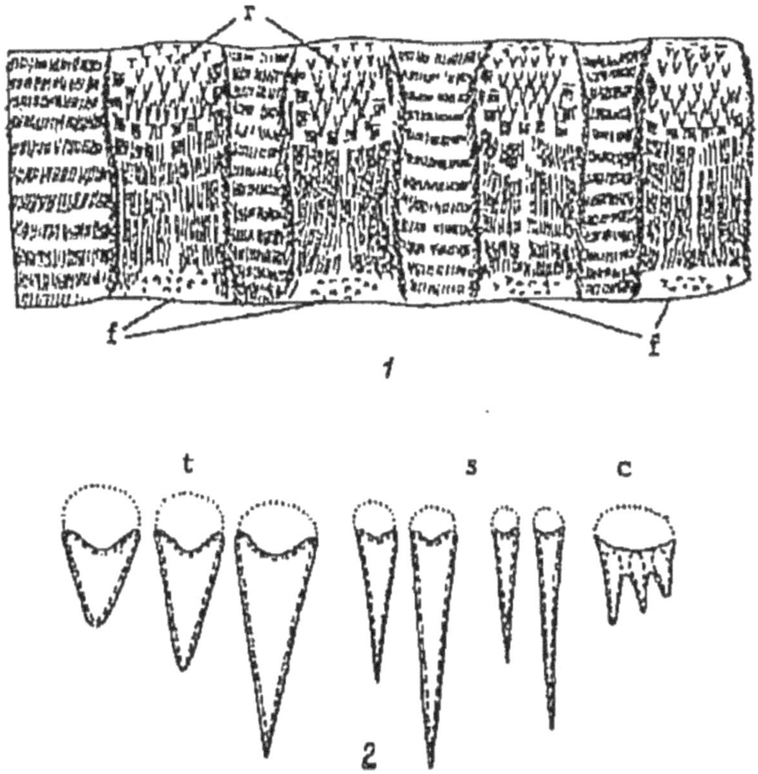

Fig. 10.2 Pattern of the structure of the proventriculus in the family Carabidae. (1) unfolded proventriculus (f: folds, r: radula with teeth); (2) elements of the chitinous structures of the radula (t: teeth, s: spines and spinules, c: combs) (Reproduced from Zhavoronkova, 1969)

A number of structures serve to limit the particle sizes that enter the midgut of ants. Ant workers are fluid feeders, and they are able to filter particles from viscous food sources (Eisner, 1957). The structures that restrict particle size are the oral opening and a specialized organ called the infrabuccal pocket (Quinlan and Cherrett, 1978; Wheeler and Bailey, 1920) (Fig. 10.3). This pocket resides within the head of the ant, and accumulates solid food particles (Eisner and Happ, 1962), such as seed material. For some species, the bolus of food that develops within the pocket is simply expelled, but in others the solid food is transferred to the larvae for digestion (Glancey et al., 1981; Wheeler and Bailey, 1920). In this way, adult workers are restricted from consuming solid portions of seeds, and the filtering mechanisms reinforce their reliance on the larval stage for the digestion of seeds, to be discussed later in the chapter. The proventriculus of some ants is also specially adapted to diet (Eisner, 1957; Eisner and Happ, 1962; Glancey et al., 1981). The two functions of the proventriculus of ants are to filter larger particles from

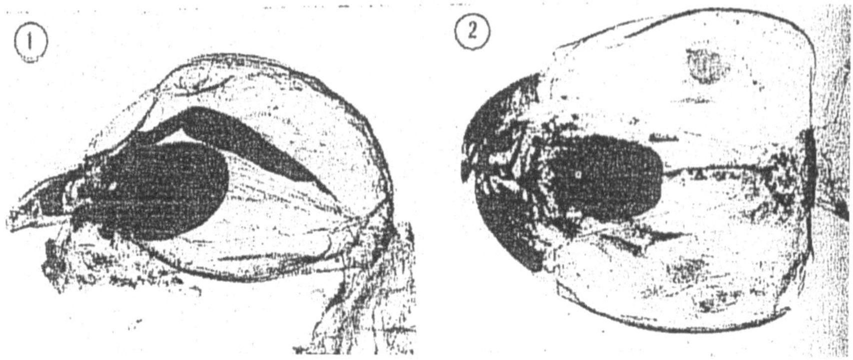

Fig. 10.3 The infrabuccal pocket in the head of Camponotus pennsylvanicus in lateral (*left*) and dorsal (*right*) aspect. The blackened area in the head is the corubundum-filled pocket (Reproduced from Eisner and Happ, 1962)

entering the midgut, and to serve as a dam that can be removed when the ants need to regurgitate fluid meals for their nestmates. Interestingly, this damming function may be reduced in the most granivorous of ants, like *Pogonomyrmex* species (Eisner, 1957). The damming function is particularly well developed for taxa that store their food internally; in harvesters the food is stored externally in granaries, and so this may partially explain why the morphological developments toward these passive dams are reduced or absent in this feeding guild.

In addition to the proventriculus, the anatomy and length of the digestive tract may be connected with food utilization (Yahiro, 1990). From the descriptions of carabid digestive tracts by Yahiro, it is difficult to draw specific conclusions that relate gut anatomy to food usage, although one trend is that many granivorous taxa (e.g., Pterostichinae, Zabrinae, Harpalinae, Panagaeinae and Zuphinae) are classified as having the Pterostichinae-type gut. This Pterostichinae-type gut is characterized with mid- and hind-guts meandering to the right and the formation of a single dextral coil (Fig. 10.4). von Lengkerken (1924; as cited in Hengeveld, 1980b) hypothesizes that the intestine of carnivorous beetles is shorter than that of phytophages. This may be because plant material is more difficult to digest than animal tissue, and so additional absorptive area is required in phytophagous carabids. The theory hasn't been well tested, but some support comes from experiments by Allen and Hagley (1982), who found that granivorous species have the slowest digestion rates of carabids.

10.1.2 Adaptations in Larval Granivores

In specialist granivorous carabid larvae, the issue of meal size is particularly acute; a preferred seed stays the same size but the larvae get bigger as they age. Thus, mandible size and shape (critical for crushing seeds) remain more constant in

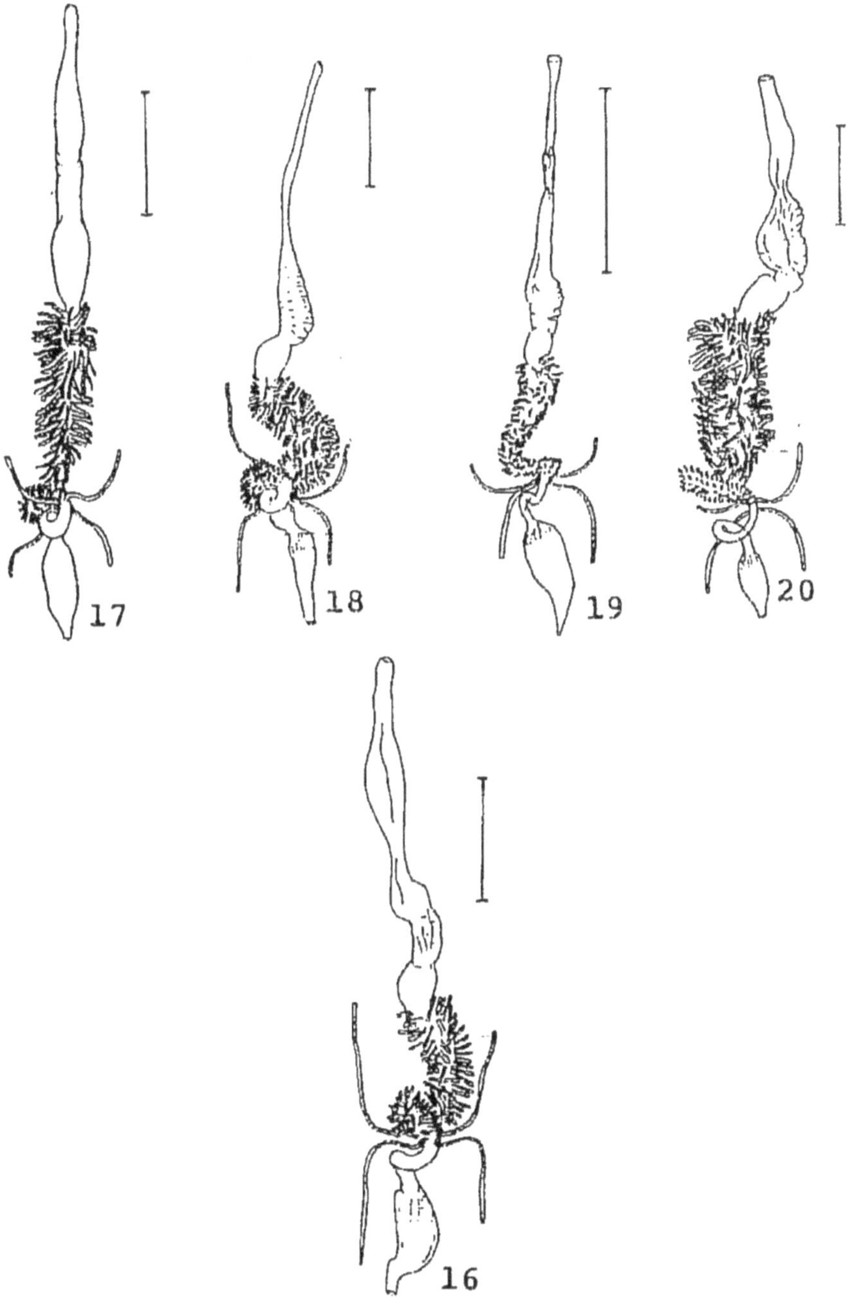

Fig. 10.4 The Pterostichine gut structure commonly seen in many granivorous Carabidae (Reproduced from Yahiro, 1990)

granivorous species than in predatory species, whose diet is more liberal (Paarmann et al., 2006; Arndt and Kirmse, 2002). Many of the mandibular traits of granivorous carabid adults transfer easily to the larval stage (mandibles of larvae are comparatively stout, with reduced terebral regions) (Brandmayr et al., 1998). Paarmann et al. (2006) also suggest that granivorous larvae are more apt to have strong terebral teeth, but these teeth are absent in the granivorous larvae of *Selenophorous* (Arndt and Kirmse, 2002).

Adult ants collect the seeds, but as will be described below, it is up to the 4th instars to process them. As such, one morphological feature of ant larvae believed to aid in triturating solid food particles like seeds is the trophorhinium (Wheeler and Bailey, 1920). This structure is located on the top and bottom of the buccal cavity, and consists of several striated ridges. Food that is dragged along these ridges is crushed and large or coarse fragments may be filtered out. This structure seems to be widespread in ants, although reports of it beyond Wheeler and Bailey's early description are scarce.

10.2 Seed Feeding Techniques

Seed predators approach the process of seed consumption differently, although each species must accomplish a similar sequence of events. These general requirements involve identifying the seed as a potential food source, manipulating and removing the seed to their dining area, and consuming the seed. The different procedures employed to accomplish each of these steps illustrate the complex physiological and behavioral adaptations of insects to a granivorous lifestyle, and ultimately underlie the dynamic exchanges between granivores and plant communities. Each of these processes also regulate seed selection by a seed predator, and will be discussed at more length in Chapters 11 and 13. Here, I discuss the physical process of consuming the seed, to shed light on how different morphological and digestive adaptations function together to macerate and digest the seed.

10.2.1 Seed Consumption Behavior

Perhaps the first thing that should be mentioned with regard to the process of feeding by granivorous entomophages is that there is a diverse array of feeding patterns among taxa. Here, the feeding styles are categorized as internal and extra-oral digestion for non-social granivores, and the colony-level feeding process of social insects, namely harvesting ants. Although these styles are disparate in many ways, certain trends among them can be drawn.

Granivorous crickets and many ground beetles consume their seeds internally, consuming solid portions of the seed. However, these entomophages are often finicky about which portions of the seed they ingest. As examples, *Amara pulpani* consumes birch seeds, beginning at the germ pole and feeding on the nutritious portions of the seed (Burakowski, 1967), and *Harpalus rufipes* eat only the endosperm of strawberry seeds, leaving the husk (Briggs, 1965). Often the external portions of the seeds, like the seed coat or pericarp are indigestible and circumvented by the insect using various means. Dick and Johnson (1958) describe seed feeding on *Pseudotsuga menziesii* by *Harpalus cautus*.

> *The beetle gnaws through the seed coat, generally starting at the flat surface of the seed, and hollows out the endosperm, leaving the seed coat intact.*

Solenopsis invicta consumes several seed species by first removing the external, protein-rich embryo. If they can penetrate the seed coat, then the rest of the seed is consumed (Morrison et al., 1997).

The portion of seeds consumed vary depending on the seed species, as has been particularly noted for grasses versus broadleaved species. *Harpalus pensylvanicus* cracked open the seeds of giant foxtail (*Setaria faberi*) and consumed the endosperm, whereas for smaller broadleaved species (i.e., *Amaranthus retroflexus*) the seed coat was crushed and consumed with the endosperm, although the seed coat is passed without being digested (Lund and Turpin, 1977). A similar behavior was seen in the field cricket *Gryllus pennsylvanicus*. For grasses, the endosperms were entirely consumed and the pericarp left intact, but with *Abutilon theophrasti* and *Amaranthus retroflexus* the pericarp was chewed and discarded (Carmona et al., 1999). Once the seeds were brought back to the nests by *Messor* ants, the involucres were removed, but this process varies with seed species (Went et al., 1972). For example, the involucre of *Chorizanthe brevicornu* was cut at one end of the fruit, and was removed like the peeling of a banana. In contrast, the involucre was cut off around the center of the woodier seeds of *C. rigida*. In other seeds (*Franseria dumosa* and *Hymenoclea salsola*), the burr was too hard to cut through, so the ants removed the cap from the seed, digested out the embryo, and discarded the remainder (Went et al., 1972).

10.2.2 Internalizing the Seed

Closer examinations of seed consumption, and how the mouthparts interact to direct the seed into the mouth, are conducted infrequently. Forsythe (1982a) describes the process for granivorous carabids, namely *Amara aulica* and *Harpalus pensylvanicus*,

> *The main work of breaking down the food is done by the mandibles; the maxillae convey the food to the mouth by dragging it over the anteriorly pointing, setae-covered surface of the ligula. These setae probably serve ... to remove and recirculate any large pieces of food or debris before they reach the cibarium. Food that reaches the hypopharyngeal membrane is held and possibly guided towards the mouth by the posteriorly projecting setae covering its surface.*

More is known of the feeding processes of the predominantly carnivorous species, such as the larvae of many carabids (Mitchell, 1963), and entomophagous adult stages (Forsythe, 1982a). In cases of carnivory the musculature to the palps is typically better developed for holding struggling prey, and extra-oral digestion is more frequently observed.

10.2.3 Digestive Enzymes

Digestive enzymes illuminate dietary range, and have been particularly well studied in carabids that digest foods extra-orally. Presumably many carabid larvae (although see Mitchell, 1963) and many adults of ground beetles are fluid-feeding. In these cases the carabids regurgitate digestive juices upon the food, or consume only the liquid contents of a food item (Hengeveld, 1980a). This undoubtedly leads to an underestimate of granivory in these taxa, since gut dissections are often the method used to determine diet under field conditions and fluid-feeding species would not have remnants of the seed coat used in diagnosing granivory. A case in point is with the genus *Carabus*, which is fluid-feeding and is generally regarded as strictly entomophagous. However, watching the feeding behavior of this beetle reveals that individuals will consume seeds (Johnson and Cameron, 1969). The digestive fluid used for pre-oral digestion in ground beetles is likely to be produced in the midgut (Hengeveld, 1980a).

Fluid-feeders aside, many carabids that consume solid foods also produce an acrid-smelling defensive regurgitate which contains digestive enzymes and may dually aid in the breakdown of food. In one omnivorous species, *Pterostichus madidus*, the defensive regurgitate contains proteases, amylases, and lipases/esterases (Forsythe, 1982b). It is remarkable that the conspicuously granivorous genera, *Amara* and *Harpalus*, do not regurgitate defensively, and it may be evidence for the lack of extra-oral digestion in some granivorous species (Forsythe, 1982b). The internal digestive proteases are characterized in one known seed-feeder, *Pterostichus melanarius,* which also does not regurgitate juices from the midgut (Forsythe, 1982b). This species possesses proteinases in their digestive tract, in particular trypsin and chymotrypsin (Gooding and Huang, 1969). In large part, the enzymatic capabilities involved in digestion by carabids, especially those that facilitate omnivory, remain unexplored.

For efficient digestion of seeds, it is important where within the digestive tract specific digestive enzymes are produced. Gryllids produce a range of different digestive enzymes, and specific nutrients appear to be digested differentially in the compartments of the digestive tract (Teo and Woodring, 1994; Thomas and Nation, 1984b). Proteins and carbohydrates are digested in the crop, and somewhat in the midgut. However, lipases are produced in the hindgut, and are shunted forward into these anterior regions in *Acheta domestica.* Sucrase and amylase, but not maltase and trehalase, are quite active in the anterior hindgut, whereas all carbohydrases are present and active in the crop. So, most of the non-lipase enzymes are produced in the midgut caecum, and absorption occurs in the midgut (Teo and Woodring, 1994). The field cricket, *Gryllus rubens*, produces amylase and lipase in the hind gut, or rather their endosymbiotic bacteria do, and absorb starches and fats here (Thomas and Nation, 1984a, b). Also, the production of digestive enzymes changes as crickets age (Koilpillai and Haniffa, 1996), and it is possible that this renders some life stages of granivorous entomophages more efficient seed consumers than others.

10.3 Seed Digestion in Harvester Ants

Ant adults are largely fluid feeding (Eisner, 1957), and the process through which they digest solid food is entirely different from solitary entomophagous arthropods. One hundred to 200 years ago, it was questionable whether seed-harvesting species of ants truly existed. Some of the support for this challenge to seed consumption in ants was raised by M. Gene in 1842 (Moggridge, 1873). He noted that although harvester ants were commonly seen bringing seeds back to their nests, he was at a loss for how these seeds were employed, since these ants could not digest hard substances and all of their food consisted of fluids or juices. Moggridge (1873) tried to maintain a colony of *Atta barbara* harvester ants in the absence of a queen or immatures to the following end:

> *On February 12 I found that all these ants, though abundantly supplied with seeds and all other kinds of food, were dead.*

He had much more success with colonies that possessed a queen and larvae, and made some observations on the behavior involved with seed consumption. In studying the workers that had been scraping the external coating of seeds in one of these colonies, he stated

> *It certainly appeared to be a bona fide meal that they were making, and not merely an act performed for the benefit of the larvae, as when they detach crumbs from a piece of bread and carry them below into the nest. However, I must own that, though I subsequently dissected ants taken in this act, which I suppose to be that of eating, I was unable by the use of the iodine test to detect starch grains in their stomachs.*

From these two observations, Moggridge came within a hair's breadth to uncovering the mechanism through which harvester ants consume seeds: workers of many species are incapable of digesting solids, and must rely on their larvae to digest the seed and pass their nutrition back to the adults through trophalaxis. Functionally, a harvester ants' digestive capacity is housed in the larval stage, and this relationship is reinforced by the larval morphology and relative abundance of digestive enzymes in the different life stages and castes. But before exploring these topics, it is illustrative to determine how nutrients are disseminated throughout the colony once workers have returned to the nest with food.

10.3.1 Nutrient Dissemination in Ant Colonies

Life stages and castes use classes of seed nutrients differently, and it is safe to say that this feeding behavior is best studied for fire ants, *Solenopsis* species. The workers consume mostly sugars, and often require a sweet liquid source of food to maintain

Table 10.1. The maximum size of food particles found within the midguts of ants. Filtration occurs by the proventriculus, infrabuccal pocket, and the mouth

Ant species	Particle size (μm)	Reference
Solenopsis invicta	<1	Glancey et al., 1981
Acromyrmex octospinosus	10	Quinlan and Cherrett, 1978
Veromessor sp.	25 × 60	Went et al., 1972
Camponotus americanus	>150	Eisner and Happ, 1962

normal activity (Tennant and Porter, 1991; Vinson, 1968). Some of these sugar sources are distributed to larvae and alates, with only a portion being retained by the workers (Vinson, 1968). Essentially, the larvae are repositories for lipids and proteins, and this is the only stage that can degrade proteins efficiently (Vinson, 1968). In fact, all protein sources are fed directly to the larvae, and none are even offered to the other adult castes by the workers when foods are initially brought to the colony (Howard and Tschinkel, 1981; Vinson, 1968). An exception is for nurse workers, that receive substantial amounts of protein-based foods, probably because they tend the larvae that are capable of digesting them (Sorensen et al., 1983). Oils are consumed by all life stages and castes (Vinson, 1968). For comparative purposes, it would be valuable for future research to examine how nutrients are distributed in the colonies of other ant species that have different natural histories.

Colony-wide nutrient usage patterns are dictated or supported by the morphology of the different life stages and castes. Liquid food sources, such as sugar solutions, are the only foods consumed by workers because a number of filtering mechanisms in the proventriculus restrict large particles from entering the midgut. Filtration is accentuated through repeated filtrations that occur during trophalaxis among workers (Eisner and Happ, 1962), and larger particles are accumulated in the infrabuccal pocket (Glancey et al., 1981). The size of filtered particles varies with ant species (Table 10.1).

10.3.2 Colony-Level Digestion of Seeds

Only the fourth instars of *S. invicta* are capable of consuming solid particles, and several adaptations are present to facilitate the feeding process (Petralia and Vinson, 1978). Specialized directional hairs and ridges in the cuticle on the antero-ventral area of the larva hold the solid food in place. This antero-ventral region on the larva is found in other ant species as well, and is called the praesaepium, or food basket. The mandibles of fourth instars are sclerotized and the head is directed such that it comes in contact with the antero-ventral region of the body. Worker ants place the solid food particles directly onto this specialized area of the larva, and do not offer solid foods to other larval instars (Fig. 10.5).

In order to digest seeds, the ants must produce enzymes that render the nutrition of the seeds suitable for digestion. The adults sometimes play a role in the initial

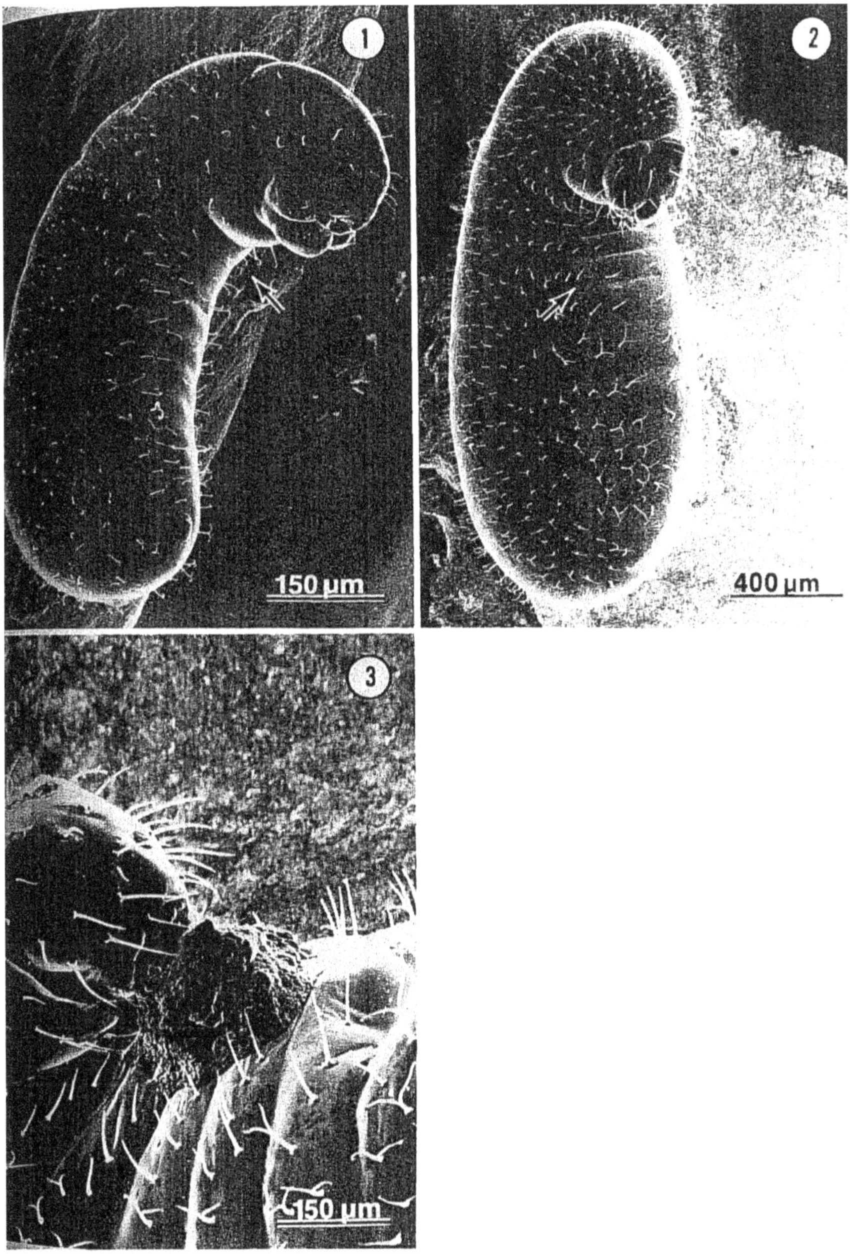

Fig. 10.5 Scanning electron micrograph of 3rd (*top right*) and 4th (*top left*) instar *Solenopsis invicta*. Note the position on the head, the straight hairs of the antero-ventral region of the 4th instar, and bifid hairs elsewhere on the body. The bottom image is of a 4th instar consuming a solid food bolus (Reproduced from Petralia and Vinson, 1978. With permission from the Entomological Society of America)

breakdown of the seeds (Wheeler and Bailey, 1920). Adults of some *Messor* species produce protopectinase in their gasters that is used to dissolve the middle lamella of the seed (Went et al., 1972). Then the larvae suck out the digested embryo from within the burr of the seed. Not all *Messor* species produce this enzyme; it has only been demonstrated in *M. pergandei*, and *M. smithi* and *M. lobognathus* do not produce it. Besides this protopectinase, no adult harvester ants are known to possess cellulose-digesting enzymes (Beattie, 1985).

Many of the enzymes required for protein digestion are present in the larval stage, which explains why the workers do not dispense proteinaceous foods throughout the colony before they pass through the larval stage. In *Solenopsis invicta*, larvae produce proteinases and actually transfer them to workers (Sorensen et al., 1983). Workers also produce proteinases, and at this point it isn't clear how the larval proteinases differ from and function in workers. Larvae of *S. invicta* use their enzymes on solid food particles that workers place upon their food basket. Although enzymatic processes are not deduced, the behavioral observations by Wheeler and Wheeler (1953) suggest that some formicine larvae also digest solid foods in this manner.

It is also worth noting that some harvester ants avoid starchy foods, possibly due to their inability to digest this carbohydrate (Wheeler and Bailey, 1920, and references therein). *Messor barbarus* may overcome their inability to digest starch by degrading it into maltose, possibly by placing seeds, or masticated portions thereof, in the sunshine (Moggridge, 1873; Wheeler and Bailey, 1920, and references therein). The simpler sugars can then be digested by the harvester.

10.4 Conclusions

In spite of vastly different natural histories among granivorous entomophages, certain morphological and physiological adaptations to consuming seeds are consistently present. Functional adaptations in these insects aid them in (1) handling of the seed, (2) crushing the external defenses of the seed, (3) protecting the midgut epithelium from the hard portions of the seed, and (4) extracting the maximum amount of nutrition from the seeds. Within solitary granivores such as carabids and crickets, many of these adaptations can be observed in individual insects. In social insects, examining only the morphology of the foraging workers does not reveal the criteria necessary for consuming and digesting seeds. However, if the colony is considered as a whole, then suddenly many of the feeding adaptations observed in individual solitary granivores become apparent. For instance, in ants the foraging workers have specializations for handling the seed but not crushing or digesting it; these tasks are reserved for other castes like the milling castes and the larval stage. This point is further illustrated by Went et al. (1972) when they describe the harvester ant *Messor*,

> *When these ants are moving their larvae, they are just carrying their stomachs around.*

Thus, certain evolutionary trends in granivorous entomophages are often present across taxa, which speak to the evolutionary importance of seed feeding in these entomophages.

Furthermore, the traits identified here may be useful in exploring or identifying putative granivory in other entomophagous arthropods and understanding the evolutionary hurdles that are encountered by entomophagous species that consume seeds.

After identifying entomophagous taxa that feed on seeds (Chapter 9) and examining the different adaptations and processes involved in digesting seeds by these insects (the current chapter), a question that remains is: *why seeds*? In the next chapter, the topic of seed nutrition as it pertains to entomophages will be addressed in full. Also, the topic of seed predation will be explored from a seed's perspective, with particular regard to nutrition and defenses as they pertain to granivorous entomophages.

Chapter 11
Seed Nutrition and Defense

The seed is an organ which performs several unrelated functions among which energy storage, dispersal, and perennation are paramount

(Levin, 1974)

To accomplish these diverse and sometimes conflicting goals, the plant must carefully balance the nutrition and structure of its seeds in order to maximize its chances for successful germination. Environmental conditions and the physiological status of the parent plant often fluctuate dramatically both geographically and temporally. Therefore, it is not surprising that there is tremendous plasticity in the nutritional and defensive traits of plant seeds within and among plant species, many of which affect their attractiveness to granivorous entomophages.

Many plants sustain very high levels of granivory (Harper, 1977). Seeds are an extremely rich source of many nutrients required by insects, and many seeds are on par with insect prey in terms of nutrition. Given that they are nutritionally rich, it should surprise no one to find that seeds are defended against granivory by insects. Indeed, predation influences the chemistry and morphology of seeds, as well as their distribution and density. By looking at the relationships among seeds and granivorous entomophages from the seed's standpoint, hopefully a better understanding of the mechanisms that operate in these interactions is gained, as well as some directions for exploiting granivorous arthropods as beneficial insects.

Given that 70% of human food consumed is directly related to seeds (Bewley and Black, 1994), it is understandable that there is a nearly endless list of scientific literature pertaining to the nutrition and chemistry of seeds. Of this body of research, the majority of information on chemical composition of seeds is for current and potential crop species (particularly within the Leguminosae and Graminae). Whether the trends in nutrition of these groups are representative of non-crop species is not clear. But these crop species give us an idea of the potential nutrients found in seeds, as well as factors that underlie the intraspecific variation in their composition. With these biases and limitations in mind, it is not the objective of this chapter to provide an exhaustive exploration of

J.G. Lundgren, *Relationships of Natural Enemies and Non-Prey Foods*,
Progress in Biological Control 7,

Published by Springer Science+Business Media B.V.

seed nutrition and defense. Rather, I will present some of the tendencies in the chemical composition of seeds as they might influence the feeding behavior of insect seed predators.

11.1 Seed Nutrition

> *The seed is dispersed from the mother plant endowed with a store of food reserves of protein, carbohydrate and fat in a more concentrated package than occurs anywhere else on the plant. Animals exploit this property when using seeds as an extremely important part of their diet.* (Bewley and Black, 1978)

Many nutrients important to insect development and reproduction can be found in large quantities within seeds, often equaling or even exceeding the levels found in insect prey. However, seed nutritional content is often plastic, with maternal nutrition, environmental factors, and genetics all playing roles in the intraspecific nutritional variation found in seeds (Bewley and Black, 1978). In fact, this variation of seed nutrition within a species is often the target of breeding programs which seek to improve crop yield and quality (Bertrand et al., 2005). Although food reserves cannot be definitively quantified for a species, it is still useful to discuss certain trends in the nutrient content of seeds from a qualitative angle, as well as explore the ranges of specific nutrients within certain species and among higher taxonomic designations of plants. To this end, several efforts are made to categorize seeds into groups based on their nutritional contents.

The first type of classification described trends in major energy storage nutrients. Protein generally comprises approximately 10–30% of food reserves for the developing seedling, and the remainder of the nutrition is present as either starch or lipid (Bewley and Black, 1978). In what can only be classified as one of the most taxonomically extensive studies on the nutrient contents of seeds, F. R. Earle and colleagues analyzed the oil and protein contents in more than 3,000 seed species comprising 113 plant families (Barclay and Earle, 1974; Earle and Jones, 1962; Jones and Earle, 1966). They also presented some qualitative studies of the presence or absence of starch, tannins, and alkaloids in the seeds of each plant species. A couple of noteworthy trends were derived from this data (Barclay and Earle, 1974; Jones and Earle, 1966), the first of which was that at the family level, protein and oil contents were positively correlated with one another. This observation was not to say that oil and protein content were positively correlated in an individual seed species, but families that had higher protein content tended to also have some species with higher oil contents. Another trend was that families low in oil and protein were inclined to have a higher number of species containing detectable levels of starch.

Reflecting on the data of Earle, his coauthors, and other research in the area, Bewley and Black (1978) describe three general approaches to energy storage in seeds. (1) Plants whose seeds rely on starch as the primary storage unit, and to a lesser extent lipids and proteins (notably species of Graminaceae). (2) Species that store energy as lipids (triacylglycerides) and protein. (3) Those species that rely on protein as their storage chemical, but contain intermediate levels of lipids and starches. An excellent illustration of the inverse relationship between oil and starch allocations in seeds are presented by Pizo and Oliveira's (2001) study of nutrient content in ant-dispersed plant seeds.

In addition to which nutrients are stored, it is important to understand where and how seeds store the nutrients that may be important food resources for granivorous entomophages. A first discussion point is the discernment of endospermic and non-endospermic seeds. As is obvious from the names of these seed classes, the major difference between them is whether or not they possess an endosperm (Bewley and Black, 1978). Nutritionally, the endosperm is quite different from the rest of the seed because it is the major site of nutrient storage for the germinating seedling. An analogous nutrient storage organ in non-endospermic seeds is the cotyledon, which are often enlarged in non-endospermic seeds (Bewley and Black, 1978). Finally, the perisperm is sometimes well developed and can function as a site for nutrient storage (Bewley and Black, 1994) (Fig. 11.1).

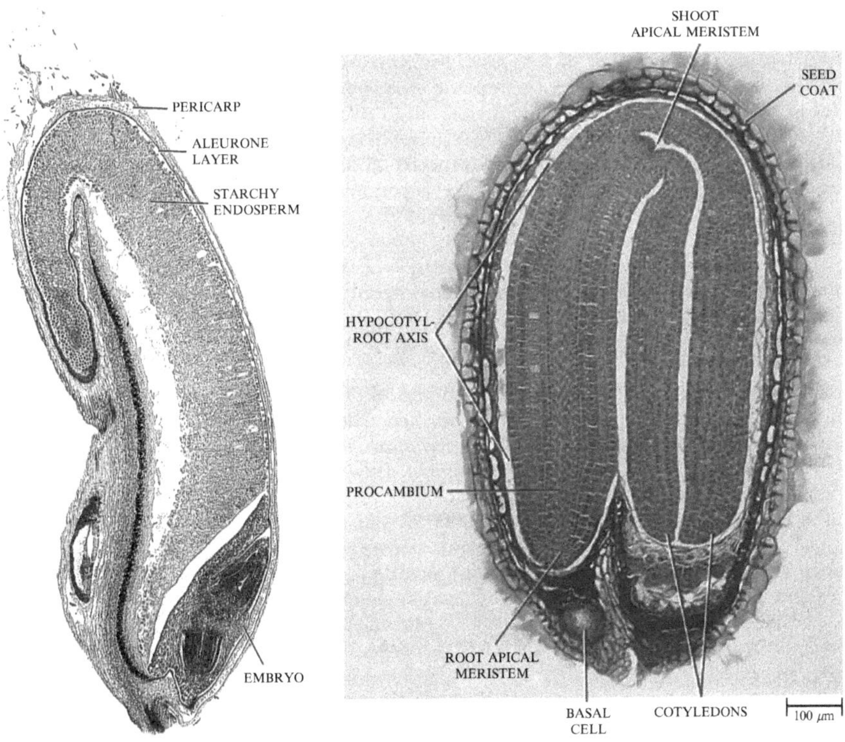

Fig. 11.1 Cross sections of endospermic (*Triticum aestivum*, *left*) and non-endospermic (*Capsella bursa-pastoris*, *right*) seeds, illustrating two major nutrient-storage strategies found in seeds (Reproduced from Raven et al., 1992)

In addition to differing in the major nutrient storage organ, species also differ in how they segregate energy sources within the seed (Bewley and Black, 1978). For instance, in certain seeds carbohydrates are stored in the endosperm, and proteins and lipids are found in the embryo (Bewley and Black, 1994). Starches frequently accumulate near the hilum of the seed (Bewley and Black, 1978). Most reserves are ultimately aggregated either in the endosperm or the embryonic tissues, especially the cotyledons. Furthermore, each storage nutrient is usually stored in discrete packages (Bewley and Black, 1978). Starches are stored as grains, proteins as protein bodies, fats as oil bodies, and minerals as phytin salts that typically occur in association with the protein bodies. The allocation pattern of specific nutrients within a seed may partially explain the feeding techniques and processes employed by granivorous entomophages to consume seeds (described in Chapter 10).

11.1.1 Carbohydrates

Simple and complex carbohydrates can account for a substantial portion of the nutrient content of seeds. Carbohydrates comprise up to 86% of seed dry weight in some species (Pizo and Oliveira, 2001). Of course, these high levels of carbohydrates are characteristic of those species that use starch as their energy reserve. For instance, free carbohydrates account for as little as 6% of the total seed dry weight in oilier species (Kelrick and MacMahon, 1985).

Typically, polysaccharides are far more abundant in seeds than mono- and oligo-saccharides (Bewley and Black, 1978). Of the carbohydrates in *Vigna radiata* var. *radiata*, starches comprise 55% of total seed weight, while reducing sugars (4.9%), stachyose (1.5%) and raffinose (0.5%) are only minor components of seed nutrition (Mubarak, 2005). This trend in carbohydrate content is found in other leguminous seeds as well (Vadivel and Janardhanan, 2005).

A number of simple sugars are present in seeds, but the major oligosaccharide in seeds is sucrose (Matheson, 1984). Trehalose is also frequently abundant in some plant groups, notably the pteridophytes. Some trisaccharides found in seeds are raffinose, umbelliferose (only in Umbelliferae), and planteose. Moreover, planteose is not found in vegetative portions of the plant (Bewley and Black, 1978).

As mentioned, complex sugars are better represented in most seeds than simple sugars. The predominant polysaccharides in seeds are starches, the composition of which generally breaks down to 20–25% amylose and 50–75% amylopectin (Bewley and Black, 1978, 1994; Matheson, 1984). In the absence of the appropriate α-amylase, starches cannot be digested by animals (MacGregor, 1983), and so enzymes are very important in insects that rely on seeds as a food source.

In addition to starch, other polysaccharides occur in seeds. In fact hemicelluloses, namely mannan, are common storage carbohydrates (Bewley and Black, 1978). In some seed species, mannans have replaced starches as the primary storage carbohydrate (Bewley and Black, 1994). Amyloids, pectins, celluloses, and musilages are other polysaccharides found in seeds, but to lesser degrees (Bewley and Black, 1994;

Kelrick and MacMahon, 1985). Lignin is another complex carbohydrate encountered in seeds. Kelrick and MacMahon (1985) found that lignin content for eight seed species ranges from 0.5–25% of total seed weight, with a mean value of 8.3%.

11.1.2 Proteins

Seeds have long been exploited as an important source of protein for both humans and livestock (Payne, 1983), and many species contain the amino acids essential for insect development and reproduction. Protein content of seeds ranges from 2–48% of seed dry weight (Barclay and Earle, 1974; Earle and Jones, 1962; Jones and Earle, 1966; Kelrick and MacMahon, 1985). But in typical seeds, protein content falls within 10–20% of total seed weight (Murray, 1984a; Pernollet and Mosse, 1983). Proteins are stored in discrete storage units called protein bodies, and storage proteins are classified into distinctive groups that vary in their solubility, structural characteristics, and amino acid constituency.

There are a number of storage proteins in seeds that all share a common function: to provide amino acids and nitrogen to the germinating plant (Pernollet and Mosse, 1983). Based on various solubility and structural characteristics, these seed proteins are classified into several groups (Pernollet and Mosse, 1983). Dicots harbor high levels of glutelin in addition to other globulin proteins, such as legumins and vicilins (Bewley et al., 2000). Monocots such as cereals frequently contain prolamines and glutelins (Bewley et al., 2000; Pernollet and Mosse, 1983). Albumins are globular water-soluble storage proteins which are also frequently encountered in seeds (Bewley and Black, 1978, 1994; Bewley et al., 2000; Murray, 1984a; Pernollet and Mosse, 1983). Globulins and prolamines tend to form protein bodies encased in starch granules near the endosperm (Pernollet and Mosse, 1983). But glutelins are a more diverse group of proteins and some can be found outside the protein body organelles (Pernollet and Mosse, 1983). Although there are similarities in the types of storage proteins found in seeds, the types of storage proteins vary among species, and even among closely related taxonomic groups (Pernollet and Mosse, 1983).

It is worthwhile to discuss these types of storage proteins found in seeds in a book on insect nutrition for at least two reasons. The first is that these different storage proteins possess different amino acid profiles (Bewley and Black, 1994), and can be inherently deficient in some amino acids that insects are otherwise unable to synthesize *de novo*. For example, prolamines are often well endowed with proline and glutamine (Bewley and Black, 1994), but deficient in other amino acids such as lysine (Bewley and Black, 1978). Legumes are generally deficient in sulfur-containing amino acids, cysteine and methionine, and some cereals are characteristically low in lysine and tryptophan (Bewley and Black, 1994). This notwithstanding, seeds often contain a diverse and abundant suite of amino acids, and frequently contain large complements of essential amino acid necessary for insect growth and development (Bertrand et al., 2005; Mubarak, 2005). In fact amino acid content in seeds can be greater (based on % dry matter) than protein-rich foods of animal origin such as egg whites (Callaway, 2004).

Another reason for an entomologist to understand the basis of seed storage proteins is that some of these storage proteins have functions other than simply providing a nutrient source for the seedling. For instance, defensive lectins can sometimes double as storage proteins, and are frequently found in protein bodies (Pernollet and Mosse, 1983). Also, albumins can be inhibitors of trypsin proteases and α-amylase in grass species (Bewley et al., 2000). These defensive proteins will be discussed more thoroughly later in this chapter.

11.1.3 Lipids

Another important storage nutrient in seeds are lipids, and oil contents range from almost 0% to 70% of seed dry weight (Barclay and Earle, 1974; Earle and Jones, 1962; Jones and Earle, 1966; Levin, 1974). As mentioned earlier, lipids and starches tend to be mutually exclusive within a single seed species, as can easily be seen in Pizo and Oliveira's survey (2001). Here, oily seeds had 62–71% lipid content, whereas less than 8% of dry mass was lipids in the starchier seeds. Triacylglycerides accumulate as lipid droplets in organelles called oil bodies. Environmental conditions such as temperature influence the abundance and classes of fatty acids that are found in seeds (Slack and Browse, 1984).

Different plant life styles or 'habits' are associated with different oil contents (Levin, 1974), and the progression of lipid contents within the Leguminosae appears in Fig. 11.2. Levin (1974) explains this relationship based on the energy

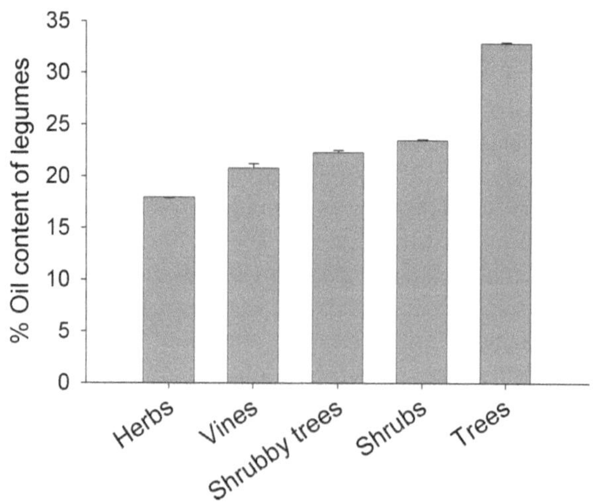

Fig. 11.2 The oil contents (mean ± SEM % of dry matter) of different 'plant habit' classifications within the Leguminosae (Reproduced using data of Levin, 1974)

efficiency of different storage nutrients and the resource needs of the seedling under different environmental conditions. Lipids contain the most energy per unit of all the storage nutrients, and so seedlings that require more energy for germination are more likely to use oil as their reserve nutrient. In harsh habitats where the seedlings experience low light or excessive competition, the extra energy provided to the seedling from the oil reserves may render it more competitive. Of course, the high energy content of oily seeds may also affect their attractiveness to granivorous entomophages.

Most seed lipids are stored in the form of triacylglycerol fatty acids (Slack and Browse, 1984). Lauric, myristic, palmitic, stearic (saturated) acids and oleic, linoleic and linolenic (unsaturated) in sum account for 95% of the fatty acids found in plants (Harwood and Russell, 1984). Within seeds, all of these but lauric and myristic acids can be found in abundance (Callaway, 2004; Slack and Browse, 1984). In many oil seed crops, oleic and linoleic acids account for up to 60% of the oils, and palmitate, stearate, and linolenate comprise less than 10% of triacylglycerides (Bewley and Black, 1994). Glycolipids and phospholipids are also encountered in seeds to varying extents (Bewley and Black, 1978).

Insects require a dietary source of sterols in order to synthesize hormones necessary for their growth and reproduction. Sitosterols, stigmasterols, and campesterols are the primary phytosterols found in plants (Bewley and Black, 1994; Harwood and Russell, 1984; Kritchevsky, 1997). Of these, sitosterols (especially β-sitosterol) are often the most abundant sterols found in seeds (Beveridge et al., 2005; Holser et al., 2004; Marcone et al., 2004; Phillips et al., 2005). In amaranth species, β-sitosterol accounts for 5.4–8.5 $\mu g\ g^{-1}$ of seed (Marcone et al., 2004), and up to 11 $mg\ g^{-1}$ of grape seed oil (Beveridge et al., 2005). In one survey, β-sitosterol is the dominant phytosterol in nearly all seeds and vegetable tissue (almonds, cashews, and walnuts) (Kritchevsky, 1997). Although extraction methods vary in their ability to quantify sterol contents, the total sterol composition of seeds typically comprises less than 5% of oils by weight (Beveridge et al., 2005; Moreau et al., 2001; Phillips et al., 2005). In order to use phytosterols in their metabolism, insects need to dealkylate the C_{28} and C_{29} sterols into the C_{27} form, which is the structure that most insects use to synthesize their molting hormones (Svoboda et al., 1978). To accomplish this, insect species use a variety of metabolic pathways (Svoboda et al., 1978), although this is not well studied for many granivorous entomophages.

11.1.4 Minerals

A large suite of inorganic minerals are contained in seeds, but once again the environment and maternal contributions influence the final mineral and micronutrient content of a seed. Inorganic minerals are largely stored as phytin, which is the salt of *myo*-inositol hexaphosphoric acid and accumulates as globoid crystals in the protein bodies of seeds (Bewley and Black, 1978; Callaway, 2004). Potassium, magnesium, phosphorus, and calcium typically represent the largest proportion of the phytin salts (Bewley and Black, 1978; Callaway, 2004; Mubarak, 2005; Vadivel and Janardhanan, 2005). Iron, manganese, copper, and sometimes sodium can also be found in phytic acid (Bewley and Black, 1994). It isn't clear whether phytin is a mineral source or sink for granivores, since it readily binds with free elements and may actually absorb more nutrients from the organism feeding on it than it contributes (Bewley and Black, 1994).

Minerals (often represented as ash content) typically comprise less than 6% of the total seed weight (Callaway, 2004; Kelrick and MacMahon, 1985; Lott, 1984; Mubarak, 2005; Pizo and Oliveira, 2001). Fenner (1983, 2004) reports an inverse linear relationship between the ash content of seeds and their weight. Furthermore, Fenner reports that a higher percentage of the internal seed is ash compared with the seed coat (Fenner, 1983), so it may be in the best interest of mineral-limited granivores to bypass the seed coat and gain access to the internal seed.

11.1.5 Vitamins

Water soluble vitamins tend to be well represented in seeds, with the exception of ascorbic acid and vitamin B12. Seeds are an excellent source of water soluble B vitamins, including niacin, pantothenic acid, Vitamin B6, riboflavin, thiamin, folate, and biotin (Grusak and DellaPenna, 1999; Lebiedzinska and Szefer, 2006). Vitamin B12 is not present in plant tissues (Grusak and DellaPenna, 1999), and its importance to insects is not entirely understood. All plant cells possess ascorbic acid (De Tullio and Arrigoni, 2003), but during dormant stages, when seeds are susceptible to post-dispersal granivory, vitamin C is often at its lowest point (De Tullio and Arrigoni, 2003). Also worth noting, ascorbic acid is a common precursor to calcium oxalate, a defensive phytochemical found in many seeds (Franceschi and Nakata, 2005). *Actinidia chinensis* is a good example of the relationship of vitamin C and oxalate in seeds. *Actinidia chinensis* has very low levels of ascorbic acid in the actual seeds (28 $\mu g\ g^{-1}$ of seeds), but the seeds had higher levels of the vitamin C metabolite, oxalate, than surrounding fruit tissues (Rassam and Laing, 2005). More information on the defensive properties of oxalate against granivorous entomophages is discussed later in the chapter.

Fat soluble vitamins, such as tocopherols (vitamin E) and carotenoid precursors of vitamin A, are also present in many seeds. Tocopherols function in protecting seed storage oils from oxidation during dormancy (DellaPenna and Pogson, 2006). The tocopherol (especially alpha-tocopherol) content of seeds is frequently higher than in other plant tissues, sometimes at 10–20 times the level found in photosynthetic tissues (DellaPenna and Pogson, 2006; Grusak and DellaPenna, 1999). In fact, vitamin E is the most abundant vitamin in hempseed (Callaway, 2004) and is also abundant in grape seed oil and corn seeds (Beveridge et al., 2005; Moreau et al., 2001). Within seeds, vitamin E tends to be specifically allocated to certain structures. For example, tocopherols are most frequently found in the axis and cotyledons of soybean seeds (Yoshida et al., 2003).

Generally, seeds are low in or devoid of carotenoids, such as vitamin A (Rodriguez et al., 1975). Carotenoids are important as components of photosynthesis, although they do occur in non-photosynthetic tissues (DellaPenna and Pogson, 2006). But some seeds have high levels of carotenoid pigments, such as those of *Momordica charantia* which are particularly high in lycopene (Rodriguez et al., 1975). It is safe to say that while carotenoids and other fat- and water-soluble vitamins have been found in seeds, their relative concentrations and functions in seeds are still poorly understood.

11.1.6 Water

A recurring observation is that seed water contents are particularly attractive to many granivorous entomophages, particularly in xeric environments where harvesting ants frequently abound. Seeds in dry habitats absorb water when relative humidity increases, thus being a reservoir of water for foraging ants (Christian and Lederle, 1984). Another trend worth mentioning is that the water content of seeds is proportionally correlated with levels of soluble carbohydrates (Crist and MacMahon, 1992). Chirstian and Lederle (1984) hypothesize that when water-limited, granivorous ants should be more attracted to seeds with high levels of carbohydrates and lower in protein and lipid content. Indeed, this is the case for *Pogonomyrmex occidentalis*, whose seed preferences correlate with the caloric content, water, and carbohydrate levels of available seeds (Kelrick et al., 1986).

11.1.7 Caloric Content

In addition to providing essential nutrients for normal metabolism, it is important that a food source have sufficient energy value that it can support the caloric needs of its consumer. Seeds are small, but their nutrient value per unit of food is quite high. From the literature, the mean energy content of 89 seed species was 4,900 cal g^{-1} of dry weight (Table 11.1). Of course, the caloric content of seeds is highly variable based on environmental conditions and the relative amounts of different macronutrients. For example, lipids yield 8,500 cal g^{-1}, whereas starch yields 4,200 cal g^{-1} (Levin, 1974), and so oil seeds are expected to have a higher energy content per unit mass than starchy seeds.

11.2 Seed Defense

While numerous aspects of plant ecology influence the consumption of seeds by granivorous arthropods and could be deemed 'defensive' in nature, here only those qualities inherent in the diaspore itself are discussed, namely the structural and

Table 11.1 The caloric contents of 89 species of seeds

Number of species tested	Range of calories	Mean calories	Citation
51	4,316–6,088	4,956 cal g^{-1}	(Kendeigh and West, 1965)
18	2,521–5,917	4,579 cal g^{-1}	(Reichman, 1976)
8	3,743–5,598	4,392 cal g^{-1}	(Kelrick and MacMahon, 1985)
11	4,301–6,575	4,832 cal g^{-1}	(Christian and Lederle, 1984)
1		5,787 cal g^{-1}	(O'Dowd and Hay, 1980)
	Average	**4,909 cal g^{-1}**	

All caloric contents listed here were collected from seed dry matter, and were determined using bomb calorimetry.

chemical defensive traits present in seeds. Specifically, the size, structure, and chemistry of the seeds are given attention. Tactics employed by plants to avoid granivory temporally and spatially, as well as other refugia from predation, will be discussed more completely in Chapter 18.

Just because a particular mechanism is effective in defending a seed from predation by insects does not mean that this is the sole purpose for the defensive adaptation. Intraspecific seed shape polymorphisms abound in nature and fulfill a number of roles in the seed's natural history including dispersal mechanisms and dormancy (Harper et al., 1970). Also, the chemical defenses of seeds frequently are limited to those that possess the dual function of providing nutrition to the developing seedling (McKey, 1979). The multifunctional aspects of seed chemistry is not surprising, given that seeds are often size limited and must prioritize their contents to maximize the success in attaining germination and establishment.

11.2.1 Seed Size

Seed size within a species is related to a number of factors, including genetic regulation, parental fitness and nutrition, the natural history of the plant, and environmental conditions (Fenner, 1992, 2004; Leishman et al., 1995). Seed size within the plant kingdom varies over ten orders of magnitude, the smallest being those of some orchids and the largest being the diaspore of the coconut palm (Harper et al., 1970). Variably sized seeds are differentially attractive to insect granivores, (a topic which will be discussed in more detail in Chapter 13). Granivorous insects frequently have an important impact on seed communities, and it is likely that relative consumption of different-sized seeds may select for those seeds which are of a size that is less attractive to ants, carabids, or crickets.

11.2.1.1 Life History Patterns and Seed Size

Research shows that certain habitats and plant life history patterns are associated with larger or smaller seeds. Specifically, seeds tend to be larger in harsher habitats.

Shady and/or dry habitats tend to have larger seeds, and high altitudes favor smaller seeded species (Baker, 1972; Davidson, 1993; Levin, 1974). For the shady or dry habitats, this is believed to be the case because seeds in these habitats require additional nutrients to establish a strong root system and be competitive with surrounding vegetation. Levin (1974) hypothesizes that high altitudes result in smaller seed sizes because of the shortened growing season in these habitats. Essentially these plants don't have the time to invest more nutrients into large-seeded progeny.

A number of lifestyles are associated with plants that produce relatively larger seeds. Persistent perennials, such as trees and shrubs, tend to produce large seeds. In one respect, these types of plants are often found in shaded habitats, and so require more nutrients. But also, long-lived organisms in general seem more willing to invest energy per capita in their offspring than species which are time-limited. Also, late-successional species tend to produce larger seeds than conspecifics that occur in earlier successions of a community (Davidson, 1993). Once again, competition is likely to be more intense in later successional communities, which could favor plants that give their seeds a leg up in the form of extra nutritional reserves.

At the end of the day, there is a finite amount of parental investment that an organism can feasibly devote to each of its progeny, and seed size is balanced with the number of seeds that can be produced. In many organisms, not just plants, two strategies of reproduction can be paraphrased as: (1) produce a large number of small seeds that require little parental investment, but whose individual chances of success are compromised, and (2) produce a small number of large seeds. These large seeds require more parental investment, but each seed is given an 'edge' in the form of nutrient reserves that can help them compete with other plants (Baker, 1972).

11.2.1.2 The Influence of Granivory on Seed Size

From an evolutionary point of view, seed size is subject to selection by granivores. One needs only look at how one granivorous species, the human race, has selected for the bulky, oversized seeds found in most crop species to see the impact that selection can have on seed size (Harper et al., 1970). Nevertheless, it is often difficult to quantify direct impacts of granivory on the selection for defensive seed sizes. Seed predation by bruchid beetles plays an important role in the seed sizes of leguminous plants (Janzen, 1969). Larvae of the bruchids feed internally on a seed. When a legume species produces many small seeds versus fewer large seeds, the small seeds suffer less predation by the beetle larvae. Essentially, the larvae cannot complete development within the small seeds. Although this system relies on a specialist insect granivore rather than an entomophagous species, it does present an argument that granivory by insects can be one aspect driving the production of small seeds in plants.

Another case that supports the idea that granivorous entomophages may be molding the sizes of seeds in seed banks deals with *Ambrosia trifida* and granivorous arthropods (*Harpalus pensylvanicus* is the only species identified) (Harrison et al., 2003). The seeds of *A. trifida* are relatively large to begin with, but a range of seed sizes are produced by each plant. Field tests on the removal rates of different size classes of *A. trifida* seeds show that small seeds are consumed at a greater rate than larger seeds. The authors hypothesize that as selection pressure on the smaller seeds of *A. trifida* persists, it is entirely possible that larger seeds will comprise a growing portion of the seed population.

11.2.2 Mechanical Defenses of the Seed

In addition to the seed itself, the diaspore is frequently covered with various external appendages that can influence the ability of granivorous entomophages to carry and manipulate the seed. At this point, it is valuable to establish that the testa (or seed coat) and the pericarp are not interchangeable terms (Bewley and Black, 1978). The testa is derived from the integument of the seed (Esau, 1977). The pericarp is an external feature of the seed that is inherited from the ovarial wall of the maternal plant (Bewley and Black, 1994). So diaspores that include the pericarp should technically be referred to as fruits; cypselas, achenes, caryopses, and samaras are all these types of seed-like fruits. Frequently, these external structures are quite nutritious and aid in attracting insects that function as dispersal agents of the seed (see Chapter 12). However, structural characteristics can also defend the seed from granivory, and it is these defensive characteristics that will be focused on here. There is an inverse linear relationship between the thickness of the seed coat and a seed's weight, heavier seeds have a disproportionately larger seed coat than smaller species (Fenner, 1983). Some of these seed coats can comprise nearly half the weight of the seed, as in *Mucuna urens* (Adebooye and Phillips, 2006).

Although the role of structural strength is believed to go hand in hand with the thickness of the external coverings of the seed, empirical data seldom puts this theory to the test. An exception is a study on the impact of the relative strength of seeds on preferences by harvester ants. Rodgerson (1998) shows that seeds which require more force to crack are less likely to be collected by ants; seeds that required more than 20 N to crack are rarely consumed by the harvester *Rhytidoponera metallica* (Fig. 11.3). Still, as is pointed out by Lundgren and Rosentrater (2007), the actual strength of a seed is related to its size, and in fact smaller seeds are generally much stronger than large ones. In this study, both the strengths of the seed coats and the internal matrices of different seeds affect their palatability to two of three granivorous insects (both carabids) under choice conditions.

11.2.3 External Structures

11.2.3.1 Appendages on Seeds

The most conspicuous features of the external covering of the seed that influence granivory are the presence of arils, hairs, awns, and spines. Some of these structures may arise from the pericarp, while others are derived from the testa. Often times, smooth-coated seeds are difficult for insects to manipulate, especially ants. The

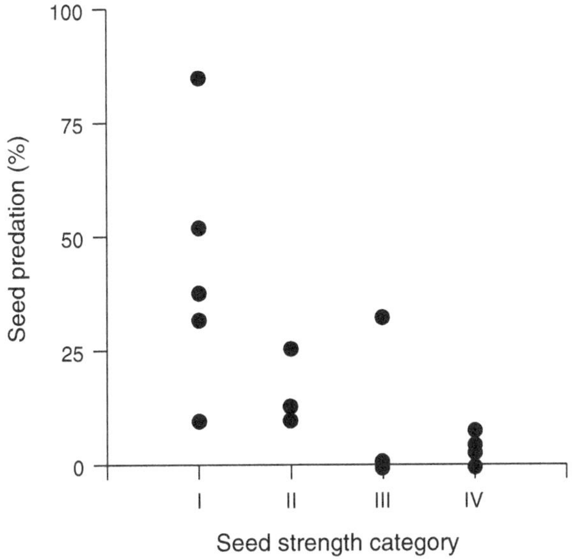

Fig. 11.3 Percentage of seeds (17 plant species) adapted for ant dispersal that were consumed by *Rhytidoponera* colonies. The seeds varied in the force necessary to crush them; categories were I: 2 < 5 N, II: 5 < 14 N, III: 14 < 41 N, IV: ≥ 41 N (Reproduced from Rodgerson, 1998. With permission by the Ecological Society of America)

presence of hairs, awns, or spines on the seeds allow the granivores to transport them, and so it is frequently the case that seeds with external projections are preferred over smooth-coated occurring in the same habitat (Azcarate et al., 2005; Pizo and Oliveira, 2001).

As a case in point, Pulliam and Brand (1975) found that *Pogonomyrmex* could not carry smooth-coated seeds back to their nests, and seeds with external projections were preferred by the ants (Fig. 11.4). The smooth-coated seeds became acceptable to the ants when they were still attached to the seed head; the ants had a 'handle' and were able to carry the seed. Based on their observations, Pulliam and Brand (1975) concluded that seed nutrition had little to do with the preference, and that seeds that did not have awns were defended against removal by the ants. The resilience of smooth-coated seeds seemed to be related to the availability of alternative foods. Of the two seeds preferred by the harvester ant, *Messor pergandei,* one was smooth-coated (Tevis, 1958). However, the preference for this smooth coated species dissipated when rains and abundance of other foods increased (Tevis, 1958).

By producing a seed population that contains a number of independent external structures, a plant can exploit several methods of seed dispersal including the ants' incapacity for carrying seeds with external features. Baker and O'Dowd (1982) describe one such system. The plant *Hypochaeris glabra* produces both beaked and unbeaked achenes, and the relative proportion of these types produced per plant is reflective of various environmental conditions (Fig. 11.5). These seed types are dispersed by different mechanisms; the beaked achenes are dispersed by the wind,

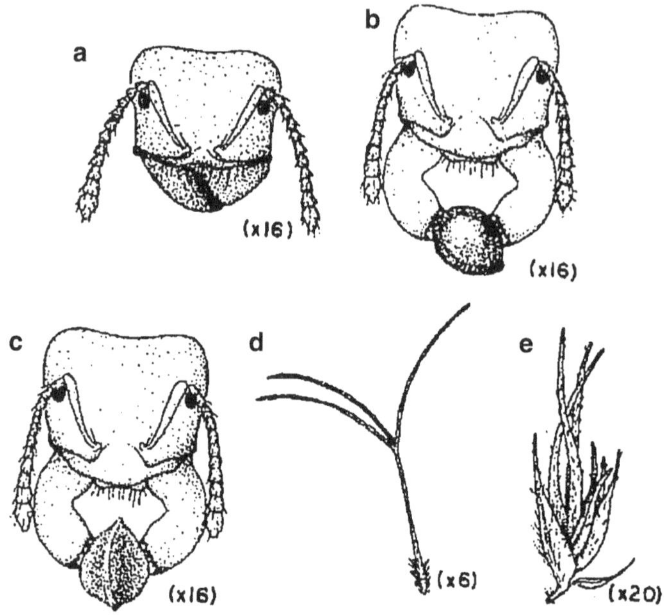

Fig. 11.4 Heads of *Pogonomyrmex* ants with closed mandibles (**a**), and mandibles grasping *Amaranthus* seeds (**b**), and *Rumex* seeds (**c**). Seeds of these species are difficult for the ants to carry, whereas seeds of *Aristida ternipes* (**d**) and *Bouteloua chondrosioides* (**e**) are carried more easily (Reproduced from Pulliam and Brand, 1975. With permission by the Ecological Society of America)

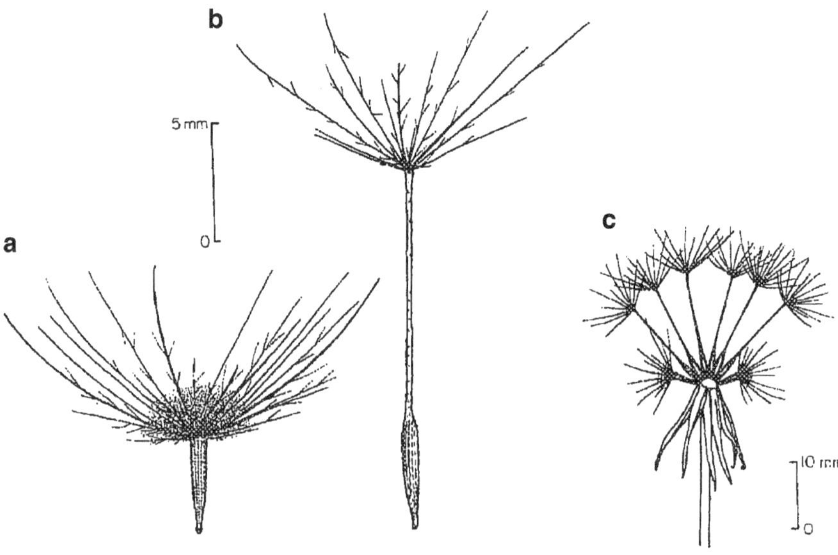

Fig. 11.5 The two achene types of *Hypochaeris glabra*: (**a**) unbeaked; (**b**) beaked; (**b**) arrangement of the achene types on the receptacle (Reproduced from Baker and O'Dowd, 1982. With permission by Blackwell)

whereas the unbeaked achenes adhere to mammalian hair. More important to the content of this book, only the beaked achenes are suitable food for colonies of *Messor andrei*. The pappus and beak are easily removed from the beaked achenes and the seeds are carried back to the nest where they are consumed. Unbeaked achenes prove bulky and typically are dropped en route to the nest. In this case, the ants may be functioning as dispersal agents of the unbeaked achenes, and this unbeaked biotype could be considered as having defensive qualities against post-dispersal granivory.

In addition to the effects of external projections on seed defense from granivory, the external structures of the diaspore can defend the seed by functioning as an impenetrable shell. In *Cryptantha flava*, the stiff and prickly calyx protects the seeds from predation by ants, specifically *Pogonomyrmex occidentalis* (Casper, 1987). When the calyx is removed, the ants selectively remove these seeds to their nests. Field observations reveal that after 24 h, 90% and 35% of hulled and intact seeds (respectively) are removed by the ants. A similar observation was seen with millet (Pulliam and Brand, 1975). *Pogonomyrmex* only removes those seeds that have their hulls removed.

11.2.3.2 Trypanocarpy

Another instance where external projections of the pericarp function defensively is in trypanocarpy (Schoning et al., 2004). The trypanocarpic diaspore contains two external components, a long awn capable of hygroscopic rotations and an upper portion that is always perpendicular to the awn (Fig. 11.6). Essentially, the seed is drilled into the soil, and after the seed is buried the awn that projects out of the soil splits off. The awns are difficult for *Messor* workers to remove, and the ants have trouble moving the seeds when the awns attached. Once the awn falls off, the seed

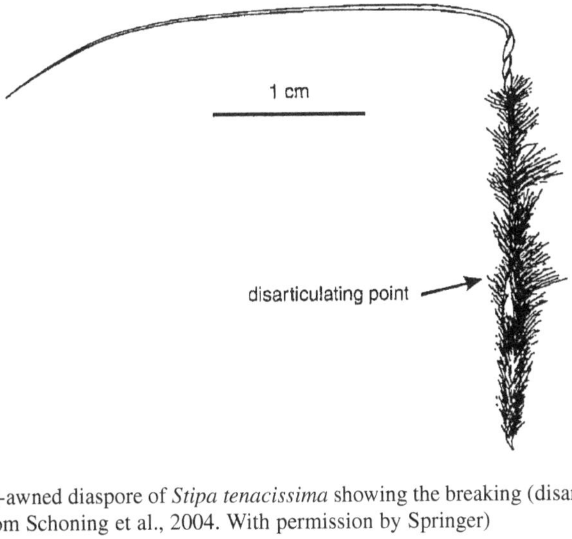

Fig. 11.6 Long-awned diaspore of *Stipa tenacissima* showing the breaking (disarticulating) point (Reproduced from Schoning et al., 2004. With permission by Springer)

is buried beneath the soil surface and is no longer at risk of predation. In the absence of granivory, it is difficult to explain why the seed would shed its awn once it has settled into the soil. For this reason, Shoning et al. believe that pressure from harvesting ants may select for seeds that shed this structure.

11.2.3.3 Parthenocarpy

An additional benefit of producing seeds with a hard external structure is that the plant can leave some of them empty, and these empty fruits can function as decoys that protect viable fruits from predation. Creating seedless fruits is called parthenocarpy, and it comes with a fitness cost to the plant. One benefit of parthenocarpy is that it may divert granivores from consuming viable seeds, although this theory is best studied in non-entomophagous species. The parsnip webworm, *Depressaria pastinacella*, feeds preferentially on empty diaspores of the wild parsnip under laboratory conditions (Zangerl et al., 1991). This is likely because there are fewer allelochemicals found in the inviable seeds (Zangerl et al., 1991). The preference for inviable seeds may be manifested in the field, where there is a lower relative abundance of inviable seeds on plants that are infested with *D. pastinacella* (Zangerl et al., 1991). Another example where parthenocarpy may reduce granivory is in *Pistacia terebrinthus* (Traveset, 1993). In this system, inviable fruits remain a red color, and viable seeds begin as red and turn green over time. A eurytomid wasp preferentially lays her eggs on red fruits, but the only red fruits present late in the season are inviable. In this way, the plant effectively avoids seed predation late in the season, although at the cost of producing the inviable fruits.

There are instances where granivorous entomophages are fooled by parthenocarpic diaspores. In *Ambrosia trifida*, there is a preference by insects and rodents for involucres that contain viable seeds, and these seeds are removed at a greater rate than unfilled involucres (Harrison et al., 2003). However, at the end of 12 months there are still some viable seeds that aren't removed, suggesting that foraging rates are less than 100% efficient for the granivores. Harrison et al. go on to speculate that one possible reason that viable involucres persist is that the unfilled involucres distract the granivores from finding the viable seeds. Also, *Messor* ants collect both viable and inviable seeds of *Franseria*, but workers distinguish between the full and empty involucres in the nest, and discard the empty involucres without opening them (Went et al., 1972). In both cases, it isn't clear whether the production of inviable seeds is actually improving the fitness of the plant, although both of these studies are in line with the hypothesis that parthenocarpic fruit are diverting granivores from viable seeds.

11.2.4 Seed Covering

The testa, or seed coat, is often a hard, protective shell surrounding the seed and is derived from the integument (Bewley and Black, 1978). The testa is coated in waxes and fats, which are embedded in several layers of thick-walled cells. Also, cells of the seed coat can contain calcium oxalate or calcium carbonate, and likely impede granivory by insects to some degree (Bewley and Black, 1994). Another conspicuous feature of the of the seed coat is the hilum, or region where the seed attaches to the funiculus (Bewley and Black, 1994). This hilum is often developed and may aid insects in transporting the seed (Fig. 11.7). Occasionally, the thickness and structural characteristics of the seed coat vary among seeds from a single plant. Harper et al. (1970) describe the different seed coats of *Chenopodium album*, a species which produces both smooth, thick-walled black seeds and reticulated, thin-walled brown seeds.

Testament to the defensive importance of the outer seed coverings are the frequent observations that seeds are only susceptible to granivory after their outer coverings have been softened by soaking in water. I have observed that seeds ordinarily unacceptable to granivorous carabids (*Harpalus pensylvanicus*, *Anisodactylus sanctaecrucis*) and field crickets (*Gryllus pennsylvanicus*) were readily consumed after soaking in water. Cardina et al. (1996) found that dry velvetleaf seeds were unacceptable to all tested insect granivores (including *Harpalus pensylvanicus* and *Amara cupreolata*) in the laboratory. However, when the seeds were soaked in water for 8 h, nearly all the granivore species consumed them. Another example is with the granivorous carabid, *Clivina impressefrons*, which in the laboratory consumed corn and soybean seeds only after they had been softened with water (Pausch and Pausch, 1980). In one study of granivory on 96 seed species by the fire ant, *Solenopsis invicta*, Ready and Vinson (1995) found that imbibing the seeds increased the rate of consumption over dry seeds in 72% of species.

In line with the hypothesis that the defensive capabilities of the seed coating is reduced through the imbibition of water, seed consumption by granivores in the field tends to be proportionally related to rainfall. Brust and House (1988b) found that seed removal was lowest early in the field season, when "*unusually hot, dry conditions*" prevailed. In another study, seed removal was tenfold higher in the week that received 5 cm of rainfall versus the two adjacent weeks in which no rainfall had occurred (Lundgren et al., 2006). Similarly, Cardina et al. (1996) found that seed predation on velvetleaf was highest during a season with particularly high rainfall, and they anecdotally associated the seasonal differences in seed removal with prevailing rainfall patterns.

Sometimes, imbibing a seed doesn't increase its susceptibility to predation, which suggests that other mechanisms are at play in seed defense beyond structural hurdles. Some seed species are not acceptable to *Solenopsis invicta* in the laboratory, even when their seeds coats are removed (Ready and Vinson, 1995)! All in all, 19% of tested seed species are resistant to attack by this fire ant. In another case, imbibing may actually protect the seed from granivory. *Amara pulpani* will not consume imbibed seeds of birch, although dry seeds are acceptable (Burakowski, 1967). A simple explanation to these studies is that the imbibing process has multiple effects on the seed's physiology. Thus, it becomes difficult to interpret whether imbibed seeds frequently become more palatable to granivores because of the softening of the seed coat, or because the phytochemical defenses are being metabolized to provide energy for the developing seedling.

11.2.5 Seed Chemistry

Of all plant tissues, secondary chemicals are frequently most concentrated in the seeds. However, the diversity of plant defensive chemicals tends to be restricted in the seeds versus in photosynthetic plant tissues. This is because the seed is a very discreet entity, and must prioritize its contents in order to maximize successful germination (it has to pack light). For this reason, seeds often rely on secondary chemicals that can either be easily translocated to the developing seedling, or possess the dual function of providing a nutrient source to the developing seedling (Janzen, 1976; McKey, 1979). McKey postulates that the secondary chemicals present in seeds should fulfill one or more of the following criteria.

> *(a) ability to be translocated to the growing seedling, which requires a degree of water-solubility and the absence of autotoxic effects during translocation. (b) ability to be readily metabolized to compounds having property (a). (c) convertibility to nutrient substances.*

Possessing defensive phytochemicals that have the dual function of nutrition for the developing seedling is a creative response by plants to the restrictions of size limitation and the need to prioritize seed contents.

The differences in physiological requirements and tolerances of insects for different phytochemicals means that a staggering array of potential defensive chemicals exists in plants (Janzen, 1979). A case in point is terpenoids, which is one of the most diverse classes of natural compounds known (30,000 have been isolated from plants), and many are reputed to have defensive properties (De Luca and St. Pierre, 2000; Mabry and Gill, 1979). But less obvious toxic compounds may act against specific predators in some situations. For example, the protein amino acids tryptophan, cystine, and methionine, normally essential to insect development, are lethal to the bruchid, *Callosobruchus maculatus*, whose host seeds contain low levels of these chemicals (Janzen et al., 1977). Thus, an understanding of the physiology of granivorous entomophages is critical to identifying putative defensive compounds in the seeds that they consume.

Undoubtedly, the chemical defenses of seeds play a role in their interactions with granivorous entomophages; however, these interactions have received disturbingly little attention from researchers. Granivorous entomophages may behave in a number of ways in order to avoid toxic seeds. First, they may consume only seeds that are non-toxic, or consume only small quantities of the toxic seeds (Carroll and Janzen, 1973). Also, insects may detoxify the defensive chemicals in the seeds (Carroll and Janzen, 1973). Finally, seed defenses are sometimes aggregated in certain compartments of the seed, and the insect may avoid the most toxic seed tissues (i.e., by consuming only the food body) (Carroll and Janzen, 1973). Steele et al. (1993) found that tannins are 12.5% and 85% higher near the embryo than in other portions of the acorns in *Quercus phellos* and *Q. laevis*, respectively. It is worth exploring whether the seed-feeding behaviors presented in Chapter 10 may represent a response to the distribution and abundance of certain plant defensive chemicals.

Because phytochemical constituents of seeds differ in their toxicity to insects, and the relationships among seed secondary metabolites and granivorous entomophages are largely unexplored, it is difficult to conduct a comprehensive review of this topic. This notwithstanding, the potential importance of these interactions for the life histories of the plants and insects are profound. Thus, an introduction to the topic is presented by highlighting what is known on the diversity and functions of some of the major classes of known seed defensive compounds, particularly with regard to insect granivores. Hopefully, this lays the groundwork for future efforts on the chemical ecology of interactions between seeds and granivorous entomophages.

11.2.5.1 Non-Protein Amino Acids

In addition to the 20 amino acids that form proteins, there are over 400 non-protein amino acids, many of which are often abundant in and readily isolated from seeds (Bell, 1984; Fowden, 1964; Rosenthal and Bell, 1979). While non-protein amino acids are present in most plants, they are particularly well studied in plant groups that are characteristically high in nitrogen, notably the Leguminoseae and Fabaceae (Baker, 1989; Seigler, 2003). Non-protein amino acids can interfere with protein synthesis, since some are structural analogues to protein-forming amino acids and function as antagonists to these molecules. As a consequence, some non-protein amino acids are toxic to a wide range of animals (Rosenthal and Bell, 1979). Mimosine, oxalyl-amino acids, canavanine, hydroxytryptophan, dihydroxyphenylalanine (L-dopa), β-cyanoalanine, and homoarginine are a few examples of non-protein amino acids with known toxicity to insects (Bell and Janzen, 1971; Bell and Tirimanna, 1965; Bewley and Black, 1994; Rosenthal and Bell, 1979; Seigler, 2003). These amino acids are readily transported from the seed to the germinating plant (McKey, 1979), and so are likely an efficient defensive compound that can be used by the developing plant for alternative metabolic functions.

Through interruption of protein synthesis and causing enzyme inhibition and dysfunction, various symptoms of toxicity of non-protein amino acids are neurological problems, musculature dysfunction, and even death (to name only a few) (Rosenthal and Bell, 1979). One non-protein amino acid that is particularly well studied with respect to insects is L-Dopa. L-Dopa occurs at concentrations of 1.5%, 2.75%, and 6–9% of dry weight in *Cassia*, *Canavalia*, and *Mucuna* species (Leguminosae), respectively (Bell and Janzen, 1971; Vadivel and Janardhanan, 2005). High concentrations of L-Dopa are associated with reduced levels of granivory by insects (Bell and Janzen, 1971), likely because this chemical interferes with proper neurological functions (Rosenthal and Bell, 1979).

11.2.5.2 Alkaloids

A large group of defensive phytochemicals, many of which have amino acid precursors, are the alkaloids (Robinson, 1979). Alkaloids are an extremely diverse group of low-molecular weight chemicals that are united by the presence of a heterocyclic ring that contains nitrogen (Robinson, 1979). Classes of alkaloids are assembled based on various similarities in their chemical structures (Robinson, 1979), and at least 12,000 alkaloids occur in plants (De Luca and St. Pierre, 2000). The physiological effects of alkaloids on insects are likely quite diverse, and their exact modes of action are not well understood. However, Robinson (1979) categorizes the possible action of alkaloids (as well as other small-molecule defensive chemicals) as interfering with:

1. Mechanisms of DNA replication, RNA transcription, and protein synthesis
2. Membrane transport processes, both active and passive
3. Enzyme inhibition and activation
4. Receptor site-binding for endogenous chemical transmitters
5. The conformation of macromolecules not included in 1–4

One effect commonly observed with alkaloids is the interference of neuroendocrine functions (Robinson, 1979). As a case in point, nicotine functions by binding to acetylcholine receptors in animal brains (De Luca and St. Pierre, 2000).

Although some plant seeds are markedly devoid of alkaloids, other species are typified by their high seed-alkaloid contents (Robinson, 1979; Seigler, 2003). Indeed, seeds of more than 500 species (of around 3,000 tested) possess alkaloids (as identified using Wagner's reagent) (Earle and Jones, 1962; Jones and Earle, 1966; Janzen, 1969; Willaman and Schubert, 1961). These surveys support the general theory that 20% of all plants contain alkaloids in at least some of their tissues (De Luca and St. Pierre, 2000). When looking at taxonomic and geographic trends in alkaloid content of plants, it appears that alkaloids are more widespread in dicotyledonous plants (Robinson, 1979), and are strongly associated with the tropics and low altitudes (Robinson, 1979). Within a plant, alkaloids tend to accumulate in

> *(1) very active tissues, (2) epidermal and hypodermal tissues, (3) vascular sheaths, and (4) latex vessels* (Robinson, 1979)

This being said, in many plants alkaloids are most abundant in seeds compared to other plant tissues (Bell, 1984). Alkaloids are readily transported through the vascular tissues of plants (De Luca and St. Pierre, 2000), and so alkaloids found in seeds are not necessarily synthesized there (Robinson, 1979). Furthermore, alkaloid contents of seeds decrease upon germination, suggesting that some may be useful in the early metabolic activities of the plant (McKey, 1979; Robinson, 1979). In addition to the physiological status of the seed, a number of factors influence the concentration of alkaloids in the seeds of a plant species. One recent example of the environmental effects on alkaloid content involves *Colchicum autumnale* (Poutaraud and Girardin, 2005). In this plant, alkaloid content of seeds varies significantly across several collection sites, ranging from 0.79% to 1.14% of seed dry weight. It appears that microelements (Ca and Co, specifically) in the soil, and subsequently in the seeds, are well correlated with alkaloid content of this plant.

11.2.5.3 Lectins

For nearly 120 years, lectins, or phytoagglutinins, have been recognized as influential protein constituents of seeds (Liener, 1979). Specifically, lectins are defined as

> *sugar binding proteins or glycoproteins of non-immune origin which are devoid of enzymatic activity towards the sugars to which they bind* (Pusztai et al., 1983)

Lectins are related to each other in that they bind to sugar residues of polysaccharide moieties frequently found on cell walls, causing the cells to agglutinate (Bewley and Black, 1978, 1994; Liener, 1979; Murray, 1984a; Pusztai et al., 1983). Lectins are very specific to the carbohydrates to which they bind, and so potentially can target very explicit groups of cells (Liener, 1979; Vaughan, 1983).

Lectins are widespread throughout the plant kingdom, and many hundreds are known (Liener, 1979; Murray, 1984a). Leguminosae are renowned for having a high diversity of lectins, although other plant families also possess these substances (Baker, 1989; Pusztai et al., 1983; Vaughan, 1983). Within a plant, lectins tend to be most concentrated in storage tissues, especially seeds (Liener, 1979; Pusztai et al., 1983). Indeed, they can comprise up to 30% of total seed proteins (Murray, 1984a). Mung bean seeds have 2,670 hemagglutinin units of lectins g^{-1} of dry weight (Mubarak, 2005).

It appears that some lectins have a dual function in plants, defending the seeds from granivory and serving as an important source of nutrients for life processes. Many lectins have an antifeedant activity for many animals, including a number of insects (Liener, 1979). For instance, the lectins of leguminous seeds repel the obligate granivore, *Callosobruchus maculatus* (Janzen et al., 1976). In addition to this defensive property, lectins appear to be an important source of nutrients for many plants. As a case in point, the lectin content of seeds readily decreases as the plant germinates, suggesting that the developing plants are using these proteins as reserves or key nutrients in their life processes (Liener, 1979).

11.2.5.4 Protease Inhibitors

> *Natural proteinase inhibitors are proteins or polypeptides that bind very specifically and tightly to enzymes that split peptide bonds of proteins, resulting in the inhibition of the proteolytic activities of these enzymes.* (Ryan, 1979)

Enzymes targeted by protease inhibitors are often endopeptidases that depend upon a serine residue in the active site, and protease inhibitors function by mimicking the substrate of the target enzyme (Haq et al., 2004; Murray, 1984a). For this reason, these inhibitors tend to be very specific for certain target enzymes (Haq et al., 2004).

Within plants, constitutive protease inhibitors can be found throughout the plant, but they are most frequently isolated from plant storage organs, where many occur in high abundance (Koiwa et al., 1997; Murray, 1984a; Ryan, 1979). The relative abundance of these protease inhibitors is usually less than 10% of total protein content of a seed (Murray, 1984a). Two examples of contents are 15.8 Trypsin inhibitor units mg^{-1} of sample in mung beans (Mubarak, 2005), and 13.5–65.4 units mg^{-1} in seven other legume species (Vadivel and Janardhanan, 2005). Because of their potential use in genetically engineered plants, inducible proteinase inhibitors that aggregate to herbivore damage receive considerably more attention than the constitutive inhibitors present in seeds (Haq et al., 2004; Kessler and Baldwin, 2002; Koiwa et al., 1997).

Three roles that these inhibitors play in seeds are (1) as a proteinaceous form of energy storage, (2) in regulating endogenous proteolytic activities, and (3) in the seed's defense against granivory (Murray, 1984a; Ryan, 1973). The most abundant proteinase inhibitors in plants appear to have little function in regulation of plant proteins (Ryan, 1979). More often, these proteinase inhibitors serve the seed by arresting the digestive enzymes of attacking organisms, notably insects (Mikola, 1983; Ryan, 1979). For instance, serine proteases (such as trypsin and chymotrypsin) are not used by plants, and so the production of large quantities of their inhibitors serves no intuitive endogenous function. However, most insects rely on serine (trypsin and chymostrypsins) and cysteine proteinases (Haq et al., 2004; Koiwa et al., 1997). In susceptible insects, inhibitors of serine proteases reduce feeding rates and result in amino-acid starvation, with symptoms including delayed growth, decreased longevity, and death (Lopes et al., 2004; Ryan, 1973). Thus, the facts that serine inhibitors, notably those of trypsin and chymostrypsin, are the most commonly described protease inhibitors found in plants (Janzen et al., 1976; Kessler and Baldwin, 2002; Koiwa et al., 1997; Mikola, 1983; Murray, 1984a; Ryan, 1973; Vadivel and Janardhanan, 2005), and that insects are frequently susceptible to their action, provide compelling evidence that the immediate primary function of these proteins is in defense. However, there is extensive evidence that, upon germination, many protease inhibitors are degraded to amino acids that serve in protein synthesis for the developing seedling (McKey, 1979).

11.2.5.5 Phenolic Compounds

There are a number of phenolic compounds that have defensive properties, including certain flavonoids, anthocyanins, tannins, and lignins (Bennett et al., 2004; Sanchez-Tinoco and Engleman, 2004; Siegler, 1979; Swain, 1979). Certain polyphenolics, such as flavonoids and tannins (or their derivatives) are the reason that many seeds are darkened (red or black) (Islam et al., 2003; Lattanzio et al., 2005).

Phenolics are quite diverse in the seeds of some plants (Sudjaroen et al., 2005). Tamarind seeds have 12 different polyphenolic compounds (Sudjaroen et al., 2005). Polyphenolic extracts comprise 0.1–0.15 g g^{-1} of the testa in different *Phaseolus vulgaris* hybrids (Islam et al., 2003). One group of polyphenolics that is particularly well studied in the case of seeds is tannins. Using the ferric chloride assay, Earle and Jones (1962, 1966) detected tannins in the seeds of more than 300 plant species from an array of families. Tannin-producing species are quite abundant in some plant communities; a case in point being the Indian cloud forest where nearly all plant species analyzed produce tannins of one fashion or another (Mali and Borges, 2003). Tannin content of seeds varies substantially among plant species. For instance, tannin contents comprise 3.3, 4.6, and 66–86 mg g^{-1} of dry seed in *Vigna radiata*, *Vigna minima*, and *Acacia* species, respectively (Falade et al., 2005; Lattanzio et al., 2005; Mubarak, 2005). *Quercus rubra* acorns contain 5–15% tannins by weight, but *Q. alba* acorns contain less than 2% tannins (Smallwood et al., 2001). Finally, in grapes the tannins are aggregated in the seeds and occur at concentrations of 2.3 g kg^{-1} of fruit (Souquet et al., 2000).

McKey (1979) postulated that many phenolics (especially tannins) are localized in the seed coats of the seed because they are not readily metabolized or water soluble, and so are not easily utilized by the germinated seedling. Indeed, several studies suggest that these compounds are more abundant in the seed coat than in the endogenous portions of the seed. For instance, tannins are restricted to the seed coat in several leguminous species (Aparicio-Fernandez et al., 2005; Vadivel and Janarhanan, 2005, and references therein). Other phenolics, including coumarin and chlorogenic acid (and their derivatives), anthocyanins, and flavonols are located at appreciable levels in the testas of a number of plant seeds (Aparicio-Fernandez et al., 2005; Bewley and Black, 1994; Briggs et al., 2005). In spite of these common trends, phenolics and tannins are not universally present at higher levels in the seed coat than in the endogenous tissues of the seed. In *Uapaca kirkiana*, all phenolic acids and tannins are present at higher concentrations in the embryo versus the seed coat (Muchuweti et al., 2006).

Phenolics function defensively through diverse means. Lignins lend structural rigidity to plant cell walls and are indigestible by many insects (Swain, 1979). Tannins bind to soluble proteins, nucleic acids, and starches, thereby interrupting enzyme functions and rendering nutrients unavailable to predators (Muchuweti et al., 2006; Swain, 1979; Vadivel and Janardhanan, 2005). Flavonoid pigments are shown to have bitter qualities that may repel insect granivores (Harbourne, 1979). Although the effects of phenolics on granivorous entomophages are not well documented, the interactions of phenolics with other seed feeding insects have been studied. For instance, in *Arabidopsis* seeds, the phenolics *p*-coumaric and ferulic acid, which were bound in the seed coat, were found to predominate and play a role in repelling attack by the *Sitodiplosis mosellana* larvae (Ding et al., 2000).

11.2.5.6 Cyanogenic Glycosides

Cyanogenic glycosides are commonly found in plants, and prove toxic to herbivores when hydrogen cyanide (HCN) is produced as a byproduct of hydrolysis (Conn, 1979). HCN is toxic to animals because it binds to metabolic proteins involved in cellular respiration (Bell, 1984; Conn, 1979; Jones et al., 2000). In order for catabolism of the glycoside to occur, both the substrate and enzyme (namely, β-glycosidase) must come in contact with one another (Conn, 1979; Seigler, 2003). Thus, by storing these different components in different cell types or organelles within a cell, the plant only produces the toxin when the damaged cell contents mix (Bell, 1984; Conn, 1979). Two other classes of plant-produced cyanogenic compounds frequently cited in the literature are cyanogenic glucosides, which are simply a derivative of the glycoside, and cyanolipids, which are only found in the Sapindaceae (Siegler, 1979).

Cyanogenic glycosides are widespread in the plant kingdom, occurring in more than 2,650 species of ferns, gymnosperms, and angiosperms (Halkier and Gershenzon, 2006; Haque and Bradbury, 2002). At least 23 cyanogenic glycosides are known from plants (Conn, 1979). Within a plant, cyanogenic glycosides are more frequently reported from the flowers, leaves and stems than the seeds (Gebrehiwot and Beuselink, 2001; Haque and Bradbury, 2002; Ruiz and Sotelo, 2001). Jones et al. (2000) note that developing seedlings often have the highest concentrations of cyanogenic glucosides in cyanogenic plants. This being said, there are some plants, such as apple and flax seed, that are renowned for having seeds that possess cyanogenic glycosides (Haque and Bradbury, 2002). In legumes, cyanogenic glycosides are typically present at less than $30\,\mu g\ g^{-1}$ of seeds (Umoren et al., 2005). A notable exception is lima beans, which have concentrations that range from $2.10–3.1\,mg\ g^{-1}$ of seed (Umoren et al., 2005).

11.2.5.7 Glucosinolates

The biosynthetic pathways for creating cyanogenic glucosides and glucosinolates have several similarities, and it is hypothesized that plants that create glucosinolates are predisposed evolutionarily to create cyanogenic glucosides (Halkier and Gershenzon, 2006). Similar to the cyanogenic glycosides, an enzyme (myrosinase) needs to hydrolyze the glucosinolate in order for the toxic compounds to be released, and plants store the enzyme and substrate independently of each other (Halkier and Gershenzon, 2006; Van Etten and Tookey, 1979). The main functional chemicals that are freed upon hydrolysis are the thiocyanate ion, nitrile, and isothiocyanate (Halkier and Gershenzon, 2006; Van Etten and Tookey, 1979). Some related compounds, the non-cyanogenic nitriles, function as enzyme inhibitors that act on cross linkages in elastin and collagen (Bell, 1984).

Glucosinolates occur almost exclusively in plants in the order Capparales, and are particularly abundant in the Brassicaceae (Halkier and Gershenzon, 2006; Van Etten and Tookey, 1979). Glucosinolates are often found at their highest concentrations in seeds (Bennett et al., 2004; Halkier and Gershenzon, 2006; Van Etten and Tookey, 1979). However, these chemicals are not synthesized in the seed, but appear to be transported there from maternal tissues (Bennett et al., 2004; Halkier and Gershenzon, 2006). Several different types of glucosinolates can be isolated from a single plant species (Bennett et al., 2004), and can occur at levels ranging from 0.4–3.0% of seed weight (Van Etten and Tookey, 1979).

11.2.5.8 Oxalate

As mentioned above in the discussion on ascorbic acid, oxalate is commonly found in seeds, and may serve a defensive role against granivory. Oxalate appears to work defensively in two ways. First, oxalate readily binds to Ca^{2+}, Fe^{2+}, and Mg^{2+}, robbing these nutrients from the granivore (Adebooye and Phillips, 2006; Franceschi and Nakata, 2005; Rassam and Laing, 2005). Once formed, calcium oxalate forms hardened crystals of various shapes and sizes. These crystals are structural defenses against herbivorous animals (Franceschi and Nakata, 2005). Oxalate was present in *Milletia obanensis* seeds at 375 μg g^{-1} (Umoren et al., 2005), and in the seeds of *Acacia* species at 22–26 μg g^{-1} (Falade et al., 2005).

11.2.5.9 Saponins

Another group of defensive phytochemicals is a class of triterpenoid glyciderides, commonly referred to as saponins. Saponins are typified as having both carbohydrate and lipid components, giving them the hydrophobic/hydrophilic tendencies for which they are known (Applebaum and Birk, 1979; Osbourn et al., 1998). Saponins are widespread in plants, occurring in more than 500 plant species in 80 families (Applebaum and Birk, 1979). Interestingly, many cereals and grasses are deficient in saponins, relative to dicotyledonous plants (an exception being oats) (Osbourn et al., 2003). Within a plant, saponins are found in many tissues, including seeds (Applebaum and Birk, 1979). Often, a suite of saponins is isolated from a single plant, with some chemicals being more abundant than others (Wei et al., 2005). One example of saponin content is with *Milletia obanensis*, which has around 100 μg g^{-1} of seeds (Umoren et al., 2005).

Saponins are implicated as having multiple properties that are repellent or toxic to insects (Applebaum and Birk, 1979). First, saponins are bitter tasting compounds, and mammalian herbivores avoid feeding on them (Bell, 1984). In addition, these chemicals reduce surface tension of fluids, and readily foam when in contact with water (Applebaum and Birk, 1979; Bell, 1984). Furthermore, some cells (particularly erythrocytes) lyse when they come in contact with saponins (Applebaum and Birk, 1979). While toxic effects of saponins are known, and the interactions of saponins with various binding sites are understood, the exact mechanisms by which these chemicals function in repelling and killing insects are still poorly understood.

11.2.6 Mucilaginous Secretions

Another type of seed defense that is observed in xeric habitats is the production of a glue-like mucilaginous coating around the seed. Eight percent of plant species collected in Eastern California and Nevada produce seeds that create mucilage

(Young and Evans, 1973). Mucilaginous species are found in the Euphorbiaceae, Labiatae, and Oanagraceae. The families with the greatest frequency of mucilaginous species are the Brassicaceae (8 of 11 genera have species that produced mucilaginous seed coats) and Plantaginaceae (the three species examined produced mucilage). Compositae and Graminae do not produce mucilage. A number of functions are attributed to this type of mucilaginous secretion. First, the coating may improve germination by allowing the seed to retain water more easily and aid in germination in the absence of soil coverage (Young and Evans, 1973). The sticky mucilage may also aid in dispersal by birds and humans (Young and Evans, 1973). However, the best documented function of the mucilaginous seed coat appears to be in deterring post-dispersal granivory by insects, especially ants.

The mucilage protects the seeds by binding to sand grains, rendering the seeds less apparent to foraging ants, and through adhering the seed to the soil surface. In the case of *Salvia columbariae*, sand binds to the mucilaginous seeds, making it difficult for *Messor pergandei* ants to recognize the seeds (Fuller and Hay, 1983). Consequently, the ants hardly touch sanded seeds until the dry, clean seeds are fully exploited (Fig. 11.7).

In a series of experiments conducted in the Negev desert, Gutterman and Shem-Tov (1997) found that the seeds of four mucilaginous species were protected from foraging ants by adhering to the soil surface. Dry seeds were collected by *Messor rugosus* within 2.25 h of being placed in the field. In contrast, when seeds were

Fig. 11.7 Seeds of *Salvia columbariae* showing dry seeds, the mucilage produced by wetted seeds, and the sand covering that adheres to the seeds after the mucilage dries (Reproduced from Fuller and Hay, 1983. With permission by the Ecological Society of America)

dampened, ants had considerably more trouble removing the seeds of three of the four species tested. For example, a mean of 63 and 142 h was required to remove 50% and 100% of the *Plantago coronopus* seeds, respectively, when they were wet. After the wetted seeds had dried, they were collected more quickly than wet seeds, although the dried mucilage slowed the removal rate compared to seeds that had never been dampened. Thus, in both of the studies discussed here, the mucilaginous seed coat would only function during years of normal rainfall, when monthly precipitation is sufficient to induce the glue secretion.

11.3 Conclusions

In a nutshell (no pun intended), seeds are an extremely rich source of nutrients, but are defended with a variety of structural and chemical features that undoubtedly influence granivory rates. However, the influence of seed nutrition and defense on the life histories of granivorous arthropods and the plants involved is still poorly understood, both from the perspective of the plant and the insects that consume them. Given the extensive list of entomophagous insects known to feed on seeds, and their importance as biological control agents of both insect pests and weeds, it is striking that more research does not focused on the relative nutrition of seeds and insect prey with respect to the biological control agents that consume them. One complication to this is the plasticity in the nutrition and morphology of seeds, even within localized plant populations. Nevertheless, insects are known to self-select their diets based on their physiological needs, and so seed nutrition and defense undoubtedly will play a role in which seeds are consumed and when they are preferred. In conclusion, the relative nutrition and defensive characteristics of weed seeds influence the range of granivorous entomophages that consume them and function as their biological control agents (an issue that will be discussed more completely in Chapter 18).

Chapter 12
Seed-Associated Food Bodies

The oily substance [of the elaiosome] is very attractive to ants, and they search for seeds and fruits so furnished, and carry them to the nest, very frequently eating off the oil-body on the way and then dropping the seed. In these cases the seed itself is not eaten or injured, so that, even if it is carried to the nest, it may soon germinate and grow there.

Ridley, 1930

A staggering number of plants produce seeds with food bodies that are attractive to insects (especially ants) which serve as potential dispersal agents. In dispersed seeds, this is believed to be a true mutualism, termed myrmecochory (myrmex = ant, kore = dispersed; in Greek) (Handel and Beattie, 1990). The seed provides a nutritious tidbit of food in exchange for being moved to a more suitable germination site, often in or near an ant nest. Myrmecochorous plant species tend to occur together within habitats, probably because they rely on the same species of ants as their dispersal agents (Beattie and Culver, 1981). And ants are choosy about which diaspores they will carry; less than 10% of seed species without food bodies are transported by ants in North American forests (Beattie and Culver, 1981). In contrast, 50% to 100% of the seeds of forest species with food bodies are carried away, clearly showing that ants perceive the nutrient-rich food body on the seeds.

Sernander (1906) was one of the first researchers to thoroughly study and classify different types of ant-dispersed seeds. His system involves 11 different classes of seed-associated food bodies, mostly based on differences in their structure (these are translated into english by Ridley, 1930). In its simplest form, the food body is an oily coating to the seed, such as is observed in *Eucalyptus* species that are transported by ants to their nests (Berg, 1975). From this rudimentary form, food bodies diversify into various shapes and sizes, sometimes growing even larger than the seed itself (Fig. 12.1).

The food body has consistently been shown to be the attractive part of the insect dispersed diaspore. Time and again, research shows that removing the food body renders a seed unattractive to dispersing ants (Brew et al., 1989; Cuautle et al., 2005; Garrido et al., 2002; Manzaneda et al., 2005) (Fig. 12.2). Interestingly, this tenet sometimes holds true even when the ants are granivorous. For example,

J.G. Lundgren, *Relationships of Natural Enemies and Non-Prey Foods*,
Progress in Biological Control 7,

Published by Springer Science+Business Media B.V.

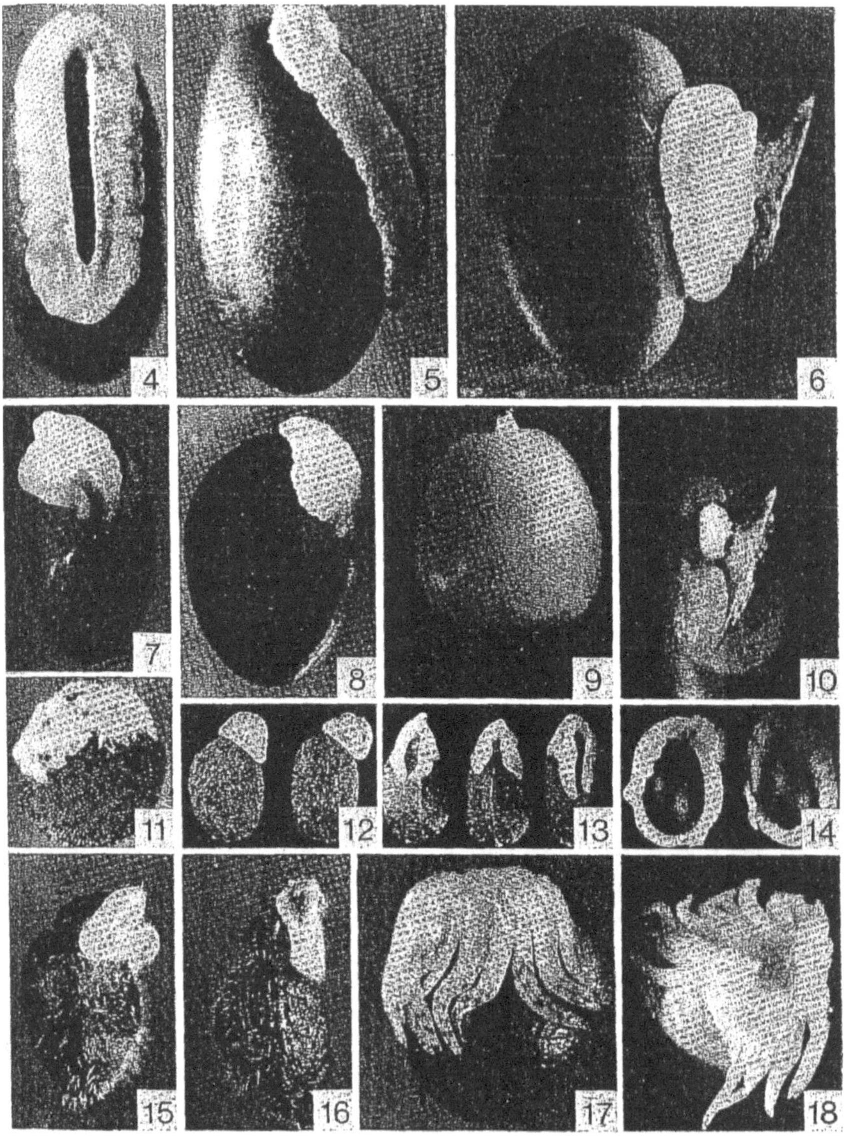

Fig. 12.1 Examples of myrmecochorous diaspores. *Hovea rosmarinifolia* (4,5), *Hardenbergia violaceae* (6,10), *Dillwynia juniperina* (7,8), *Monotoca scoparia* drupe (9), *Caesia vittata* (11), *Rulingia pannosa* (12), *Thomasia petalocalyx* (13), *Hibbertia serpyllifolia* (14), *Tetratheca* (15, 16), *Hibbertia obtusifolia* (17, 18) (Reproduced from Berg, 1975. With permission of CSIRO Publishing)

Messor pergandei, ordinarily a harvesting species, only collects the seeds of *Datura discolor* when the food bodies are present (O'Dowd and Hay, 1980). What's more, 99.9% of these seeds from these diaspores are placed on the ant middens unharmed after the food body was removed.

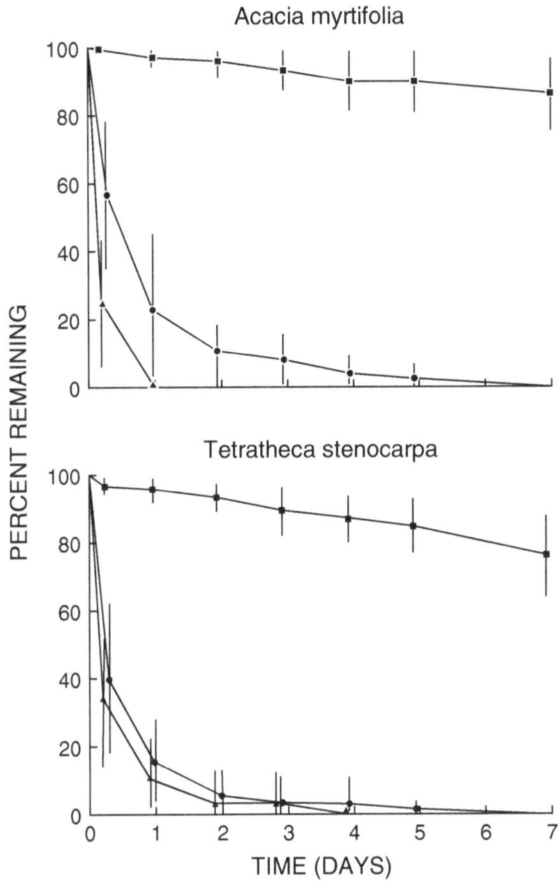

Fig. 12.2 The mean (± SEM) percent of intact diaspores (•), seeds with elaiosomes removed (▪), and eliasomes alone (▲) of two myrmecochorous plant species that were removed from seed dishes by foraging ants (Reproduced from Brew et al., 1980. With permission by Springer)

The nutrition of food bodies mimics that of insect prey, and so these structures are attractive to a range of entomophagous insect species. Beattie (1985) explains that providing food as a reward to beneficial insects and dispersal agents is a risky business. First the food has to be of nutritional suitability for a range of potential dispersal agents, which vary interspecifically as to their nutrient requirements. Preferably, the nutrients are less attractive to thieves. Second, different ontogenic stages of a dispersal agent differ with regard to their nutritional requirements. For instance, the nutritional requirements of an ant colony changes over a growing season, which challenges the development of an 'ideal' nutritional profile for a particular dispersal agent. However, once accomplished, insect-dispersed seeds are clearly strongly selected for within certain habitats. Harper et al. (1970) says it well:

> *It is only a short step from seed predation which reduces the reproductive success of a plant to the predation of fleshy receptacles and pericarps in which a seed escapes predation and gains dispersal.*

The topic of seed-associated food bodies is elevated to the chapter level rather than being included as part of Chapter 11 for two reasons. First, the food body is often nutritionally distinct from the associated seed, arguably making it a distinct type of non-prey food from the seed. Second, a host of entomophagous insects consume food bodies associated with seeds that are otherwise entirely uninterested in the seeds by themselves. Thus, the insects associated with insect-dispersed seeds overlap with, but are not synonymous with granivorous entomophages.

In this chapter, the diversity and distributional trends in plants that produce food bodies will be discussed, as will some of the structural and chemical aspects of foods bodies and the plants that produce them. Furthermore, the entomophagous species that are attracted to and consume the food bodies of these seeds will be showcased. Finally, the evolutionary implications for ant-dispersed seeds will be discussed, including why these relationships may have developed in the first place and how invasive species are disrupting their delicate balance.

12.1 Diversity of Plants that Produce Seed-Associated Food Bodies

Worldwide, more than 3,100 plant species in more than 87 genera and 80 families are known to be ant-dispersed (Handel and Beattie, 1990; Holldobler and Wilson, 1990). Most myrmecochorous species occur in the southern hemisphere; Australia (1,500 species) and Southern Africa (1,300 species) to be precise (Beattie and Hughes, 2002; Berg, 1975; Davidson and Morton, 1981). The northern hemisphere, notably U.S.A. and Europe, have around 300 myrmecochorous species (Beattie and Hughes, 2002; Berg, 1975). The plants residing in the two hemispheres have very different life history syndromes (Table 12.1).

Myrmecochorous plants are a diverse guild, and this method of dispersal has evolved multiple times within the plant kingdom. Indeed, it appears that habitat

Table 12.1 Life history syndromes of myrmecochorous plants in the northern (USA and Europe) and southern (Australia and Southern Africa) hemispheres (Adapted from Beattie and Hughes, 2002)

	Northern hemisphere	Southern hemisphere
Number of species	~300	~2,800
Typical growth form	Understory herbs	Woody shrubs
Vegetation type	Deciduous forests	Dry sclerophyll woodland and heath (Australia), fynbos (South Africa), often fire-prone, on low-nutrient soils
Elaiosome	Soft, collapsible, desiccates and becomes unattractive within a few days	Hard, long-lived, may retain attractiveness for several years
Seed presentation	Clumped on ground as peduncle bends, some species ballistic	Seeds dropped singly, often preceded by ballistic expulsion

type drives the creation of ant-attractive food bodies rather than the phylogenetic placement of a plant species (Westoby et al., 1991). Plants that produce seeds with food bodies are particularly well represented in temperate mesic forests, desert shrublands, and sclerophyll forests (Holldobler and Wilson, 1990). Also, a growing body of literature shows that tropical forest species also produce seeds with food rewards. For example, tropical *Calathea* seeds bear food bodies and are attractive to a host of ant species (Horvitz and Beattie, 1980).

Within a habitat, species that produce food bodies can account for a substantial proportion of the plant community. For instance, 30–40% of species produced seeds with food bodies in the understory community of temperate forests (Beattie and Culver, 1981; Gorb and Gorb, 1999; Wolff and Debussche, 1999), the community surrounding *Fagus sylvatica* (Wolff and Debussche, 1999), and arid shrublands of Australia and South America (Holldobler and Wilson, 1990). The consistency in this upper limit to the number of myrmecochorous plant species across localities and habitats suggests that there is a maximum number of ant-dispersed plants that a habitat can carry (Beattie and Culver, 1981). There also may be a successional trend in the dominance of ant-dispersed plants (Wolff and Debussche, 1999). From Wolff and Debussche's work, it appears that early successional habitats favor dispersal by harvesting ants and are less likely to provide food rewards to non-granivorous species for their dispersal services. As the habitats in this Mediterranean system age and become more forested, the number of plant species that provided food rewards increases.

12.2 Physical Characteristics of Food Bodies

A number of morphological structures associated with the seed can function as an insect-attracting food body. Affiliated with the seed itself are organs like the aril, hilum, strophiole, and caruncle, which all give rise to food bodies in various plant groups (Berg, 1975, 1979; Harper et al., 1970). Arils are simply outgrowths of the seed coat, and can come in many shapes and sizes. Sometimes arils contain chemicals that are found nowhere else on the plant (Bewley and Black, 1978). A caruncle is an aril emanating from the micropyle. A strophiole is a warty outgrowth of the hilum, which is itself an outgrowth of the seed where the funicle was attached (Bewley and Black, 1978). The seed appendages, in addition to functioning as food bodies for dispersal, may also be important in the detachment of the seed from the fruit (Berg, 1979). Affiliated with the fruit, food bodies can be derived from the receptacle, the style-base, or the glandular disc surrounding the base of the ovary (Berg, 1975; Harper et al., 1970). Also, the fruit itself can function as the food body (Berg, 1975; Harper et al., 1970).

The food body frequently is light colored in contrast with the darkened true seed (Berg, 1975; Soukup and Holman, 1987), and its resistance to desiccation varies among habitats. The food bodies associated with seeds in dry habitats of the southern hemisphere tend to be hard and resistant to desiccation (Berg, 1975). However, those myrmecochorous seeds of temperate forests have fleshy food bodies that dry out and lose their function within a few days of being released from the parent plant (Gorb and Gorb, 1999) (Table 12.1). For these seeds, time is of the essence. A common misconception is that myrmecochorous seeds have a harder seed coat than other seeds in order to dissuade consumption by ants once the food body is gone. Although seed strength is certainly well-correlated with the ability of an ant to consume a seed, myrmecochorous seeds do not require more strength to crush than seeds without food bodies (Rodgerson, 1998). Of course, only a single study explores this topic, and the hypothesis that myrmecochorous seeds have superior mechanical defenses to normal seeds merits further attention.

The size of the food body can be quite variable, and this attribute influences the attractiveness of the seed to ants. The size and shape of the food body can even vary substantially among seeds from a single plant, as well as among conspecifics from different localities (Garrido et al., 2002). It is often the case that seeds with high food-body to seed size ratios are more attractive to dispersing ants (Lanza et al., 1992; Manzaneda et al., 2005; Mark and Olesen, 1996; but see Garrido et al., 2002). The food body itself may be useful as a type of handle for manipulating the seed, i.e., which facilitates carrying it back to the nest (Horvitz, 1981).

12.3 Chemical Composition of Food Bodies

The food bodies associated with seeds are a nutritious resource, containing a range of nutrients. The most extensive survey on the nutritional composition of food bodies is that of Bresinsky, who found that 76% of seed species (41 tested) contain at least some sugars, including glucose, fructose, saccharose, and xylose (summarized by Beattie, 1985). Protein, starch, vitamin B1, and vitamin C are found in 16, 9, 19, and 20 of the food bodies, respectively. Other work indicates that food bodies are a source of water; *Acacia myrtifolia* and *Tetratheca stenocarpa* are 8.7% and 14% water by weight, respectively (Brew et al., 1989). But the most prevalent nutrient class associated with food bodies, and the best studied, is lipids. Lipids are present in 93% of the species tested by Bresinsky, and this lipid fraction can account for a large proportion of the dry weight of seeds. Although it varies among species, up to 51% of the dry weight of food bodies associated with ant-dispersed seeds are lipids (Beattie, 1985; Brew et al., 1989). The abundance of oils in seed-associated food bodies, and the critical function of these oils in attracting insects, has led to these appendages being dubbed elaiosomes (Brew et al., 1989).

A growing body of literature supports the claim made by Carroll and Janzen (1973) that the food body of seeds is an analogue to dead insects. Although the nutritional composition of food bodies varies substantially among congeners (Lanza et al., 1992), one fascinating convergent pattern that is possessed by nearly all food bodies across phylogenies and geographic regions is in their fatty acid profiles. The fatty acid profiles of food bodies contain high fractions of palmitic, oleic, and sometimes linoleic acids (Bebawi and Campbell, 2004; Hughes et al., 1994;

Kusmenoglu et al., 1989; Lanza et al., 1992; Pizo and Oliveira, 2001; Soukup and Holman, 1987). Interestingly, the associated seeds typically have a different profile, which are consistently lower in palmitic acid and higher in arachidic or linoleic acids (Bebawi and Campbell, 2004; Hughes et al., 1994; Soukup and Holman, 1987). Comparing the fatty acid profiles among insects, food bodies, and their associated seeds, reveals that the fatty acid distributions of the food bodies more closely mirror those of insect prey than those of the seeds (Hughes et al., 1994). In fact, some food bodies are statistically indistinguishable from insect prey. Table 12.2 gives a summary of the fatty acid profiles of insects, and 22 species of food bodies and their accompanying seeds.

A number of researchers have determined that the lipid fraction of food bodies function as the attractants used by the ants that disperse them (Brew et al., 1989; Midgley and Bond, 1995). Ricinoleic acid was one of the first chemicals found to elicit a feeding response in ants, specifically *Lasius fuliginosus* (Bresinsky, 1963, as presented in O'Dowd and Hay, 1980). Subsequent work has not duplicated the attractiveness of this substance in other ant-food body interactions. Rather, research has focused primarily on the diglyceride fraction of the food body (Kusmenoglu et al., 1989; Skidmore and Heithaus, 1988), particularly in the outermost tissues (Marshall et al., 1979). Specifically, 1, 2-diolein was frequently encountered in the food bodies of seeds and has been repeatedly shown to elicit carrying behavior in a number of ants (Brew et al., 1989; Hughes et al., 1994; Marshall et al., 1979). Oleic (Brew et al., 1989; Marshall et al., 1979; Skidmore and Heithaus, 1988) and linoleic (Lanza et al., 1992; Skidmore and Heithaus, 1988) acids have also been shown to have attractiveness to some ants, although not at the same intensity as the 1, 2-diolein. Previous research has found that a number of these fatty acid chemicals elicit other behaviors in ants. For instance, ricinoleic acid was found in ant larvae, and oleic acid elicited corpse removing behaviors (Brew et al., 1989; Holldobler and Wilson, 1990). The diglyceride 1, 2-diolein has been identified as a major neural lipid found in most insects (Hughes et al., 1994). Thus, it seems probable that the plants evolved to mimic these behavior-eliciting chemicals in ants (and other

Table 12.2 The fatty acid profiles of seeds and eliasomes (n = 22), compared with various insect prey (156 species in 7 insect orders)(From Thompson, 1973)

	Fatty acid profiles (%)		
	Insect prey	Eliasomes[a]	Seeds[a]
Palmitic (16:0)	23.7	18.66	6.34
Palmitoleic (16:1)	7.8	9.95	0.87
Stearic (18:0)	4.54	2.52	1.35
Oleic (18:1)	33.3	36.99	42.74
Linoleic (18:2)	10.2	25.12	22.70
Linolenic (18:3)	8.6	1.78	0.67
Arachidic (20:0)	No data	1.26	25.57

[a]Data summarized from the seed-elaiosome diasporas evaluated by Soukup and Holman (1987), Hughes et al. (1994), and Bebawi and Campbell (2004).

entomophagous insects) in order to exploit them as dispersal agents. Confirming this notion, ants tested from numerous geographic locations and phylogenetic placements have been shown to be universally attracted to these chemical cues, even under new associations (Midgley and Bond, 1995).

12.4 Diplochory and Seed Cleaning

A single plant species takes advantage of multiple methods of dissemination, and myrmecochory is frequently regarded as a secondary dispersal mechanism (Berg, 1975). For example, Baker and O'Dowd (1982) describe the influence of plant density in *Hypochaeris glabra* on the production of two seed types that have different dispersal agents. At low densities, the plants produce beaked achenes, each with a long pappus, that are wind disseminated. When plants become more crowded, they produce a greater proportion of unbeaked achenes that adhere to animal fur and feathers. Ants are only able to carry and disseminate the unbeaked achenes. So this plant has multiple methods for seed dispersal that are plastic according to the density of the parent plants.

Vertebrates are often messy eaters, and fruit that attracts birds and bats is frequently dependent on ants as secondary dispersal agents that promote seedling germination. Seeds of these vertebrate-disseminated fruits that are discarded beneath the parent tree are extremely susceptible to degradation by microorganisms. The fate of these fruits is in the control of ants, which respond to fruit resources in a number of ways. In a case study that focuses on the fleshy arils (with high lipid content) of *Cabralea canjerana*, the largest of the ant species bring entire fruits to their nests, or deposit the fruits beneath the leaf litter (Pizo and Oliveira, 1998). Other ant species clean the fleshy portions of this fruit from the seed. A third group buries the seeds before removing the food body. The ants take several hours to clean the seeds, but in the process they improve the germination of the cleaned seeds over those with their fruits unhampered (Pizo and Oliveira, 1998). For example, removal of the aril of *C. canjerana* by ants increases the germination rate from 42% to 100% (Pizo and Oliveira, 1998), a trend observed in other tropical fruit-fall systems as well (Guimaraes and Cogni, 2002; Horvitz, 1981; Ohkawara and Akino, 2005). The cleaned seeds are much less susceptible to microbial infection. In fact, it appears that the ants actually treat the seeds in a way that improves germination over simply mechanically removing the fruit tissue (Ohkawara and Akino, 2005). Research on *Nephelium lappaceum* and a suite of ants suggests that the omnivorous ants apply some anti-fungal substance to the seed (Ohkawara and Akino, 2005). Even after all of the fruit pulp is removed, the ants will continue to lick the surface of the seed (Ohkawara and Akino, 2005). The ant-tended seeds have less fungal hyphae with fewer spores (Fig. 12.3). So, not only do ants frequently avoid consuming the myrmecochorous seeds, but they may actually be promoting germination through treating the seeds against seed-degrading microorganisms.

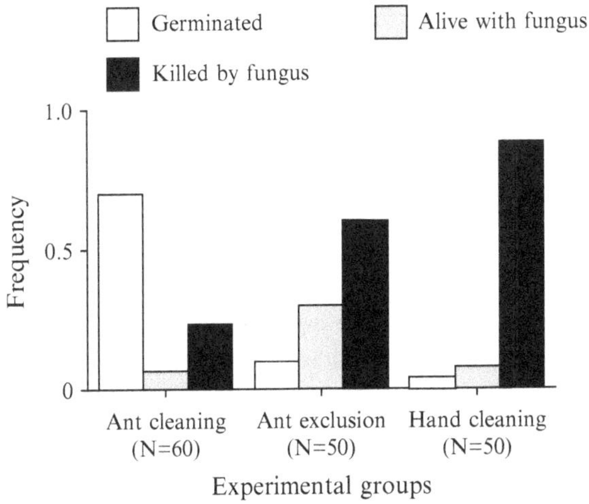

Fig. 12.3 Effect of seed cleaning (by *Pheidole plagiaria* or by hand) on *Nephelium lappaceum* seed performance. Data are frequency of germination, alive, and dead with fungus infection (Reproduced from Ohkawara and Akino, 2005. With permission by the Japan Ethological Society)

12.5 Ants

The attractiveness and suitability of food bodies for entomophagous insects are best documented in ants. A wide range of ant species collects diaspores with food bodies, including many carnivorous species. Indeed, often a large proportion of an ant community is attracted to seeds with food bodies. From the plant's perspective, these ants are of varying quality as dispersal agents, transporting the seeds to different lengths and endpoints and collecting the seeds at different rates.

The ants that consume and collect diaspores with food bodies or an oily covering are a diverse lot, both behaviorally and phylogenetically (Carroll and Janzen, 1973). Ants in the subfamilies Ponerinae, Mirmicinae, and Attini consume the food bodies associated with seeds and aid in the dispersal of tropical seeds (Pizo and Oliveira, 2001). Specific genera associated with seed dispersal are *Rhytidoponera*, *Pheidole*, *Iridomyrmex*, *Odontomachus*, *Formica*, *Myrmica*, and *Aphaenogaster* (Berg, 1979; Handel and Beattie, 1990). An assorted group of feeding guilds is also attracted to and consumes the food bodies of seeds. Some entomophagous ants that are not attracted to any other seeds will collect those of myrmecochorous plants (Handel and Beattie, 1990; Horvitz and Beattie, 1980). Even harvester ant species that normally consume the seeds will disseminate myrmecochorous diaspores (Handel and Beattie, 1990; O'Dowd and Hay, 1980), sometimes leaving the seeds unharmed. For example, *Solenopsis geminata geminata* consumes the food body associated with *Calathea* seeds, but does not damage the seeds themselves (Horvitz, 1981).

The communities that collect seeds with food bodies are generally quite speciose. Often the number of ants found in association with a given diaspore species ranges from five to more than 20 species (Boulay et al., 2005; Brew et al., 1989; Cuautle et al., 2005; Horvitz, 1981; Ohara and Higashi, 1987). Often it is the case that a minority of ant species within these communities accounts for the majority of seed removal. For instance, *Aphaenogaster japonica* and *Myrmica ruginodis* account for more than 93% of the observations of seed removal for *Trillium* species in Japan (Ohara and Higashi, 1987), and *Rhytidoponera metallica* removes 57% of all *Acacia terminalis* and *Dillwynia retorta* seeds in Australia (Hughes and Westoby, 1992). Hughes and Westoby (1992) classify the ant community associated with seed depots of *Acacia terminalis* and *Dillwynia retorta* into two guilds. Guild 1 consists of small ant species that are best described as harvesters. Guild 1 species don't disperse the seeds very far, and they frequently consume the seeds that they collect. Guild 2 species are larger and better suited as dispersal agents. These species tend to specialize only on the food bodies attached to the seeds. From these studies, community structure associated with a given seed species clearly influences its ultimate success in germination.

First, not all seeds are equally attractive to different members of an ant community. For example, the removal rates of *Calathea* seeds range from 0% to 100% among eight species of ants evaluated in the laboratory (Horvitz, 1981). It appears that in certain habitats, different ant tribes specialize on different classes of food bodies; Ponerinae species remove arillate seeds, and Attini specialize on pulpy seeds (Pizo and Oliveira, 2001). The case of *Iridomyrmex purpureus* colonies, which deposit an estimated 334,000 seeds of bellyache bush per hectare in their middens, demonstrates that some ant-seed interactions are quite strong (Bebawi and Campbell, 2004). Once a source of diaspores is identified by an ant colony, the efficiency with which it collects and disperses the seeds fluctuates depending on the ant species involved.

As mentioned above, time is of the essence for a seed that relies on ants as a dispersal agents. In temperate habitats, the food body quickly desiccates and becomes unattractive to ants. But regardless of habitat, the longer the seed remains on the surface beneath the parent plant, the more exposed it is to a range of mortality factors. Although it differs greatly among ants, seeds, and environmental conditions, myrmecochorous seeds are generally removed within several hours of their deposition (Gorb and Gorb, 1999; Hughes and Westoby, 1990). Berg (1975) found that most patches of myrmecochorous seeds are entirely or largely exploited within minutes of their placement. Ants with large colonies, especially carnivorous species, tend to be the most efficient dispersal agents (Gorb and Gorb, 1999).

Once collected, different ant species vary in their dispersal capabilities, which in turn affect the final dispersion pattern of the next generation of plants. In a comprehensive analysis of dispersal distances for ant-dispersed seeds (2,524 seed species), Gomez and Espalader (1998) reveal several patterns of ecological relevance. First, the worldwide mean dispersal distance of myrmecochorous seeds is 0.96 m (Fig. 12.4), however there are significant effects of geography and vegetation on seed dispersal distances. Seeds in the northern hemisphere are dispersed about half the distance as those in the southern hemisphere. Also, dispersal distances are significantly farther in sclerophyllous than in mesophyllous vegetation. These trends may be redundant, since sclerophyllous and mesophyllous vegetations predominate in

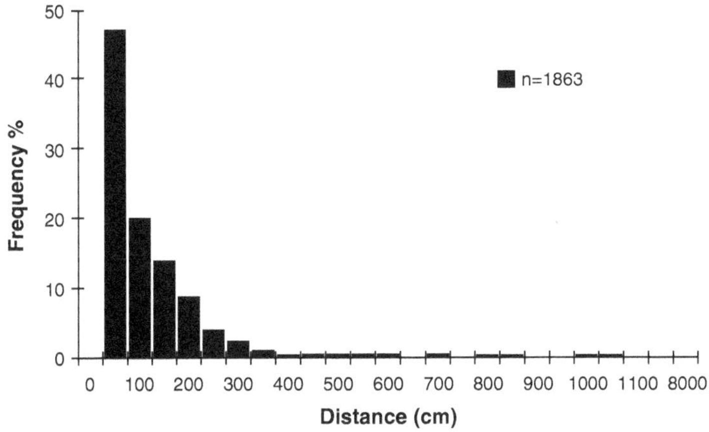

Fig. 12.4 The frequency of dispersal of myrmecochorous diaspores from worldwide reports (Reproduced from Gomez and Espalader, 1998. With permission from Blackwell)

the southern and northern hemispheres, respectively. Furthermore, different ant subfamilies carry seeds different distances; Formicinae, Ponerinae, Dolichoderinae, and Myrmicinae carry seeds mean distances of 2.26, 1.41, 0.87, and 0.79 m, respectively (Gomez and Espadaler, 1998). This may be a reflection of the relative size of these different ant taxa (Horvitz and Schemske, 1986). Larger ponerine ants (longer than 1.0 cm) move the seeds of *Calathea* species up to 10 m (Horvitz, 1981; Horvitz and Beattie, 1980), but smaller ants (0.2–1 cm) struggle to move seeds 10 cm from the nest (Horvitz and Beattie, 1980). The smallest ants of this community (shorter than 0.2 cm) are entirely unable to move the seeds, but will consume the food bodies on the spot (Horvitz, 1981; Horvitz and Beattie, 1980). This raises the idea that some ants may be predisposed to be better dispersers of seeds. A case in point is the twelve ant species that disperse the seeds of *Acacia terminalis* and *Dillwynia retorta* (Hughes and Westoby, 1992). In this study, two dispersal agent 'classes' exist, that of *Rhytidoponera metallica* which remove seeds a mean of 1–1.5 m (and a maximum of 11 m), and the rest of the ant species, which primarily move the seeds less than 0.5 m.

Ant-selected seeds are frequently returned to the nests, which vary in their suitability as germination sites. Specifically, nest characteristics that fluctuate among ant species include the colony architecture, nest depth, and levels of specific nutrients. Nest traits may explain the relative germination rates of a seed species collected by different ant species. For example, significantly fewer *Viola* seeds germinate on the nests of *Lasius* (5%) compared with those placed on *Formica* nests (65%) (Culver and Beattie, 1980). Also, once the seed reaches the nest, dispersing ants may decide to consume the seed as well as the food body, particularly when other foods are limited (Boulay et al., 2005). Experimentally limiting the resources available to ants shows that some species are inherently more likely to consume the seed portion of the myrmecochorous diaspore. Confining three ant species in their nests with *Hardenbergia violacea* diaspores reveals that *Aphaenogaster longiceps* and a *Pheidole* species are prone to consuming the seeds in their nests, whereas *Rhytidoponera metallica* do not (Hughes and Westoby, 1992).

Hughes and Westoby (1990) point out that in habitats where myrmecochorous plants abound, most seeds are readily collected by ants. So, it does not do for the mother plant to worry about whether or not her seeds will be collected. Rather, it may be more important for the mother plant to influence *which* ant species collect her seeds, since not all ants are created equally. Given the importance of dispersal to the success of the developing seedling, and the intense variability among ant species as dispersal agents, it is remarkable that plants are not more ant-specific in the timing of seed shed or the morphology and chemistry of their seeds (Handel and Beattie, 1990). Future research will hopefully address the measures that plants take in order to influence which dispersal agents will ultimately collect their progeny.

12.6 Other Entomophagous Insects

With all of the attention paid to the interactions among ants and seeds with food bodies, the importance of seed-associated food bodies to other carnivorous insects is largely overlooked. I do not challenge the importance of ants in the dispersal of these seeds. After all, ants are frequently the most abundant insects occurring in habitats that abound with insect-dispersed seeds. However, it is worth pointing out that the mechanisms that render these food bodies attractive to ants also function in haling down other entomophagous insects, particularly in temperate ecosystems. These other entomophages can be superior dispersal agents to the ants, but they can also interfere with seed dispersal. In either case, the importance of these species in the dispersal of seeds merits considerably more attention.

Yellow jackets, (including *Vespula flavopilosa, V. maculifrons*, and *V. vulgaris*) are thought to be important dispersal agents of the seeds of several northern hemisphere plants that produce seeds with food bodies. In fact, Jules (1996) goes so far as to coin a name for this type of seed dispersal: vespicochory. Yellow jackets disperse the seeds of *Trillium catesbaei, T. cuneatum, T. ovatum, T. undulatum*, and *Vancouveria hexandra* (Jules, 1996; Pellmyr, 1985; Zettler and Spira, 2001). The wasps quickly locate seeds of these species, often chewing into the fruits to collect the seeds still in the pods. With seed in tow, the wasps alight on a nearby tree, and cut the food body from the seed which falls to the ground. In other cases the wasp carries the entire diaspore out of sight. The wasps are highly efficient at foraging for these seeds, collecting 94% of them within 1 h of their placement (Zettler and Spira, 2001). Also, the wasps recruit nestmates to seed sources, in order to fully exploit the patch. When the wasps alight on a tree, the mean dispersal distance of the seed is around 1.4 m (Zettler and Spira, 2001). However, a large proportion of the seed groups are carried more than 30 m out of sight of the parent plant (Jules, 1996; Pellmyr, 1985; Zettler and Spira, 2001). Yellow jackets nest in the ground, so it seems likely that any seeds that are returned to the nest will be buried.

Furthermore, the nests of yellow jackets are abandoned in the autumn, and so uneaten seeds may remain in the nest below ground (Jules, 1996). The suitability of these wasp nests as germination sites is unknown. Pellmyr (1985) believes that seed pods of *Vancouveria hexandra* may dehisce while the seeds are still green and on the plant as an adaptation to dispersal by wasps. The phenomenon of vespicochory appears to be fairly common in western North America, and may also occur in parts of Europe (Jules, 1996). More research is needed before the importance of this type of dispersal to plants, and the importance of food bodies to yellow jackets, can be fully assessed.

Another entomophagous group that is attracted to the food bodies associated with seeds is the Carabidae. *Synuchus impunctatus* acts quite 'ant-like' in its response to the diaspores of *Melampyrum lineare* (Manley, 1971). The beetle forages slowly along the soil surface until in comes across the *M. lineare* diaspore. Then, it manipulates the diaspore with its mandibles, and if a spongy white caruncle is associated with the seed, the beetle quickly runs the entire diaspore beneath a nearby branch or log. Manley postulates that the beetle cannot carry seeds without the food body. Under cover, the beetle consumes the food body, leaving the seed intact. It appears that the beetle has some fidelity to the same seed cache, returning multiple times to the same location. Also, Manley points out that the beetle must come in contact with the seed, and that no volatile chemical cues appear to be operating in this system. If the seed cache can be empirically established as a "safe site" for seed germination, then the *Synchus-Melampyrum* relationship is a mutualistic one.

In other instances, carabids may actually hinder the dispersal of insect-dispersed seeds (Higashi and Ito, 1991; Ohara and Higashi, 1987). At night, a suite of carabids are found consuming the food bodies of *Trillium* seeds, including *Carabus arboreus*, *C. arcensis conciliator*, *C. opaculus*, *C. granulatus yezoensis*, *C. japonicus*, *Cychrus morawitzi*, *Damaster blaptoides rugipennis*, *Pterostichus thunbergi*, *P. subovatus*, *P. orientalis*, *Synuchus melantho*, *S. nitidus*, and *Harpalus ussuriensis*. Of these, *Carabus arboreus* is the dominant consumer of the food bodies of *Trillium kamtschaticum* (43 of 57 observations), and *Pterostichus thunbergi* and *Carabus japonicus* are dominant consumers of *Trillium tschonoskii* (28 and 23 of 91 observations, respectively). It takes the *Carabus* species less than 20 s to consume the food body, and carabids destroy approximately half of the food bodies available at a site in a single night. As a result of this carabid feeding, the seeds of *Trillium* are found to be mostly clumped at the sight of their deposition near the plant. Only 15% are dispersed by local ants, which are not attracted to the seeds when the food bodies are removed. Thus, other entomophagous species are attracted to the food bodies designed to draw in ants, and can interfere with the ultimate dispersal of these plants.

With this in mind, another explanation for the wide disparity in the number of myrmecochorous plant species between the northern and southern hemispheres that has not been previously pointed out may exist. Specifically, the epigeal insect community, particularly with regard to entomophagous species, may include disproportionately more non-ant omnivores in temperate forests. If non-dispersing entomophagous species and ants are equally attracted to the enticing chemistry of the food body, then one would not expect this mode of dispersal to proliferate here. However, where ants (or other nest-building social insects) dominate the insect communities, one might expect plants that specialize on this form of dispersal to accumulate and diversify.

12.7 Ants as Dispersal Agents

> *It frequently happens on the wild hillside, the position of a nest of* Atta barbara *is indicated by the presence of a number of plants growing on or round the kitchen midden, which are properly weeds of cultivation, and strangers to the cistus- and lavender-covered banks of the garrigue.* (Moggridge, 1873)

There are a number of testable hypotheses that may explain the evolutionary development of ant-dispersed seeds (Beattie, 1985; Holldobler and Wilson, 1990). These include factors inherent to the feeding and foraging behaviors of ants, and the intrinsic properties of ant nests. As Moggridge pointed out more than 130 years ago, ant nests are often easily identified in the landscape as being much different from surrounding vegetation. Indeed, even species that do not produce food bodies (and hence are not considered 'ant-dispersed') benefit from germinating upon ant nests. In one study, Rissing (1986) describes that six of the 17% of plant species (none of which produce food bodies) present in close association with the nests of *Messor pergandei* or *Pogonomyrmex rugosus* are more abundant in this local area than in surrounding vegetation. Furthermore, individuals of these favored species on ant nests grew larger and were capable of producing more seeds. Rissing (1986) goes on to postulate that this type of food-bodiless interaction with ant dispersal agents is the first evolutionary step toward the production of specialized ant-attracting seed structures. So, it appears that ant nests are better places to germinate. But why? Below are summarized some of the hypotheses on this topic. Certainly, these theories are not mutually exclusive, and in fact multiple mechanisms may be operating within individual systems.

12.7.1 Ant-Treated Seeds

One hypothesis is that the ants themselves are treating the seeds in such a way that promotes their germination. The simple act of removing the food body may be sufficient to improve the germination of ant-dispersed seeds (Bebawi and Campbell, 2004) (Fig. 12.5). However, removing the food body mechanically is not always sufficient to obtain the high levels of germination observed on ant nests, suggesting that the ants are doing something else to the seed that fosters germination. Germination of *Turnera ulmifolia* seeds was 26% in ant-treated seeds, whereas only 6% of those seeds with their food bodies mechanically removed germinated

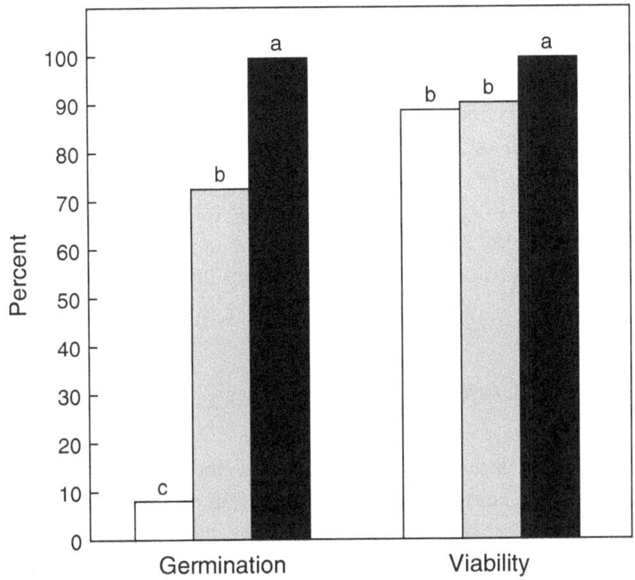

Fig. 12.5 Germination and viability of intact (open bars), non-carunculate (shaded bars) and *Iridomyrmex spadius*-discarded (solid bars) seeds of *Jatropha gossypiifolia*. Within seed variables, bars with the same letter are not significantly different (Reproduced from Bebawi and Campbell, 2004. With permission by CSIRO Publishing)

(Cuautle et al., 2005). Some seeds require scarification in order to germinate (Culver and Beattie, 1980; Karban and Lowenberg, 1992), and ants are known to apply anti-microbial substances to seeds that may also promote germination (see the section on seed-cleaning, this chapter).

12.7.2 Escaping Seed Mortality

Seeds within the shadow of the parent plant are susceptible to multiple sources of mortality, including predation and fire. Several studies indicate that seed predation is lower upon ant nests. A case in point is *Datura discolor*, which experiences an 8-fold reduction in predation when deposited in the middens of *Messor pergandei* and *Pogonomyrmex californicus* compared to when they are left beneath the plant canopy (O'Dowd and Hay, 1980). *Miribalis hirsuta* seeds also suffer much greater predation near the original plant. At a distance of 0.7 m from the parent plant, seed predation is virtually non-existant (Platt, 1976). Still, seeds upon ant middens are not immune to predation, and there is considerable variability in the rate of predation upon these 'safe sites' (Manzaneda et al., 2005). For instance, seed predation of *Helleborus foetidus* surrounding ant nests is no different from undispersed seeds at two sites in

Spain, is greater on ant nests than at one site, and is reduced at another (Manzaneda et al., 2005).

The propensity of burying seeds is well documented as reducing predation, but ants must bury the seeds above a specific depth (Manzaneda et al., 2005), or germination will be delayed or fail entirely (Cavers, 1983). Also, burying the seeds reduces mortality from fire, but only at depths of more than 1–2 cm (Berg, 1975; Hughes and Westoby, 1992). An ideal depth for burying seeds that balances the costs associated with reductions in germination with the benefits of fire and predation avoidance appears to be around 3 cm (Hughes and Westoby, 1992).

12.7.3 Avoiding Competition

Competition from the parent plant, conspecifics, and other flora is intense within the seed shadow, and being transported even a short distance may improve the chances of successful seedling establishment. Seedling survival of *Mirabilis hirsuta* is lower beneath the parent plant, and so being transported even a short distance to ant nests may be sufficient to improve chances of survival (Platt, 1976). Also, when the seedlings of *Carex pedunculata* are raised with other common woodland sedges, their fitness is reduced (Handel, 1978). This suggests that being moved to an ant nest where competitors are reduced has positive effects on this plant species. One final point is that in order for plant competition to be sufficiently reduced on ant middens, the ants must change nest sites periodically so that competition near their nests does not exceed that of the surrounding habitats (Beattie, 1985).

12.7.4 Providing Favorable Germination Sites

A large body of literature suggests that the soil nutrient profile of ant nests is considerably different from the surrounding habitat, and that these conditions may be favorable to the germination and establishment of ant-dispersed plants (Culver and Beattie, 1983). Ant nests differ from surrounding soils in particle sizes, and nutrient and water contents (Culver and Beattie, 1980). These conditions frequently promote plant germination and growth compared to surrounding soils (Bebawi and Campbell, 2004; Culver and Beattie, 1980).

Further evidence for the 'nutrient hypothesis' of Beattie (1985) comes from the fact that myrmecochorous species (especially shrubby plants, less than 2 m tall) are particularly abundant in habitats (deserts and xeric shrublands) that typically have infertile soils, particularly those low in potassium (Hughes et al., 1993; Westoby et al., 1991). Ant nests are often higher in nutrients critical to plant development (i.e., phosphorous and nitrogen) than surrounding soils (Beattie and Culver, 1981). Other types of animal dispersed seeds produce fleshy fruits with a relatively high ratio of potassium to nitrogen compared with ant-dispersed seeds

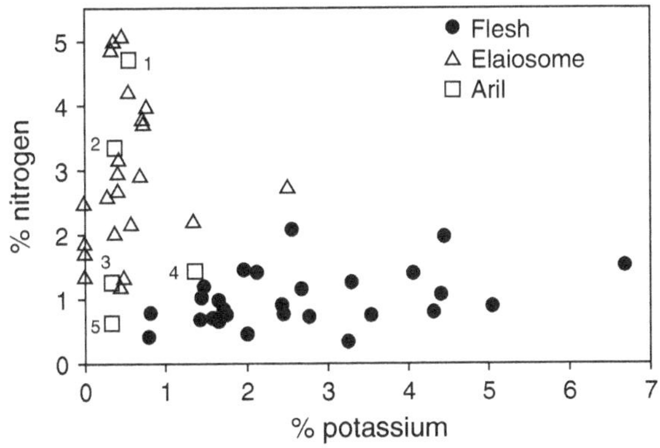

Fig. 12.6 Nitrogen and K content (% dry weight) in diaspores of 56 species (N = 22, elaiosome bearing species, n = 29 fleshy-fruited species, n = 5 arillate species) collected in Australia (Reproduced from Hughes et al., 1993. With permission by Blackwell)

(Hughes et al., 1993) (Fig. 12.6). Thus, seeds with food bodies are energetically cheaper to produce, and plants that display other forms of animal dispersal may not be able to acquire the nutrients necessary to excel in myrmecochorous plant-dominated habitats.

12.8 Invasive Species and Myrmecochorous Plants

The ecological interactions of invasive species with native flora and fauna are varied, but often the end result is at the expense of local biota. The interactions among myrmecochorous plants and ants are often affected by invasion processes. These interactions under two scenarios are documented, one where the myrmecochorous plant is the invader, and one where the endemic ants are displaced by exotic ones. Both situations demonstrate the fragility of plant-ant interactions and how a single species can upset long-standing ecosystem processes.

One important facet to the disruption of ant-dispersed plant communities is that the cues contained in the food body of myrmecochorous seeds appear to be universally attractive to ant species. For instance, South African and Australian ants are both attracted to the food bodies associated with Australian *Acacia* seeds (Midgley and Bond, 1995). Thus, invasive myrmecochorous plants are liable to find dispersal agents in recipient biotas. Such is the case with the *Jatropha gossypiifolia*, which is invasive in Australia (Bebawi and Campbell, 2004). A native *Iridomyrmex spadius* is involved in the population increase of this exotic plant, and facilitates the spread of its seeds across ordinarily inhospitable habitats (Bebawi and Campbell, 2004).

Invasive ants are well documented as disrupting native ant communities, and myrmecochorous plants suffer when the invasive ant is not as efficient in dispersing seeds as the native fauna. In a review of this topic, Ness and Bronstein (2004) found that invasive ants adversely affect these plants in a number of ways. First, the invasive ants may be slower at collecting the seeds or carry them a shorter distance than native species. In eight of nine cases examined, the invasive ants transport seeds a shorter distance than native species (Ness and Bronstein, 2004). Second, exotics may be seed predators, actually consuming the seeds in addition to the food bodies. Third, the invasive species may not bury the seeds after carrying them, leaving them prone to predation or unsuitable germination conditions. Finally, they may consume the food body without moving the seed from beneath the parent plant's seed shadow. Another important consideration is that invasive ants tend to be relatively small, and this limits which foods can be tackled individually (Ness and Bronstein, 2004). As a result, small seeded species tend to predominate the myrmecochorous communities in ant-invaded habitats (Ness and Bronstein, 2004). An illustration of the important impacts of invasive ants on ant-dispersed plants is found with the red imported fire ant, *Solenopsis invicta*, which is facultatively granivorous. Under field conditions, *S. invicta* move 100% of five of six myrmecochorous seed species (*Trillium undulatum*, *T. catesbaei*, *T. discolor*, *Sanguinaria canadensis*, and *Viola rotundifolia*, but only 33% of *Iris cristata*) to their nest within 12 h of seed rain (Zettler et al., 2001). Of these seeds, only 5–70% of seeds are recovered on the surface of ant nests, suggesting that many are destroyed by the ants. Furthermore, the majority of these collected seeds are scarified (Zettler et al., 2001). Thus, the invasion of *S. invicta* may have important deleterious effects on these North American ant-dispersed plants.

12.9 Conclusions

Clearly, the food bodies associated with seeds are an important source of non-prey foods for entomophagous insects. However, as is evidenced by the literature presented here, ants are almost the sole focus of the interactions among food bodies and entomophagous species. Attracting a single class of entomophagous insect to act as a dispersal agent is a risk-filled endeavor for a plant, particularly in habitats rich in insect species. More work should be conducted on the importance of food bodies to the life histories of entomophagous species other than ants, and how these alternative species function in the life processes of the plant.

The provision of food bodies, and reliance on ants for seed dispersal, also raises some questions concerning the science of the biological control of weed seed banks. As will be discussed in the Chapter 18, successful weed biological control is dependent on the spatial and temporal occurrence of seeds, as well as the granivorous insects that may be present in an agricultural system. To my knowledge, the influence of seed-associated food bodies on seed preferences by granivores and the level of granivory in managed habitats has yet to be studied, though would merit examination for some regions.

Chapter 13
Seed Preferences of Natural Enemies

The preferences for seeds by granivorous entomophages are a clear indication that these arthropods are able to sense differences in the seed and make selections based on their physiological and morphological circumstances. For example, harvesting ants often 'major in', or prefer, seeds of a certain species (Holldobler and Wilson, 1990; Rissing, 1981). For example, 90% of seeds consumed by *Messor pergandei* are of three species: *Mentzelia, Malvastrum*, and *Oenothera claviformis* (Tevis, 1958). Yet these three seed species account for only 8% of the seeds available in the habitat.

In fact, a staggering number of studies have been conducted on the preferences of granivores, both in the laboratory and in the field. Surprisingly, there have been very few attempts to synthesize the different factors responsible for these preferences. Many of the topics discussed in the preceding four chapters were invariably involved in the preferences of granivorous entomophages. Although sensory cues used in the decision-making process for granivory remain poorly understood, granivorous entomophages have been shown to respond to visual, tactile, and chemical cues derived from the seed. Seed acceptability and suitability for granivores form the basis for understanding the importance of seeds as a food source for predators of arthropods, as well as the utility of arthropod granivores as modifiers of plant communities.

13.1 Sensory Cues Involved in Seed Selection

It is plain from a number of studies that granivorous arthropods recognize specific seeds, although it is not always clear what senses are being alerted by the seed. For instance, granivorous arthropods can detect whether a sealed involucre of *Ambrosia trifida* contains a seed or not, and they prefer to remove those with seeds, but how they know remains a mystery (Harrison et al., 2003). Another case involves *Messor pergandei*, which manipulates the seeds of *Datura discolor* before preferentially selecting seeds with food bodies (O'Dowd and Hay, 1980). The senses employed by the ants that allow them to recognize seeds with and without food bodies

J.G. Lundgren, *Relationships of Natural Enemies and Non-Prey Foods*,
Progress in Biological Control 7,

Published by Springer Science+Business Media B.V.

remains a matter of speculation. Visual, tactile, and chemical cues are all implicated in distinguishing preferred seeds for granivorous insects.

Visual cues may be important in seed recognition, but there is little empirical evidence to bear this claim. The utility of visual cues for granivorous entomophages that forage primarily at night (i.e., many ground beetles, Brust et al., 1986) may be of limited value. Similarly so for immature stages that have simple eyes more suited for detecting light intensities rather than distinct images. However, numerous granivorous entomophages forage during the day, such as desert-dwelling harvester ants, and vision may play a role in the identification of seeds.

Tactile cues are used to assess the size and structure of a seed and determine its palatability. Ants will often antennate and handle a number of seeds before selecting one to carry back to the nest (Abramsky, 1983). Tactile cues may also operate in the larvae of carabids, *Trechus quadristriatus* and *Bembidion lampros*, which detect food through contact with food particles. These larvae walk with their mandibles spread apart and use their palps to feel potential food particles. The mandibles close on an object, and it is then torn with labial and maxillary spines. The palps then direct the food particles into the mouth (Mitchell, 1963). It seems likely that other cues, namely chemical, work in concert with tactile signals in detecting immobile foods (Mitchell, 1963).

Surface chemistry as a primary seed-selection signal by post-dispersal granivores has the greatest body of evidence in the literature. The chemistry of the seed covering can harbor chemicals that function as attractants to or in repelling granivores. An example of this occurs with seed extracts offered to *Solenopsis invicta* (Ready and Vinson, 1995). Three seed species (*Lactuca scariola*, *Panicum vargatum*, and *Triticum aestivum*) are damaged more frequently after they are rinsed with hexane and acetone, but *Pyrrhopappus pauciflorus* is removed less frequently after washing. Thus, some species rely on repellent chemicals to ward off fire ants, while others possess surface chemicals that are attractive to the ants.

Another simple yet creative study that further supports the importance of seed chemistry to recognition by granivores was conducted by Nickle and Neal (1972). Different seed species were of varying attractiveness to *Pogonomyrmex badius* in the laboratory. However, when the attractive seed (in this case, *Triticum*) was ground and placed on previously unattractive seeds (*Paspalum notatum* and *Carex scoparia*), these were avidly collected by the ant. This suggests that the attractive factor of wheat seed could be easily transferred to unpreferred species.

A similar story occurs with food bodies of seeds. Ricinoleic acid, oleic acid, and 1,2-diolein are all associated with attractiveness to different ant species (Bresinsky, 1963; Marshall et al., 1979). As mentioned in Chapter 12 on myrmechory, seeds may actually produce these attractive chemical signals to elicit ant tending. For instance, ricinoleic acid is produced by ant larvae, and oleic acid is a chemical given off by ant corpses, both of which require ants to carry them to other locations. Thus, by exploiting ant chemical elicitors, the seed is carried to more favorable germination sites.

In some ways insects are simple creatures and are easily fooled. An instance of mistaken identity in ants is recorded by Tevis (1958). He found that up to 22% of *Messor pergandei* returning to their nests carried inedible objects such as twigs, dried flowers, stems, and peduncles. Went et al. (1972) hypothesize that this unexpected assemblage of items was related to the seed type under collection by the foraging column, and may smell like the target seed. Thus they are accidentally collected, and it isn't until the more perceptive workers in the colony examine them that they are recognized as non-food items and surrendered to the midden outside the nest. Another example is noted by Moggridge (1873), who found cynipid galls within the granaries of *Messor barbarus barbarus*;

> *It seems difficult to understand how it comes that these galls are systematically placed among the seeds, for it was evidently no chance occurrence, and I can only conjecture that the worker ants may have brought them in and stored them under the impression that they were really seeds!*

These stories of mistaken identities by ants reinforce the importance of chemical signals in food recognition by seed predators.

13.2 Seed Traits Influencing Seed Selection

Characteristics of both the granivore and seed play a role in seed preferences. Sensory capabilities of the insects discussed above are clearly important in restricting the seed range that can be exploited, but what characteristics in the seeds are the granivores looking for? A wide range of different seed attributes have been identified as influencing the preferences of granivores. In the end, granivore-seed interactions are seldom simple, and are frequently regulated simultaneously by multiple seed traits and predator physiologies.

13.2.1 Seed Size

Seed size is one of the more important factors restricting the dietary breadth of seed consumers, and all else being equal, larger seeds should be preferred over smaller seeds. This is predicted by optimal foraging theories which state that an organism should try to maximize their net energy intake (Schoener, 1971). Thus, larger seeds should contain the most nutrition and theoretically be preferred over smaller seed species. Still, there is an upper limit to the size of seeds that a granivore can realistically manipulate and consume, and any trends in the size-specific preferences for seeds need to be qualified by stating the largest seed that is acceptable to a given granivore (Lundgren and Rosentrater, 2007). For example, Honek et al. (2007) found that, of 28 dicotyledonous seed species, the mean maximum sized seeds consumed by 30 species of granivorous carabids are less than 0.7 mg (the largest carabid in this study is 36 mg). Also in this study, the heavier beetles preferred to eat the heavier seeds. In sum, the working theory for granivorous insects is that they will prefer the largest seeds that fall below the threshold for acceptable size.

Some examples of this theory in action begin with the harvester ant, *Messor arenarius*, which prefers larger seeds to small ones in one desert ecosystem (Abramsky, 1983). Also, some ants will work together in order to exploit larger seeds than they could individually (Brown et al., 1979). In general, there appears to be a correlation between seed preferences and worker size (Holldobler and Wilson, 1990). This is easily seen in work by Pizo and Oliveira (2001) which shows that one ant species' ability to disperse medium sized seeds is much less (15 cm) than their ability to disperse smaller seeds (29 m). Thus, the ants seem to prefer the seeds that they can easily carry. Kaspari (1996) found that seeds carried were well correlated with head capsule size in five of seven harvester ant species, indicating that ants would frequently collect the largest seeds that they could handle (Fig. 13.1). Another concept illustrated with Kaspari's study is that the largest ants could consume a wider range of seed sizes, so the idea of collecting the largest seed possible may be best applied to smaller ant species whose food options are more limited. For instance, Tooley and Brust (2002) hypothesize that the upper size limit of seeds consumed by carabid beetles is related to their ability to carry the seed. Indeed, the largest seeds offered to *Solenopsis invicta* (more than 50 mg) are not easily manipulated and are left untouched in the laboratory (Ready and Vinson, 1995). A complication with the 'largest seed first' theory is when large seeds contain more indigestible material.

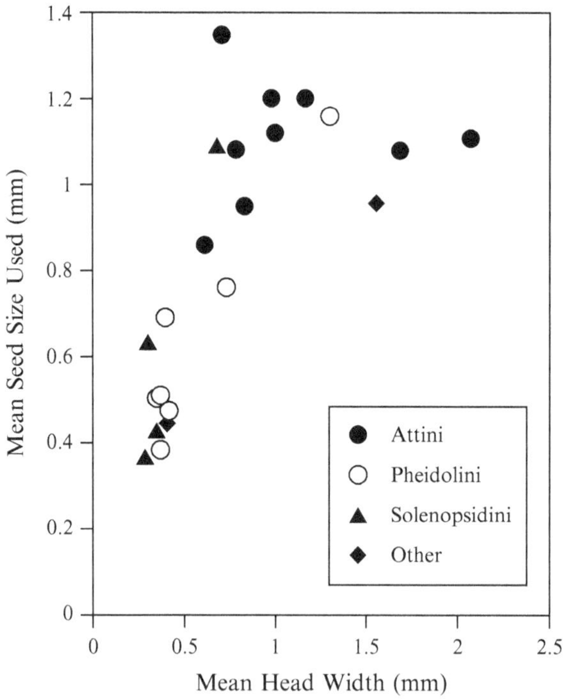

Fig. 13.1 The correlation between mean seed size and species mean size (Reproduced from Kaspari, 1996. With permission by Springer)

For instance, *Pascopyrum* seeds are large but the nutritious material comprises only a portion of the seeds' size (Whitford, 1978). In this case, *Pascopyrum* seeds are unpreferred by ants, in spite of their large size (Kelrick et al., 1986).

In differentially sized seeds, the assumption that 'all else is equal' is seldom realized, and as a result it is commonly observed that smaller seeds are preferred over the larger ones. In the carabid, *Amara cupreolata*, small seeds of *Agrostis*, *Poa annua*, *P. trivialis*, *P. pratensis*, *Festuca rubra* var. *commutata*, and *Lolium perenne* are consumed before larger seeds in the laboratory (Johnson and Cameron, 1969). In old-fields, ants favor the smallest seeds (Mittelbach and Gross, 1984). Larvae of the carabid *Harpalus rufipes* also prefer to consume small seeds over larger ones (Hartke et al., 1998). Hölldobler (1976) found that *Pogonomyrmex rugosus* consistently favors smaller seeds over larger ones, based on recruitment intensity. The preference is the same whether different sizes of the same seed or different seeds of varying size are offered, suggesting that seed size is the trait examined by the ants. In another study, Crist and MacMahon (1992) found that harvester ants prefer to remove seeds that range from 0.26–0.5 mg size range, particularly those of *Poa* species and *Alyssum desertorum*. This is in spite of 75% of seeds (namely *Bromus tectorum*) in this habitat falling within the size range of 1.51–2.50 mg.

Preferences for seed sizes are one distinguishing factor that allows multiple niches to coexist within a single habitat, and thus permits diverse granivore communities to arise. A case in point is an extensive series of experiments on seed preferences of carabids recently conducted in Eastern Europe (Honěk and Martinkova, 2003; Honěk et al., 2003). Two species of ground beetles, *Harpalus rufipes* and *Harpalus affinis* were offered the seeds from 65 dicotyledonous plant species in the laboratory in multichoice tests. These studies revealed that *H. rufipes* preferred seeds of medium size, and less so those of large and small-seeded species. *Harpalus affinis*, a smaller beetle, preferred small- and medium-weighted species, and heavy seeds were less preferred. Seed size described 25–33% of the variation in seed consumption, but seed hardness may also have been involved. Brust and House (1988b) presented data on seed removal by two groups of carabids that were grouped by size, small (less than 15 mm) and large (greater than 15 mm). Small carabids fed mostly on the small *Ambrosia artemisiifolia* and *Amaranthus retroflexus* seeds, whereas larger beetles consumed the larger seeds of *Triticum*, *Datura stramonium*, and *Senna obtusifolia*.

Another factor worth mentioning is that some granivores can assess the density of a seed, in addition to its size. *Anisodactylus sanctaecrucis* and *Harpalus pensylvanicus* base their seed preferences on seed density, this being the prevailing physical selection criterion for *A. sanctaecrucis* (Lundgren and Rosentrater, 2007). *Harpalus pensylvanicus* also responds to the seeds' mass, size, and strength in addition to density. Finally, none of the measured physical properties of seeds influence the preferences of *Gryllus pennsylvanicus* under choice conditions.

At the end of the day, spurious results stemming from examinations of seed size preferences highlight that there is no single reason that one seed is consumed while another is passed over. *Centaurea* and *Dipsacus* seeds are similar in size. But when these seeds are placed in the field concurrently, only the *Centaurea* seeds are acceptable to harvesting ants (Mittelbach and Gross, 1984). There may even be a taxonomic predisposition to in some granivores to consume seeds of a certain size, regardless of the other available resources (Honek et al., 2007). So while seed size is undoubtedly one component influencing granivory, other features besides seed size play a role in the seed preferences of granivorous entomophages.

13.2.2 External Features

The external topography of seeds also influences the relative predation rates of seeds. Many granivorous entomophages have difficulty manipulating smooth seeds, and thus only collect and consume those seeds with external 'handles'. For instance, *Pogonomyrmex* ants display clear preference for seeds with awns, hairs, or other projections (e.g., *Andropogon*, *Aster*, and *Diodia*), and smooth seeds are generally not attractive to the ants in the field (Pulliam and Brand, 1975). In the laboratory, *Solenopsis invicta* prefers small seeds with external morphological features (Ready and Vinson, 1995). It is also worth mentioning that food bodies, while being a source of nutrition, also function as structures for manipulating the seeds (O'Dowd and Hay, 1980). Even in the presence of handles, ants express different abilities to carry seed species (Culver and Beattie, 1978). For example, *Myrmica* drops myrmecochorous *Viola* seeds 19% of the time, whereas *Aphaenogaster* drops them on only 4% of foraging bouts. This affects the relative importance of these ant species as dispersers of *Viola* seeds. Similar to seed size, preferences for seeds with external structures can be influenced by competition. In desert and grassland systems, ants specialize on seeds with external structures, while rodents and birds consume many of the smooth-coated seeds (Pulliam and Brand, 1975).

13.2.3 Seed Covering

Other seed traits that influence the relative preference of granivorous entomophages are defensive properties of the seeds, the physiology of which are discussed in Chapter 11. One of these defenses, relative seed coat thickness or hardness, has been shown to influence the ability of a granivore to successfully access the internal nutrients of some seeds. Thus, the seeds that are preferred are those in which the nutritional content can be accessed by the granivore. When the seed coat is compromised through cracking (Morrison et al., 1997), softened by imbibition with water (Brust and House, 1988b), or in young seeds (Briese and Macauley, 1981), formerly unsuitable seeds become more attractive to the granivore.

Brust and House (1988b) observed the restrictive capabilities of the seed covering to a large granivorous carabid.

> "Harpalus caliginosus *was unable to penetrate the pericarp of wheat or sicklepod seeds*."

After struggling with the seeds for several minutes, the authors examined the seeds under the microscope.

> *Numerous scratches and depressions were observed in the pericarp, but most seeds were still intact, hard, and presumably viable.*

Observations during subsequent weeks revealed that this large carabid species eventually could crush the seed covering of the seeds. So, apparently the beetle learned how to crack the seed covering over time. A similar story involved larvae of *Notiobia flavicinctus* feeding on fig seeds. This granivorous larva could not crack the seeds of *Ficus hebetifolia*, and actually scored its mandibles in attempting to do so (Paarmann, 2002b).

Further evidence for the importance of the seed coat in determining seed preferences came from Honěk et al. (2003). The authors noted that seeds of similar weights, but different seed coat thicknesses, were consumed at different rates by granivorous carabids; the thicker the seed coat, the lower the consumption rate. Seed preferences of ants were also dictated by the thickness of the seed coat. When comparing the preferences of *Solenopsis invicta* for different crop seeds, the thick seed coats of soybean and cotton seed probably reduced consumption of these species relative to corn, wheat, and sorghum seeds (Morrison et al., 1997).

13.2.4 Nutrition

Nutritionally, not all seeds are created equally and granivores should theoretically select seeds of high nutrition preferentially. *Chenopodium album* seeds support the greatest fecundity in the carabid, *Harpalus rufipes*, in the laboratory. Furthermore, *C. album*, *Senecio vulgaris*, *Agrostis tenuis*, *Festuca ovina*, and *Phleum pratense* seeds foster a faster larval growth rate than seeds of cereals and ryegrass in the laboratory (Briggs, 1965). In this case, the laboratory preferences manifest themselves in the field by the spatial distribution of *H. rufipes* larvae, which are aggregated in stands of *C. album*. *Amara similata* lay more eggs when fed a seed mixture than seed species offered individually (Jorgensen and Toft, 1997b). Thus different seeds possess different levels of individual nutrients, and the beetles self-select their optimum diet from the seed mixture. When seed species are offered individually to *A. similata* females, *Poa annua* support less egg production than seeds of *Taraxacum* and *Tripleurospermum inodorum*. Another physiological requirement that is critical to granivores, especially in desert ecosystems, is water (Crist and MacMahon, 1992). Morrison et al. (1999) found that water content of seeds has an effect on the relative consumption rates of different seeds by the fire ant, *Solenopsis invicta*.

13.2.5 Seed Viability

Granivores can distinguish viable seeds from non-viable ones, and generally prefer the viable seeds. In one study, Crist and MacMahon (1992) found that 88% of seeds that ants carry from a seed source are viable. But only 50% of the seeds in the seed source are viable, suggesting that the ants preferentially select viable seeds for removal.

13.2.6 Grass Versus Broadleaf Species

It is worth mentioning that grass seeds are frequently preferred over broadleaf species (Saska et al., 2008; Heggenstaller et al., 2007 but see Brust, 1994). This preference is particularly well noted under field conditions. Carroll and Risch (1984) found that the fire ant, *Solenopsis geminata geminata*, caches large quantities of grass seeds in their nests, and these ants prefer grass seeds over broadleaf seeds. Similarly, under choice conditions in the field, Lundgren et al. (2006) found that grassy weed seeds are preferred over broadleaf species by granivorous arthropods in three different cropping systems. Of 16 weed seed species tested, *Harpalus pensylvanicus* prefer those of *Setaria viridis* (Lund, 1975), and adults and larvae of this ground beetle are at high abundances in association with *S. viridis* patches in agricultural fields (Kirk, 1973). Lundgren and Rosentrater (2007) found that *Gryllus pennsylvanicus* is unaffected by the physical characteristics of a range of broadleaf and grass seeds in the laboratory, but consumes the seeds of *Digitaria sanguinalis* more than twice as frequently as any of the eight other species tested. In another trial with *G. pennsylvanicus*, *Panicum miliaceum* is the most preferred seed among six tested, but this preference is not statistically different from two *Amaranthus* species (O'Rourke et al., 2006). Given the severity of preference for grass species in the laboratory, it is hypothesized that this very abundant granivore in North American cropland is fairly oligophagous in its granivory in the field. One explanation for the observations of grass seed preferences may be that grass seeds are lower in defensive chemicals (Janzen, 1971), unless symbiotic fungi are present.

13.3 The Dynamics of Preferences

Finally, seed preferences are a plastic process, changing spatially and temporally with the circumstances of the environment and physiological statuses of the seeds and granivores. Some individuals within an ant colony may specialize on a certain seed species, but this specialization may be a temporary arrangement (Nickle and Neal, 1972; Rissing, 1981). *Pogonomyrmex rugosus* individuals specialize on different seed species, but the preference of an individual ant worker shifts from *Poa pratensis* to *Lolium perenne* during a single foraging period. In their examination of two granivorous carabids, Honěk et al. (2006) found that the seed preferences remain constant throughout the growing season, although the level of seed consumption varies considerably over the season. Preferences by harvester ants also shift with proximity to the ant nest. Essentially, ants get choosier the further they get from their nest, as is expected for central-place foragers (Kelrick et al., 1986). In one study, Crist and MacMahon (1992) detail how preferences of *Pogonomyrmex occidentalis* are based primarily on seed size when the seeds are within 5 m of the nest. However, when the same species combinations are offered

10 m from the colony, seed preferences are correlated with seeds that have small cell walls, high energy content, and greater percent soluble carbohydrates and water content. So ants become finicky when they have to travel to get the seeds.

Another complicating factor is that the results of laboratory seed preference studies do not necessarily transfer to the field. Briese and Macauley (1981) found that the most commonly collected seeds in the field by harvester ants are the least preferred in the laboratory. They conclude that in this instance, foraging in the field is mitigated largely by the abundance of seeds.

13.4 Conclusions

As is evident in this chapter, a tremendous body of literature on seed preferences of post-dispersal seed predators is available, although attempts to synthesize these studies are scarce. The extensive amount of information on granivorous entomophages that is uncovered in the review of literature leads to the conclusion that relationships between granivores and seeds are extremely complex, and a suite of factors are operating on each granivore-seed interaction. It appears that seed preferences by granivorous entomophages are a means to two ends: (1) securing the necessary resources for physiological processes at minimal energetic and physiological costs to the granivorous species, and (2) a way for granivore communities to partition a class of highly nutritious foods in such a way that numerous species can coexist on a limited resource.

In conclusion, the current chapter and its four predecessors on seed and granivore natural histories set the stage for questioning applications of granivory in biological control and IPM, as well as giving a springboard for probing the evolutionary development of granivory in entomophagous arthropods. Given the background knowledge generated thus far, applied questions that remain to be addressed revolve around the importance and potential of biological control of weed seed banks by seed predation, and the importance of non-prey foods such as seeds to biological control of insect pests.

Section IV
Fungi and Microorganisms

Microbes are seldom ostentatious, and efforts to explore and understand their significance to the nutritional ecology of entomophagous arthropods are only just beginning to gather steam. This is a group of incredible diversity and abundance, and as might be expected their contributions to insect nutrition are also diverse. In this section, I primarily focus on three areas; the extent of mycophagy in entomophagous taxa, and the symbiotic associations of microbes with both non-prey foods and the entomophagous arthropods that consume these foods. Fungi are a source of food for numerous entomophagous species, but the full extent of mycophagy in natural enemies will only be realized when more scientists focus their attention on these organisms during gut content analyses and laboratory feeding assays. Microbes are omnipresent in nature, and frequently form close relationships with non-prey foods and arthropods. These symbioses influence the dietary breadth of entomophagous species and provide a further example of just how truly complex a process omnivory is in insects. Admittedly the microorganisms section of this book is comparatively short, but this is more reflective of how little work has been done on the interactions between microbes and natural enemies rather than evidence for the unimportance of these interactions.

IV.1 Fungi as Food for Arthropods

There are likely 1.5 million fungal species (140,000 of which produce mushrooms) scattered throughout the terrestrial world (Chang and Miles, 2004). Yeasts are some of the most pervasive fungi. Up to 20,000 yeast cells are present in each gram of soil, but the densest populations of yeasts are found in association with plants (on the phylloplane, flowers, and fruits) (Do Carmo-Sousa, 1969). Even mushrooms have yeast associates (Do Carmo-Sousa, 1969). The biomass of fungus present in many habitats is overwhelming; one estimate is that 80,000 t of edible fungus (by human standards) is produced annually in Finnish forests (Hackman and Meinander, 1979). Fungal fruiting bodies are an extremely variable resource in space and time, sometimes springing up seemingly from nowhere, and quickly disappearing once their spores are dispersed.

Most fungi produce spores which are consumed by numerous arthropods. Whether spores are sexual or asexual (termed conidia), a major reason for their existence is dispersal (Garraway and Evans, 1984). Spores are transported by animals or water, or are blown to new substrates by the wind (Garraway and Evans, 1984). Yeasts are also frequently found floating in the air (Do Carmo-Sousa, 1969), such that any insects that learn to exploit fungal material as food should have little trouble encountering a snack (Hanski, 1989).

Most fungi spend at least part of their lives saprophytically in the soil, and quickly respond to newly available resources. Sooty molds, a polyphyletic group of fungi that quickly colonize sugary secretions, are an excellent example of the ability of fungus to colonize ephemeral resources. Hughes (1976) reports that scale populations on some trees in New Zealand can produce so much honeydew that sooty mold forms a thick mat on the forest floor. An exception to the saprophytic lifestyle in fungi are the rusts (Uridinales) which are adapted to live almost entirely on plants (Savile, 1976).

Fungi support a diverse range of arthropod associates, including numerous entomophagous species. Ninety-eight beetle species are found in association with the fruiting bodies of *Polyporus squamosus* (Klimaszewski and Peck, 1987). This is in part because the life of a fungal sporocarp is host to a succession of insects, as the nutritional and habitat quality provided by the fruiting body changes over its life (Kukor and Martin, 1987).

Numerous entomophagous species are associated with the fruiting bodies of mushrooms. Anthocorids, carabids, staphylinids, spiders from nine families, predatory mites, phalangiids, clerids, tachinids, and numerous parasitoid Hymenoptera from an array of families are all found in association with *Piptoporus betulinus* fruiting bodies (Pielou and Verma, 1968). Similarly, Matthewman and Peilou (1971) found numerous predators living in association with the polyspore fungus *Fomes fomentarius*, including anthocorids, predaceous mites, staphylinids and other predatory beetles, ants, centipedes, phalangiids, and spiders. A diverse array of parasitoids (hymenopterans, phorids, and tachinids) are also captured on this species of mushroom. But this is not to say that these entomophagous species are mycophagous; many are feeding on the mycophagous residents of the fungi (Lewis and Worthen, 1992). For instance, staphylinids are some of the most common residents of sporocarps (Klimaszewski and Peck, 1987), but typically their diet is not fully known. As Ashe (1984) puts it

> *A wide variety of staphylinids visit fresh mushrooms. ...However, most mushroom visitors appear to be predaceous on other arthropods which occur there.*

Arthropods generally fall into one of three types of associations that vary by their degree of dependence on fungus (Hammond and Lawrence, 1989; Klimaszewski and Peck, 1987). *Mycetobionts* have an obligate dietary relationship with fungus. *Mycetophiles* are more polyphagous insects that can complete their life cycle in the absence of fungus (many staphylinids fall into this category). Finally, *mycetoxenes* are occasionally found in association with fungi, but are fully able to carry on in the absence of fungi. Those arthropods that feed on spores are termed microphagous, and those that consume the mycelium and context of the fungus are macrophagous mycophages (Lawrence, 1989). Most mycophagous natural enemies can be categorized as mycetoxenes or mycetophiles, and although it is seldom substantiated whether the diversity of entomophagous species found in fungus actually eats their surroundings, the next chapter will illustrate that this is likely the case for many species.

For facultative mycophages, fungi are unique from animal and plant material in several respects (Hanski, 1989). Structurally, fungi rely primarily on chitin and non-cellulosic β-(1, 3)- and β-(1–6)-glucans rather than cellulose, lignin, and pectin (as in plants). These storage polysaccharides necessitate a unique set of digestive enzymes in insects hoping to tap into the energy potential found in fungi. Morphological adaptations to microphagy are a bit more diagnostic than those of macrophagous arthropods (Lawrence, 1989). To collect spores and conidia, brush-like and toothed regions of the mouthparts have evolved in some entomophagous species; crushing the spores is accomplished using expanded denticulate regions of the mandible. These adaptations to mycophagy in entomophagous species underline the importance of this non-prey food to the fitness of natural enemies.

IV.2 Symbioses

> *By the term endosymbiosis we mean well-regulated and essentially undisturbed cooperative living between two differently constituted partners. It is usually a far more highly organized partner which shelters another within its body, and the mutual adaptation is so complete as to justify the assumption that the arrangement is useful to the host.* (Buchner, 1965)

Thus read the first lines of Buchner's amazing synthesis on endosymbioses between animals and microbes. Symbioses between microorganisms and insects were first uncovered with enthusiasm in the late 1800s, and their significance to the feeding ecology of the insects was substantiated largely after 1910 (Buchner, 1965). Relatively recent development of molecular tools has facilitated the study of microbial ecology, and the current state of knowledge concerning endosymbiosis vastly transcends what was presented nearly 50 years ago in Buchner's book. See Vega and Dowd (2005) for a list of insect-yeast associations.

Protozoa, yeasts, higher fungi, bacteria, and plasmids are just a few of the organisms to hold nutritional significance in their symbioses with arthropods. In addition to residing within arthropods, environmental microorganisms quickly change the quality and acceptability of most non-prey foods. Thus, through the close association with both insects and the foods they consume, microbes play an inextricable role in the nutritional ecology of entomophagous species.

Chapter 14
Mycophagy

Facultative mycophagy is likely the one of the most pervasive and least documented forms of omnivory within entomophagous species. The simple fact is that when scientists actually search for fungal material during gut analyses of natural enemies, they typically find it to some degree. The evolutionary development of mycophagy in entomophagous species is best described for beetles, notably the Coccinellidae and the Staphylinidae. But each group of natural enemies under study has at least a few scattered reports of mycophagy in species normally regarded as predators or parasitoids. Although fungus contains a rich assortment of nutrients critical to the nutrition of entomophagous species, the importance of mycophagy to the natural history of most natural enemies is poorly understood relative to other non-prey foods.

14.1 Fungi as Food for Natural Enemies

The diversity of fungi is staggering, and for several reasons it is to be expected that we know very little of the nutritional quality of most fungi. First, the nutritional value of different tissues within the same fungus can vary widely; hyphae/mycelium, conidia, fruiting bodies, and spores of a single species all have different nutritional profiles (Garraway and Evans, 1984), and support very different arthropod communities (Lawrence, 1989). Moreover, fungi are very sensitive to environmental fluctuations, and

> *within the bounds set by the metabolic capacities of the species, the composition* [of a fungus] *varies widely with the environment. The quantities of fat, carbohydrate, ash, wall material, and total nitrogen are all more or less responsive to the culture medium.* (Cochrane, 1958)

Also, as mushrooms age, they change substantially with respect to nutrition. The current state of knowledge with respect to the nutritional value of fungi stems largely from commercially produced species (mushrooms) of interest to humans. Although many of the nutrient analyses presented below are derived from commercial mushrooms, and are questionably applicable to most fungus-entomophage interac-

J.G. Lundgren, *Relationships of Natural Enemies and Non-Prey Foods*, Progress in Biological Control 7,

Published by Springer Science+Business Media B.V.

tions, the nutrient analyses presented at least give us a ballpark range of the nutrition of this non-prey food.

14.1.1 Water Content

One fairly consistent nutritional component of fungi is that they have a fairly high water content. Typically, more than 85% of the fresh weight of vegetative tissues of fungi is comprised of water (Chang and Miles, 2004; Cochrane, 1958; Kalberer, 1990; Kurtzmann, 1997; Leschen and Beutel, 2001). Nonetheless, environmental conditions affect the water content of fungi, and spores are lower in water than other fungal tissues. For instance, Todd and Bretherick (1942) found uredospores of unknown origin to be composed of only 15% water.

14.1.2 Carbohydrates

The major nutritional component of fungi is carbohydrates, primarily in the form of polysaccharides (Griffin, 1994). Proximate analysis usually shows that carbohydrates comprise around half the dry weight of fungi, but range from 3–85% (Chang and Miles, 2004; Cochrane, 1958; Griffin, 1994; Kurtzmann, 1997; Mueller et al., 2001). Very little of this carbohydrate is in the form of mono- or oligosaccharides. Chitin, cellulose, and glucan are the major structural polysaccharides found in the cell walls of fungi, and glycogen is the major storage polysaccharide (Garraway and Evans, 1984; Griffin, 1994; Kurtzmann, 1997; Phaff et al., 1966). Cochrane (1958) discusses that 5.5–10.6% of fungal dry weight is composed of chitin. Cellulose and lignin are other structural polysaccharides found in fungi, but many insects are unable to digest these polysaccharides (Garraway and Evans, 1984). The glycogen found in fungi is very similar to that found in animal tissues (Cochrane, 1958), and so is likely readily digestible by entomophagous arthropods. It is interesting to note that while most fungi have few simple sugars, the mutualistic fungi symbiotic with attine ants tend to have high levels of trehalose, in addition to protein-bound amino acids, that make the nutrients more accessible to ants (see below for more discussion of this relationship) (Swift et al., 1979). Finally, spores are nutritionally disparate from most other fungal tissues, and carbohydrate content is no exception to this pattern. Todd and Bretherick (1942) found that uredospores contain approximately 26% of their dry weight as carbohydrates, the majority of which were reducing sugars and only 0.78% being starch.

14.1.3 Proteins

Proteins, including their amino acid precursors, are the next most abundant nutrient found in fungi. Relative to some non-prey foods, fungi are an excellent source of protein. Typically, from 20–40% (actual range from 4.6–61%) of fungal dry weight is protein-based nitrogen (Chang and Miles, 2004; Cochrane, 1958; Griffin, 1994; Kurtzmann, 1997; Mueller et al., 2001; Todd and Bretherick, 1942). In their survey, Mueller et al. (2001) determine that mean (±SD) protein content of 49 species of Basidiomycetes is 21.37% ± 10.26% of dry weight. Fungi are also an appreciable source of amino acids essential to insect growth and development (Chang and Miles, 2004; Kurtzmann, 1997). The least concentrated amino acid is tryptophan in many fungi; the most abundant is often lysine (Chang and Miles, 2004).

14.1.4 Lipids

A variety of lipids are present in fungi, including sterols that are a dietary requirement for insects. Lipids typically comprise less than 10% of fungal dry weight (Chang and Miles, 2004; Cochrane, 1958; Mueller et al., 2001), but can reach as high as 87% of tissues in some species (Griffin, 1994). Fatty acids in fungi are typically unsaturated, and all fungi have palmitic and stearic acids (Chang and Miles, 2004; Harwood and Russell, 1984). Myristic, palmitoleic, stearic, oleic, arachidonic, and linoleic acids are also found in many fungi (Chang and Miles, 2004; Cochrane, 1958; Harwood and Russell, 1984). Longer-chained fatty acids may be common in certain species, but generally are minor components of fungi (Harwood and Russell, 1984). Sterols are invariably present as ergosterol, but other C27, C28, and C29 sterols are also present in fungi (Chang and Miles, 2004; Griffin, 1994; Harwood and Russell, 1984; Kurtzmann, 1997). An exception to this is the Uridinales, which replace ergosterol with C29 sterols (Harwood and Russell, 1984). Sterols represent 4.0–5.4% of total lipids in fungi (Cochrane, 1958), but some yeasts can contain 10% of their dry weight as sterols (Harwood and Russell, 1984). Ultraviolet light converts ergosterol into vitamin D (Chang and Miles, 2004; Kurtzmann, 1997), and it isn't clear how this vitamin functions in entomophagous insects. Inositol and choline, two important dietary requirements of insects, are also important to the physiology of fungi and are thus ubiquitously present in this non-prey food (Cochrane, 1958; Garraway and Evans, 1984; Griffin, 1994; Harwood and Russell, 1984).

14.1.5 Vitamins and Minerals

Up to 12% of the dry weight of most fungi is ash (Chang and Miles, 2004; Cochrane, 1958; Griffin, 1994), and fungi are a good source of many minerals and vitamins necessary for insect fitness. Potassium is invariably the most abundant mineral found in fungi, followed by the elements Na, Ca, P, Mg (Chang and Miles, 2004; Cochrane, 1958; Garraway and Evans, 1984; Griffin, 1994; Kurtzmann, 1997; Todd and Bretherick, 1942). Cu, Zn, Fe, Mn, Mo, Cd, S, Ca, Co, and Al are also found in many fungi, but at much lower concentrations (Chang and Miles, 2004; Cochrane, 1958; Garraway and Evans, 1984; Kurtzmann, 1997). In addition to minerals, fungi are a good source of B-vitamins, including thiamine, biotin, pyridoxine, riboflavin, and vitamin B12 (Cochrane, 1958; Garraway and Evans, 1984; Kurtzmann, 1997). Finally, β-carotene is the predominant carotenoid found in fungi, and while γ-carotene is not infrequent in fungi, α-carotene has yet to be detected (Harwood and Russell, 1984).

14.1.6 Defensive Properties of Fungi

Finally, just as with the preceding non-prey foods, fungi possess a whole range of non-nutritive secondary chemicals that presumably defend the fungus from mycophages. Although the toxicity of mycotoxins has made a strong impression on the human race for hundreds of years, the importance of secondary chemicals in anti-predator defense is surprisingly understudied (Rohlfs et al., 2007). In addition to deterring predators, the secondary chemicals found in fungi may have important implications for the suitability, apparency, and acceptability of a given fungus to a mycophagous insect (Kukor and Martin, 1987). Although there are a few examples where natural enemies use fungal secondary metabolites as kairomones for finding mycophagous hosts or prey (Dicke, 1988a; Kukor and Martin, 1987), the direct role of these chemicals in the nutrition of facultative mycophages is unknown. In addition to these protective secondary chemicals, fungi (especially spores) also possess a range of structural defenses that are presumed to deter unwanted mycophagy. Spores often have rigid and protective walls that must be overcome by mycophagous entomophages in order to access the nutrients (Lawrence and Newton, 1980; Savile, 1976). In fungus-like myxomycetes, these spore walls have calcium carbonate that further lend to their rigidity (Lawrence and Newton, 1980). The spore walls of some rust fungi have warts or spines that have a dual function of defense and dispersal (Savile, 1976). Moreover, the spore-bearing sori of certain rust fungi, (examples occur in *Puccinia*, *Uridinopsis* and *Uromyces*, as well as many others) are protected against mycophagy by spikes or spines of various origins (Savile, 1976). Exploration of the interactions between the defenses of fungi and natural enemies seems a fruitful branch of research heretofore untouched.

As detailed above, fungi are composed of a series of substances that are simply not encountered in entomophagy and phytophagy. Consequently, mycophagous arthropods require a series of digestive adaptations that enable them to exploit the maximum nutrition from this food source. Most notably, many mycophagous insects possess digestive enzymes that allow them to digest β-1, 3-glucans, α-1, 4-glucans, and β-1, 6-glucans, and chitin, which are major structural polysaccharides fairly distinct from plants and insects (except for chitin in insects, of course) (Hanski, 1989; Kukor and Martin, 1987; Martin et al., 1981). The gut pH may also be indicative of diet in that it may support the enzymatic and metabolic reactions unique to mycophagy; the guts of herbivores tend to be alkaline, whereas mycophagous species tend to have a neutral gut pH (Martin et al., 1981). Fungi have a high caloric content for those species that can unlock fungal nutrients. One estimate for mushrooms is that energy content ranges from 2,760–3,920 calories g^{-1} of dry weight (Chang and Miles, 2004).

14.2 When Mycophagy Benefits the Fungus

Not all cases of mycophagy by arthropods are at the expense of the fungus, and several instances of fungi capitalizing on facultative mycophagy as a means of spore dispersal are documented. A range of fungi, especially within the rust fungi (Uridinales) produce sugary secretions to attract mycophagous entomophages (Stoffolano, 1995). Spores that attach to the bodies of insects attracted to the sugary secretions are then transferred to other fungi, thereby encouraging the outbreeding of the fungus (Webber and Gibbs, 1989). Gilbert and Jervis (1998) mention that at least members of the Syrphidae, Phoridae and Tachinidae are attracted to the sugary secretion of rust fungi and ergot (*Claviceps purpurea*). But the 'honeydew' of these fungi may be spiked with numerous secondary chemicals with unknown effects on visiting insects (Todd, 1967). Paracelsus recognized the delusional side effects of consuming the honeydew of ergot, and it may be that Coleridge was referring to this as he describes the delusional visions of paradise held by Kubla Khan (Todd, 1967).

And all who heard shall see him there,
And all should cry, Beware! Beware!
His flashing eyes, his floating hair!
Weave a circle round him thrice,
And close your eyes with holy dread,
For he on honey-dew hath fed,
And drunk the milk of Paradise
(Kubla Khan, S. T. Coleridge, 1798)

Also worth mentioning is that being consumed by an entomophagous arthropod is not always a death sentence for a spore, which may use beneficial insects as vectors to reach new hosts. *Hippodamia convergens* is an effective vector of *Discula destructiva*, a pathogen of *Cornus florida*. In addition to carrying spores of this pathogen on their bodies (Colby et al., 1995), *H. convergens* consumes the fungus, and transmits spores in its frass to new plants. A substantial number of spores (10^8 per beetle) of *D. destructiva* can survive in the digestive tract of *H. convergens* for up to 96 hours in the laboratory (Hed et al., 1999), and fewer can survive for up to 16 days (Colby et al., 1996). Also, the plant-pathogenic yeast, *Nematospora coryli*, is vectored to new *Brassica* plants through the mouthparts of the nabid, *Nabis alternatus* (Burgess et al., 1983; Lattin, 1989).

14.3 Mycophagous Taxa

14.3.1 Arachnida: Araneae

In addition to trapping pollen grains (discussed in Chapter 6), spider webs often trap substantial quantities of fungal material, and these spores likely provide nutrition to web-building spiders when they consume their webs. Bera et al. (2002) found that up to 13% of organic material recovered from spider webs is fungal, including spores of *Alternaria*, *Curvularia*, and Microthyriaceae. Still, pollen tends to be intercepted by spider webs more frequently than fungal material (Bera et al., 2002; Linskins et al., 1993).

14.3.2 Arachnida: Acari

Mites got their start evolutionarily as predators, but the diets of many taxa designated as predaceous also include fungus to varying degrees (Krantz and Lindquist, 1979; OConnor, 1984). Indeed, some tydaeid mites initially regarded as predators of herbivorous mites in orchards are determined to be exclusively mycophagous under closer examination in the laboratory (McCoy et al., 1969). The most mycophagous of mites attract interest as potential biocontrol agents of powdery mildews in some crops, especially grapes (English-Loeb et al., 1999; Norton et al., 2000). Another mite, in this case presumed to be mostly mycophagous, is the cheese mite, *Tyrophagus putrescentiae*. This mite is able to detect volatile extracts, namely cis- and trans-octa-1, 5-dien-3-ol, produced by numerous species of fungi (Vanhaelen et al., 1979, 1980). Vanhaelen et al. speculate that these volatiles, which give fungus its 'mushroomy' scent, may be broadly important in mitigating mite-fungus interactions. Interestingly, this same soil-dwelling mite is a key predator of *Diabrotica undecimpunctata howardi* eggs. *Tyrophagous putrescentiae* is able to detect rootworm eggs from up to 5 cm away (quite a distance for a mite!), and inflict a heavy toll on *D. u. howardi* eggs in the field (Brust and House, 1988a). The bottom line is that the feeding behavior of mites with respect to mycophagy is seldom as clear as it seems upon superficial examination.

Within the Phytoseiidae, a number of species can complete development and even reproduce on food of fungal origin, although these seldom are ideal foods for predaceous mites (Huffaker et al., 1970; McMurtry et al., 1970). Mites in the genus *Typhlodromus* receive the most attention for their mycophagous habits, but unsurprisingly, not all fungal species are equally suitable for these mites (Chant, 1959; Putnam, 1962; Zaher and Shehata, 1971; Zemek and Prenerova, 1997). In one set of experiments, powdery mildew (*Plasmopara viticola*) densities are correlated with increased abundance of two phytoseiids, *Amblyseius andersoni* and *Typhlodromus pyri* in vineyards (Duso et al., 2003; Duso et al., 2005). When powdery mildew populations are controlled, these predaceous mites suffer. Gut analysis revealed that 39 of 40 individuals of *A. andersoni* had consumed the fungus. Ultimately, the prevalence of fungus on the phylloplane may facilitate the persistence of predaceous mites in cropland in the absence of prey, as first postulated by Chant (1959).

14.3.3 Coleoptera: Carabidae

Mycophagy in Carabidae is fairly widespread, although its importance to their life history is entirely unexplored. To date, all records of mycophagy in carabids stem from gut dissections of field-collected specimens, and nearly all records can be ascribed to one of two researchers, Stephen Forbes and Michael Davies. In addition to the fungal species listed in Table 14.1, Hammond and Lawrence (1989) mention that carabids will consume Sphaeriales (Ascomycotina), Aphyllophorales, Agaricales, Russulales (Basidiomycotina).

Fungal material is known from the guts of 41 species, but taxonomically, mycophagy involves a slightly different subset of carabids than of those that are granivorous

Table 14.1 Species of Carabidae recorded as mycophagous under field conditions

Species	Reference	Notes
Abax parallelepipedus	(Davies, 1953)	
Abacidus permundus	(Forbes, 1883)	
Agonum gratiosum	(Davies, 1953)	
Amara aenea	(Davies, 1953)	
Amara aulica	(Davies, 1953)	
Amara carinata	(Forbes, 1883)	About ¼ of the food ingested was fungal in origin (*Peronospora*)
Amara familiaris	(Davies, 1953)	
Amara impuncticollis	(Forbes, 1883)	Had traces of fungi in their diet
Amara montivaga	(Davies, 1959)	
Amara plebeja	(Davies, 1953)	
Amara sp. (listed as *A. vulgaris*)	(Davies, 1953)	
Anchomenus dorsalis	(Davies, 1953)	
Anisodactylus rusticus	(Forbes, 1883)	Ate *Cladosporium* and *Peronospora* fungi
Bembidion guttula	(Davies, 1953)	
Bembidion lampros	(Davies, 1953; Sunderland, 1975)	
Bembidion lunulatum	(Davies, 1953)	
Bembidion obtusum	(Davies, 1953)	
Bembidion tetracolum tetracolum	(Davies, 1953)	
Calathus fuscipes	(Davies, 1953)	
Calathus melanocephalus	(Blunck, 1925; Larochelle, 1990)	
Chlaenius	(Forbes, 1881, 1883)	In one field, around 10% of their diet consisted of fungi, *Coprinus* and *Dematei*. Around 5% of food at another site was fungal. Spores and fungal material were found in the stomachs of *C. platyderus*
Cratacanthus dubius	(Forbes, 1881)	A few fungal spores were found in their otherwise empty gut
Cyclotrachelus faber	(Freitag, 1969)	Spores of fungi were found in the gut contents
Cyclotrachelus sodalis colossus	(Forbes, 1883)	A few fungal spores were found in their stomachs (2 of 5 specimens)

(continued)

Table 14.1 (continued)

Species	Reference	Notes
Eurycoleus	(Erwin and Erwin, 1976)	Adults are ectoparasitoids of fungus beetles, and consume fungal spores during the adult stage, as well as host-feeding of sorts. They like smooth, polyspore fungi, and by passing spores in their feces, they probably help to disseminate the host fungus to new trees
Harpalus	(Forbes, 1883)	
Harpalus affinis	(Davies, 1953)	
Harpalus attenuatus	(Davies, 1953)	
Harpalus caliginosus	(Forbes, 1881, 1883)	Ate a little *Helminthosporium* at one site
Harpalus herbivagus	(Forbes, 1881, 1883)	About a third of their diet was fungal. In 1881, a fleshy, cartilaginous fungus was found in their stomachs.
Harpalus pensylvanicus	(Forbes, 1881, 1883; Kirk, 1973)	At one site, 8% of food was fungal, *Peronospora* (mostly) and *Helminthosporium*. A few spores were found in some of the guts.
Harpalus rufipes	(Larochelle, 1990)	
Lebia atriventris	(Forbes, 1881, 1883)	Spores of *Phoma* and *Helminthosporium* were found in guts (five of seven specimens; the others hadn't eaten anything).
Notiophilus rufipes	(Davies, 1953)	
Notiophilus substriatus	(Davies, 1953)	
Paradromius linearis	(Davies, 1953)	
Poecilus cupreus	(Davies, 1953)	
Pterostichus	(Forbes, 1883)	Around 10% of the food taken by half the beetles contained fungal spores. *Helminthosporium* was the only fungus noted.
Pterostichus madidus	(Davies, 1953)	
Pterostichus oblongopunctatus	(Davies, 1953)	
Stenolophus	(Forbes, 1883)	Only 2% of their diet was fungal, split equally between *Helminthosporium* and *Peronospora. Ramularia* and *Coleosporium* were also found in *Stenolophus* stomachs.

(discussed in Chapter 9). For instance, species of *Cyclotrachelus*, *Notiophilus*, and *Bembidion* consume fungal material, but seldom seeds. Meanwhile, the granivorous members of Harpalini and *Amara* are broadly represented in the list of mycophagous taxa. Another set of observations gives some additional information on the potential importance of mycophagy for carabids. As mentioned in Section III, *Harpalus pensylvanicus* and *H. eraticus* larvae create burrows in which they reside, overwinter, and cache seeds (Kirk, 1972). The larvae of *H. pensylvanicus* also make shallow trenches outside the entrance to their burrows, where Kirk (1973) hypothesizes that they consume soil microorganisms as food. Also striking is that Kirk noted no evidence of actual seed consumption within the burrows of the larvae, even though the larvae are confined with the seeds all winter. It would be fascinating if these seed caches serve as a substrate for microorganisms that are then consumed by the *Harpalus* larvae, essentially constituting a similar nutritional system to that seen in leaf-cutting ants.

14.3.4 Coleoptera: Coccinellidae

Mycophagy pervades many clades of this family. On one extreme, the Halaziini (formerly the Psylloborini) are a tribe of exclusively mycophagous coccinellids (Hodek and Honěk, 1996). Like mycophagous mites, coccinellids in the genera *Psyllobora* and *Ileis* receive attention as a potential source of biological control of powdery mildews (Davidson, 1921; Takeuchi et al., 2000). The most mycophagous species possess morphological adaptations to the mouthparts that facilitate the collection and consumption of fungal spores, similar to those rakes and combs used in pollinivory. Specifically, the mandibles of members of the Halaziini have two tips and a series of spines or teeth on their inner, ventral margin that help to scrape spores from fungal material (Kovar, 1996; Samways et al., 1997) (Fig. 14.1). *Tytthaspsis sedecimpunctata* is a great example of this adaptation (Ricci, 1982; Samways et al., 1997). Polyphagous species typically lack these morphological adaptations to fungal feeding, but this is not to say that fungi are not important to their life histories.

In several published gut analyses, even the best appreciated of aphidophagous coccinellids consumed fungal spores as an important component of their diet. In his examination of the gut contents of agricultural coccinellids, Forbes found fungal material in the guts of all eight species examined (31 of 39 individuals), and this class of food comprised 45% of the food that these beetles had consumed (Forbes, 1881, 1883). In fact, 90% of food found in the guts of *Coccinella novemnotata* was *Ustilago helminthosporium* spores. Subsequent gut dissections concur with the early findings of Forbes that predaceous ladybeetles in agriculture are frequently mycophagous (Anderson, 1982; Hagen et al., 1976; Lundgren et al., 2004; Ricci et al., 2005). Putman (1964) found that nearly all of the four most abundant predaceous cocci-nellids found in peach trees consisted of 'detritus', in other words plant material, fungal spores and pollen (507 guts dissected in total). Indeed, 50% of these

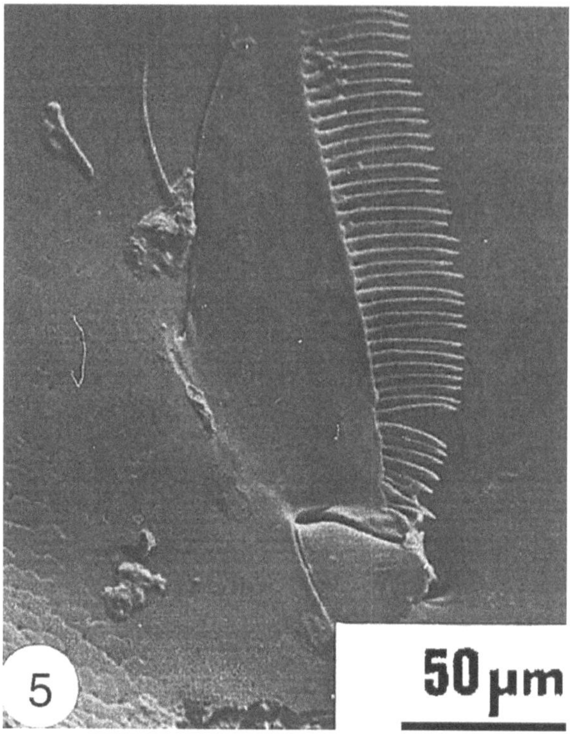

Fig. 14.1 Ventral view of the right mandible of *Tytthaspis sedecimpunctata*, showing the distinct comb-like prostheca (Reproduced from Samways, 1997. With permission from Taylor and Francis)

beetles had no other solid food in their guts besides this detritus. Nearly all adults of *Hippodamia notata*, in addition to consuming aphids, also consumed *Cladosporium* in one observation series (Ricci and Ponti, 2005); larvae were also very mycophagous. Quite a range of fungal groups have been found in the guts of coccinellids, including tissues or spores of *Helminthosporium*, *Ustilago*, *Cladosporium*, *Discula*, *Septoria*, *Uredo*, *Coleosporium*, *Menispora*, *Stemphylium*, *Sphaeronemei*, *Myxogastres*, *Macrosporium*, *Oidium*, *Peronospora*, *Alternaria*, *Monilinia fructicola*, and *Puccinia* (Anderson, 1982; Forbes, 1883; Hed et al., 1999; Putnam, 1964; Ricci, 1986a; Ricci et al., 1983; Ricci and Ponti, 2005; Ricci et al., 2005; Triltsch, 1997, 1999).

In addition to supporting prolonged survival in the absence of prey, mycophagy frequently coincides with two critical life processes in ladybeetles, diapause and reproduction. In part, the importance of mycophagy to overwintering success in coccinellids may be related to the fact that other foods become scarce late in the growing season, whereas fungi sometimes persist into this time of year. Regardless of why, it is a fact that numerous ladybeetles rely on fungi the most late in the growing season, building up nutrient reserves for dormancy (Anderson, 1982; Ricci et al., 1983) In *Coccinella septempunctata*, adults increase their consumption of *Alternaria* and *Puccinia* spores during pre-dormancy, even when aphids are abundant.

This suggests that self-selection of nutrients present in fungi may be necessary for overwintering (Triltsch, 1997, 1999). Spring is another critical time for mycophagy in ladybeetles, when ladybeetles require nutrition for dispersal and reproduction (Anderson, 1982; Triltsch, 1997) (see Chapter 1). Mycophagy in coccinellid larvae is not well explored and solid conclusions on the importance of fungi in larval diets really can't be drawn at this point. However, what little information is available suggests less reliance on fungal material by larvae compared with adults (Hukusima and Itoh, 1976; Ricci, 1986a; Triltsch, 1999).

14.3.5 Coleoptera: Staphylinidae

Given that the Staphylinidae is such a large and diverse family for which we have comparatively little biological understanding, it is surprising that staphylinids are probably the best understood group of natural enemies with regard to their adaptations to facultative mycophagy. The three subfamilies where many economically important natural enemies reside (Aleocharinae, Staphylininae, and Tachyporinae), all have members that are facultatively mycophagous to varying degrees (Hammond and Lawrence, 1989; Leschen, 1993). Only a few staphylinid lineages are obligate mycophages (the subtribe Gyrophaenina within Aleocharinae, and the genus *Oxyoporus*) (Ashe, 1984). Evolutionarily, the Aleocharinae as a whole arises from primitively predaceous ancestors (Ashe, 1984, 1993). However, those tribes within the Aleocharinae with a more mycophagous lifestyle originate from ancestors with distinct brushes on their mandibles, the function of which is adapted to collecting fungal spores (Ashe, 1984).

The structure of the mouthparts is well correlated with diet in many staphylinids, making this a useful group in understanding the adaptations diagnostic for mycophagy in natural enemies. A first key point in relating mouthpart morphology with the consumption of fungus is that different fungal tissues present unique challenges for consumption. For example, feeding on the context of a sporophore is much different than grazing the surface of the hymenium or eating free spores or conidia. Lawrence (1989) classifies these two processes as macrophagy and microphagy, respectively, and notes that feeding on fungal context is a difficult process likened to feeding on wood. Consequently, there are few instances of facultative context feeders, and many of the predaceous species of interest to this book are best classified as microphagous mycophages. Another caveat worth discussing with regard to structure and function in staphylinid feeding behavior is that feeding adaptations are often more evolved in the larval stage, since this is the primary feeding stage and populations are potentially more limited by the larvae's ability to feed efficiently (Lawrence, 1989). Moreover, mouthpart structure is highly conserved in staphylinid adults (Lawrence, 1989; Leschen, 1993), further blurring the relationship between structure and diet.

In summarizing the evolution of mouthpart structure in relation to mycophagy in staphylinids, Ashe (1993) figures that most adaptations to this lifestyle occur on the maxillae, mandibles, and epipharynx of the insects. Various components of the maxillae have combs or rakes that are useful in harvesting spores, or grazing the hymenium of fungi (Ashe, 1984, 1993; Lawrence, 1989). Once the spores are collected by the maxillae, the mandibles are used to grind the material into digestible matter. The molar lobe of the Staphylinidae is largely lost in the predaceous ancestors of the group, but in mycophagous species a pseudomola has resurfaced secondarily to accomplish the work tackled by the molar lobe in other beetles (Newton, 1984). Specifically, the pseudomola grinds the fungal spores, and often this structure has denticles or teeth that allow the trituration of fungal material (Ashe, 1984, 1993; Lawrence, 1989; Leschen, 1993; Leschen and Beutel, 2001) (Fig. 14.2). The larvae of staphylinids (at least the Aleocharinae) have simplified, sickle-shaped mandible with a small subapical tooth. This tooth is one aspect of the larval mouthparts that varies little in response to diet (Ashe, 1993). Infrequently, the larval mandible of mycophagous species have a bifid tip or set of spines that assist in filtering spores from a substrate (Ashe, 1993; Leschen and Beutel, 2001) as in Fig. 14.3. Within larvae of *Sepedophilus*, Leschen and Beutel (2001) speculate that different mandibular tips have evolved in response to feeding on different fungal structures. Those species whose larvae feed on persistent and tough fruiting bodies of mushrooms have a chisel-ended or serrate mandible. Those species that specialize on softer fungi are associated with a more robust mandible containing fine asparites (filters). Finally, the epipharynx is adapted to mycophagy in several respects. In microphagous staphylinids, the epipharynx has denticles that further triturate spores (Ashe, 1993), whereas a few genera that feed by juicing the context of a fruiting body tend to have a number of epipharyngeal tubes that facilitate the drinking of fluids squeezed from

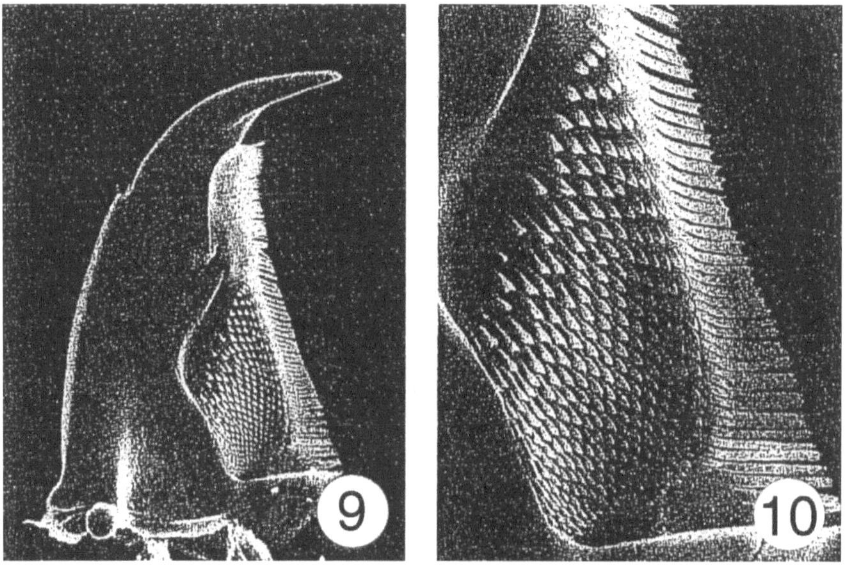

Fig. 14.2 Details of the denticles on molar surface of the adult mandibles of *Bolitochara lunulata* (Reproduced from Ashe, 1993. With permission from the Entomological Society of America)

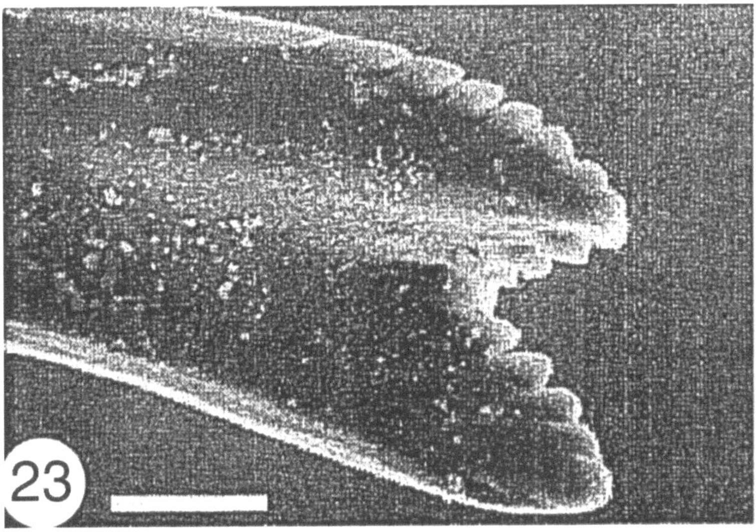

Fig. 14.3 Lateral view of bifid mandible of *Sepedophilus* type C (Reproduced from Letschen and Beutel, 2001. With permission from Blackwell)

the fungal cells (Leschen and Beutel, 2001). Finally, although partially diagnostic, mouthpart structure is fairly conserved within certain taxonomic groups, and it is advisable to back any presumptions on diet based upon morphology with gut analysis.

Several agriculturally important staphylinids consume fungal material in cropland, and mycophagy in these staphylinids is an important consideration when designing biological control programs. In cereal fields, *Tachyporus hypnorum* and *T. chrysomelinus* adults, and *Tachyporus* larvae consume substantial quantities of powdery mildew (*Erysiphe*) spores (more than 85% of meals) in addition to consuming cereal aphids (Sunderland, 1975). Under no-choice conditions in the laboratory, each beetle can consume 300–450 spores per day (Dennis et al., 1991). Another key staphylinid predator in this system, *Philonthus cognatus*, does not eat the fungus even under no-choice conditions. The mass production of staphylinids for augmentation biological control depends on understanding the mycophagous nature of some species, and incorporating fungi into the rearing regimens of these agents (Birken and Cloyd, 2007).

14.3.6 Neuroptera: Chrysopidae

The predaceous and non-predaceous adults of Chrysopidae consume fungal material, especially yeasts (Canard, 2001). In a comprehensive survey on the nutrition of Hungarian chrysopids, Bozsik (1992) found that yeasts (and some spores) are commonly found in the stomachs of *Chrysoperla carnea*, *Dichochrysa prasina*, *Chrysopa formosa*, *C. pallens*, *C. perla*, and *C. viridana*. Although present in all species examined, the predaceous species are less likely to have yeast in their stomachs. The potential role of symbiotic yeasts in the nutrition of lacewings is discussed more extensively in the Chapter 15. A final piece of evidence that suggests the importance of fungi, especially yeasts, to the nutrition of chrysopids is that these species are commonly reared on Wheast (a commercial formulation of *Saccharomyces fragilis* with a milk whey substrate) and yeast hydrolysates in the laboratory (Hagen and Tassan, 1966, 1970; Sheldon and MacLeod, 1971), and are attracted to cropland sprayed with this artificial food (Hagen et al., 1976).

14.3.7 Heteroptera

Reports of mycophagy in predaceous bugs are rare, but some anthocorids are adapted to consuming fungi as part of their diet. Chu (1969) presents that two predaceous anthocorids, *Lyctocoris beneficus* and *Xylocoris galactinus*, are capable of completing development on a diet of only moldy corn seed. Although not as suitable as prey for normal predator fitness, these bugs sustain themselves for long periods of time on fungus, but cannot lay eggs on this diet by itself. Also, the anthocorid *Anthocoris nemorum* can detect the presence of the entomopathogenic fungus, *Beauveria bassiana*, on nettle leaves (Meyling and Pell, 2006). Although they are deterred by this entomopathogenic fungus, the same sensory mechanisms used to recognize *Beauveria* may be useful in identifying other fungi as food. Clearly, the topic of mycophagy in predaceous bugs is ripe for exploration.

14.3.8 Diptera

Generally speaking, fungus is a minor component of the diets of predatory or parasitoid species of flies. It is interesting that some species of entomophagous Diptera share their familial designation with mycophagous species. These include Stratiomyidae, Scenopinidae, Syrphidae, Phoridae, and Empididae (Hackman and Meinander, 1979; Hammond and Lawrence, 1989; Maier, 1978). The nutritional ecology of syrphids that are mycophagous as larvae is not entirely understood. In addition to consuming fungi, it is conceivable that closer examination will reveal that some of these species are predaceous on other insects they encounter (as suggested by Hackman and Meinander, 1979). Indeed, ancestral syrphids are mycophagous, and entomophagy is a derived state for the family (Rotheray and Gilbert, 1999), so it should not be surprising to see overlap in the dietary ranges of mycophagous and entomophagous guilds to some degree. Closer examination of the cyclorrhaphous flies may reveal a greater degree of mycophagy than is currently realized. Many of the adaptations by flies to pollinivory that are noted in earlier chapters are transferable to the consumption of fungal material. Broadhead (1984) mentions that the prongs or scoops of the labellum are used in snipping fungal material, in addition to the width of the pseudotracheal canals (and reduced number), are diagnostic of mycophagy in non-predaceous lauxaniid flies. Some of these same adaptations are described to aid pollinivory in entomophagous species of Diptera.

14.3.9 *Parasitoid Hymenoptera*

Although numerous parasitoids are associated with mycophagous hosts (Hammond and Lawrence, 1989; Rotheray, 1990), I am unable to report even a single instance of direct mycophagy in parasitoid Hymenoptera. In laboratory feeding trials, yeast does not prolong the lives of *Trichogramma* wasps (Ashley and Gonzalez, 1974; Leatemia et al., 1995). Given the importance of nectar and pollen to the fitness of many parasitoid wasps, it would not be surprising to find fungal material in the guts of some parasitoid wasps, especially in the larger ichneumonoid species.

14.3.10 *Formicidae*

The best documented case of mycophagy in ants is inarguably the case of Attini (subfamily Myrmicinae) ants and their symbiotic fungus (Beattie and Hughes, 2002; Cherrett et al., 1989). All ants within this group, the main diversity of which occurs in *Atta* and *Acromyrmex*, are obligate mycophages on *Leucocoprinus* or *Leucoagaricus* species. The intricacies of the mutualism between these two groups of organisms are amazing, and the relationship results in leaf-cutter ants dominating many Neotropical habitats. In leaf-cutter genera, the ants harvest only the plants that promote fungal growth (Carroll and Janzen, 1973; Hubbell et al., 1983; Ridley et al., 1996; Swift et al., 1979), and remove microbial competitors from their nests (Swift et al., 1979). The fungus is capable of accessing nutrients from vegetation that are otherwise unavailable to these insects, and it packages the nutrients into a mycelial food body that is nutritionally complete for developing ant larvae (Beattie and Hughes, 2002; Hartzell, 1967; Stradling, 1987). The relationship likely evolved from ant and fungal forerunners associated with the same nest cavity (i.e., wood inhabiting ants). Another option is that the fungus initially relied on the ants as dispersal agents, as seen with myrmecochory and seeds (Mueller et al., 2001; Sanchez-Pena, 2005) (Chapter 12). These fungus-growing associations likely evolved 45–60 million years ago (Mueller et al., 2001).

Even though Attini-fungus interactions involve strict mycophagy in the ant, there are a few aspects of this relationship that may improve our understanding of how entomophagous insects use fungi as food. Facultative mycophagy in ants is rare at best (Mueller et al., 2001), although it would not be unexpected to find instances of this phenomenon in some ant species. Some evidence for more wide-spread facultative mycophagy in ants come from Torres (1984), who reports that 15 of 21 ants species carry "fungi or feces" (a somewhat contrived food category) back to their nests. The infrabuccal cavity of ants is frequently loaded with fungus, but it appears that most ants discard fungal material to the midden rather than provide it to the fourth instar for digestion (Mueller et al., 2001). Also, the physiological adaptations to mycophagy found in leaf-cutter ants may shed light on this feeding behavior in other ant groups (many of these adaptations are listed in Cherrett et al., 1989). In conclusion, this is an excellent group of insects with which to transition to the next chapter, which deals extensively with the nutritional symbioses between entomophagous species, non-prey foods, and microbes.

14.4 Conclusions

Numerous entomophagous species consume fungus routinely, and under-represented taxa in this chapter are likely to yield more reports under closer scrutiny. Fungus presents a diverse suite of tissues and organs that can be exploited by entomophagous arthropods, but each presents structural and chemical obstacles that need to be overcome before the rich nutrition found in fungus can meet the energetic needs of the arthropods (Leschen, 1993). The importance of this food to natural enemies is evident in the morphological and physiological adaptations that are expressed in those entomophagous species that have come to rely the most on this fungus. It is interesting to note that the anatomical features that employed to facilitate spore-feeding in insects (a series of brush-like abrasive features on the mouthparts for collecting and crushing the spores) are similar to those seen in pollinivorous species. A closer look at these relationships may reveal how isolated morphological adaptations can have wide implications for the dietary breadth of an omnivorous species. Finally, although numerous instances of mycophagy are noted for natural enemies, very little is known about the function of fungus in the nutritional ecology of these arthropods.

Chapter 15
Symbioses with Microorganisms

Microorganisms are an inescapable presence in most biotic interactions, and they influence the nutritional ecology of natural enemies in at least two major ways. First, their interactions with the food items themselves often change the quality and attractiveness of these substances for natural enemies. Presented in this chapter are three such interactions: when microorganisms (especially fungi) affect seeds, nectar, and honeydew for natural enemies. The microbial community of insect guts plays an important and often underestimated role in the nutritional ecology of entomophagous species, and internal nutritional symbionts are the focus of the second half of this chapter. Clearly, as a discipline we are only just beginning to understand how microbes render the nutritional ecology of entomophagous species more complex, and it is hoped that this short review will stimulate more research in this expanding area of biology.

15.1 Contaminated Non-Prey Foods

15.1.1 Endophytes and Seeds

Want ye corn for bread?
I think the Duke of Burgundy will fast
Before he'll buy again at such a rate.
'Twas full of darnel: Do you like the taste?

Henry VI: Act III, Sc. 2

This passage by Shakespeare describes darnel (*Lolium temulentum*), a plant often found within cereal fields that is best appreciated for its mammalian toxicity and bitter taste. Leemann (1933) presents a compelling case that the toxicity of this plant stems not from the seed itself, but rather from endophytic microorganisms that produce defensive chemicals. Although Leemann believes the endophyte to be

J.G. Lundgren, *Relationships of Natural Enemies and Non-Prey Foods*,
Progress in Biological Control 7,

Published by Springer Science+Business Media B.V.

a fungus, more recent research suggests that a nonfungal endophyte may be responsible for the observed toxicity (Faeth, 2002). Indeed, endophytic microorganisms, particularly fungi, are quite widespread in plants, and the functions of these often mutualistic (from the plant's perspective!) symbionts are being revealed.

15.1.1.1 Patterns in Endophytic Infections

Endophytes by definition live internally within the plant, but are not pathogenic to the host in a traditional sense. Evolutionarily, these endophytic mutualists are likely derived from plant pathogens (Clay and Schardl, 2002). Both bacterial and fungal endophytes are known, although the insect-resistant properties are better appreciated for the fungi. Nevertheless, endophytic bacteria can be quite abundant within plants (Vega et al., 2005), and the plant-insect relationships of bacterial endophytes merit further attention. Endophytic fungi generally stem from the Ascomycota, but are widespread throughout this order (Carroll, 1988). Most endophytes are transmitted horizontally, and are not contained in the embryo of the seed (Faeth, 2002). Exceptions to this are the fungal endophytes of grasses, of which some are passed vertically to the seeds (Faeth, 2002).

The fungal endophytes associated with grasses have been the target of substantial research efforts, in part due to the importance of cereal crops and rangeland forage, and the utility of these plants as turfgrasses (Brem and Leuchtmann, 2001; Clay and Schardl, 2002). The best understood endophytic fungi reside within the Clavicipitaceae (Clay, 1988), which form very close relationships with their grass hosts, nearly attaining organelle status in some cases (Clay, 1992; Clay and Schardl, 2002). One estimate is that 30% of pooid grasses harbor fungal endophytes (Faeth, 2002). In grasses, endophytes display one of three classes of reproductive strategies, (1) exclusive sexual reproduction, (2) exclusive vegetative reproduction, or (3) a little of both (Clay and Schardl, 2002). The endophytes that reproduce sexually are infrequently found in the seeds, and produce fruiting bodies. These species either prevent or reduce seed production in the plant, and so are better described as pathogens. The species that reproduce asexually are dependent on the host plant for their persistence, and are transmitted vertically to the plant's progeny via the seed. Although genetic recombination does not occur in these asexual species, they can hybridize with other strains. The beneficial qualities of endophytes are best appreciated for these asexual forms of endophytes of grasses (Clay and Schardl, 2002; Faeth, 2002; Hill et al., 2005), particularly for those in the genus *Neotyphodium*.

Although endophytes are widespread within the plant kingdom, there is considerable variability in the infection rates and within-plant distribution of endophytes. It is frequently the case that different populations of the same plant species have very different endophytic infection rates (Jensen, 2005). Within plants, many endophytic fungi live intercellularly and are present at different levels in the various tissues. In one case, the endophytic fungus, *Phomopsis casuarinae*, of *Casuarina equisetifolia* is present throughout the entire plant, except for the cotyledon and embryo (Bose, 1947). But the hyphae of the fungus are in the testa of the seed (Bose, 1947). In addition to variability in the infection rates within and among plants, the level of defensive chemicals produced by these endophytes is extremely plastic (Clay and Schardl, 2002; Leuchtmann et al., 2000).

15.1.1.2 Endophyte Function

One way that endophytes benefit their host is through the production of mycotoxins that are allelopathic to other plants, phytopathogens, and herbivores. In particular, many endophytes produce alkaloids (loline, lolitrem, and ergot peramine) (Clay, 1988; Clay and Schardl, 2002), normally not present in aposymbiotic grasses. Mycotoxins produced may also reduce plant competitors and microbiological pathogens. One test shows that 10 of 18 tested grass species produce alkaloids, presumably because of their endophyte mutualists (Leuchtmann et al., 2000).

Although not all grass endophytes alter feeding by herbivores (Saikkonen et al., 1999), several instances where endophyte infections have a strong influence on herbivore–plant interactions are in the literature. In one study, 9 of 11 endophyte-infected grasses have some inhibitory effects on insect feeding (Clay and Schardl, 2002). In one case, herbivores (mammalian and insects) affect the level of infection in tall fescue grasses under field conditions (Clay et al., 2005). There are often reproductive costs and reductions in host competitive ability inherent in endophyte infections (even in asexual grass endophytes) (Clay and Schardl, 2002; Richmond et al., 2003), but these costs are overcome by the contributions to plant fitness made by the endophytes. Thus, the endophyte-plant relationships persist (Clay, 1988; Clay and Schardl, 2002). Still, the interactions between endophytes and plants are complex and the ability of endophytes to reduce herbivory is by no means universal within plants (Faeth, 2002).

Endophytic microorganisms influence granivory of grasses by entomophages, which is understandable given that grasses are the only plants known to transmit their endophytes vertically to their offspring. The alkaloids produced by endophytes are present at their highest concentration in the seeds of ryegrass (*Lolium perenne*) (Ball et al., 1997), and in tall fescue, *Festuca pratensis* (Justus et al., 1997) (Fig. 15.1). Within the seed, the embryo of *F. pratensis* has two fold higher alkaloid contents compared with the rest of the seed (Justus et al., 1997). Endophytes of grass seeds reduce seed consumption by *Pogonomyrmex rugosus* and *Acheta domestica*. In *P. rugosus*, the ants collect infected and uninfected seeds equally, but discard the infected seeds to their middens, which incidentally are an excellent site for germination (Knoch et al., 1993). The endophyte *Acremonium loliae*, in association with ryegrass, is toxic to *Acheta domestica* (Ahmad et al., 1985). Essentially, the mycotoxins produced in the seeds interfere with the membrane permeability of the midgut epithelium, resulting in the dissociation of the gut lining from connective tissues. Ultimately, complete digestive failure occurs in the crickets, which soon die. Given that the seed is such a critical point in the life stage of the plant and that endophytic toxins are expressed at their highest levels in the seeds of grass species (Knoch et al., 1993), it is surprising that more interest isn't given to the importance of endophyte defenses against post-dispersal granivory.

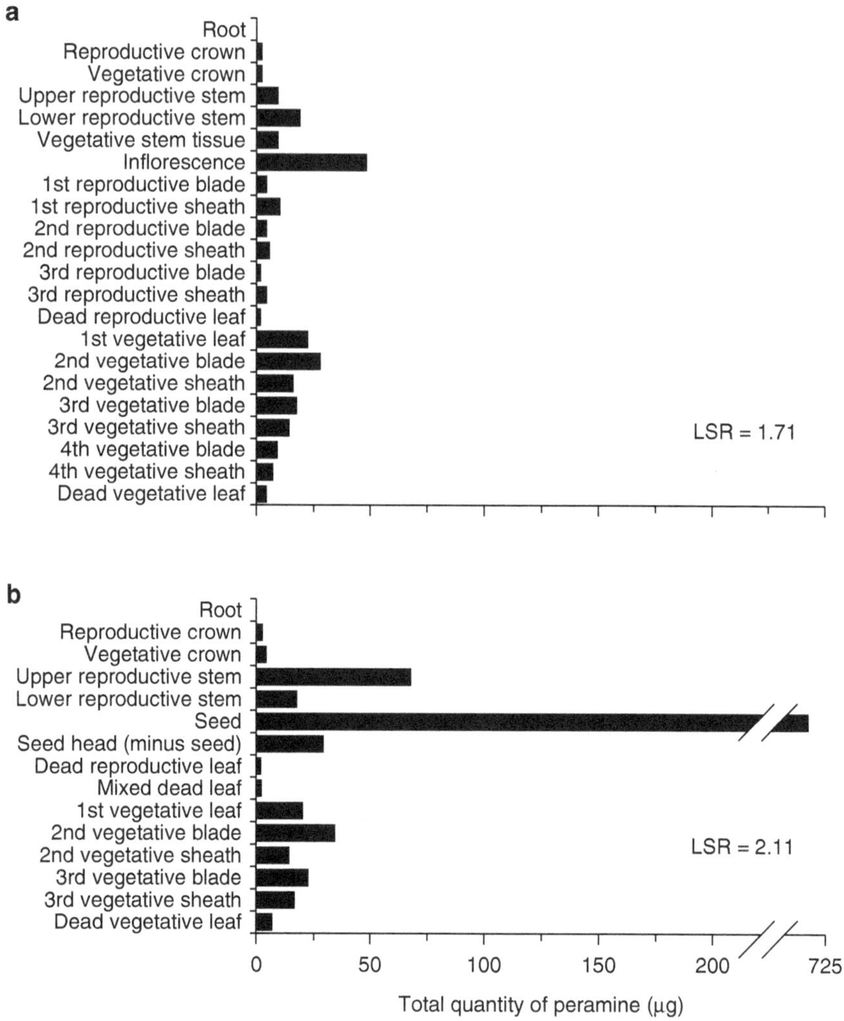

Fig. 15.1 Total quantities of peramine in different components of vegetative and reproductive tissues from endophyte-infected (*Neotyphodium lolii*) *Lolium perenne* plants (Reproduced from Ball et al., 1997. With permission from Springer)

15.1.1.3 Other Seed–Microbe Interactions

Finally, many non-endophytic fungi influence granivory rates by entomophagous insects. For example, seeds uninfected with fungus are removed at twice the rate of moldy seeds by the harvester ant *Pogonomyrmex occidentalis* (Crist and Friese, 1993). Also, some fungal associates may actually make seeds more attractive to granivorous entomophages. Grain infested by fungi may be more easily digested by natural enemies (Dicke, 1988a). Finally, ergot infections of *Paspalum dilatatum* render the seeds more attractive to the fire ant, *Solenopsis invicta* (Vinson, 1972). This is likely because the fungus produces a sugary secretion to which the ants are attracted (Vinson, 1972). The magnitude of interactions of microbial seed associates and granivory rates are still poorly understood, but could be very important to the ecology of both parties.

15.1.2 Nectar and Yeasts

> *It is a fact, which does not appear to be widely known, that the nectar of some flowers is frequently infected with one or more species of yeast, sometimes to such an extent that it is visibly fermented.* (Betts, 1920)

To think of nectar nutrition in the absence of microbes provides an incomplete view of things. Most floral nectars have a diverse microbial community, especially containing yeasts in the genera *Candida*, *Torulopsis*, and *Metschnikowia* (Ehlers and Olesen, 1997; Eisikowitch et al., 1990b; Lachance et al., 2001; Last and Price, 1969; Phaff et al., 1966; Sandhu and Waraich, 1985). In all, dozens of yeast species are isolated from nectar (Last and Price, 1969; Phaff et al., 1966); the nectars in one survey contain 36 yeast species from 12 genera (Sandhu and Waraich, 1985). It might be expected that yeasts should be quite pervasive in floral nectar (Phaff et al., 1966); after all, nectar presents a rich nutritional source for yeasts, and flower-visiting insects provide an excellent mechanism for transferring yeasts from flower to flower (Betts, 1920; Corbet et al., 1979; Eisikowitch et al., 1990b; Phaff et al., 1966). For example, in their examination of the yeasts associated with cactus flowers, Lachance et al. (2001) describes the strong associations between beetle pollinators and *Candida cleridarum*; 19 of 22 collected beetles carried the yeast.

The reality of the situation is that yeasts in nectar are not as ubiquitously present in flowers as one might expect. Although up to 90% of flowers in some habitats have yeasts (Eisikowitch et al., 1990a), and single flowers can support millions of yeasts (Phaff et al., 1966), typical yeast-nectar associations are somewhat less impressive. In most situations, far fewer (usually less than 50%) flowers are inhabited by yeasts (Ehlers and Olesen, 1997; Gilliam et al., 1983; Phaff et al., 1966; Sandhu and Waraich, 1985), and some flower species are entirely devoid of microbial inhabitants (Gilliam et al., 1983). The microclimate found in the flower, presence of antimicrobial secondary chemicals present in the nectar, and the efficiency of transmittal may all contribute to why some flowers have a rich microbial community, while others are practically pristine.

Microbial residents of nectar play an important role in the dynamic exchanges between flowers and insects. From a flower's perspective, troublesome nectar-inhabiting yeasts may reduce self pollination by inhibiting pollen germination, as seen in *Asclepias syriaca* and *Metschnikowia reukaufi* (Eisikowitch et al., 1990b). But, insects with greater residence times on a flower and that fly shorter distances once they have picked up a flower's pollen make the best pollinators from the plant's perspective.

Thus, when some nectar-microbes ferment nectar sugars into alcohols, they cause a 'drunken pollinator' syndrome that may promote more efficient pollination (Betts, 1920; Corbet et al., 1979; Ehlers and Olesen, 1997). Ehlers and Oleson (1997) implicates a *Cladosporium* species as the ethanol-producing yeast found in nectar of *Epipactis* flowers that both intoxicates and is transferred from flower to flower by fruit wasps. From the insect's perspective, yeasts can alter the odor and concentrations of nectar as well as its nutritional composition, possibly affecting its attractiveness and suitability for glucophagous species (Corbet et al., 1979; Kevan et al., 1988). However, one of the only studies to examine this theory found that foraging bees are unaffected by the presence of microorganisms found in *Asclepias syriaca* nectar (Kevan et al., 1988). Regardless, it appears that glucophagous entomophages regularly consume yeasts along with sugar-meals, and the implications of this inadvertent mycophagy for the physiology of glucophagous entomophages remains to be examined.

15.1.3 Sooty Molds and Honeydew

Microorganisms affect the nutrition and attractiveness of honeydews in several ways. First, endosymbionts of sternorrhynchans alter the sugar content of honeydew before it is secreted. More specifically, bacterial symbionts synthesize some of the trisaccharides frequently encountered in honeydews (Bates et al., 1990; Davidson et al., 1994; Tarczynski et al., 1992). As discussed in the glucophagy section, these trisaccharides have important implications for the interactions with natural enemies. Once the honeydew enters the environment, it soon is colonized by a broad microbiological community, the most conspicuous of which are termed sooty molds (Hughes, 1976).

> *The term sooty mould has been and continues to be variously applied. In its broadest sense it has included not only superficial saprophytes but also certain parasitic fungi whose dark, conspicuous, superficial hyphae insert a variety of absorption mechanisms into the plant tissues.* (Hughes, 1976)

All sooty molds are members of the Ascomycetes; other fungal groups do not possess the same pigmentation characteristics present in the sooty molds. Although best appreciated for their symbiosis with honeydew, these fungi also occur in association with other sugar sources like those from glandular trichomes and EFNs (Hughes, 1976). Sooty molds affect the nutrition of honeydew-feeding natural enemies in at least two ways. First, once established, the sooty moulds alter the nutritional landscape of the phylloplane by exuding sugars, pectic acids, and sugar alcohols (Hughes, 1976). Also, the entomophagous species that consume honeydew also eat the sooty molds alongside the sugars. Sheldon and MacLeod (1971) describe the fruiting bodies of the sooty molds, *Helminthosporium*, *Alternaria* (mostly), *Piricauda*, and *Fumago* in the guts of *Chrysoperla carnea* and *Chrysopa nigricornis* collected in the field. Coccinellids are another natural enemy that frequently has sooty mold in its gut. At least 26 coccinellid individuals (many of which were *Coleomegilla maculata*) from one survey ingest spores of

Alternaria, a sooty mold found in peach orchards (Putman, 1964). Although this may be incidental consumption along with the desired honeydew, data from Triltsch (1997, 1999) indicate that coccinellids are more mycophagous than many of us give them credit for.

In fact, feeding on fungi associated with honeydews may have pre-dated actual entomophagy of sternorrhynchan prey that produce the honeydew, and may have led to the evolution of dietary specialization in predaceous beetles, especially within the Coccinellidae (Leschen, 2000). Briefly, those clades that were ancestrally predaceous have not specialized on aphids. But many predaceous beetles with mycophagous ancestors have tended to evolve into aphid specialists. In examining feeding records and phylogenies, Leschen concluded that specialization on sternorrhynchan prey from mycophagous ancestors has likely occurred in 10 out of 11 specialized predatory beetle groups. Thus, sooty molds may have helped to form the current suite of aphidophagous predators.

15.2 Nutritional Symbionts of Entomophagous Species

> *For the most part it was some insufficiency in food sources which led to the establishment of symbiosis, or better stated, certain food sources became available to the animals only after they had symbionts at their disposal to compensate for the deficiencies.* (Buchner, 1965)

Microbial symbioses contribute to the nutrition of an arthropod when (1) the diet of the arthropod lacks specific nutrients entirely, and (2) when required nutrients are present in the diet but are unavailable because of a lack of metabolic tools in the insect or when the nutrient is bound to indigestible compounds. Given that at first glance natural enemies consume nutritionally robust foods for at least part of their lives, it is easy to understand Buchner's early proclamation that predatory species are disinclined toward nutritional symbioses

> *...types of symbiosis which play a role in the metabolism of animals are lacking, above all, in predators...*

But as is repeated in the book you are reading, the diets of arthropods are much less defined than many would prefer. Non-prey foods are extremely heterogenous in their nutrition and defense, and these traits restrict which organisms can consume them. Understanding how nutritional symbioses contribute to the acceptance of non-prey foods by entomophagous species will undoubtedly advance our understanding of how ordinarily carnivorous organisms can make a living as vegetarians, and ultimately how these mutualisms have driven, or at least facilitated, the evolution of herbivory within arthropods. Still, most research only scratches the surface as to how nutritional symbionts contribute to facultative phytophagy in entomophagous species.

A tremendous diversity of microorganisms have formed tight relations with arthropods, including plasmids, protozoa, bacteria, yeasts, and higher fungi (Campbell, 1989, and numerous references therein). Insects are often born with sterile guts, and consume a range of microbes during their lifetimes (Chapman, 1998). Koch (1960) states that

> *Without doubt, the majority of symbionts were first taken up with nourishment by way of the mouth.*

Thus, when the microbial community of an insect's gut differs from that of the surrounding habitat, it may be worth exploring for mutualistic symbioses. Because microbes found on a food are likely adapted to digesting this particular substance, it makes sense that an insect benefits from harnessing the talents of these microbes (Martin, 1992). The diversity of feeding modes, even over the ontogeny of an insect, supports symbioses among a diverse range of arthropods and microbes, and even different life stages of the same insect species may develop distinct symbioses (Jones, 1984).

15.2.1 Physiological Adaptations to Symbioses in Insects

There seems to be no end to the diversity of internal physiological structures where microbes of nutritional relevance to insects may reside, but close proximity to some portion of the digestive tract is commonly a prerequisite for these adaptations (Koch, 1960). The most obvious location of microbes of nutritional relevance to arthropods is within the gut lumen itself. Chapman (1998) points out that those insects with straight, simple digestive tracts have fewer nooks and crannies for symbiont communities to persist. Of the three main portions of the gut, the hindgut most frequently houses symbionts in insects (Bignell, 1984; Chapman, 1998; Koch, 1967). Within the digestive tract, microbial populations are often maintained and spatially restricted by the physiological conditions of the gut, especially pH (Bignell, 1984; Haas and Konig, 1988). For instance, in the detritivore, *Tipula abdominalis*, a narrow region of the midgut is maintained at a high pH (near 11). This is where Martin et al. (1980) believe symbiotic bacteria with strong proteolytic capabilities might reside and assist in the digestion of proteins bound to indigestible molecules (like tannins).

Any caecae, pockets, or structural anomalies of the gut may be specializations for housing symbiotic microbes (Koch, 1960). Chrysopid adults house yeasts within a large diverticulum that attaches to the posterior end of the foregut (Hagen and Tassan, 1966). Forbes (1892) provides a nice description of the diversity of caecae found in association with the midguts of various heteropterans, all of which house bacteria. Lygaeidae, Pentatomidae, Scutelleridae, Pyrrhocoridae, and Coreidae all have a series of gastric caecae that vary in their complexity and arrangement (see also Koch, 1967). In the cinch bug, between five and eight caecae radiate from a single point just before the end of the midgut. It is fascinating that even closely related genera within Heteroptera can vary dramatically in the arrangement (and even presence or absence) of these gastric caecae. Finally, the strongest associations between microbes and insects are manifested in specialized cells that house intracellular microorganisms, termed mycetocytes and mycetomes (Campbell, 1989; Chapman, 1998; Tanada and Kaya, 1993).

15.2.2 Nutritional Functions of Microbial Symbioses

Jones (1984) rightly highlighted the fact that microbes are seldom ubiquitous in their taxonomic and functional associations with arthropods. This is to say that no single microbe occurs widely in insects that remedies a taxonomically widespread nutritional deficiency. The functional outcome of each microbe-insect symbiosis needs to be considered on a case-by-case basis. Still, it is a useful exercise to briefly summarize the known contributions made by symbiotic microorganisms to their host's nutrition. The nutritional roles of symbiotic microbes can be typified as either contributing specific key nutrients required by arthropods, or enhancing or augmenting nutritional processes in the host. Studies on the contributions of microbes to insect nutrition should focus mainly on the nutrients that are consistently lacking from an insect's food source, but are critical to the fitness of the insect (Jones, 1984).

Symbiotic microorganisms can provide key nutrients to arthropods by serving as a food themselves, or by producing specific nutrients from substrates which insects cannot metabolize. In the former case, the microbes serve as the sole food source for the arthropod, or as a nutritional supplement to a suboptimal diet (Vega and Dowd, 2005; Jones, 1984). One example of this is observed in Diplopoda, in whom the yeast symbionts typically associated with Malpighian tubules venture into midgut, where they are digested (Byzov et al., 1993). More frequently reported are instances where the microbe provides some nutrient that arthropods cannot synthesize on their own. Most notable are B-vitamins (Akman et al., 2002; Campbell, 1989), sterols (Campbell, 1989; Morales-Ramos et al., 2000; Wetzel et al., 1992), and amino acids created from non-essential amino acid precursors (Campbell, 1989; Gil et al., 2003; Prosser and Douglas, 1991; Shigenobu et al., 2000).

In addition to the direct provision of nutrients to insect hosts, symbiotic microorganisms also augment normal digestive processes, allowing insects to make better use of what they eat. This result may arise from more efficiently extracting dilute nutrients from a food source, or providing the necessary enzymes to metabolize foods or byproducts more efficiently (Jones, 1984). Microbes digest molecules that many insects cannot, especially some of the structural and storage polysaccharides found in plant material (Martin, 1992). Specifically, microbes produce cellulases, pectinases, ligninases, and chitinases that enable arthropods to extract the most energy from their foods (Breznak and Brune, 1994; Campbell, 1989; Hogan et al., 1985; Howard et al., 1985; Hungate, 1938; Martin, 1984). In the omnivorous *Acheta domestica*, the hindgut microflora contributes a range of carbohydrases that broadens the suite of nutrients that can be extracted from low quality food

(Kaufman and Klug, 1991) (Fig. 15.2). Pectin, amylose, xylan, raffinose, and locust bean gum are all digested more efficiently in crickets with their symbionts intact versus in aposymbiotic conspecifics. When the quality of the dietary carbohydrates are deliberately changed periodically over nymphal development, the aposymbiotic crickets only grow when the high quality carbohydrates are offered; those with their symbionts intact grow even when times are nutritionally tough (Kaufman and Klug, 1991). It is worth noting that yeasts are unable to digest cellulose (Phaff et al., 1966), and therefore these symbionts do not fulfill this role in insects. In addition to providing digestive enzymes to arthropod hosts, microbes also contribute to the storage and recycling of nitrogen (Byzov et al., 1993; Campbell, 1989; Cochran, 1985; Douglas, 1998; Potrikus, 1981), sulfate assimilation (Douglas, 1998; Shigenobu et al., 2000), and fatty acid metabolism (Campbell, 1989 and references therein) in their hosts.

A final function of microbial symbionts is the detoxification of plant allelochemicals (and insecticides) harmful to the arthropod hosts (Vega and Dowd, 2005; Jones, 1984). In one early research system, a bacterial symbiont of *Rhagoletis pomonella*, *Pseudomonas melophthora* detoxifies six different insecticides under in vitro culture (Boush and Matsumura, 1967). Whether this happens in the host, or how the host accommodates the insecticide breakdown products is unknown. Still, the detoxification of plant secondary compounds (e.g., terpenoids) is commonly accomplished in insects by bacterial associates (Campbell, 1989 and references therein), and it is conceivable that these detoxification capabilities facilitate the degradation of pesticides in insects as well (Berenbaum, 1988). Of relevance to this book is the detoxification capabilities of microbe-symbiotic natural enemies that

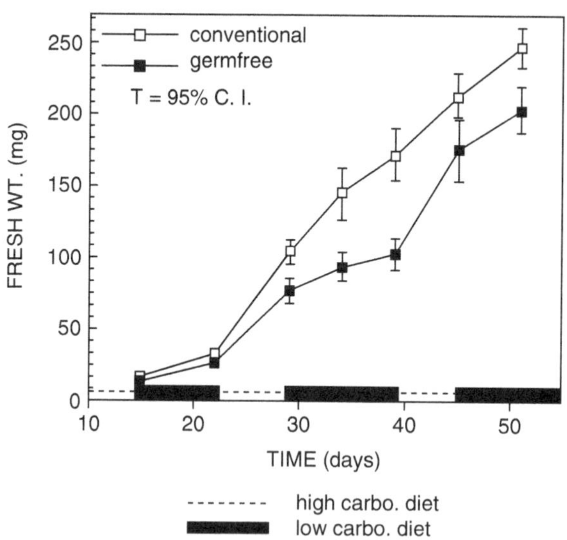

Fig. 15.2 Growth of aposymbiotic (germfree) and symbiotic (conventional) *Acheta domesticus* reared on an alternating diet regime with high and low quality foods (Reproduced from Kaufman and Klug., 1991. With permission by Elsevier)

allow them to consume chemically defended non-prey foods, but I know of no reports of this in the literature.

15.2.3 Natural Enemies and Microorganism Associations

> *Although there are many scattered reports on the bacteria associated with insects, knowledge concerning the bacterial flora of insects in general is markedly scant. Since there are over 600,000 known species of insects, it is apparent that a considerable amount of work will be necessary before even a limited survey of this field can be accomplished. In the meantime it seems logical that a study of a few representative species of Hexapoda might be worthwhile.* (Steinhaus, 1941)

15.2.3.1 Chrysopidae

Since first discovered 85 years ago (Cowdry, 1923), and popularized nearly 50 years ago (Hagen and Tassan, 1966), the microorganisms (especially yeasts) associated with green lacewings have been the focus of a considerable body of research that provided an excellent example of the microbial promotion of dietary breadth in arthropods best appreciated for their entomophagous tendencies. In spite of a wealth of information that has been acquired regarding the symbiotic relationships between these two sets of organisms, the microbial contributions to the nutrition of lacewings has not yet been empirically substantiated.

Ken Hagen and colleagues prompted a long-lived line of research that continues to this day with the following statement:

> *Budding yeasts were found in the diverticulum of the adult foregut of* [Chrysoperla] carnea. *The yeast was found in both the laboratory stock as well as field collected specimens. It is speculated that the yeast may play a mutualistic role in the synthesis of essential metabolites in the host which are often lacking in the natural adult diet, honeydew.* (Hagen and Tassan, 1966)

Adult lacewings was shown to possess a large diverticulum that joined the digestive tract directly anterior to the midgut (Fig. 15.3). This diverticulum was often filled with yeasts, initially assigned to the genus *Torulopsis* (Hagen and Tassan, 1972; Hagen et al., 1970). Investigations of larval guts did not produce yeasts, and Hagen et al. concluded that the relationship between lacewings and yeasts was a loose one, perpetuated when adults consumed the yeasts in contaminated honeydew or nectar which were then passed among a lacewing community through trophalaxis. Perhaps even more interesting, female lacewings were shown to possess tracheae of broader diameter that presumably provided additional oxygen to the symbiotic yeasts (Hagen et al., 1970).

A number of recent explorations validate that lacewings are frequently host to a diverse community of microorganisms, many of which are yeasts in the genera *Candida* and *Metschnikowia* (= *Torulopsis*, in part) (Chen et al., 2006; Nguyen et al., 2006; Suh et al., 2004; Woolfolk and Inglis, 2004). Although Hagen et al. (1970) suggest that European and North American *Chrysoperla carnea* have different

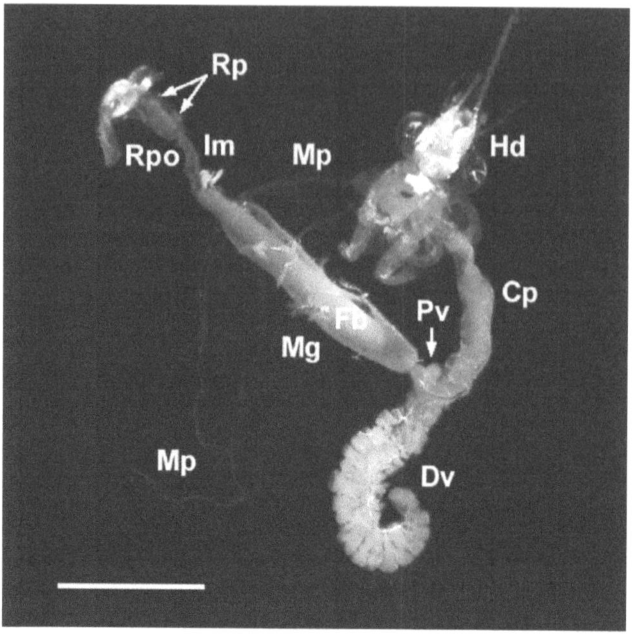

Fig. 15.3 Alimentary canal of an adult *Chrysoperla rufilabris* where symbiotic yeasts reside. Cp, crop; Dv, diverticulum; Fb, food bolus; Hd, head; Im, ileum; MG, midgut; Mp, Malpighian tubules; Pv, proventriculus; Rp, rectal pads; Rpo, rectal pouch (Reproduced from Woolfolk et al., 2004. With permission from the Entomological Society of America)

yeast symbionts, more recent descriptions reveal a fairly strong fidelity between lacewings and certain yeast species (Suh et al., 2004). Specifically, the same yeasts *Candida pimensis* and *C. picachoensis*, are found in the guts of lacewings collected in geographically distant populations (Arizona and Louisiana), suggesting that either these yeasts are extremely widespread, or that there is a strong relationship between yeast and host (Nguyen et al., 2006). *Metschnikowia pulcherrima* are found in the guts of all *Chrysoperla rufilabris* that have yeast symbionts (71% of individuals), with cell counts on the order of 5×10^2–10^5 (Woolfolk and Inglis, 2004). Transient filamentous fungi and bacteria are also found in the guts of this lacewing species. Recently eclosed lacewings do not possess yeast symbionts (Woolfolk and Inglis, 2004), and likely have to acquire them from the environment. Not found in Hagen's early work is that the larvae of *Chrysoperla carnea* also host a diverse community of microorganisms, but few yeasts (Chen et al., 2006). Chen et al. suggest that because the larval midgut is closed at the junction to the hindgut, the dense populations of free-living bacteria in the midgut lumen may be important in maximizing the digestion of the food material. Finally, Gibson and Hunter (2005) describe how *Chrysoperla* mothers transfer yeasts to their egg surfaces, but lack of yeasts in the larvae of lacewings and newly eclosed adults seems to preclude vertical maintenance of the symbiosis.

Several physiological adaptations promote the symbioses between yeasts and lacewings. Canard and colleagues (Canard, 2001; Canard et al., 1990) expound upon the initial observation of Hagen et al. (1970) that yeast-bearing *Chrysoperla carnea* have increased tracheation to the diverticulum. Specifically, Canard et al. (1990) provide data that clearly show that glucophagous lacewing adults have tracheae of greater diameter near the diverticulum compared to predaceous lacewing species. These morphological adaptations to symbiosis with yeasts suggest a strong physiological benefit to the lacewing. Within the digestive system of the adults, the yeasts are restricted to the foregut and diverticulum (Woolfolk and Inglis, 2004), but some make their way into the midgut lumen and hindgut (Chen et al., 2006; Woolfolk and Inglis, 2004). This residency pattern may be reinforced by the presence of forward-pointing hairs that line the proventriculus of *Chrysoperla* (Woolfolk et al., 2004) (Fig. 15.4). Little absorption occurs within the diverticulum itself, which is lined with cuticle (Woolfolk et al., 2004).

The repeated demonstration of diverse yeasts in the guts of lacewings, and the morphological adaptations to housing the yeasts in the lacewing diverticulum suggest that these yeasts are providing some beneficial function to the lacewing. The honeydew diet of chrysopids with the strongest yeast associations does not contain the requisite nutrients for maximum oogenesis (Hagen and Tassan, 1972). Honeydews have minimal amino acid contents, and feeding trials suggest that the yeast symbionts may be supplementing the diets of glucophagous lacewings with the key amino acids, valine, threonine, and phenylalanine (Hagen and Tassan, 1972; Hagen et al., 1970). These authors also point out that the yeasts do not improve oviposition rates when fed certain honeydews, and recognize that the interactions between yeasts and lacewings may be more complex. Many of Hagen et al.'s assertions regarding the function of yeasts in lacewings are recently challenged by Gibson and Hunter (2005), largely based on the inability to replicate the results of these early experiments. First, Hagen et al. use sorbic acid to cure their lacewings, which does not produce aposymbiotic individuals in Gibson and Hunter's experiments. Gibson and Hunter produce aposymbiotic lacewings using cycloheximide, but they don't

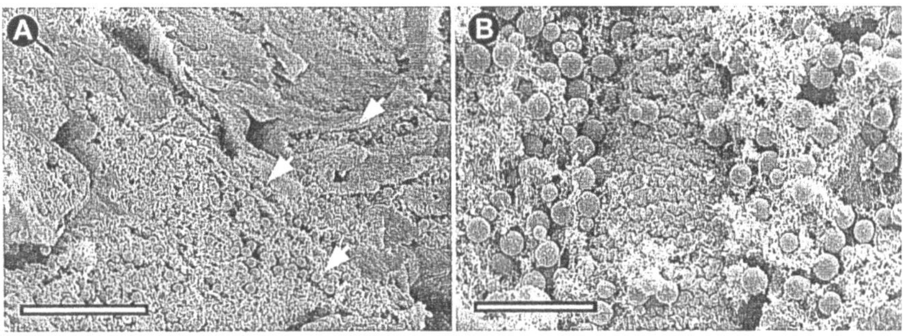

Fig. 15.4 Scanning electron micrographs of *Chrysoperla rufilabris* adult foregut with high populations of yeast (arrows) in between fold structures within cryofractured foregut. Bar 50 μm (Reproduced from Woolfolk et al., 2004. With permission from the Entomological Society of America)

use this treatment to explore the nutritional contributions of the symbionts. The latter work shows that the number of yeasts is significantly correlated with total fecundity in the females, but the authors challenge the notion that this improvement is based on amino acid content in the diet. In replicating Hagen et al.'s amino acid supplementation experiments, a consistent trend in Gibson and Hunter's work is that inclusion of the amino acid valine seems to actually reduce both yeast count and realized fecundity in the lacewings. The idea that lacewings consume honeydew and nectar in part to feed their yeast symbionts, which then form part of the diet in the lacewing is suggested by Woolfolk et al. (2004) and Chen et al. (2006), but no empirical studies test this hypothesis. The bottom line is that the contribution of yeast symbionts to the nutrition of lacewings remains to be imperviously established.

15.2.3.2 Crickets

Crickets are one of the 'lab rats' used to understand the contributions of symbiotic microorganisms to insect nutrition. A number of species have bacterial residents in their guts (Kaufman et al., 1989; Steinhaus, 1941), and these symbionts are maternally inherited in some species (Koch, 1960). In *Acheta domesticus*, microorganisms residing in the hindgut enhance the diversity of polysaccharides that can be consumed by the insect (Kaufman et al., 1989). These microorganisms allow the cricket to adapt to a changing diet more quickly, thereby speeding development and improving fecundity over aposymbiotic crickets when both groups are faced with a suboptimal diet, or a suboptimal diet periodically alternated with a more digestible one. Gryllids possess the same carbohydrase profile in their hindgut as *A. domesticus*, suggesting a similar symbiosis with microorganisms (Kaufman and Klug, 1991). Of reference to granivory, removal of microbial gut populations with antibiotics is associated with a 40% reduction in seed (*Chenopodium album*) consumption by *A. domesticus* (J. G. Lundgren , unpublished data 2006).

15.2.3.3 Heteroptera

A number of heteropterans have symbiotic microorganisms that facilitate herbivory, and although there are numerous reports on the most entomophagous heteropterans hosting a diverse gut community, the functions of these symbionts remain a mystery. Forbes (1892) is one of the first to document specialized organs (gastric caecae) that house bacteria in several families of Heteroptera, but especially in the cinch bug. He even notes that the caecae of the Coreidae have increased tracheation, similar to the diverticula of lacewings described above. After identifying the bacterial associate of the chinch bug as *Micrococcus insectorum*, he basically ceases this line of research.

> *I have no present desire to speculate concerning the meaning of the bacterial contents of these glands, but limit myself to this preliminary account...* (Forbes, 1892)

Glascow reveals some of Forbes' later unpublished findings, summarizing them as

> [Forbes] *also established later, by the examination of a great variety of insects of different orders, and especially of Heteroptera, that the chinch bug was not unique in this regard, but that the same phenomenon also occurred in a number of other species of Lygaeidae as well as in representatives of several other families of Heteroptera, and that wherever the caeca were present in this group, they were always filled with specific bacteria.* (Glascow, 1914)

Glascow advances these initial observations by determining that microorganism species vary among hosts, but are fairly consistent within a host species. Moreover, he found that the bacteria, which cannot be cultured, can be maternally inherited by offspring. A number of other researchers have since frequently found microbial associates of herbivorous heteropterans (Haas and Konig, 1988; Hosokawa et al., 2007; Kikuchi et al., 2007; Koch, 1967; Martin et al., 1987; Prado et al., 2006; Ragsdale et al., 1979), and blood-sucking triatomine reduviids (Buchner, 1965; Koch, 1967).

The symbiotic microbes of entomophagous species inarguably receive less attention from biologists, although there are clues that point to dynamic relationships worth considering in these natural enemies. Of the most predaceous species, Nabidae, Reduviidae, *Geocoris uliginosus* and *Podisus maculiventris* do not have the caecae necessary for harboring bacterial symbionts (Forbes, 1892; Glascow, 1914). This doesn't preclude them from having nutritional symbionts. Indeed, Glascow describes the same species of bacteria in the guts of *P. maculiventris* (present in 50% of individuals) as is seen in the caecae of the herbivore *Holcostethus limbolarius*. And Cowdry (1923) describes fungi and bacteria living in the guts of a *Nabis* species. Other predaceous heteropterans have at least transient bacterial residents in their guts as well; *Sinea diadema* (Reduviidae) and *Lygus pratensis* (a facultatively entomophagous Lygaeidae) both have two bacterial species residing in their guts (Steinhaus, 1941). Also, gerrids have a number of gut symbionts, which do not exceed 36% infection of surveyed insects and are not pathogenic to the bugs (Klingenberg et al., 1997). Still, the absence of caecae does suggest that the relationships are weaker in entomophagous species than they are in herbivorous and blood-sucking heteropterans, and the contribution of these resident microorganisms to the nutrition of entomophagous insects is entirely unknown.

15.2.3.4 Coleoptera

With the recent exception of the Carabidae, very little is known of the microbial residents of predatory beetles. One report states that flower-visiting cantharid adults (*Raxonycha* species) have yeasts (*Metschnikowia corniflorae*) in their guts (Nguyen et al., 2006), and another species has fungi in its gut (Cowdry, 1923). A number of unidentified yeasts were isolated from the stomachs of basidiocarp-dwelling carabids and staphylinids (Suh and Blackwell, 2005). Buchner (1965) discusses how mycetomes are present in the Malpighian tubules of pollinivorous *Dasytes* females, but not the males. In his broad survey for microbes in the digestive tracts of insects, Steinhaus (1941) determines that the lampyrid, *Photinus pyralis*, *Coccinella novemnotata*, and a coccinellid larva have bacteria in their guts. *Adalia bipunctata* is known to have a *Rickettsia* inhabitant in its intestinal lumen (Cowdry, 1923). Although Forbes (1892) mentions that carabid beetles have gastric caecae similar to the Heteroptera, no bacteria are associated with these organs according to his notes. Cowdry (1923) describes the presence of bacteria in the gut lumen of *Cicindela punctulata punctulata* and numerous *Rickettsia* within the digestive tract of *Anisodactylus agricola*.

Recent explorations by myself and colleagues reveal that carabids have a rich bacterial community associated with their digestive tracts, and one function of these bacteria may be in facilitating seed consumption by facultatively granivorous species. Direct cell counts of the bacterial community in the guts of field-collected (and primarily predatory) *Poecilus chalcites* reveal 1.5×10^8 bacteria ml^{-1} of gut (Lehman et al., in press), and reducing the dietary breadth of the insects through lab culture reduces the bacterial diversity within this species. In another study, the guts of two granivorous carabids, *Harpalus pensylvanicus* and *Anisodactylus sanctaecrucis*, both harbor simple yet consistent bacterial communities that are distinct between the two carabid species (Lundgren et al., 2007). Perhaps what is more interesting is that curing these two species of their bacterial associates with antibiotics reduces the consumption of *Chenopodium album* seeds by 40% (Lundgren et al., 2007). The exact contributions of these gastric bacteria to digestion of seed material remain to be established.

15.2.3.5 Formicidae

Blochmann's early discovery that *Camponotus ligniperdus afer* and *Formica fusca fusca* harbor endosymbionts was the first report of an insect-microbe mutualistic symbiosis (Buchner, 1965; Koch, 1960) (note that another *Camponotus* also has bacterial gut residents; Cowdry, 1923). Based on embryonic development, Koch (1967) believed that congeners *F. rufa rufa* and *F. sanguinea* may have had similar bacterial symbioses to those of *F. fusca fusca* that have secondarily been lost. More recent work shows that nutritional symbiosis between microbes and omnivorous ants may be much more pervasive and integral than early workers could have imagined. In *Camponotus floridanus*, bacterial symbionts in the genus *Blochmannia* (especially *floridanus*) provide a series of amino acids to the workers, most notably tyrosine and the essential amino acids phenylalanine and methionine (Feldhaar et al., 2007; Zientz et al., 2006). Also important, these bacteria aid in nitrogen metabolism and reducing sulfate to sulfide for their host (Feldhaar et al., 2007; Zientz et al., 2006). Aposymbiotic workers that are amino-acid limited produce fewer brood than symbiotic workers and aposymbiotic workers supplemented with a dietary source of amino acids (Feldhaar et al., 2007). These bacteria are intracellular, living within specialized baceteriocytes housed in the midgut epithelium and in the ovaries, and they are likely transmitted vertically to the brood (Wolschin et al., 2004). Several other genera within the Formicinae are known to harbor this bacterium, and Feldhaar et al. (2007) speculate that this association has persisted for 30–40 million years. Indeed, these symbiotic interactions may have facilitated the current species diversity present in this genus.

Other ants also harbor gut bacteria that may contribute to their nutrition. *Tetraponera* species have a number of bacterial associates in their guts (Stoll et al., 2007), as do *Solenopsis* colonies. In an elegant study pertaining to nutritional symbionts of fire ants, Ba and Phillips (1996) determine that colonies of *Solenopsis invicta* in the southern U. S. are associated with several *Candida* species and *Debaryomyces nasenii* var. *hansenii*. The most abundant microbes (90% of the microbial community) are *Candida parapsilosis* and *C. lipolytica*, and these species are geographically widespread and prevalent throughout the season. Very few of the adults (3.27%) and third instars harbor the yeasts, but 80% of fourth instars in 100% of the colonies are symbiotic, mostly with *C. parapsilosis*. In that this life stage is where most of the digestion of solid food material occurs within the colony, it seems likely that the yeast is providing some nutritional function. Ba and Phillips (1996) discuss evidence that points toward the yeasts' contribution of the sterols, ergosterol and zymosterol, to the nutrition of the colony.

15.2.3.6 Other Natural Enemies

A handful of explorations describe the microbiota of the digestive tracts of a range of other natural enemies. Yeast-like intracellular symbionts are apparently transferred vertically to the progeny of the ichneumonid *Pimpla turionellae* (Middledorf and Ruthmann, 1984). These yeasts are found in the hemolymph and fat body of the adult wasp, and may contribute to the wasp's nutrition through the frequent passage of large vacuoles across the cell membrane. For the most part, the function of yeasts associated with parasitoids is largely unknown (Vega and Dowd, 2005). The ichneumonids, *Casinaria infesta* and *Echthromorpha maculipennis*, also are associated with bacteria (Cowdry, 1923). Predatory wasps also have bacterial associates (Cowdry, 1923). Of the entomophagous Diptera, two out of four syrphid adults (*Eristalis* species) tested have bacteria in their guts (Cowdry, 1923).

The Arachnida also have known bacterial associates. *Enterobacter* was isolated from *Galendromus occidentalis* (Hoy and Jeyaprakash, 2005), and the bacterial symbiont *Aranicola proteolyticus* (= *Serratia proteamaculans*) contributes the broad-spectrum protease, arazyme, that may be useful in the digestive processes of the spider *Nephila clavata* (Bersanetti et al., 2005). The jumping spider, *Salticus scenicus*, may have maternally inherited microbes associated within its haemocoel (Cowdry, 1923). A crab spider, *Misumena vatia*, has many protozoa-like microbes in its gut, and *Leiobunum vittatum dorsatum* has intestinal bacteria as well (Cowdry, 1923). Also, Cowdry found some microbes in the guts of the centipedes *Scolopendra subspinipes subspinipes*, and *Scutigera forceps*. Beyond these superficial descriptions, little is known concerning the functions of these relationships between other natural enemies and their gut microbes.

15.3 Conclusions

While there are numerous descriptions of the microbial diversity symbiotic with non-prey foods or the digestive systems of entomophagous species, we are only just beginning to understand the importance of these interactions. Nevertheless, some putative ecological functions and processes are assigned to microbes (alterations in the palatability of non-prey foods to natural enemies, and increasing the dietary breadth of entomophagous species) but their extent needs to be further resolved. It is clear that microorganisms play an important part in the lives of natural enemies, and how symbiotic interactions drive the evolution of omnivory in entomophagous species and their relationships with non-prey foods will likely prove to be a fruitful branch of research in the near future.

Section V
Applied Aspects of Non-Prey Foods for Natural Enemies

This book has set about describing the physiological and evolutionary importance of omnivory in entomophagous species. As mentioned in the introduction, understanding the biology and ecology of key top-down sources of pest mortality is a crucial first step in managing natural enemy populations. The next three chapters will discuss how knowledge regarding the breadth of diet within most natural enemies affects the ecological services provided by these insects, and how these characteristics inherent in many natural enemies can be used to manipulate their function as biological control agents. Moreover, omnivory by entomophagous species opens up new linkages within food webs that have important implications for higher trophic levels. Specifically, omnivory on non-prey foods reveals a whole suite of ecological pathways whereby farming decisions pertaining to pest management and cultural techniques can affect biological control agents.

Recognizing the nutritional importance of non-prey foods to the ecology of natural enemies, and that omnivorous natural enemy populations track non-prey resources, allows cropping systems to be manipulated in ways that encourage more predictable outcomes of biological control. Also, recognition of their true feeding behavior allows us to tailor IPM systems that minimize the direct effects of herbivore resistance in crop plants to higher trophic levels. Finally, it gives us the opportunity to use omnivorous arthropods to manage pests that are themselves a form of non-prey food, namely weed seeds and fungal phytopathogens.

Chapter 16
Non-Prey Foods and Biological Control of Arthropods

> *Thus, the view is unrealistic that a given species is a poor predator only because it accepts alternate foods. Only if the preference of alternate foods results in neglect of the prey is this view necessarily valid.*
>
> Huffaker and Flaherty, 1966

Non-prey foods are an inextricable part of the diet of most entomophagous species, and thus provide a way in which biological control practitioners can manipulate the efficacy of natural enemies. The unique nutritional ecology of natural enemies comes to play in improving classical and augmentative releases of biological control agents, and providing for the nutritional needs of natural enemies can be a powerful tool that allows land managers to conserve and promote endemic natural enemies within designated areas. Although providing for the nutritional needs of natural enemies intuitively enhances biological control, there are sometimes complex and unintended repercussions on entomophagous communities resulting from this practice. While this fact should not preclude the incorporation of non-prey foods into cropland, both the benefits and limitations of this strategy need to be considered when applying non-prey foods in an IPM context.

16.1 Improving Biological Control of Arthropods Using Non-Prey Foods

16.1.1 Improving Natural Enemy Releases

Recognizing and providing for the distinct physiological needs of species under consideration for classical or augmentative biological control programs may influence whether releases of these natural enemies establish (in classical programs) and the duration of their effectiveness (in augmentation programs). Indeed, among arthropods, predators and parasitoids are some of the poorest colonizers of highly

J.G. Lundgren, *Relationships of Natural Enemies and Non-Prey Foods*,
Progress in Biological Control 7,

Published by Springer Science+Business Media B.V.

disturbed areas where pests are most abundant (Landis and Menalled, 1998). This concept of combining classical and augmentation programs with strategies often associated with conservation biological control is termed 'integrated biological control' (Gurr and Wratten, 1999). Combining habitat management, particularly as it pertains to the provision of non-prey foods, with the initial release and subsequent integration of the natural enemy into a recipient region will eliminate one causative factor for the occasional failure of biological control.

16.1.1.1 Classical Biological Control

> No organism can survive in a new environment where any individual factor essential to its existence is lacking. It is the neglect of this principle, which may be considered almost a biological axiom, that in many cases has prevented other parasites from becoming equally effective in the control of their specific injurious host. (Wolcott, 1942)

Numerous scientists seeking to introduce natural enemies into new biota notice that failure to accomodate the nutritional ecology of natural enemies may partially explain the low rate of establishment commonly associated with classical biological control (Beirne, 1962, 1975; Bugg and Pickett, 1998; Gurr and Wratten, 1999; Jervis et al., 1996b; Townes, 1958; Wolcott, 1942). Indirect evidence that supports this hypothesis comes from the greater level of successful introductions experienced in more stable habitats (Hall and Ehler, 1979); in addition to numerous other qualities beneficial to natural enemies, these types of habitats have a greater diversity and abundance of non-prey foods. With few exceptions, efforts in establishing natural enemies focus primarily on interactions of the natural enemy and potential host/prey and largely ignore the requirements of non-prey foods experienced by the natural enemy. Non-prey foods can extend the lives of released natural enemies, and thereby facilitate the synchronization of released natural enemies with the optimal stage of their host or prey. Moreover, non-prey foods fulfill the metabolic requirements of maintaining reproductive potential and dispersal critical to the establishment of a predator or parasitoid. Several programs examine the post-introduction feeding behavior of natural enemies, and attribute the ability of introduced natural enemies to find and exploit non-prey foods as one reason for their effective establishment (Beggs, 2001; Gallego et al., 1983; Watson and Thompson, 1933; Wolcott, 1942).

Several authors note that providing the right type of non-prey food can be as critical to successful introductions as the availability of general non-prey resources within a recipient biota (Box, 1927; de Charmoy, 1917; Wolcott, 1941). Certain parasitoids have very distinct feeding ecologies entirely unrelated to their hosts, and ignoring this aspect of parasitoid life history delays or dooms a biological control introduction. For example, certain scoliid and tiphiid wasps feed only on honeydew, spending little time foraging at flowers (Gardner, 1938; Hocking, 1967). In the absence of honeydew, the scoliids disperse rapidly from the release area, and may have difficulty in subsequently finding potential hosts. Hagen and Tassan (1972) point out that understanding nutritional relationships between entomophagous species and microbial symbionts that facilitate omnivory reduces the risk that a natural enemy is released into a new region without the requisite nutritional symbioses. Clearly, recognizing the nutritional ecology of the natural enemy, and adapting the post-release strategy to address their needs through selecting optimal release sites or managing the release site appropriately (as in Box, 1927; Pickett et al., 1996), will facilitate natural enemy establishments.

16.1.1.2 Augmentative Biological Control

Similar to classical biological control programs, the effectiveness of augmentation programs can be improved using non-prey foods that prolong the lives of the released natural enemies, and arrest their emigration from the targeted area. Within enclosed systems, such as stored grain or greenhouse environments, the availability of non-prey foods is drastically reduced even beyond that experienced in monoculture farmland. Moreover, dispersal within these artificial systems is severely reduced, and inundative releases of natural enemies can be very effective in managing these pests. Still, there are costs associated with releasing natural enemies, and prolonging the efficacy of the biological control agents would improve their economy. An example of non-prey foods improving natural enemy efficacy in a highly managed system is presented by Wäckers (2003). In this study, he shows that *Anisopteromalus calandrae*, a parasitoid of the stored grain pest *Callosobruchus chinensis*, cannot establish in small experimental arenas with hosts and beans alone. When honey is added to this same system, the parasitoid drives the pest population to extinction. Within greenhouses, the use of banker plants is proposed to improve the efficacy of released natural enemies. Banker plants typically provide non-pest alternative hosts or prey that allow natural enemy populations to build up before moving onto the greenhouse crop (Yano, 2006). However, there is also potential for using banker plants to build up omnivore populations in the same fashion by selecting plants that possess nutritional characteristics favorable to natural enemies (Matteoni, 2003). Also, a number of retailers of commercially produced natural enemies advocate providing sugar solution to the natural enemies at the time of release. Non-prey foods for application in greenhouses should be selected with extreme care, since many greenhouse pests can also use these resources.

The benefits of providing non-prey foods along with inundative releases of commercially produced biological control agents under field conditions is of more questionable value, although more research is needed on this topic. Two studies examine the effects of non-prey foods on *Trichogramma* augmentations. Begum et al. (2006) try to conserve augmented *T. carverae* within orchards using flowers of known value to the parasitoid. Lundgren et al. (2002) try to promote parasitism of *Pieris rapae* eggs by augmented *T. brassicae* using weekly sugar sprays. Neither of these supplements improve the generally low parasitism rates inflicted by the released *Trichogramma* wasps. Smith et al. (1986) show that feeding *T. minutum* honey upon emergence is sufficient to increase parasitism of spruce budworm in enclosed cages, and this simple procedure may go far in improving local parasitism rates. In line with this result, Lundgren et al. (2002) also determine that egg parasitism is greater in cabbage plots where point releases of honey-fed *T. brassicae* are practiced compared to those plots where unfed, *Trichogramma*-parasitized hosts are broadcasted over the plants. Feeding the natural enemies before their release also makes sense when one considers the literature on the foraging behavior of hungry versus satiated parasitoids (Lewis et al., 1998). While this is a promising area, more research is needed on tactics of integrating non-prey foods with augmentation programs before clear recommendations can be made to producers.

16.1.2 Conservation Biological Control

This approach to biological control seeks to ensure that the populations of entomophages that occur naturally in and near cropland overlap with critical resources that ultimately lead to greater natural enemy densities (Barbosa and Benrey, 1998). Relative to classical biological control, conservation biological control as a discipline is late in blossoming (excuse the pun). van Emden (2003) cites several reasons for why this may be, including:

1. Insecticides are very effective.
2. Farmers are adverse to incorporating potential weeds into cropland.
3. The concept of biodiversity is too complex to be useful on farms.

To this list I also add that a much wider array of disciplines than just entomology/applied ecology need to be involved in implementing conservation biological control efforts on farms. Perhaps resulting from this, biological control as a science is excellent at documenting the numerical and functional responses of natural enemies to conservation efforts, but much less effective from a crop production and agro-economic perspective. Also, the effects of non-prey foods on top-down regulation of pest populations are much more complex than is initially apparent, and understanding the multi-trophic cascades resulting from inputs of non-prey foods on the outcome of biological control on farms needs more attention. Thus, while a large base of literature indicates that conservation biological control holds great potential for reducing pest pressure under the correct circumstances, the practicalities of putting this knowledge to use in agriculture are less transparent.

In the simplest of terms, a population is affected by four general parameters: death rate, birth rate, immigration, and emigration. Many of the attempts to conserve natural enemies using non-prey foods focus on one or more of these components to population growth, and this conceptual framework is used here to discuss the utility of non-prey foods in conservation of natural enemies.

To begin, it is well documented that non-prey foods improve the survival and longevity of natural enemies. This is particularly true when prey is scarce or absent from a habitat. Many studies tend to focus on the influence of non-prey foods on individual natural enemies, often under controlled situations. But, by simultaneously improving the survival of numerous predators and parasitoids, the degree of intraguild interactions are increased in conservation biological control programs. The implications of these intraguild interactions on the net survival of individual species will be discussed more at length later in this chapter.

In addition to improving survival, many non-prey foods influence the movement of natural enemies in and out of cropland. Natural enemies are well known to disperse in search of non-prey foods, and the aggregation of natural enemies to sources of natural resources is discussed repeatedly throughout this book (see also Coll, 1998a for more discussion). Some specific examples include carabids that aggregate to areas with high numbers of preferred seeds (Honěk and Martinkova, 2001, 2003; Kirk, 1973), anthocorids that migrate to spring-pollinating catkins of various trees (Anderson, 1962b), predatory mite populations more sensitive to the local availability of pollen than prey on the same trees (Grafton-Cardwell et al., 1999; McMurtry and Johnson, 1965), syrphids that visit distinct nectar-bearing flowers when they are hungry (Pontin et al., 2007), and green lacewings that are attracted to certain volatile chemicals associated with honeydews (Hagen et al., 1976). Moreover, many natural enemies base their decision to leave a farm field on the availability of plant-based resources, which are often more stable in time and space then ephemeral prey-based resources (Fig. 16.1). By affecting the movement pat-

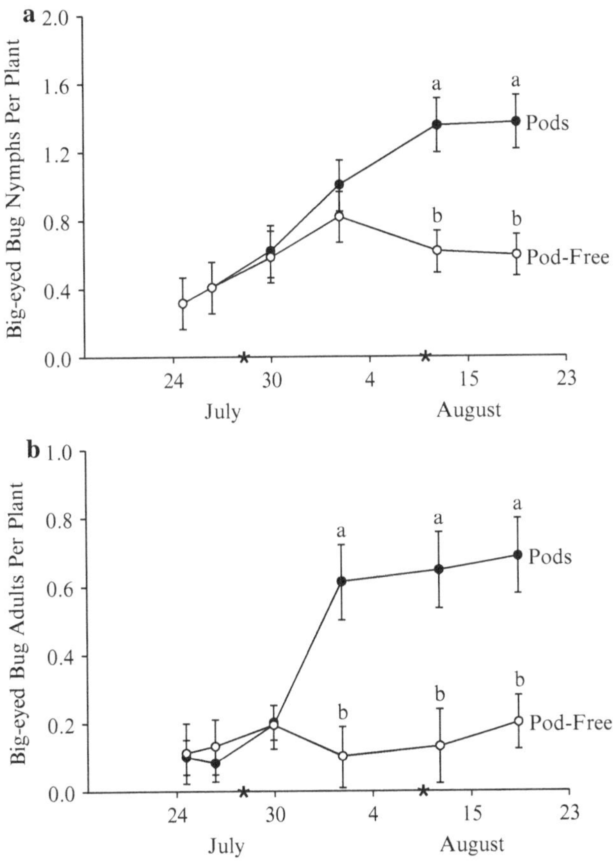

Fig. 16.1 Density of *Geocoris punticeps* (**a**) nymphs and (**b**) adults in lima bean plots with and without pods, an important non-prey food for *G. puncticeps* (mean ± SEM) Means with different letters are significantly different. Asterisks on the x axis indicate dates of fruit-thinner applications used to establish the treatments (Reproduced from Eubanks and Denno 1999. With permission by the Ecological Society of America)

terns of entomophagous arthropods, either through arresting dispersal from a designated area or attracting them to it, non-prey foods increase the local abundance of biological control agents.

Another way that conservation efforts associated with non-prey foods affect natural enemy populations is through increasing their reproductive capabilities or provoking oviposition within a localized area. Many natural enemies lay their eggs in response to plant-based cues and the local availability of non-prey foods (Griffen and Yeargan, 2002a, b; Lundgren et al., 2004; Seagraves and Yeargan, 2006; Shaltiel and Coll, 2004). *Orius insidiosus* is an excellent case in point; females of this species select which plants to oviposit upon based on the relative nutritional suitability of these plants for their developing offspring both in the laboratory and in the field (Lundgren et al., 2008a, b). The ability of non-prey foods to elicit reproduction becomes particularly important in the case of synovigenic life history omnivores, whose immatures are the only entomophagous life stage. For these species, biological control will not be maximized without non-prey resources because the mothers require these extra nutrients for ovigenesis.

While aggregating natural enemies within cropland is necessary, this is not sufficient to increase biological control; natural enemies need to eat more pests, pest populations need to be reduced, plant damage needs to reduced, and yields improved (Wratten et al., 2003). Much more research focuses on the effects of conservation efforts on the abundance of natural enemies than on the latter experimental endpoints. To encourage growers to implement non-prey foods as a means for conserving natural enemies, realistic strategies that are economical and cost effective need to be provided in a transparent format.

16.2 Strategies for Incorporating Non-Prey Foods into Cropland

The ability of non-prey foods to improve survival, enhance reproduction, reduce dispersal from the crop field, and attract natural enemies from outside of crop fields raises the methodological conundrum of how best to incorporate non-prey foods into crop production systems and biological control programs. Non-prey foods within cropland can originate from (1) vegetational diversity, (2) food sprays, and (3) the crop itself (the latter topic will be discussed in Chapter 17), and non-prey foods accompany several agronomically proven practices currently available to farmers. The end result is that there are numerous exciting opportunities for manipulating natural enemy populations in ways that make biological control programs more predictable and reliable, and many of these tactics are based on understanding the omnivorous habits of most natural enemies.

16.2.1 Land- and Farm-Scape Diversity as a Source of Non-Prey Foods

given the meta-population...characteristics of many groups [of natural enemies], refugia next to very intensive cultivation may be thought by some to be the agricultural equivalent of establishing a kindergarten next to a busy freeway. (Wratten et al., 1998)

There have been a plethora reviews, especially over the past ten years, which tout the benefits of incorporating biodiversity into cropland and the positive effects it has on natural enemy populations (Altieri and Whitcomb, 1979; Bugg and Waddington, 1994; Coll, 1998a; Jervis et al., 1993; Nentwig et al., 1998; Zandstra and Motooka, 1978). Indeed, more often than not, vegetational diversity in and around cropland leads to greater abundance of entomophagous arthropods (Andow, 1991; Russell, 1989). Although there are numerous reports of the association between biodiversity and reduced pest populations, this is far from universally the case (Andow, 1991; Bugg et al., 1987; Gurr et al., 1998; Kemp and Barrett, 1989; Risch et al., 1983; Russell, 1989). The myriad ways in which complex food webs generate unpredicted trophic interactions are likely to blame for the periodic breakdown in top-down regulation of pests, and this necessitates that farmland be manipulated in directed ways that promote biological control, rather than just incorporating biodiversity for its own sake. Certainly, more complex habitats provide a greater diversity and abundance of non-prey foods that can bolster the fitness of entomophagous species. But there are several practical considerations that must be factored into integrating vegetational diversity into cropland.

Numerous characteristics of habitat complexity are theorized and experimentally shown to promote the abundance and diversity of natural enemies, and the beneficial effects of habitat diversity to natural enemies is observed on a range of spatial scales. More complex habitats may support greater natural enemy populations by offering a greater diversity of microclimates, sources of shelter, alternative prey, overwintering sites, more refugia for herbivores that reduce the risk of prey becoming locally extinct, more preferred oviposition sites, and of course, a greater abundance and diversity of non-prey foods (Coll, 1998a; Landis et al., 2000). The benefits of habitat complexity are observed at the landscape, farm, and within-farm spatial scales (Altieri and Letourneau, 1982; Gurr et al., 2003; Landis et al., 2000). Enhanced habitat diversity within a landscape is associated with greater natural enemy populations and pest suppression on specific farms within the landscape matrix (Landis and Menalled, 1998; Landis et al., 2005), in addition to providing numerous other benefits to farming operations. Moreover, providing a greater diversity of habitats within a farm can also promote biological control in specific fields. But several characteristics of entomophagous arthropods support the argument that the best way to implement non-prey foods as a tool to promote natural enemies within specific fields is to integrate these non-prey foods as intimately and uniformly as possible within the crop that requires protection.

Foraging decisions by natural enemies reflects the need to balance their time among several key life history processes, including meeting their immediate nutritional requirements and maximizing reproduction (Barbosa and Benrey, 1998; Lewis et al., 1998). Biological control agents unfortunate enough to find themselves in an expansive monoculture of a particular crop may find plenty of potential prey or hosts, depending on the time of year, but will often need to migrate to at least the field borders in order to locate non-prey foods of sufficient quantity. Moreover, although natural enemies are able to migrate throughout a landscape in a short amount of time (30–180 m per day are not uncommon for the better dispersers) (Corbett, 1998; Gurr et al., 1998; Lavendero et al., 2005), and nutritional benefits of non-prey foods offered by field borders do spill over into adjacent crop fields (Wratten et al., 1998), these benefits seldom extend throughout the field and lead to edge effects in the abundance and function of natural enemies (Grout and Richards, 1990, 1992; Hausammann, 1996; Heimpel and Jervis, 2005; Houssain et al., 2002; Landis et al., 2005; Lavendero et al., 2005; Rogers and Potter, 2004; Shaltiel and Coll, 2004; Tylianakis et al., 2004; van Emden, 1965; White et al., 1995) (Fig. 16.2).

While many entomophagous species are able to disperse in search of non-prey foods, many others can not. For many predators, the larval stage is relatively or entirely immobile and only those non-prey foods integrated into the crop are accessible. Moreover, many parasitoid and some predator adults are at the mercy of the wind when they decide to fly in search of non-prey foods (depending on the architecture of the crop), and those that choose to leave a field in search of non-prey resources may never re-enter. While landscape and farm-level diversity meet several needs of natural enemies (e.g., overwintering habitats, necessary microclimates for key life processes) and should be encouraged, incorporating non-prey foods directly into cropland will minimize the need for natural enemies to emigrate from cropland and maximize both their reproductive success and impact on pest populations. Of course, this then begs the question of how to incorporate biodiversity and non-prey foods into cropland without sacrificing the primary goal of agriculture, which is to produce the crop itself.

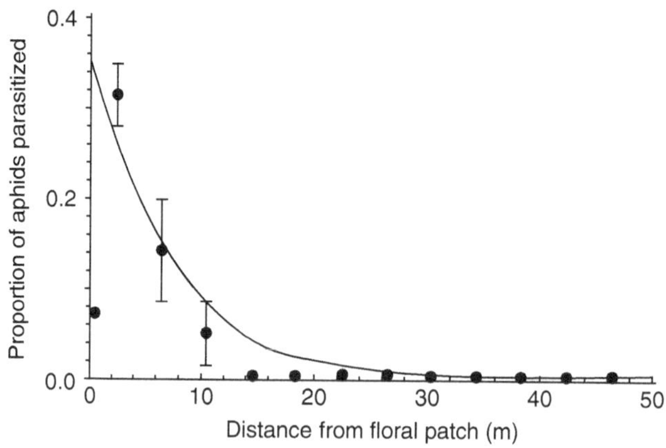

Fig. 16.2 Mean (±95% CL) proportion of experimentally placed *Metopolophium dirhodum* parasitized by *Aphidius rhopalosiphi* at increasing distances from the nearest floral resource patch (Reproduced from Tylianakis et al., 2004. With permission by the Ecological Society of America)

16.2.2 Integrating Vegetational Diversity Within Fields

Incorporating non-prey foods into crop fields will likely maximize the efficiency and impact of natural enemies on incipient pest populations (Coll, 1998b; Wilkinson and Landis, 2005). This is certainly the case for large acreages, and applies less to smaller or partitioned fields (Coll, 1998b; Wratten et al., 1998). From the seed literature, Harper (1977) presents that seed aggregations can be thought of as islands, and the larger and closer they are the more likely they are to be found by natural enemies. From this perspective, the uniformity and widespread incorporation of non-prey foods into cropland will more easily enable their discovery by hungry natural enemies.

Numerous agronomically sound practices currently exist to incorporate vegetational diversity into cropland without overtly sacrificing the integrity of the farming operation (Bugg and Pickett, 1998; Landis et al., 2000; Speight, 1983). Nearly a dozen such tactics available to producers are listed in Table 16.1. From this table it is conspicuous that vegetational diversity can be integrated into crop fields either as strips or more uniformly incorporated throughout the field. Using strips as a strategy does not compete directly with the crop, but necessitates that the strips are spaced at sufficient intervals as to allow natural enemies to move evenly into the adjacent crop. For instance, Nentwig (1998) advocates planting weed strips every 50–100 m so that natural enemy communities will disseminate evenly throughout the cropped areas of the field. More uniformly incorporating vegetation does not necessitate taking specific areas of farmland out of production, but with this strategy, non-crop plants compete directly with the crop.

Ultimately, for vegetational diversity to serve its role as a source of non-prey foods, it needs to provide nutritional resources at key periods when pest suppression is most important. The timing of when the non-prey nutritional resources are most important will vary with the crop and the pest or pest complex involved, although often the non-prey food will be of particular importance early in the growing season before pest populations outbreak. A tactic for deciding which types of plants may be most beneficial to natural enemies in a particular cropping system may be extracted from trends in the results of previous research. For instance, Coll (1998b) found that incorporating cereal crops had a more frequent positive effect on parasitoids than when legume crops were included in intercropped systems. Care should also be taken to preferentially apply native species over exotics when selecting non-crop vegetation for incorporation into cropland (Fiedler and Landis, 2007). It may also be that the best strategy for providing non-prey foods is implementing a combination of the tactics listed in Table 16.1. For instance, sowing a winter cover crop that is removed from the field as the crop germinates, and reducing herbicide applications during the remainder of the season or sowing internal beetle banks or weed strips might be a composite solution that meets the nutritional needs of biological control agents. It should also be noted that dead vegetation supports an entire detrital food web that supplies both prey and non-prey foods to higher trophic levels. Finally, the decision of when to eliminate vegetation from a field can be manipulated to encourage sequestered natural enemies in these areas to migrate onto the remaining crop. For instance hay fields can be harvested in strips, such that non-prey foods produced by the crop remain available to beneficial insects throughout the season (Houssain et al., 2002).

Table 16.1 Agronomic practices that encourage plant diversity within farm fields

	Practice	Description	Reference
Weeds	Reduce herbicide applications	Allow low levels of weeds throughout the field by reducing the dose or number of herbicide applications	(Altieri and Letourneau, 1982; Zandstra and Motooka, 1978)
	Directed herbicide applications	Target the most competitive weeds with selective herbicides, leaving the others as a source of diversity	(Zandstra and Motooka, 1978)
	Herbicides applied in strips	Applying herbicides in strips, so that weeds are present in portions of the field throughout the season	(Bugg and Waddington, 1994)
	Cultivated weed strips	Selected weed mixtures sown along strips throughout the field	(Hausammann, 1996; Landis et al., 2000; Nentwig et al., 1998; Zandstra and Motooka, 1978)
Abandon monoculture	Simulta-neous intercrop	Traditional polyculture	(Altieri and Letourneau, 1982; Coll, 1998b; DeLoach and Peters, 1972; Wilkinson and Landis, 2005)
	Relay intercrop	Planting the next crop directly into the standing crop, never leaving bare soil	(Altieri and Letourneau, 1982; Coll, 1998b; Landis et al., 2000; Slosser et al., 2000)
	Harvest in strips	Hayfields that are harvested in a staggered pattern, always leaving habitat for natural enemies	(Gurr et al., 2003; Houssain et al., 2002; Summers, 1976)
Cover crops[a]	Fall planted	Traditional use of cover crops, planted in the fall and removed before the next year's crop matures	(Clark, 1998; Wilkinson and Landis, 2005)
	Ground cover/living mulch	Planting an understory crop, which is not killed before the primary crop is planted. This strategy is particularly relevant to orchard systems	(Altieri and Letourneau, 1982; Bugg and Waddington, 1994; Flaherty et al., 1971; Landis et al., 2000; O'Neal et al., 2005; Prasifka et al., 2006; Wilkinson and Landis, 2005)
Diversification within mono-culture		Plant several varieties of the same crop with different characteristics	(Gurr et al., 2003)
Relax the mono-culture		Plant several closely related crop species	(Gurr et al., 2003)
Within-field uncultivated areas	Beetle banks	Uncultivated areas of the field	(Landis et al., 2000; MacLeod et al., 2004; Sotherton, 1995)
	Hedgerows		(Wratten et al., 1998)

[a] This is a difficult category since there are a wide array of uses and definitions of the term "cover crop". I include two of the more commonly used applications of cover crops in this list. This complexity is well described for orchard systems by Bugg and Waddington (1994).

Coll (1998a) advocates this strategy of killing off non-crop vegetation at key points during pest development to encourage the natural enemies to emigrate to the crop (see Perrin, 1975; Sluss, 1967 for specific examples of this tactic in action). Thus, in addition to carefully selecting which vegetation to encourage, deciding how to manage this vegetation once planted also influences the function of non-prey foods in promoting biological control.

16.2.3 Food Sprays

A promising tactic that incorporates non-prey foods directly into cropland is the application of artificial food supplements, and a substantial body of research shows the positive effects of this practice on biological control under an array of circumstances. Almost invariably, food supplements take the form of sugar sprays, often enhanced with yeast-based proteins, and intend to mimic the nutrition provided by honeydew. The application to cropland of mass produced pollen (Flaherty et al., 1971; Kennett et al., 1979) and crushed sunflower seeds (Tamaki and Weeks, 1972) are two exceptions to this trend. More often than not, food supplements increase the abundance of targeted groups of natural enemies and reduce pest populations over untreated controls (Fig. 16.3).

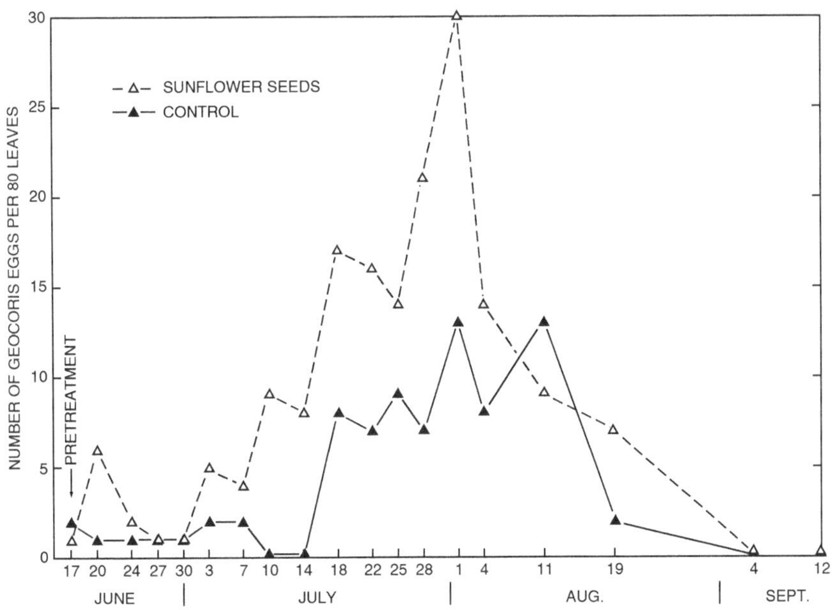

Fig. 16.3 Number of *Geocoris* eggs found on sugarbeet leaves in plots baited with sunflower seeds (an important non-prey food for this species) and unbaited plots (Reproduced from Tamaki and Weeks1972)

Although applying food supplements seems like a very straightforward process and there is considerable evidence of its benefits to pest management, several factors sometimes lead to unpredictable outcomes that are discussed later in this chapter.

Published research strongly supports the utility of food supplements as a way to increase natural enemy abundance and function in a variety of production systems and for a diversity of natural enemies. The effects of food supplements on a wide array of different natural enemy taxonomic and functional groups have been studied in 13 crops, and turf and forest systems (Table 16.2). Applying food supplements to managed lands at least partially increases the abundance of natural enemy communities over untreated fields in 91% of studies (Table 16.2). In 80% of studies, at least part of the pest community is reduced through the application of food supplements, and in 86% of studies, herbivore damage is reduced through the application of food sprays (but only seven studies followed their trials through to this endpoint; Table 16.2). It should be pointed out that there are some biases associated with the numbers presented here. First, there is likely an inherent publication bias toward publishing the positive results of food sprays on natural enemies. Duelli (1987) mentions anecdotally at least two unpublished unsuccessful food spray programs. Also, the studies presented here sometimes focus on more than one pest or natural enemy; if there are beneficial effects of the food spray on any of these species (increased abundance of any natural enemies, or decreased abundance of one of the targeted pests), then the study is considered a success in this analysis. A more comprehensive meta-analysis of food supplement studies may reveal a less extreme level of success, but the trends in the published literature will certainly be similar to the ones presented here. Although current evidence strongly suggests the benefits of food sprays for conserving natural enemies in cropland, more work is needed before farmers can implement this strategy to improve crop yields within agronomically relevant systems.

16.2.3.1 Disseminating Food Sprays

There are a number of ways to optimize the contributions of food supplements to natural enemy populations. Application technologies such as hand sprayers, fly-over sprays, point source applications, and even paint ball guns are used to disseminate food supplements throughout target areas. Several studies indicate that entire fields do not need to be coated in the food sprays in order to experience beneficial effects (Liber and Niccoli, 1988), which should make the use of this technique more applicable and affordable under normal farming operations. For instance, Evans and Richards (1997) found that applying sucrose to exterior rows of a crop field is as effective in aggregating natural enemies within the field as applying the spray throughout the entire field. Moreover they show that sugar sprays can be used to aggregate predators within the field center while the rest of the field is sprayed with insecticide, thereby reducing natural enemy exposure to the chemicals. Hagen and colleagues found that applying cards

Table 16.2 Summary of the research involving the application of non-prey food supplements to manipulate the abundance and function of natural enemies. Often, the studies were multi-year, involving multiple pests and multiple natural enemies. Thus, considerable variation is sometimes present in the studies, and a particular characteristic is designated 'yes' if at least part of the pest or natural enemy community responded to the food treatments. '-' indicates that the study did not provide the designated information

Crop	Food supplement[a]	Targeted natural enemies	Target pest	Improve natural enemy numbers?	Improve reproduction in predator?[b]	Reduce pest densities?	Reduce plant damage?	Study
Alfalfa	Yeast hydrolyzate + sucrose + choline chloride	*Chrysoperla carnea*, syrphids, and coccinellids	–	Yes	Yes	Yes, at least in one year when aphids and three species of lepidopterans were measured	–	(Hagen et al., 1976; Hagen et al., 1971)
Alfalfa	Sucrose; Wheast; or sucrose + Wheast	Predators	*Acyrthosiphon pisum*	Yes (sugar did more widely than protein supplements)	–	yes	–	(Evans and Swallow, 1993)
Alfalfa	Protein hydrolysates (blood, casein, yeast)	*C. carnea* and some coccinellids	–	Yes	–	–	–	(Hagen et al., 1976)
Alfalfa	Sucrose	*Bathyplectes curculionis* and coccinellids	*Hypericum postica* and *A. pisum*	Yes	Yes	Yes, only *A. pisum* measured	–	(Jacob and Evans, 1998)
Alfalfa	Sucrose	*B. curculionis* and *Coccinella septempunctata*	*H. postica* and *A.pisum*	Yes	See text	Yes	–	(Evans and England, 1996)

(continued)

Table 16.2 (continued)

Crop	Food supplement[a]	Targeted natural enemies	Target pest	Improve natural enemy numbers?	Improve reproduction in predator?[b]	Reduce pest densities?	Reduce plant damage?	Study
Alfalfa	Sucrose	Coccinellids (three species)	–	Yes	–	–	–	(Evans and Richards, 1997)
Apple trees	Sucrose + Formula 57 yeast product	Predators	A number of pests; unspecified green fruitworm was most abundant	No	No	–	Yes	(Hagley and Simpson, 1981)
Artichoke	Wheast + sucrose	*Hemerobius pacificus*	*Myzus persicae* and *Platyptila carduidactyla*	Yes	Yes	Yes	–	(Neuenschwander and Hagen, 1980)
Cabbage	Sucrose	*Trichogramma brassicae*	*Pieris rapae*	No	No	No	No	(Lundgren et al., 2002)
Corn	Sucrose	Coccinellids (three species)	–	Yes	–	–	–	(Ewert and Chiang, 1966)
Corn	Sucrose + molasses	Predators	*Ostrinia nubilalis*	Yes	Yes, for *C. carnea*	Yes	yes	(Carlson and Chiang, 1973)
Corn	Sucrose	Predators	*Spodoptera frugiperda*	Yes	No	Yes	Yes	(Canas and O'Neil, 1998)
Corn	Sucrose	Predators	*Rhopalosiphum maidis*	Yes	No	Yes	–	(Schiefelbein and Chiang, 1966)
Corn, cotton	Wheast + sucrose	*Coleomegilla maculata*	–	Yes	–	–	–	(Nichols and Neel, 1977)
Cotton	Food Wheast + honey	*C.carnea*	*Helicoverpa zea* and *Heliothis virescens*	Yes	No	–	–	(Butler and Ritchie, 1971)

Cotton	Yeast hydrolysate; Food Wheast + sucrose	*C.carnea*	*H. zea*	Yes	Yes	Yes, at least in one year when aphids and three species of lepidopterans were measured	Yes	(Hagen et al., 1976; Hagen et al., 1971)
Cotton	Food Wheast + sucrose	*C.carnea*	Heliothine pests	Yes	Yes	Yes	Yes	(Hagen et al., 1970)
Cotton	Envirofeast®	Predators	Heliothine pests	Yes	–	–	–	(Mensah and Singleton, 2003)
Cotton	Sucrose + *Torula* yeast	Predators	*H. zea* and *Aphis gossypii*	Yes, as long as it was early in the season and not near relay crops	–	Yes for *A. gossypii*; no for *H. zea*	–	(Slosser et al., 2000)
Cotton	Envirofeast®	Predators	Heliothine pests	Yes	–	–	–	(Mensah, 1997)
Grapes	*Typha* spp. pollen	*Galendromus occidentalis*	Tetranychidae	Yes (1 of 2 locations)	–	–	–	(Flaherty et al., 1971)
Navel orange trees	*Typha latifolia* pollen	*Euseius hibisci*	*Panonychus citri*	Yes (during winter and fall, but not during spring)	No	Yes	–	(Kennett et al., 1979)
Oak-maple forest	Sugar sprays	Formicidae (mostly *Formica subsericea* and *Aphaenogaster fulva fulva*)	*Lymantria dispar*	Yes, compared with ant excluded treatment	–	Yes, compared to ant-excluded treatment	–	(Weseloh, 1993)

(continued)

Table 16.2 (continued)

Crop	Food supplement[a]	Targeted natural enemies	Target pest	Improve natural enemy numbers?	Improve reproduction in predator?[b]	Reduce pest densities?	Reduce plant damage?	Study
Olive trees	Tryptophan	*C.carnea*	–	Yes	–	–	–	(McEwen et al., 1994)
Olive trees	Brewer's yeast, sucrose, & tryptophan	*Chrysoperla*	*Prays oleae*	Yes	–	Yes	Yes	(Liber and Niccoli, 1988)
Peppers	Wheast + sugar or yeast hydro-lyzate + sugar	*C.carnea*	*Myzus persicae*	Yes	Yes	Yes	–	(Hagen and Hale, 1974)
Potato	Feed-Wheast + honey or molassess	*C. carnea*, coccinellids, syrphids	*M. persicae*, *Lygus hesperus*, *Leptinotarsa decem-lineata*, and *Autographa californica*	Yes	Yes, *C. carnea*	Yes	–	(Ben Saad and Bishop, 1976b)
Potato	Molasses, honey, and tryptophan and/or Wheast	*C.carnea* and coccinellids	–	Yes	–	–	–	(Ben Saad and Bishop, 1976a)
Potato	Protein hydro-lysate	*C. carnea* and coccinellids	*M. persicae* and *Macrosiphum euphorbiae*	No	No	No	–	(Shands et al., 1972)
Sugarbeets	Sunflower seeds	*Geocoris*	–	Yes	Yes	No	–	(Tamaki and Weeks, 1972)
Turf	Sucrose	*Tiphia*	*Popillia japonica* and *Cyclocephala*	Yes (one species)	Yes, for *Cyclocephala* parasitoid	–	–	(Rogers and Potter, 2004)

Wheat	Acacia honey	Predators	*R. maidis*	Yes: carabids; no: staphylinids	–	No	–	(Monsrud and Toft, 1999)
Wheat	Hydrolysed and oxidized L- and D-tryptophan	*C. carnea*	Aphididae	Yes	–	–	–	(Dean and Satasook, 1983)

[a]Dissolved in water unless otherwise noted.

[b]Includes parasitism rate.

coated with food supplements at point sources throughout the fields is effective in augmenting certain natural enemies within cropland (Hagen et al., 1971). Most recently, Mensah and Singleton (2003) detail how band-applications of EnviroFeast® is as effective in aggregating natural enemies within cotton fields as spraying the entire field. Mensah also shows that the beneficial effects of food sprays on natural enemy abundance dissipates within 20 m of sprayed areas (Mensah, 1997), so spacing of these food spray bands is critical. There is also some flexibility with the exact concentration of sucrose and protein sources; in other words, adding more food to the solution above a certain concentration doesn't seem to improve the function of the food spray (Carlson and Chiang, 1973; Tassan et al., 1979, but see Hagen et al., 1976; Slosser et al., 2000). Finally, the effects of food sprays are typically short-lived. Sugar sprays are easily washed away by rain, and yeast-based foods lose their attractiveness over time. Generally, the beneficial effects of food sprays entirely dissipate within a week of their application (Butler and Ritchie, 1971; Evans and Swallow, 1993; Ewert and Chiang, 1966; McEwen et al., 1994).

16.2.3.2 Desired Qualities in Food Sprays

Some food supplements are primarily useful in arresting foraging behavior of natural enemies, but are not effective in attracting natural enemies from outside of treated cropland (Ewert and Chiang, 1966; Hagen, 1986; Hagen et al., 1976; Hagen and Hale, 1974; Hagen et al., 1971). Plants get around the challenge of attracting beneficial insects to sugar solutions by advertising the presence of nectar with floral displays, and possibly through the inclusion of traces of semiochemicals like amino acids; honeydews are also advertised to natural enemies with volatile chemicals. Ken Hagen and colleagues reason that advertising the application of sugar resources with accompanying semiochemicals known to attract natural enemies should augment their densities over applying arrestants by themselves. With this in mind, they show that including tryptophan, an amino acid that is associated with honeydew, to sugar sprays attracts lacewings from outside the targeted cropland (Ben Saad and Bishop, 1976a; Hagen et al., 1976; Hagen et al., 1971; Liber and Niccoli, 1988) (Fig. 16.4). Volatile chemicals associated with protein hydrolysates also attract certain natural enemies better than sucrose alone (Hagen et al., 1971, 1976; Nichols and Neel, 1977). Because many natural enemies, especially those that consume non-prey foods primarily as adults (like syrphids, parasitoids, some lacewings, etc.) balance their time between meeting physiological and reproductive needs, the semiochemicals included with the food spray should be selected with care. Advertising the presence of hosts/prey with host/prey-derived chemicals, and providing primarily non-prey foods instead, seems a bit like playing dirty pool and may ultimately reduce the effectiveness of these 'duped' natural enemies. Thus, it seems best to mimic the signals indicating the presence of non-prey foods rather than hosts/prey when applying food sprays.

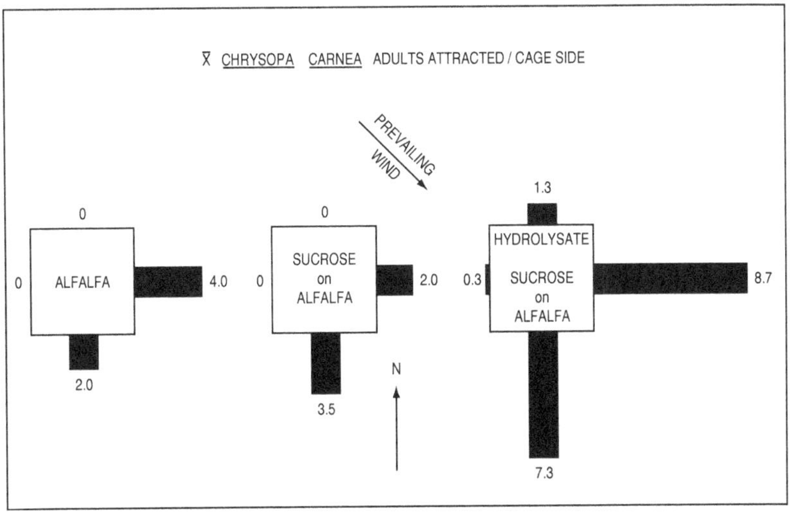

Fig. 16.4 Mean number of *Chrysoperla carnea* adults entrapped per side of cages (cages treated with Tanglefoot). The results indicate a volatile attractant given off by the yeast hydrolysate. Note the apparent upwind movement of adult *Chrysoperla* (Reproduced from Hagen et al. 1971)

Aggregating large numbers of the phytophagous stages of a predator or parasitoid are not sufficient to enact pest control. The impact of the natural enemy population on the pest is improved if the former responds to the food supplement by reproducing, a phenomenon that has been noted in 59% of studies (N = 17; Table 16.2). Sugar sprays typically do not elicit an oviposition response in many predators (Schiefelbein and Chiang, 1966); for example, Hagen and colleagues deduce that *C. carnea* won't lay eggs on sucrose alone, but when Food Wheast (a yeast, *Kluyreromyces fragilis*, produced on cheese whey) is added added to the spray, lacewings respond by laying eggs (Ben Saad and Bishop, 1976b; Hagen, 1986; Hagen and Hale, 1974; Tassan et al., 1979). In spite of the early promise suggested by Food Wheast as an attractant and an elicitor for predator reproduction, other studies do support the benefits of supplementing sucrose applications with Food Wheast (Ben Saad and Bishop, 1976a; Butler and Ritchie, 1971; Dean and Satasook, 1983; Duelli, 1987). In one 2-year study, coccinellids respond to the more diverse diet during one of the two growing seasons, but other natural enemies (*Geocoris*, *Chrysoperla*, syrphids, and *Bathyplectes curculionis*) are generally at equivalent numbers in fields treated with sucrose or sucrose and Food Wheast (Evans and Swallow, 1993). Also, there is no evidence that Food Wheast attracts natural enemies in this study; it only arrests their foraging. Since green lacewings are most attracted to food volatiles during the 3–4 days post-eclosion, one explanation for the differential success rates in these Wheast programs may have to do with the relative number of generations experienced in different geographies (Dean and Satasook, 1983).

16.2.3.3 Timing of Food Sprays

Because food sprays are short-lived under field conditions, optimizing their application to coincide with when pest management is needed the most will improve their utility for farmers. Research has consistently shown that the successful outcome of applying food sprays and supplements is strongly influenced by the general availability of non-prey foods within a habitat, and this gives some insight when application of food sprays would be most useful. When aphids or other non-prey nutritional resources are widely available in the environment, natural enemies are typically less or un-affected by food sprays in the targeted crop (Carlson and Chiang, 1973; Hagen et al., 1970, 1976; Kennett et al., 1979; Nichols and Neel, 1977; Slosser et al., 2000; Tassan et al., 1979). Thus, it will often be the case that food sprays are best applied early in the season, before crop-incorporated non-prey foods like honeydew, many flowers, and pollen are widely available. Moreover, application of non-prey foods can help to stabilize predator populations within cropland before pests become abundant (Kennett et al., 1979), thereby reducing the booms and busts of insect communities and helping to suppress or delay pest outbreaks (Hagen and Hale, 1974; Hagen et al., 1971; Schiefelbein and Chiang, 1966).

16.2.3.4 Additional Benefits of Food Sprays

The benefits of food sprays to pest management may not necessarily come from the omnivorous habits of the higher trophic levels. Mensah shows that the food supplement, EnviroFeast®, deters oviposition by some lepidopteran pests (Mensah, 1996; Mensah et al., 2000). Also, the attractiveness of the food supplement to other insects may further support higher abundance of natural enemies (Monsrud and Toft, 1999). For instance, Flaherty et al. (1971) describe how non-pest tydeid mites respond positively to field-applied pollen in grapevines, and predatory phytoseiids built up on these pollinivorous mites. Because of the pollinivorous alternative prey, the predatory mites are better able to suppress the target pest, spider mites. Thus, by fostering more biodiversity in cropland, food supplements can function to indirectly encourage biological control or they may function to reduce pests in ways other than top-down suppression.

16.3 Complications with Utilizing Non-Prey Foods in Pest Management

While there are numerous instances that demonstrate that non-prey foods in the forms of foods sprays and within-field vegetational diversity can improve a biological control program, the intricacies of multi-trophic interactions sometimes lead to unintended consequences for pest management that need to be considered. These complications come in many forms, from directly distracting natural enemies from eating the target prey, to inadvertently sending perturbations throughout the food web. By no means should this prevent biological control practitioners from employing non-prey foods, but the issues detailed below should be addressed as new programs are implemented.

16.3.1 Are Non-Prey Foods a Sink for Biological Control?

Much discussion over the years revolves around whether omnivory on non-prey foods is a beneficial aspect for biological control. After all, if a biological control agent is consuming things like pollen or seeds, then it is not eating the target prey. Although there is ample literature discussed above that indicates that the availability of non-prey foods often improves biological control, it is not instantly apparent as to how this can be. The current train of thought, theoretically and empirically reinforced, is that while per capita consumption of pests may be reduced by predators when non-prey food is abundant, the availability of non-prey foods improves the population growth rate of these predators such that the population or community of predators is better able to suppress pests. This notwithstanding, instances exist where non-prey food availability distracts from entomophagy within cropland.

Several examples, especially from the laboratory, indicate that the consumption of non-prey foods distracts natural enemies from consuming the targeted prey. In the lab, brown lacewings (Robinson et al., 2008), phytoseiid mites (James, 1989; McMurtry and Scriven, 1966; Nomikou et al., 2004; Skirvin et al., 2007), praying mantises (Beckman and Hurd, 2003), anthocorids (Skirvin et al., 2007), and coccinellids (Spellman et al., 2006) all consume fewer prey when non-prey foods are offered. Similarly, parasitoids are less apt to search for and sting hosts when they are hungry and sugar is available (Siekmann et al., 2004). Many of the studies conducted in the laboratory on single or few generations of the predator should be interpreted with great care. Initially pointed out by McMurtry and Scriven (1966), and better explored with theoretical models by van Rijn et al. (2002) is that although prey consumption may be reduced under laboratory settings, predators eating a diversified diet frequently experience higher levels of fitness (Fig. 16.5). Consequently their populations will grow faster and larger, and the net level of pest suppression will ultimately be greater when non-prey foods are available. In an elegant set of experiments, Eubanks and Denno (2000) demonstrate this process experimentally. In laboratory and field cage studies, *Geocoris punctipes* fed lima bean pods and prey consume fewer prey than those receiving prey alone. This leads to the hypothesis that less predation will occur in lima bean fields with bean pods than those without. But under true field conditions, the availability of lima beans helps to dramatically reduce pest populations over fields without non-prey foods (Fig. 16.6).

Under field conditions, there are a few instances where non-prey foods distract from predation. First, hungry parasitoids are more apt to search for food than search for and sting hosts within cornfields (Takasu and Lewis, 1995) (Fig. 16.7). Among predators, phytoseiid, staphylinid and coccinellid omnivores are distracted by non-prey foods from consuming key crop pests. During anthesis in cornfields, Cottrell and

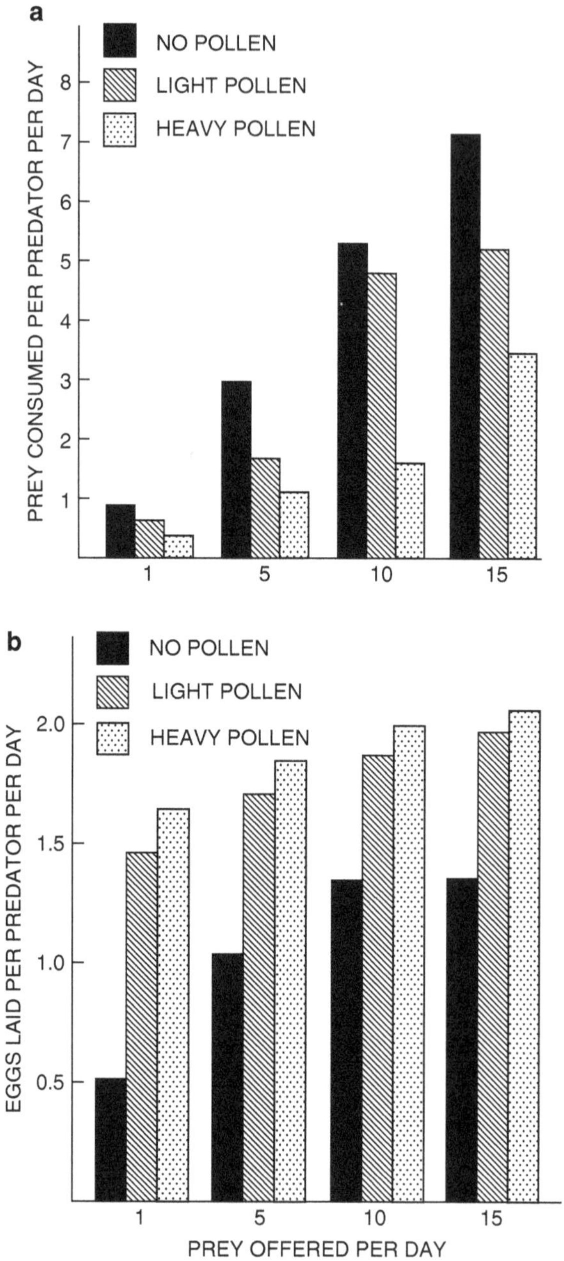

Fig. 16.5 Average number of *Oligonychus punicae* consumed (top) and number of eggs laid (bottom) per day by adult female *Amblyseius hibisci* fed varying levels of pollen. Differences between the treatments were significantly different (Reproduced from McMurtry and Scriven, 1966. With permission by the Entomological Society of America)

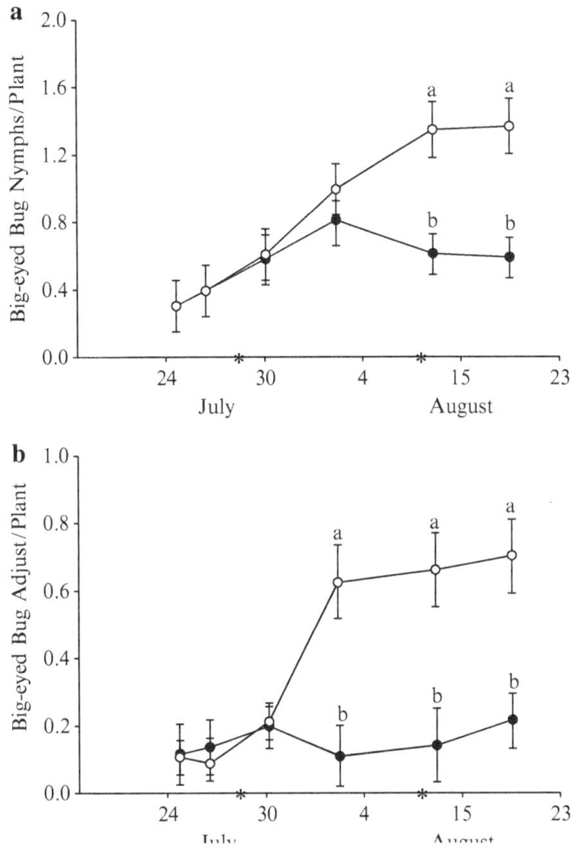

Fig. 16.6 The number of *Geocoris punticeps* adults (**a**) and nymphs (**b**), and aphids (**c**), caterpillars (**d**), and other herbivores (**e**) in plots that provide lima beans (an important non-prey food for *G. punticeps*) (•) and in plots with non-prey resources reduced (○). Seasonal mean predator and herbivore populations (**c–e**) were significantly different between treatments. Asterisks indicate the dates when fruit thinner was applied to the plots to maintain the treatments (Reproduced from Eubanks and Denno, 2000. With permission by the Ecological Society of America)

Yeargan (1998) show that predation on *Helicoverpa zea* eggs by *Coleomegilla maculata* is significantly reduced in pollen shedding cornfields, and Lundgren et al. (2004) go on to show through gut analysis that *C. maculata* larvae are much more pollinivorous than predaceous during this period of the field season. In wheat fields, fungi can detract key staphylinid predators from foraging for aphids (Dennis et al., 1991; Dennis and Sotherton, 1994). In this system, at least two dominant staphylinids, *Tachyporus chrysomelinus* and *Philonthus cognatus*, are attracted to fungi and preferentially feed on this over aphids. In the field, the lower leaves of wheat plants are frequently contaminated with a diverse flora of fungi, and foraging staphylinids are arrested on these lower leaves and never encounter the aphids on the upper foliage of the plants. Finally, only 63% of the *Typhlodromus pyri* population in apple orchards

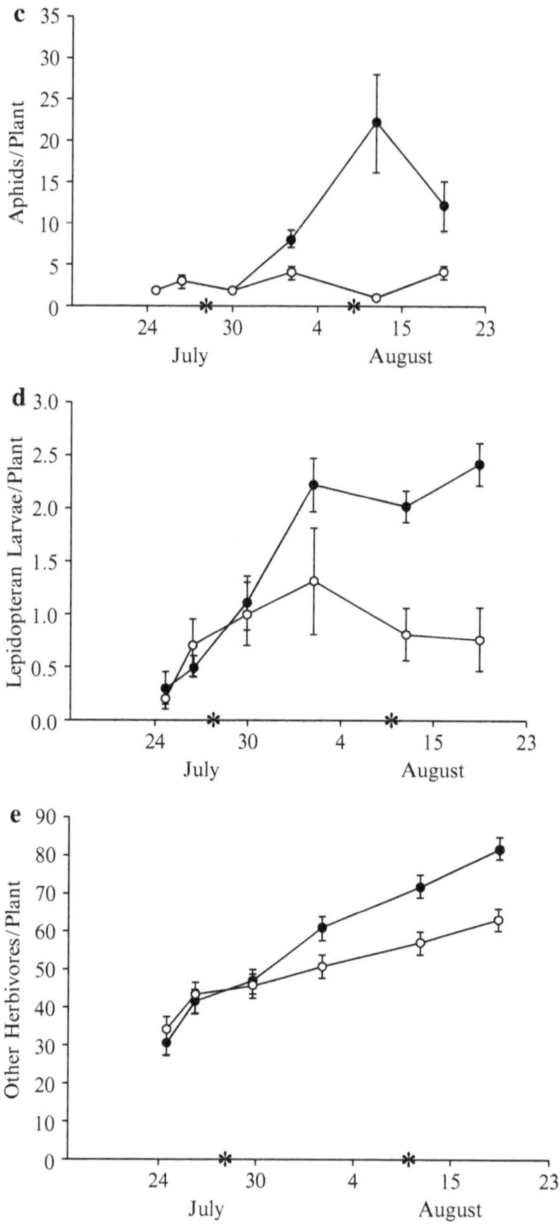

Fig. 16.6 (continued)

have prey remains in their guts, even when prey is widely available. Chant (1959) feels that this is largely due to the availability of pollen during the sampling.

This discussion is not to downplay the importance of non-prey foods in conserving natural enemies within cropland, but rather to illustrate one potential unintended outcome where the non-prey foods actually function as sinks for natural enemies.

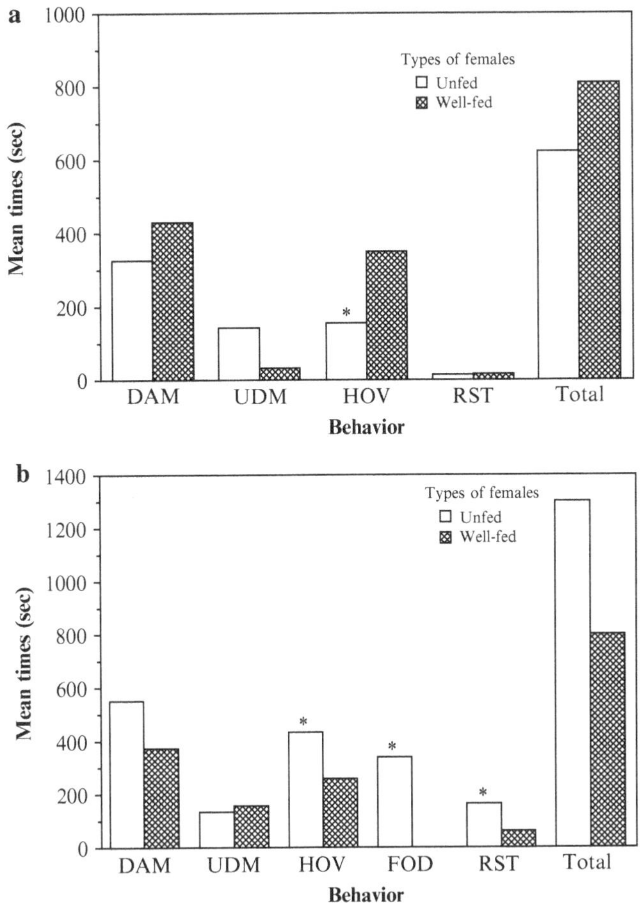

Fig. 16.7 Time allocation by female *Microplitis croceipes* in a corn plot with (top) and without (bottom) food. DAM, searching damaged leaves; UDM, searching on undamaged leaves; HOV, hovering; RST, resting; FOD, searching on food plants. Asterisks indicate significant differences between treatments (Data reproduced from Takasu and Lewis, 1995. With permission by Elsevier)

Most instances where non-prey foods distract from predation under field conditions involve the most omnivorous of entomophages, and some of the more nutritious non-prey foods. Thus, the outcome of a biological control program will depend both on the nutritional ecology of the natural enemy as well as the quality of the non-prey food (Eubanks and Styrsky, 2005).

16.3.2 Caveats to Vegetational Diversity

The effects of plant diversity on higher trophic levels are often much more dynamic than many biological control practitioners would like, and not all of these multi-trophic interactions are good for pest management (Bugg and Pickett, 1998; Gurr et al., 2003).

Sometimes, providing non-prey foods via vegetational diversity increases the abundance of natural enemies, but their function as pest control agents remains unaffected (Berndt et al., 2007; Braman et al., 2002). In the case of life history omnivores, where primarily the immature stages are entomophagous, a failure to reproduce in response to the availability of non-prey foods may be to blame (Chandler, 1968a). But more often than not, cases where non-prey resources fail to increase natural enemy function defy explanation. In other situations, vegetational diversity actually disrupts biological control. Essentially, while the beneficial aspects of vegetational diversity for natural enemies are undeniable, vegetational diversity comes along with a lot of extra baggage that may adversely and directly impact both crop production and pest management.

In addition to providing non-prey foods, non-crop vegetation can promote adverse multitrophic effects that ultimately impede pest management. First, increasing within-field habitat complexity may impede the foraging efficiency of natural enemies (Coll, 1998b). Also, non-prey foods may promote populations of pests, hyperparasitoids, or parasitoids of predators (Bugg and Pickett, 1998; Gurr et al., 1998; Stephens et al., 1998), thereby possibly impeding pest management. Also, alternative prey that occurs on non-crop vegetation may be preferred over the target prey, thereby functioning as a sink for biological control (Coll, 1998b). Finally, additional vegetational diversity does not necessarily lead to greater density of non-prey foods over monocultures. Andow and Risch (1985) showed that although pollen (and prey) resources are more diverse and evenly distributed in space and time in a polyculture system, pollen density is greater in maize monocultures. To *Coleomegilla maculata*, this greater pollen density overcomes the more even availability of non-prey foods in the polyculture, and more predation on ECB eggs within the monoculture system is the result.

Taxonomic and functional guilds of natural enemies respond very differently to vegetational diversity. Literature reviews indicate that parasitoids are typically favored by plant diversity (Coll, 1998b). But heteropteran predators are promoted by diversity in only 27% of studies (Coll, 1998a). This same pattern with heteropterans is observed in a number of food spray studies (see below), and could be the result of the attraction of more natural enemies to the abundance of non-prey foods, and the subsequent intraguild predation contests among species. Essentially, by promoting one natural enemy with non-prey foods, one may inadvertently remove or reduce the abundance of another natural enemy (Gurr et al., 2005). The end result of these complications with vegetational diversity is that while herbivore populations are increased in diversified cropland over monocultures in only 15% of the cases, they are unequivocally reduced in diversified crop fields in only 50% of studies (Andow, 1991).

16.3.3 Troubles with Food Sprays

The response of different species to food sprays varies considerably, and the effects of this practice on key pests and natural enemies need to be evaluated in each study system. For instance, several studies report that food sprays increase pest populations

(Cranshaw et al., 1996; Hagen and Hale, 1974; Slosser et al., 2000). This may be the result of the pest using the food supplement as part of their diet, or through unintended and overall deleterious interactions among higher trophic levels. For instance, *Orius*, *Nabis*, and sometimes *Geocoris* populations are either unaffected by food sprays or occur at reduced densities in some fields where food sprays are applied (Ben Saad and Bishop, 1976b; Evans and Swallow, 1993; Hagen et al., 1971), presumably through the increase in the occurrence of intraguild predation. One theory is that more omnivorous species are less competitive in intraguild interactions than more predaceous species (Coll and Guershon, 2002). In several studies, Ted Evans and colleagues suggest that application of sucrose can have complex interactions among natural enemies and the pests they suppress in alfalfa fields (Evans and England, 1996; Jacob and Evans, 1998). In this system, the parasitoid of alfalfa weevil, *Bathyplectes curculionis*, feeds on honeydew of the pea aphid; the more pea aphids, the more weevil parasitism. When sugar sprays are applied, this trophic cascade is disrupted, or perhaps replaced is a better word. Coccinellids aggregate to sugar-sprayed areas, where they consume pea aphids. But ladybeetles also eat alfalfa weevil larvae, and there is no short-term net difference in weevil suppression between sprayed and unsprayed plots. But given that the effects of food sprays are short-lived, what are the consequences of pea aphid predation on the long term populations of weevil parasitism if sugar sprays are ceased?

Finally, natural enemies often have very specific nutritional ecologies that need to be considered when implementing food sprays (Monsrud and Toft, 1999). For example, *Tiphia vernalis*, a scoliid parasitoid of Japanese beetle, only responds to sugar sprays when they are applied to tree trunks, rather than the turf where their hosts reside. This is because this wasp typically feeds on honeydew produced by arboreal aphids. Meanwhile, a congener, *Tiphia pygidalis* (a parasitoid of *Cyclocephala*), is easily manipulated through applying sugar sprays to turf (Rogers and Potter, 2004). Finally, several food sprays are known to be phytotoxic to certain crops (Hagen et al., 1970, 1971, 1976;Hagley and Simpson, 1981). For all of these reasons, the use of food sprays need to be evaluated on a case by case basis before they should be widely advocated for adoption.

16.3.4 Are Omnivorous Natural Enemies Pests?

Omnivory sometimes leads to problems when the natural enemies damage the crop itself. In general, damage is inflicted to either the fruit or the crop seeds, but this damage is seldom of great concern to producers. The most conspicuous exception to this is damage inflicted to fruit by omnivorous heteropterans, which are important pests in crops like cotton and strawberries (Schaefer and Panizzi, 2000). Another recent problem accompanies the invasion of North America and Europe by *Harmonia axyridis*. Presumably in search of water and possibly a sugarmeal, this ladybeetle damages apples, pears, peaches, cherries, blueberries. Perhaps more importantly, when *H. axyridis* are harvested with wine grapes, they alter the flavor of the resulting product and the wine industry is expressing concern over this problem (Koch, 2004; Kovach, 2004).

The perception of granivorous entomophages as beneficial to crop production is contingent on them not consuming the crop itself. Generally, this is not an issue for granivorous insects, since crops are selected to produce relatively seeds one to three orders of magnitude larger than those of weeds (Murray, 1984b). Nevertheless, a few instances of granivorous entomophages acting as pests of crop seeds are worth mentioning here. The strawberry seed beetle, *Harpalus rufipes*, is of great importance to strawberry producers in England during outbreak conditions (Briggs and Tew, 1965). Also, *Zabrus tenebrioides* is a notorious pest of grain in Europe (Bassett, 1978). In North America, the seedcorn beetles, *Stenolophus lecontei* and *Clivina impressefrons*, sporadically cause problems (Bigger and Blanchard, 1959). Likewise, *Clivina fossor* is considered a pest of sugarbeets in Northern Europe. However, the impact of these insects on cornfields is seldom well documented, and the beneficial services they provide may outweigh the occasional crop damage inflicted. Occasionally, seed feeding by harvester ants becomes problematic for agriculture and forest ranges (Andersen, 1990, 1991; Holldobler and Wilson, 1990; Morrison et al., 1997). This is particularly well documented in Australia, which has a large cohort of harvester ant species that inhabit most landscapes. In this country, insecticidal sprays and baits are used to reduce seed losses to harvesting ants (Andersen, 1990). Non-chemical management tools are also available for controlling harvesting ants. Coating the crop seeds with substances repellent to harvesters, or sand that reduces the ability of the ants to harvest the seeds are two such strategies. Also, compacting the soil after planting, timing the planting when ant activity is low, and increasing the seeding rate to satiate the ants are other non-chemical methods for managing harvester ants. These few examples notwithstanding, the vast majority of granivorous entomophages have a minimal impact on crop seed mortality.

16.4 Conclusion

conservation biological control will have the highest likelihood of success if we recognize both the constraints and opportunities afforded by agroecosystems. (Barbosa, 1998)

The deployment of non-prey foods to conserve natural enemies within cropland and enhance natural enemy releases holds great promise as an IPM tool so long as the limitations posed by this strategy are recognized. A recurring theme throughout this chapter is that while providing non-prey foods within cropland consistently *changes* the natural enemy community, it does not necessarily *improve* it from a functional standpoint. Concerted efforts to provide the correct type of non-prey food for the circumstances will reduce undesired outcomes (Barbosa, 1998; Gurr et al., 1998, 2005). Of course, this then begs the question of which are the 'right' kinds of non-prey foods for natural enemies, a topic which will be discussed at length in the concluding chapter of this book. Still, Landis and Menalled (1998) raise an excellent point in their discussion of the more directed approach to conservation of natural enemies commonly put forth in today's literature. They argue that common agricultural practices exclude a large portion of the natural enemy community, and practices that are directed toward conserving only members of the subset which resides in cropland may further exclude the natural enemies most fit for the job. This makes a strong case for a broader approach to conservation biological control that targets a wider suite of natural enemies, assuming that it does not aggravate pest problems.

> *In most of these situations there would be a reduction in total land in crop production, since space is needed for weed growth. For this system to be economically feasible therefore, the resulting reduction in yield would have to be compensated for by decreased crop damage by insects* (Zandstra and Motooka, 1978)

In spite of the intuitive benefits of providing non-prey foods to encourage biological control, only a small segment of the farming community intentionally implements these strategies. Zandstra and Motooka recognized this fact in describing different methods for integrating weeds into cropland. What is currently lacking is the relative efficacy of non-prey foods in reducing pests from an economic standpoint (but see Sotherton, 1995). Only when the economic feasibility of non-prey foods can be demonstrated to producers will they begin to adopt these tactics on a large scale (Risch et al., 1983). To this end, the conservation biological control community needs to shift the endpoints of their research from simply measuring natural enemy abundance or impact on the pest to addressing how non-prey foods impact crop yields and the relative profitability of different tactics. Finally, augmenting non-prey foods within cropland to conserve natural enemies is addressing a symptom of a much larger problem: the dearth of habitat complexity and over-emphasis on monoculture production practices (Landis et al., 2000).

Chapter 17
Plant-Incorporated Pest Resistance and Natural Enemies

It is short-sighted to develop a chemical control program for the elimination of one insect pest and ignore the impact of that program on the other arthropods, both beneficial and harmful, in the ecosystem.

Stern et al. 1959

Plant breeders and pest managers seek to make crops resistant to herbivores through selective breeding of resistant lines and by actively placing insecticidal compounds into plants in the form of systemic insecticides or insecticidal transgene products. These strategies (host plant resistance [HPR], systemic insecticides, and genetically modified [GM] crops) differ from each other in the nuances of their efficacy and persistence, but all have the similarity in that the theoretical (and often actual) outcome of their use reduces environmental exposure to insecticides over that experienced with insecticidal sprays. Moreover, since the insect resistance factor is expressed solely in plant tissue, herbivorous pests are theoretically much more exposed to the insecticidal chemicals than higher trophic levels. Commonplace omnivory in entomophagous arthropods complicates this widely held perception.

There are a number of ecological pathways through which plant-incorporated defensive chemicals can alter the abundance, fitness, and function of higher trophic levels. A natural way for these chemicals to affect natural enemies is by eliminating or reducing the quality of prey or hosts within cropland. After all, the raison d'être for these chemicals is to kill or disturb herbivores. Many studies examine prey-mediated effects of resistant plants on natural enemies. Some studies experimentally describe synergistic effects of plant-incorporated resistance with biological control (Bell et al., 2001; Pair et al., 1986; Wyatt, 1970) (Fig. 17.1), while other studies measure no tri-trophic effects (Dogan et al., 1996; Lundgren and Wiedenmann, 2005; Pimentel and Wheeler, 1973; Riddick and Barbosa, 1998); under these circumstances it is concluded that HPR and biological control are compatible. But instances where HPR (Kauffman and Flanders, 1985; Kennedy et al., 1975; Orr and Boethel, 1986; van Emden, 1999), and systemic insecticides

J.G. Lundgren, *Relationships of Natural Enemies and Non-Prey Foods*,
Progress in Biological Control 7,

Published by Springer Science+Business Media B.V.

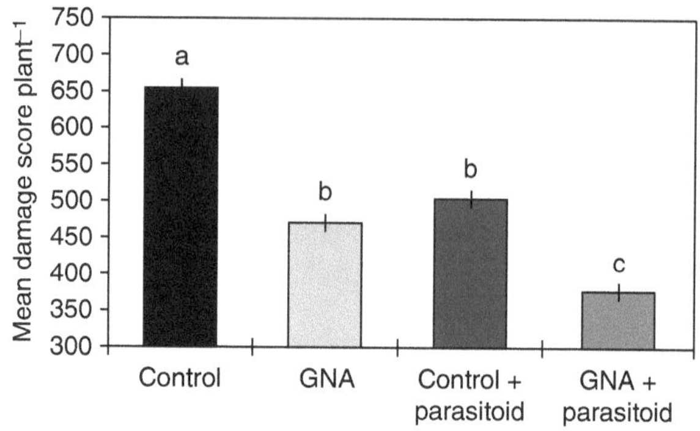

Fig. 17.1 The effect of GNA expression and the presence and absence of *Eulophus pennicornis* on the damage incurred by potato plants from *Lacanobia oleracea*. The damage rating system was applied 17 days post-infestation with lower scores indicating less damage. Columns capped by different letters are significantly different (Reproduced from Bell et al., 2001. With permission by Springer)

(Ahmed et al., 1954; Feese and Harlan, 1975; McClanahan, 1967) contribute to declines in natural enemy abundance or fitness through the alteration of prey populations are also documented. And within GM crops, events that produce Bt (Bernal et al., 2002b; Dutton et al., 2002; Hilbeck et al., 1998a, 1999; Ponsard et al., 2002; Romeis et al., 2004), or lectins (Birch et al., 1999; Couty et al., 2001a, b; Down et al., 2000; Setamou et al., 2002a) sometimes adversely affect natural enemies by reducing prey/host quality or abundance. Even when adverse tri-trophic effects on predators are noted in laboratory tests, these effects do not necessarily transfer to the field.

However, the central theme of this book is omnivory, and most natural enemies will consume plant tissue and exudates, especially when prey is scarce. Thus, when herbivore prey are reduced through the action of plant-incorporated toxins, conditions for a perfect storm involving *direct* interactions between protected crop plants and natural enemies exist. While these direct interactions tend to be less well-studied relative to prey-mediated interactions, numerous reports indicate that they are an important consideration when discussing the dynamics between HPR, systemic insecticides, or GM crops and biological control.

17.1 Host Plant Resistance

> *It is as an aid to other control measures that insect resistance is most important and may be most commonly used.* (Painter, 1951)

Plant defense to herbivores is attained through selective crop breeding programs, and in most overviews of the topic it is generally concluded that HPR and biological control are compatible pest management strategies (first described with models by van Emden, 1966) (Bergman and Tingey, 1979; Clement and Quisenberry, 1999; Maxwell and Jennings, 1980; Singh and Singh, 2005; Smith, 2005). While there are several aspects of HPR that make it very amenable to IPM programs, it is dangerous to assume that pest-resistant crops are going to have no impact on higher trophic levels (Groot and Dicke, 2002; van Emden, 1995; Verkerk et al., 1998). In a review of the literature, Hare (1992) points out that in 40% of studies (N = 16), HPR is antagonistic toward parasitoids. The other 60% of studies show no effect on biological control or there are synergistic interactions between HPR and parasitoid performance. In spite of this, many examples exist that document how HPR and biological control lead to lower pest densities than either strategy on their own (Kartohardjono and Heinrichs, 1984; Kuo, 1986; Riggin et al., 1992; Starks et al., 1972). The earliest attempts to explain how HPR may upset biological control dealt with prey-mediated interactions (Painter, 1951). Painter's main conclusion is that natural enemies can be affected tri-trophically when prey/host densities are decreased in resistant crops or when prey/host quality is reduced.

Bergman and Tingey (1979) raise the notion that there may be direct, or bi-trophic, interactions of HPR with natural enemies through differences in volatile production, nutrition or toxicity, and plant architecture and mechanical defenses (i.e., trichomes). Many reviews since put particular emphasis on the contributions of plant architecture to foraging success in natural enemies and the prey-mediated effects of HPR on biological control (Bottrell et al., 1998; Hare, 1992; Smith, 2005). Although often mentioned, the importance of differential emission of synomones, particularly in attracting natural enemies to plant-based foods, and variability of the quantity and quality of non-prey foods produced within a crop species receive less attention from scientists. As these two ecological pathways are related to the omnivorous tendencies of natural enemies, the topic merits a discussion here.

17.1.1 Nutritional Suitability of Resistant Plants to Natural Enemies

Plant species differ in their nutritional suitability for natural enemies (examples in Lundgren et al., 2008a; Naranjo and Stimac, 1985), but varieties within a crop species also vary in their effects on higher trophic levels when these arthropods feed on them directly. There are very few studies that examine direct bi-trophic interactions between herbivore-resistant crops and facultative phytophagy by natural enemies, with most focus given to *Geocoris punctipes*. When nymphs of this bug are reared on *Helicoverpa zea*-resistant and susceptible soybean lines, nymphal development time and adult size are significantly hindered in the resistant lines. Fitness degradations are observed both in the laboratory and the field, and experimental methods suggest that the resistant phytochemistry may involve an induced plant response to insect feeding (Rogers and Sullivan, 1986, 1987). A study comparing the effects of high gossypol (a terpenoid) cotton with susceptible cotton on *G. punctipes* survival does not report any deleterious effects (De Lima and Leigh, 1984). Clearly, more research is needed to determine whether facultative phytophagy by natural enemies, particularly under conditions of prey limitation, interferes with integrating HPR and biological control.

17.1.2 Intraspecific Variation in Synomone Production

Many predators and parasitoids use synomones produced by plants to locate habitats where prey or hosts are likely to be found (Barbosa and Wratten, 1998; Cortesero et al., 2000; Dicke et al., 1990; Hagen, 1986; Verkerk et al., 1998; Vinson, 1977, 1981). Some examples within cropland include *Orius insidiosus*, which is attracted to volatiles from corn silks when looking for prey (Reid and Lampman, 1989) and the parasitoid *Diaeretiella rapae*, which is attracted to the allyl isothiocyanate produced by collards, their host's plant (Read et al., 1970). Moreover, sometimes plant volatiles that are entirely unrelated to prey or hosts inexplicably elicit a flight response in natural enemies, even when the natural enemy has experience with the volatiles from the prey/host (Shahjahn, 1974 Nentwig, 1998; Agelopoulos and Keller, 1994; Herrebout and van der Veer, 1969; Powell and Zhi-Li, 1983). Given that entomophagous species locate non-prey foods using chemo-sensory adaptations, it is conceivable that plant-based volatiles that are unrelated to prey/hosts are one way in which natural enemies find non-prey/host resources.

Researchers have only skimmed the surface of intraspecific variation in allelochemical production within a crop, but there are a few examples that suggest that intraspecific allelochemical variation may affect natural enemies. In 1966, Franklin and Holdaway show that the tachinid parasitoid, *Lydella thompsoni*, is attracted to the volatiles from certain corn hybrids, but not others (Franklin and Holdaway, 1966). This relative attraction leads to differential levels of parasitism of *Ostrinia nubilalis* on the different corn hybrids. Maize hybrids also display varying attractiveness to the parasitoid *Campoletis marginiventris*, a parasitoid of *Spodoptera littoralis*, under laboratory conditions (Fritzsche Hoballah et al., 2002). Similarly, the greenbug parasitoid, *Lysiphlebus testaceipes*, is more attracted to volatiles emitted from aphids on resistant oat cultivars than from those on susceptible ones, leading to three times higher parasitism rates on resistant plants (Schuster and Starks, 1975). When infested with the same number of spider mites, volatiles from bush beans are unattractive to predatory mites, whereas volatiles from pole beans are (Dicke et al., 1990). Also, there is intraspecific variation in the attraction of resistant and susceptible host plants infested with *Nilaparvata lugens* to the predators *Micraspis hirashimai* and especially *Cyrtorhinus lividipennis* (Rapusas et al., 1996). Evidence also suggests that our artificial selection of crop cultivars has dramatically reduced the amount and diversity of volatiles produced in commercial crops compared to the native landraces (Fritzsche Hoballah et al., 2002; Loughrin et al., 1995) (Fig. 17.2), and there may be additional chemical cues that can be reincorporated into crop germplasm by enthusiastic and open-minded plant breeders that will improve the attractiveness of these crops to natural enemies. Another consideration is that environmental conditions may affect the differential attractiveness of crop cultivars to natural enemies (Halitschke et al., 2000). Finally, not all natural enemies are attracted to the same chemical profiles, and selecting to attract one natural enemy may very well exclude another. As a case in point, Elzen et al. (1986) showed that the parasitoid

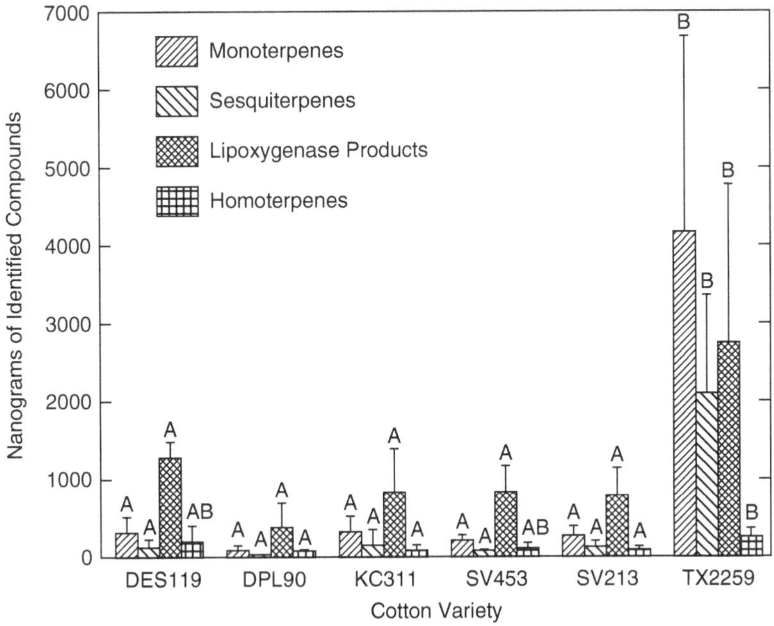

Fig. 17.2 Total identified volatiles, divided by biosynthetic class, from cotton varieties damaged by *Spodoptera exigua* larvae. TX2259 is a naturalized variety, the rest are commercial varieties. Data represent the means of six experiments ± SD. Within biosynthetic class, bars with the same letter are not significantly different (Reproduced from Loughrin et al., 1995. With permission by Springer)

Campoletis sonorensis, is attracted to volatile terpenoids from cotton, and flies more frequently to cotton cultivars possessing internal leaf glands than to glandless cultivars. The exact opposite trend is observed for *Microplitis croceipes*, which is more attracted to glandless cotton (Navasero and Elzen, 1989). Whether these variations are driven more by the pest or by the plant, the end result is that intraspecific variation in the emission of plant volatiles used by natural enemies may be a way to integrate natural enemies into cropland by exploiting their capacity for using plant-based cues.

Entomophagous arthropods are attracted to allelochemicals of crops that signal the availability of non-prey foods, and variability of the volatile profiles present in crops suggests that the chemical cues associated with non-prey foods are at least equally variable among different crop cultivars. Understanding the importance of different non-prey foods to natural enemies and the cues that they use to find these foods may allow the development for crop cultivars more conducive to biological control, or at least allow the screening existing cultivars for their compatibility with key natural enemies. Also, this variability in infochemicals associated with non-prey foods provides one more explanation for the variability in the outcome of biological control programs.

17.1.3 Intraspecific Variation in Non-Prey Food Production

The ability to select crop lines that favor biological control is difficult because many plant traits are polygenic and isolating particular phenotypes that favor biological control without affecting many other aspects of crop performance is challenging and not always easy to assess in breeding programs (Bottrell et al., 1998; Poppy and Powell, 2004). For these same reasons, as crop varieties are developed for a range of different growing conditions and interests, the phenotypic outcome of the selection process of new germplasm should consider unintended consequences in the direct interactions of these crops with biological control agents. A case in point involves how aphids respond to partially resistant Ommid wheat by producing 40% less honeydew than on susceptible lines of wheat (van Emden, 1995). In the case of the aphid parasitoid, *Aphidius rhopalosiphi*, less honeydew on the resistant wheat improves foraging, since the parasitoid spends less time cleaning themselves. There is also substantial intraspecific variation in the quantity and quality of non-prey foods naturally present in commercially competitive crop varieties that may affect the efficacy of natural enemies in cropland.

Two veins of research suggest that quantity and quality of non-prey foods vary intraspecifically in commercially available crop varieties, and that these traits influence natural enemies. Perhaps the best-studied example of intraspecific variability in non-prey food availability within a crop comes from cotton varieties with and without EFNs. The presence of EFN in cotton fields is correlated with natural enemy abundance (especially heteropteran predators) (Agnew et al., 1982; Henneberry et al., 1977; Naranjo and Gibson, 1996; Schuster and Calderon, 1986; Schuster et al., 1976; Scott et al., 1988; Yokoyama, 1978, but see Stone et al., 1984), and the relative efficacy, behavior, and longevity of parasitoids and predators on nectaried cotton cultivars suggests that natural enemy abundance is caused in part because of the presence of the nectar (De Lima and Leigh, 1984; Lingren and Lukefahr, 1977; Schuster and Calderon, 1986; Treacy et al., 1987; Yokoyama, 1978). Indeed, the most abundant predator in many cotton fields is *Solenopsis invicta*, and activity patterns of this predator are closely tied to seasonal nectar production in the EFN-bearing fields (Agnew et al., 1982).

The nutritional quality of non-prey foods also varies among crop varieties. Selecting for different flower cultivars in lucerne sometimes accompanies changes in the nutrition of pollen as well (Karise et al., 2006). In field corn, intraspecific variation in the nutrition of corn pollen influences the survival of *Coleomegilla maculata* (Lundgren and Wiedenmann, 2004), *Orius insidiosus* (Chapter 8, this volume), and possibly the mite *Neoseiulus cucumeris* (Obrist et al., 2006b). The quantity of pollen produced by different corn hybrids also varies substantially. To verify this, ask anyone who has the displeasure of walking into a sweet corn field during anthesis and compare the degree of pollen coverage with another walking into a field corn plot during the same developmental stage. Certainly, a large proportion of natural enemies occurring in cornfields during anthesis consume corn pollen (Corey et al., 1998; Lundgren et al., 2004, 2005), but whether selecting corn hybrids that produce more or higher quality pollen enhances the relative abundance of biological control agents or their efficacy is untested. At the same time, it is important to consider the dual effect of variation in non-prey foods on the herbivore community (Naranjo and Gibson, 1996). For example, EFN in cotton also increases populations of tarnished plant bugs (Scott et al., 1988) and nectaried cotton may accompany increases in total insect populations (Adjei-Maafo and Wilson, 1983). Another consideration when developing these lines is that heterogeneity in the behavior and physiological requirements of different natural enemies makes it necessary to target the selection of plant traits with respect paid to only the most key natural enemies in the pest management system (Bottrell et al., 1998).

17.2 Systemic Insecticides

> *Many compounds with systemic properties are now known and it appears probable that any insecticide which is sufficiently water soluble and stable may possess some degree of systemic action.* (Metcalf, 1955)

Systemic insecticides have been used to focus insecticidal action against lepidopteran pests for more than 50 years. The first insecticides specifically applied for their systemic capabilities were organophosphates (Metcalf, 1955). Today, neonicotinoids (imidacloprid, thiamethoxam) and pyrethoids (deltamethrin) have joined (and often replaced) the use of systemic organophosphate and carbamate insecticides (e.g., disulfoton, aldicarb, acephate, and phorate). Often, systemics have been applied to the soil near the plants early in development, and more recently as seed treatments before planting even occurs. In either case, the outcome has been very little or no direct exposure of foliar-dwelling natural enemies to the pesticide through application.

17.2.1 Systemic Insecticides in Non-Prey Foods

Once applied, the insecticide is translocated throughout the plant, often rapidly. Systemic insecticides are found in the nectar of treated plants within 6 h of application (Davis et al., 1988). Many systemic insecticides have use restrictions emplaced to ensure that the chemical is not applied close to when the crop is flowering in order to reduce exposure to non-target species (Johanson et al., 1983). In sweet clover (*Melilotus alba*), systemics are found in floral nectar, but Johansen and colleagues found that when systemic insecticides are applied according to label instructions, they pose little risk to foraging honeybees. Although systemics dissipate from non-prey foods after a week of application, some of these insecticides persist in non-prey foods produced by plants for weeks (and sometimes years!) after application (Barker et al., 1980; Cate et al., 1972; George and Rincker, 1982; Glynne Jones and Thomas, 1953; Jaycox, 1964; Lord et al., 1968; Sclar et al., 1998; Tasei et al., 1994; Waller and Barker, 1979; Waller et al., 1984). Recently, many seed companies treat their crop seeds with systemic insecticides (particularly imidacloprid and thiamethoxam) to encourage higher yields. In sunflower and corn, imidacloprid seed treatments are consistently shown to be translocated to pollen and nectar during flowering, thereby exposing predators and parasitoids to the insecticides for weeks after they are initially applied (Bonmatin et al., 2005; Halm et al., 2006; Laurent and Rathahao, 2003; Schmuck et al., 2001). Thus, systemic insecticides are persistent within plants, and natural enemies that consume non-prey foods containing these insecticides are directly exposed to insecticides for potentially prolonged periods of time.

17.2.2 The Compatibility of Systemics and Natural Enemies

A number of factors affect the ultimate level of exposure that natural enemies experience when they consume non-prey foods contaminated with systemics. First, different insecticides behave divergently when they enter a plant. For instance, in some organophosphate chemicals, the initial active ingredient is degraded into metabolites, some of which are equally toxic to non-target beneficial species (George and Rincker, 1982; Thomas and Glynne Jones, 1955). Additionally, crop species react differently to systemic insecticides (Smith and Krischik, 1999). Nectary structure, particularly whether nectaries are vascularized, is of little importance in whether a plant will exude systemic insecticides in their nectar (Davis et al., 1988). The idiosyncrasies in systemic behavior aside, a consistent observation is that many of the systemic insecticide classes (i.e., neonicotinoids, pyrethroids, organophosphates, and carbamates) are found in at least one type of plant-based food consumed by natural enemies (Table 17.1), and these chemicals change the quantity and quality of plant-based resources (Table 17.2). In a review, systemic insecticides are found in the nectar of treated plants in 71% of 34 studies (11 plant species and 13 systemic insecticides are detailed) (Davis et al., 1988).

Exposure to systemic-tainted plant tissues is deleterious to many natural enemies through direct toxicity within laboratory-based studies. Hymenopteran parasitoids fed nectar from insecticide-treated plants may experience drastic reductions in survival for a substantial amount of time after initial treatment (Cate et al., 1972; Krischik et al., 2007; Stapel et al., 2000) (Fig. 17.3). In one study, the flight capabilities of *Microplitis croceipes* wasps are also impaired when these wasps eat insecticide-tainted nectar from systemically treated cotton plants (Stapel et al., 2000). Coccinellid and chrysopid predators also experience reduced survival when enclosed with systemically treated flowers versus untreated flowers (Rogers et al., 2007; Smith and Krischik, 1999) (Fig. 17.4). *Coleomegilla maculata* adults fed from flowers of plants treated with imidacloprid have reduced mobility and fecundity as well (Smith and Krischik, 1999). Heteropterans, *Nabis, Geocoris* and *Orius*, feed extensively on vegetation, particularly when prey is scarce, and thus it is not surprising to find that these predators are adversely impacted by systemic insecticides (Ridgway and Jones, 1968; Ridgway et al., 1967; Sclar et al., 1998; Torres and Ruberson, 2004). In a creative twist on the unintended consequences of systemics on higher trophic

Table 17.1 Systemic insecticides found in non-prey foods and the doses at which they occur in these tissues

Insecticide trade name	Insecticide class	Crop species	Initial applica-tion rate	Non-prey food contami-nated	Maximum dose found	Reference
Schraden	Organophosphate	*Brassica*; *Sinapis alba*		Anthers	0.0038 ug per anther	(Glynne Jones and Thomas, 1953)
Deltamethrin	Pyrethroid	*Brassica napus* var. *oleifera*		Anthers	0.047–0.605 ppb	(Tasei et al., 1994)
Aldicarb	Carbamate	*Gossypium*		EFN	Toxic to *Campoletis perdistinctus*	(Cate et al., 1972)
Aldicarb (Temik 15G)	Carbamate	*Gossypium*	1 mg per plant	EFN	Toxic to *Microplitis croceipes*	(Stapel et al., 2000)
Imidacloprid (Provado)	Neonicotinoid	*Gossypium*	0.005 mg/plant	EFN	Toxic to *M.croceipes*	(Stapel et al., 2000)
Acephate (Orthene 75S)	Organophosphate	*Gossypium*	0.0019 mg/plant	EFN	Toxic to *M. croceipes*	(Stapel et al., 2000)
Shell SD-9129 (dimethyl phosphate ester with 3-hydroxy-*N*-methal-*cis*-crotonamide	Organophosphate	*Gossypium*		EFN	85 ppm	(Ridgway and Lindquist, 1966)
Dimethoate (Cygon) and carbofuran (Furadan 480)	Organophosphate and carbamate (respectively)	*Vicia faba*	Two treatment rates for each; 5 and 50 ug/l	EFN		(Davis et al., 1988)
Aldicarb	Carbamate	*Medicago sativa*		Floral nectar	5.74 ppm	(George and Rincker, 1982)

(continued)

Table 17.1 (continued)

Insecticide trade name	Insecticide class	Crop species	Initial applica-tion rate	Non-prey food contami-nated	Maximum dose found	Reference
Imidacloprid (Guacho)	Neonicotinoid	*Helianthus annuus*	700 g/kg	Floral nectar	1.9 ppb	(Schmuck et al., 2001)
Dimethoate	Organophosphate	*Phacelia campanularia, Borago officinalis, Brassica napus*	0.1% applied to soil	Floral nectar	Toxic to bees	(Jaycox, 1964)
Dimethoate (Cygon)	Organophosphate	*Allium*	300 ppm AI	Floral nectar	7 ppm	(Waller and Barker, 1979)
Dimethoate (Cygon)	Organophosphate	*M. sativa*	304 ppm AI per plant	Floral nectar	0.5 ppm	(Barker et al., 1980)
Dimethoate (Dimethogon)	Organophosphate	*Citrus limon*	1.12 kg of AI per ha	Floral nectar	0.1 ppm	(Waller et al., 1984)
Tricholofon	Organophosphate	*M. sativa*	Foliar spray	Floral nectar		(George and Rincker, 1982)
Dimethoate (Cygon) and carbofuran (Furadan 480)	Organophosphate and carbamate (respectively)	*Ajuga reptans*	Two treatment rates for each; 5 and 50 ug/l	Floral nectar		(Davis et al., 1988)
Dimethoate (Cygon) and carbofuran (Furadan 480)	Organophosphate and carbamate (respectively)	*B. napus*	Two treatment rates for each; 5 and 50 ug/l	Floral nectar		(Davis et al., 1988)
Dimethoate (Cygon) and carbofuran (Furadan 480)	Organophosphate and carbamate (respectively)	*V. faba*	Two treatment rates for each; 5 and 50 ug/l	Floral nectar		(Davis et al., 1988)
Deltamethrin	Pyrethroid	*B. n.* var. *oleifera*		Floral nectar	0.012–019 ppb	(Tasei et al., 1994)

Phosphamidon		*P. campanularia, B. officinalis, B. napus*	0.8% applied to soil	Floral nectar	Toxic to bees	(Jaycox, 1964)
Imidacloprid	Neonicotinoid	*Fagopyrum esculentum*	Label rates	Floral nectar	Toxic to *Anagyrus pseudococci*	Krischik et al., 2007
Dimethoate	Organophosphate	*V. faba*	25–30 mg	Floral nectar	Toxic to fruit flies	(Lord et al., 1968)
Dimethoate	Organophosphate	*Nasturtium (Tropaeolum)*	25–30 mg	Floral nectar	Toxic to fruit flies	(Lord et al., 1968)
Dimethoate	Organophosphate	*Fuschia*	25–30 mg	Floral nectar	Toxic to fruit flies	(Lord et al., 1968)
Schraden	Organophosphate	*Brassica, S. alba*		Floral nectar	5.5 ppm (21 ppm?)	(Glynne Jones and Thomas, 1953)
Schraden	Organophosphate	*B. officinalis*		Floral nectar	2.5 ppm (21 ppm?)	(Glynne Jones and Thomas, 1953)
Shell SD-9129 (dimethyl phosphate ester with 3-hydroxy-*N*-methal-*cis*-crotonamide	Organophosphate	*Gossypium*		Floral nectar	85 ppm	(Ridgway and Lindquist, 1966)
Imidacloprid	Neonicotinoid	*F. esculentum*	Label rates	Flowers	Toxic to *Chrysoperla carnea*	(Rogers et al., 2007)
Imidacloprid	Neonicotinoid	*Asclepias curassavica*	Label rates	Flowers	Toxic to *C. carnea*	(Rogers et al., 2007)

(continued)

Table 17.1 (continued)

Insecticide trade name	Insecticide class	Crop species	Initial applica-tion rate	Non-prey food contami-nated	Maximum dose found	Reference
Imidacloprid	Neonicotinoid	*H. annus*	Label rates	Flowers	Toxic to *Coleomegilla maculata*	(Smith and Krischik, 1999)
Imidacloprid	Neonicotinoid	*Chrysanthemum morifolium*	Label rates	Flowers	Toxic to *C. maculata*	(Smith and Krischik, 1999)
Imidacloprid	Neonicotinoid	*Taraxacum officinale*	Label rates	Flowers	Toxic to *C. maculata*	(Smith and Krischik, 1999)
Imidacloprid	Neonicotinoid	*H. annuus*		Pollen	11 ppb	(Bonmatin et al., 2003)
Imidacloprid	Neonicotinoid	*Zea mays*		Pollen	3 ppb	(Bonmatin et al., 2003)
Imidacloprid (Guacho)	Neonicotinoid	*H. annuus*	700 g/kg	Pollen	3.9 ppb	(Schmuck et al., 2001)
Imidacloprid (Guacho)	Neonicotinoid	*Z. mays*	700 g/kg	Pollen	2.1 ppb	(Bonmatin et al., 2005)
Imidacloprid (Guacho)	Neonicotinoid	*Z. mays*	1 mg/seed (30% higher than label rates)	Pollen	13 ppb	(Laurent and Rathahao, 2003)
Dimethoate (Cygon)	Organophosphate	*M. sativa*	304 ppm AI per plant	Pollen	0.5 ppm	(Barker et al., 1980)
Tricholofon	Organophosphate	*M. sativa*	Foliar spray	Pollen		(George and Rincker, 1982)

Table 17.2 Ecological pathways through which plant-incorporated insecticides and natural enemies interact. Specifically, a summary of the chapter that indicates whether the pesticidal properties change the non-prey food quantitatively or qualitatively, and whether natural enemies are affected as a result. References for specific interactions can be found throughout the chapter

Pathway		Intraspecific variability/HPR		Systemic insecticides (including seed treatments)		GM crop	
		Variability in observed trait[#]	Effects on natural enemies[&]	Variability in observed trait[#]	Effects on natural enemies[&]	Variability in observed trait[#]	Effects on natural enemies[&]
Floral nectar	Quantity of nectar	?	?	X	?	X	?
	Quality of nectar	X	?	X	–	X	?
EFN	Quantity of EFN	X	–	?	?	?	?
	Quality of EFN	?	?	X	–	X[*]	?
Honeydew	Quantity of honeydew	X	+	?	?	X	+
	Quality of honeydew	?	?	?	?	X	–[%]
Pollen	Quantity of pollen	X	?	?	?	?	?
	Quality of pollen	X	–	X	–	X	0
Seeds	Quality of seeds	?	?	X	–	X	0
Leaves	Toxicity in vegetative tissues	X	–	X	–	X	?
	Variation in allelochemicals	X	?	?	?	X	0

[*]I was unable to find studies that specifically examined whether the transgene product was found in EFN, but several instances indicate that transgene toxins can be found in the phloem.
[%]This study simulated the effects of honeydew resulting from GNA-expressing plants on parasitoid feeding using a spiked sugar solution.
[&]'–' indicates that variability has reduced natural enemy fitness in at least one instance, '0' indicates that the interaction has been tested and no deleterious results have resulted, and '+' means that the only results measured to date have revealed a positive effect on the natural enemy.
[#]An 'X' indicates that research has shown variability in the given trait.
'?' indicates that no research could be found on this topic.

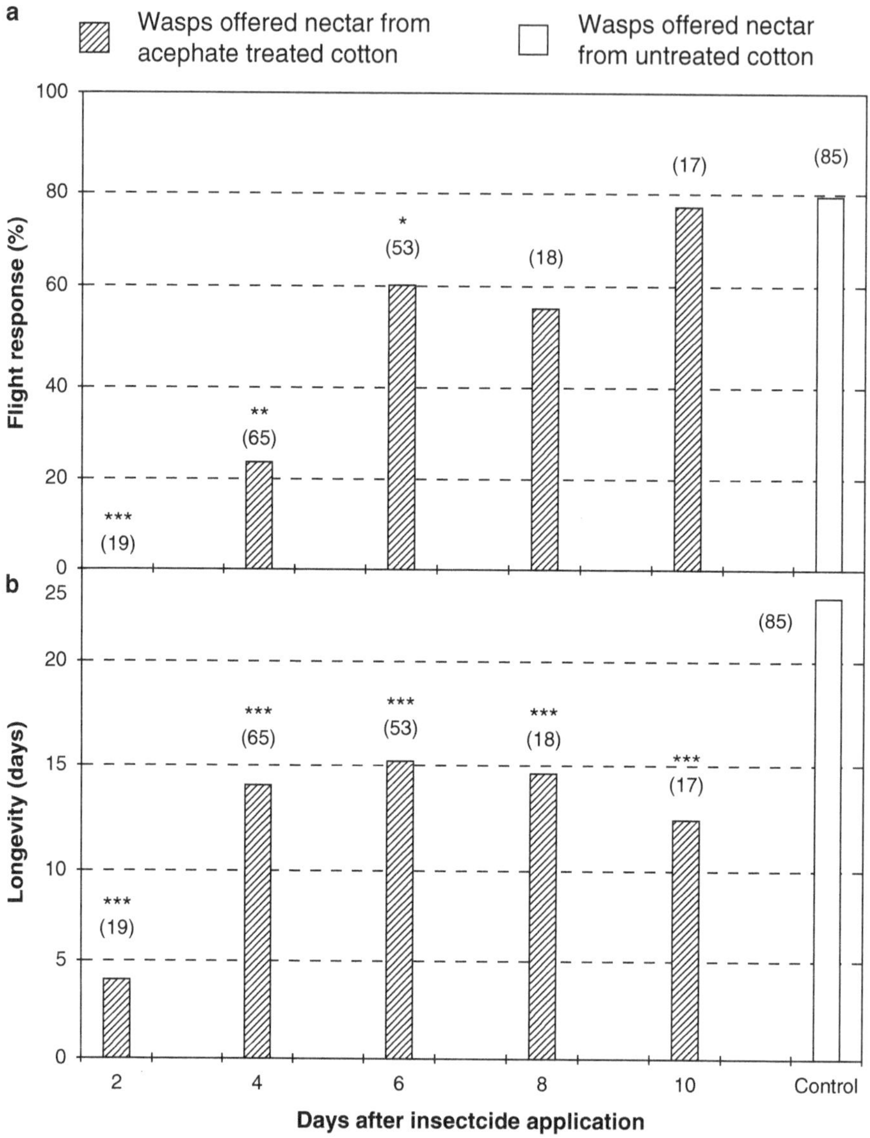

Fig. 17.3 Flight response to *Helicoverpa zea*-damaged cotton leaves in a wind tunnel **(a)** and longevity **(b)** of *Microplitis croceipes* females offered nectar from acephate-treated cotton. Nectar was collected 2–10 days after insecticide application and fed to wasps. Asterisks indicate significant differences from control at levels (*) $P = 0.05$, (**) $P = 0.01$, (***) $P = 0.001$ (Reproduced from Stapel et al., 2000. With permission from Elsevier)

levels, systemic insecticides are used to determine whether predatory phytoseiid mites consume plant material as part of their diet (Magalhaes and Bakker, 2002; Nomikou et al., 2003a). By observing differential levels of mortality in several

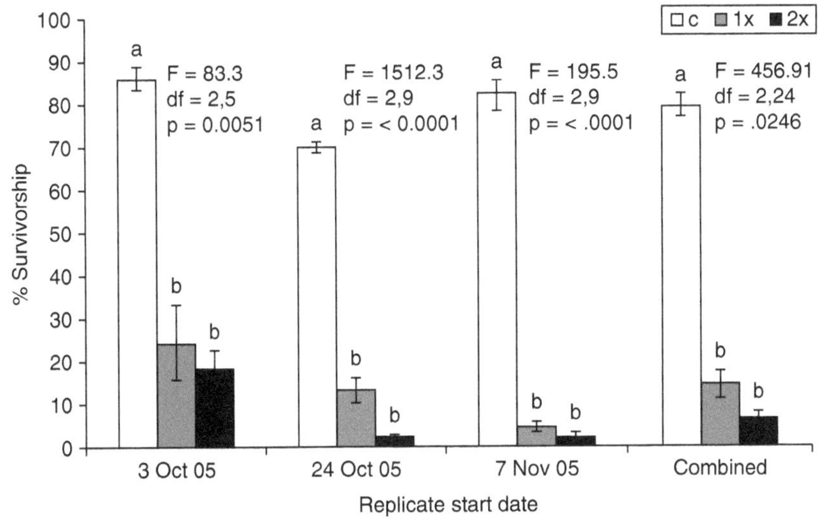

Fig. 17.4 Survival at 10 days of adult *Chrysoperla carnea* exposed to *Fagopyrum esculentum* flowers treated at 0 (c), 1x or 2x the label rate of Marathon 1% G (Reproduced from Rogers et al. 2007. With permission by Elsevier)

predatory mite species raised on systemically treated plants, these authors assess whether each species relies on vegetation in their diet.

The results of field studies are inconsistent with regard to the impact of systemic insecticides on natural enemy populations under field conditions. At one end of the continuum, treating cotton fields with a side-dress of aldicarb virtually eliminates natural enemy populations compared with untreated fields (Timmons et al., 1973). Other studies also record reductions in natural enemies (Cate et al., 1973), especially heteropterans which often rely on plant-based material in their diet (Albajes et al., 2003; de la Poza et al., 2005; Morrison et al., 1979; Ridgway and Jones, 1968; Ridgway et al., 1967; Sclar et al., 1998; Torres and Ruberson, 2004). Al-Deeb et al. (2001) show that adverse effects of imidacloprid seed treatments are most severe when *Orius insidiosus* is prey limited, and likely have to rely more on treated plant material to make up for their diet. In the same study, thiamethoxam killed *O. insidiosus* adults whether prey is available or not. It is also notable that in at least three field studies, pests that are under regulation by predatory heteropterans have higher populations when the systemic insecticide reduce the densities of their primary predators (Albajes et al., 2003; Morrison et al., 1979; Sclar et al., 1998). Other studies do not report consistent reductions in the numbers of heteropterans or other predatory taxa exposed to systemics when they are applied as seed-treatments (Al-Deeb and Wilde, 2003; Bhatti et al., 2005b; Charlet et al., 2007; Reed et al., 2001). In general, when systemics are applied as seed treatments, they seldom disrupt epigeal predator communities (Albajes et al., 2003; Duan et al., 2004). This is surprising in that this guild of predators is highly exposed to the insecticide delivery method, experiences high levels of mortality when enclosed with treated seeds in the laboratory (Mullin et al., 2005), and a number of species consume seeds under field conditions. To conclude, systemic insecticides have direct effects on natural enemy populations, especially on predatory heteropterans, in the laboratory and in the field, but these dynamics do not always behave as predicted.

17.3 Insecticidal GM Crops

Dozens of crops are now transformed to express resistance to insect pests, but the only commercialized events currently available are those producing Cry proteins from the entomopathogen, *Bacillus thuringiensis*. Specific Cry proteins (Cry1Ab, Cry1Ac, Cry2Ab2, Cry 9C, Cry1F, Cry3A, Cry3Bb1, a binary compound of Cry34Ab1 and Cry35Ab1, to name a few that have been used in commercialized GM crops) are toxic only under very distinct physiological conditions present in some insect guts, and phylogenetic uniformity in gut conditions within insect orders allow the creators of GM crops to target specific pest groups with the insecticide (e.g., Lepidoptera, Coleoptera). Of course, gut conditions of pests and natural enemies are not entirely exclusive, and although Bt crops have a relatively narrow spectrum of activity, there is a potential for these crops to affect natural enemies with similar gut physiologies to the targeted pest.

Other toxins engineered into non-commercialized crops have a broader spectrum of insecticidal activity. A number of other crops are engineered to express genes that code for lectins (the most famous of which being snowdrop lectin, or *Galanthus nivalis* agglutinin; GNA) (e.g., Couty et al., 2001a; Kanrar et al., 2002), trypsin inhibitors (e.g., Hilder et al., 1987), and other defensive toxins (e.g., Bell et al., 1999; Gatehouse et al., 1996; Picard-Nizou et al., 1995) (see Chapter 11 for more details on these secondary compounds). The proteins produced, crop species transformed, and gene promoters that are integrated into the transformed crop all influence the degree of hazard and exposure experienced by natural enemies through bi-trophic interactions.

17.3.1 Transgenic Toxins in Non-Prey Foods

Expression of the transgene within crop tissues is regulated in great part by the promoter that is included in the transgene package. The most commonly used of these promoters in commercial crops is CaMV 35S, from the cauliflower mosaic virus. CaMV 35S is a constitutive promoter, but transformed crops operated by solely by this promoter tend to express the transgene at much lower levels in vascular tissue and pollen than in other tissues of the plant. Because of these characteristics of the CaMV 35S promoter, controlling phloem feeding pests (sternorrhynchan species) with GM crops usually requires a different promoter that will produce the toxin within the cells of the phloem. Phloem-specific promoters are incorporated into uncommercialized GM crops, and the transgene product is found in nectar and honeydew that result from these crops. This being said, crops and transgenic events behave differently even when under control of the same gene promoter, and extrinsic factors like weather, seasonal changes, and geography all influence transgene expression (Grossi-de-Sa et al., 2006). Thus, while some trends exist in how different promoters affect the presence of toxins in GM non-prey foods, expression levels in different tissues is not entirely predictable and each transgenic event needs to be assessed individually (Malone and Pham-Delegue, 2001).

Of the non-prey foods produced by transgenic crops, leaves and stems, pollen, and sometimes seeds have the highest levels of Bt in events under the influence of the CaMV 35S promoter. In Bt cotton, Bollgard I and II (events 531 and 15985, respectively) both produce Cry1Ab in pollen (around 0.6 ppm dry weight); what is interesting is that in Bollgard II, two Bt toxins are produced by the plant (Cry1Ac and Cry2Ab2), but only the Cry 1Ac is found in the pollen (Grossi-de-Sa et al., 2006). Seeds of Bollgard I contain 0.86 ppm dry weight (Grossi-de-Sa et al., 2006), but Cry1Ab is absent from the kernels of corn (event 176) (Fearing et al., 1997). Both lepidopteran- and coleopteran-specific Bt corn produce the toxin in the pollen; two estimates being 1.2–2.4 ppm Cry1Ab fresh weight (0.3 ppm dry weight) in event 176 (this event employs a leaf-specific promoter, PEPC, the constitutive CaMV 35S promoter, and a maize specific pollen promoter) and 77.1 ppm Cry3Bb1 fresh weight in event MON863 hybrids (Duan et al., 2002; Fearing et al., 1997; Obrist et al., 2006a). Thus, regardless of whether a pollen-specific promoter is employed in Bt crops, the pollen can contain the transgene product.

Literature on the production or presence of toxins in the floral nectar, EFN, and honeydew produced by or from GM plants is inconsistent. In two Bt corn events that produce Cry1Ab, one operated by CaMV 35S and the other being event 176, the phloem contains at most 1.1 ppb Cry1Ab, but there is no measurable toxin detected in the honeydew of *Rhopalosiphum padi* reared on these plants (Raps et al., 2001). A higher and more consistent level of Cry1Ac (2.7 ppb; about 27 times less abundant than in the leaves) is found in the phloem sap of Bt oilseed rape (again, the CaMV 35S promoter) grown in the greenhouse (Burgio et al., 2007), but these results are far less consistent in field-grown corn plants. In Bt rice lines (expressing Cry1Ab or Cry1Ab/Cry1Ac) transformed to contain one of several constitutive or leaf-specific promoters, the honeydew of *Nilaparvata lugens* contains the toxin to varying degrees; for example planthoppers reared on the lines operated by the CaMV 35S and rice actin promoters have 5.0 and 1.3 ppb of the toxin in their honeydews, respectively (Bernal et al., 2002a) (Fig. 17.5). Not surprisingly, non-Bt GM crops that are operated by phloem specific promoters (e.g., rice sucrose synthase 1; RSs1) contain the transgene product in their floral nectar and EFN at relatively consistent levels (Couty et al., 2001a; Hilder et al., 1995; Rao et al., 1998; Shi et al., 1994; Wang et al., 2005; Wu et al., 2006). In fact, the successful transformation and phloem-specific expression of GNA tobacco plants is actually validated by testing the honeydew of *Myzus persicae* for GNA (Shi et al., 1994).

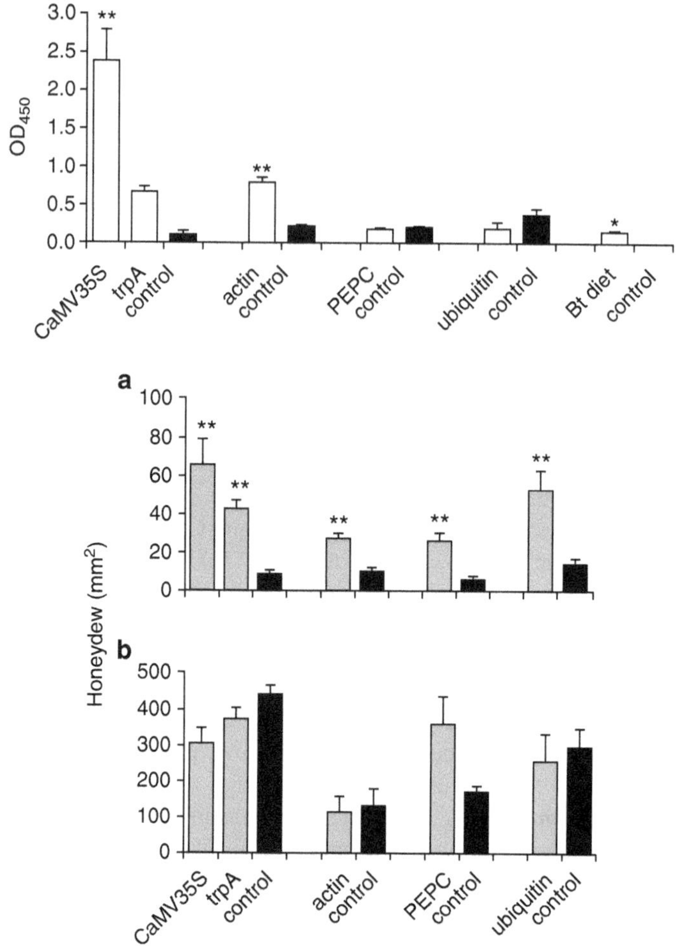

Fig. 17.5 Optical density (indicating of the presence of Cry toxin) (*top*), and quantity (*bottom*) of honeydew produced by *Nilaparvata lugens* when reared on GM rice plants and their respective control line. In the bottom figure, (**a**) white spots produced by acidic honeydew derived from xylem feeding; (**b**) blue spots produced by basic honeydew derived from phloem feeding. Asterisks indicate significant differences between the GM and control plants (Reproduced from Bernal et al., 2002. With permission by Blackwell)

In addition to potentially contaminating phloem-based non-prey foods with the insecticidal transgene product, GM crops may affect natural enemies by reducing the quantity or quality of these phloem-based resources. Genetic modification of oilseed rape plants (transformed to produce bean chitinase using a CaMV 35S promoter) changes the quantity and sugar content of the crop's nectar (Picard-Nizou et al., 1995). Honeydew production is reduced in *Nilaparvata lugens* feeding on GNA-rice (Rao et al., 1998), and in *Lipaphis erysimi* fed transgenic canola that produces wheat germ agglutinin (WGA) (Kanrar et al., 2002). In contrast, *N. lugens* produces more honeydew when reared on Bt rice, but its feeding behavior changes to consume more xylem fluids versus phloem fluids, and thus the honeydew of planthoppers fed Bt rice is quantitatively and qualitatively different from nontransgenic plants (Bernal et al., 2002a) (Fig. 17.5). Another instance is lepidopteran-resistant Bt corn, which appears to favor *Rhopalosiphum maidis* populations (possibly through improved amino acid availability in Bt plants), resulting in greater overall honeydew quantity (but not quality) on Bt plants than on non-GM corn hybrids (Faria et al., 2007). These changes documented in nectar production and aphid feeding and honeydew production are far from ubiquitous among transgenic crops (Pierre et al., 2003; Shieh et al., 1994), and more research is needed to better understand when and where changes may occur. The impacts on natural enemies of unintended changes to honeydew and nectar production resulting from crop transformations appear trivial at first glance, but no research has substantiated this one way or the other.

17.3.2 Bi-trophic Interactions of Natural Enemies and GM Crops

Spray applications of Bt (reviewed by Groot and Dicke, 2002; Obrycki et al., 2004), and diet-incorporated assays involving Cry toxins (Hilbeck et al., 1998b, but see Duan et al., 2005; Romeis et al., 2004; Sims, 1995) are deleterious to some natural enemies. Moreover, natural enemies residing in and near Bt crops contain Cry toxins (Harwood et al., 2007; Ludy and Lang, 2006b; Torres et al., 2006; Zwahlen and Andow, 2005). In a survey of natural enemies collected from Bt cornfields, Harwood et al. (2005) found several predators that frequently had appreciable amounts of Cry1Ab in their stomachs. *Nabis roseipennis* is the most frequently contaminated predator (77% of 30 individuals), containing a mean of 1.85 ppm Cry1Ab, and various spiders, *Coleomegilla maculata* and other coccinellids, *Orius insidiosus*, and *Photinus pyralis* also test positive for the protein. During anthesis in European Bt cornfields, several field-collected predators contain Cry1Ab in their stomachs, including *Orius*, *Stethorus*, and *Chrysoperla* (Obrist et al., 2006a). *Chrysoperla* collected 3 m from Bt fields still test positive (though the other predators do not). In laboratory assays, Obrist and colleagues determine that the two greatest sources of the Cry1Ab toxin in Bt corn fields (event 176) during the sample period are corn pollen and the herbivorous mite, *Tetranychus urticae*, and when *O. majusculus* is fed these foods in the laboratory, Cry1Ab can be detected in the predator (Obrist et al., 2006a). Thus, ingesting Bt corn pollen explains in part why certain field-collected natural enemies contain Bt toxins during anthesis.

Given that GM non-prey foods sometimes contain transgenic toxins, and that Cry toxins are found in the guts of natural enemies collected from Bt cropland, several studies attempt to ascertain whether these toxins affect the fitness of natural enemies under controlled conditions. By far, most laboratory studies on GM non-prey foods focus on prey-mediated effects, but a few studies have examined whether omnivory on pollen, seeds, and sugar sources may function as a possible pathway for GM crops to impact higher trophic levels. Pilcher et al. (1997) were the first to examine whether consuming Bt maize pollen (in this case, Cry1Ab) in the laboratory affects the fitness of predators in the laboratory; *Chrysoperla carnea*, *Coleomegilla maculata*, and *Orius insidiosus* were the predators addressed. With this study, Pilcher and colleagues top a list of studies that demonstrate that *Bt* pollen of several events (Cry1Ab, Cry3A, Cry3Bb1) have no consistent measurable effects on predators and parasitoids representing a diverse array of taxonomic groups and feeding ecologies (Bai et al., 2005; Duan et al., 2002; Ferry et al., 2007; Geng et al., 2006; Ludy and Lang, 2006b; Lundgren and Wiedenmann, 2002, 2004; Obrist et al., 2006b). Feeding on Cry3Bb1 maize seed in the laboratory is similarly benign to five carabid species, some of which are known granivores (*Pterostichus melanarius*, *Harpalus pensylvanicus*, and *Chlaenius tricolor tricolor*) (Mullin et al., 2005), and consuming foliage of Bt potatoes has no adverse effects on a number of predatory heteropteran species (Armer et al., 2000), nor is any Cry1Ac found in the heteropterans allowed to feed directly on Bt cotton plants (Torres et al., 2006).

To date, few studies examine the effects of honeydew resulting from sternorrhynchans fed GM crops or the nectar from these plants on natural enemies. But in an attempt to artificially approximate toxic honeydew as a pathway for interactions of GM crops and natural enemies, Romeis et al. (2003) offer sugar solutions containing GNA to three parasitoids *Aphidius colemani*, *Trichogramma brassicae*, and *Cotesia glomerata* in the laboratory. Feeding time, longevity, survival rate, and fecundity is affected in at least some of the parasitoids relative to those fed untreated sugar solutions. Thus, if a GM crop was to contain GNA in its phloem, then sugar feeding by natural enemies may affect the compatibility of biological control and this GM crop. But when the fitness of aphid populations is improved on lepidopteran-resistant Bt corn, and honeydew is more abundant, this can lead to greater populations of honeydew-consuming natural enemies (Faria et al., 2007). Under confined laboratory conditions, the increased availability of honeydew produced on Bt-hybrids leads to greater parasitoid longevity and parasitism rates of *Spodoptera littoralis* by *Cotesia marginiventris* (Faria et al., 2007). Whether the function of natural enemy populations in Bt corn are similarly improved under field conditions, especially given the reduced quality of *Spodoptera* hosts on Bt maize, is a question well worth pursuing.

Another factor that may affect the performance of natural enemies, but one that is not well tested for natural enemies, is whether predators and parasitoids can detect and avoid GM plants. Essentially, do GM crops affect natural enemies through changes to allelochemicals, as indicated with HPR? Certainly, herbivores can detect GNA in transgenic plants (Foissac et al., 2000), and *Chrysoperla carnea* assess the relative prey quality of Bt and non-Bt fed *Spodoptera littoralis* larvae and preferentially feed on non-Bt prey (Meier and Hilbeck, 2001). One study suggests that GNA-expressing and non-transgenic sugarcanes are equally acceptable to *Cotesia flavipes* (Setamou et al., 2002b). Schuler et al. (1999) points out that by reducing feeding by herbivores on GM crops, there are fewer herbivore-induced plant volatiles, which may in turn reduce exposure of natural enemies to the effects of the transgene product. Whether natural enemies can also detect dissimilarities in pollen or phloem-based foods from GM and non-GM crops remains to be explored.

Dozens of studies investigate the field-scale impacts of Bt crops on non-target organisms (Al-Deeb and Wilde, 2003; Bhatti et al., 2005a, b; de la Poza et al., 2005; Duan et al., 2004; Ludy and Lang, 2006a; Meissle and Lang, 2005; Obrycki et al., 2004; Orr and Landis, 1997; Pilcher et al., 1997; Reed et al., 2001; Sisterton et al., 2004, to name a few). The wealth of field data on predators and parasitoid species gives us the opportunity to determine whether the trends in the bi-trophic interactions between Bt crops and higher trophic guilds measured in the laboratory stand up under more realistic conditions. Two recent meta-analyses on non-target impacts of Bt crops record these general at the taxonomic (Marvier et al., 2007), and functional group levels (Wolfenbarger et al., 2008). Barring a few exceptions, there are no consistent non-target effects of Bt crops on natural enemies. When Marvier et al. (2007) compare the abundance of non-target arthropods in Bt cotton versus untreated controls, hemipterans are one of the groups significantly impacted by GM crops. But this same trend is not observed in Bt maize fields. But in maize fields, hymenopteran abundance is significantly reduced in both Lepidoptera- and Coleoptera-specific fields. Wolfenbarger et al. (2008) substantiate this trend with parasitoids in lepidopteran-specific Bt cornfields, and point out that nearly all of the studies pertain to the specialist parasitoid of *Ostrinia nubilalis*, *Macrocentrus grandii*. Thus in this case, tri-trophic interactions appear to influence specialist parasitoid abundance. Other effects of Bt crops on the abundance of predators, parasitoids, and omnivores are also brought to light in these studies, but they are not consistent across all crops. For instance, there is a slight negative effect of Bt cotton on predators compared with untreated cotton fields, apparently driven by relatively low numbers of coccinellids and nabids in the Bt cotton. Predators are actually more abundant in Bt potatoes over untreated fields. Generally speaking, spraying insecticides on the crop is more deleterious to natural enemies than Bt crops are, exceptions being omnivorous ants in Bt cornfields. In this case, Bt crops have a moderate deleterious effect on ants over other insecticide-treated controls. To summarize, GM crops can express insecticidal compounds in and change the production levels of non-prey foods, but these changes do not apparently have deleterious effects on natural enemies except for phloem-targeted events. Meta-analysis of existing field data point out only a few inconsistent negative effects of Bt crops on higher trophic levels.

17.4 Conclusions

Examining the diversity of ways in which non-prey foods are affected qualitatively and quantitatively when anti-herbivore properties are selected for or actively placed into crop plants gives a more complete view of how complex bi-trophic interactions among natural enemies and pest-resistant crops can be (Table 17.2). Listing the pathways whereby individual technologies change the crop, and bi-trophically the fitness of natural enemies, reveals conspicuous knowledge gaps. For example, the role of intraspecific variability of plant-based allelochemicals in natural enemy attraction is only investigated in depth for conventional germplasms. Although systemic insecticides often change the physiology of the crop (Gupta and Krischik, 2007; Rebek and Sadof, 2003), and unintended physiological effects of transgenesis have been recorded (Birch et al., 2002; Saxena and Stotzky, 2001; Faria et al., 2007), the impact of these changes to the plant in terms of nutrition and attractiveness to natural enemies is seldom investigated. Another trend worth noting is that although considerable research investigates where insecticidal toxins show up in GM crops, very little work explores the diversity of pathways in which natural enemies are actually affected by this exposure (a conspicuous exception is feeding on Bt pollen by natural enemies). Another observation worth noting is that the effects of plant-incorporated insecticides on fungi and microorganisms, both as food (and thus vectors of the insecticide) and as nutritional symbionts, are entirely overlooked as pathways for these technologies to influence higher trophic levels. Finally, ants are some of the most influential of natural enemies within ecosystems, but the impacts of plant-incorporated insecticides on this group escape the notice of scientists almost entirely. The few studies that examine this influential group suggest that formicids are potentially affected by plant traits (EFN-bearing cotton) (Agnew et al., 1982) and Bt crops under some circumstances (Wolfenbarger et al., 2008). Clearly, more work is needed on these understudied interactions.

A question that is posed at the beginning of this chapter is whether or not plant-incorporated resistance and biological control are compatible, particularly from the perspective of bi-trophic interactions. After carefully reviewing the topic, I am very comfortable with giving the answer: 'sometimes'. Several factors complicate our ability to make general predictions regarding the risk posed by individual technologies to higher trophic levels. The idiosyncratic way in which individual crop cultivars and species interact with insecticides and transgene expression, and the environmental effects on the degree and type of resistance ultimately shown can have dramatic effects on the ultimate outcome of these interactions (Cortesero et al., 2000). One take home message from Table 17.2 is that when pest resistant properties are incorporated into plants, it should be presumed that higher trophic levels will be directly exposed to the toxins through omnivory unless it can be experimentally shown that specific pathways are defunct. This is not to say that natural enemies will ultimately be harmed by the insecticidal properties. As illustrated with Bt plants, the level of toxicity of the particular pest resistance factor ultimately plays a role here as well.

> *Thus low levels of plant resistance have value in integrated control, especially as they have the added advantage of delaying the breakdown of resistance which can occur with strongly resistant plant varieties.* (van Emden, 1966)

Biological control of the target pest and plant-incorporated resistance are shown to be most compatible when the crop exhibits only partial resistance (Adkisson and Dyck, 1980; Gould et al., 1991; Hare, 1992; van Emden, 1986, 1995), which raises concern over the current high-dose strategy that is applied with GM crops (Gould, 1998), and systemic insecticides (Dively et al., 1998), a topic reviewed more at length by van Emden (1999). The long term sustainability of GM crops and systemic insecticides depends in part on how well they fit into IPM frameworks. Application and marketing of Bt crops or systemic seed treatments as silver bullets that will eradicate pest populations is a practice that society has experienced before and we are still living with the consequences. Current GM technology, involving Bt crops, seems to have few direct adverse effects on natural enemies, but it is only a matter of time before a GM crop is commercialized that *does* have direct insecticidal effects on higher trophic levels. Seed treatments, HPR, and GM crops offer tremendous opportunities for pest management, but they should be viewed and designed as tools that can be integrated with other control strategies for the most economical and sustainable solutions. As such, the diversity of ways that these technologies interact bi- and tri-trophically with biological control agents should be evaluated on a case by case basis. It is hopeful that many of the proposed frameworks for considering the risk of newly developed GM crops shine attention on bi-trophic interactions of natural enemies and crops into their non-target risk assessments (Andow and Hilbeck, 2004; Romeis et al., 2006). Whether newly developed HPR lines and systemic insecticides should be upheld to the same pressures as GM crops is a question worth asking.

Chapter 18
Biological Control of Weed Seeds in Agriculture Using Omnivorous Insects

Some seeds fell by the wayside...some fell upon stony places...and some fell upon thorny places...but others fell upon good ground

The Gospel Matthew (Chapter 13)

The christian apostle Matthew falls into a broad group of nature observers that fail to recognize the importance of seed predation on seed demography.

Seeds differ from other forms of non-prey food in that the seed can itself give rise to a pest. For this reason, weed seeds are the target of biological control using granivorous entomophages. Indeed, the seed stage is particularly vulnerable to biological control, since granivory ultimately leads to the death of a plant (Hulme, 1996). Granivory of weed seeds on farms is typically quite substantial (Table 18.1). The majority of mortality incurred by plant populations occurs during the seed stage. Consequently, feeding on dispersed seeds shapes the density and dispersion of plant communities. Many of the topics discussed in previous chapters of the book (e.g., seed preferences, the feeding behaviors of granivorous taxa, and seed nutrition and defense) directly influence the outcome of biological control programs of weed seeds in agricultural fields. The topic of weed biological control is a very broad one (McFayden, 1998), and the current chapter will focus on a very specific (but relevant) type of biological control, namely post-dispersal seed predation of agricultural weeds by granivorous entomophagous arthropods.

Within agricultural systems, entomophagous insects comprise a large component of the post-dispersal granivore community (Gallandt et al., 2005; Hulme, 1998; Lundgren, 2005; Lundgren et al., 2006; Mauchline et al., 2005; Menalled et al., 2000; Nystrand and Granstrom, 2000). Rodents are another important group of granivores (Hulme, 1998, 2002; Kollmann et al., 1998). Some work shows that rodents do not prefer highly disturbed habitats like agricultural fields (Cardina et al., 1996), but many other studies show that rodents consume seeds at equal or even superior rates compared to granivorous insects (Westerman et al., 2003a, Brust and House, 1988; Cardina et al., 1996; Harrison et al., 2003; Menalled et al., 2000; Watson et al., 2003). Guilds of granivorous organisms can be categorized by their spatial and temporal occurrence and their physiological capabilities to ingest different weed seed species. For example, in

J.G. Lundgren, *Relationships of Natural Enemies and Non-Prey Foods*,
Progress in Biological Control 7,

Published by Springer Science+Business Media B.V.

Table 18.1 Field studies of weed seed removal rates in cropland. The removal rates presented here are generally the top predation levels discovered in each experiment, to illustrate the potential for weed seed biological control under the best circumstances

Removal rate	Seed species	Crop	Granivores	Reference
4.8% day^{-1}	Five weed species	Corn	Ground beetles and *Gryllus pennsylvanicus*	(Brust and House, 1988)
6.4–21%	*Alopecurus myosuroides*, *Bromus sterilis*, and *Avena fatua*	Cereal fields (margins)	Vertebrates excluded	(Povey et al., 1993)
13–43% 48 h^{-1}	*Abutilon theophrasti*	Different field crop rotations	Vertebrates and invertebrates	(Westerman et al., 2005)
32–60% 48 h^{-1}, higher for *S. faberi*	*Setaria faberi* and *A. theo-phrasti*	Three different field crop rotations	Vertebrates exluded; invertebrates	(Heggenstaller et al., 2007)
Up to 70–100% 48 h^{-1}, after mid-August in corn fields. Lower earlier in the season and in other crops late season	*Setaria faberi*	North American field crops	Vertebrates excluded; prima-rily *G. pennsylvanicus* and *Allonemobius allardi*	(O'Rourke et al., 2006)
Up to 90% 5 day^{-1} during August and September	*Panicum dichotomiflorum* and *Chenopodium album*	Three cropping systems	Carabids targeted; vertebrates excluded	(Menalled et al., 2007)
65% $year^{-1}$	*Carex pilulifera*	Forest	*Harpalus solitaris*, vertebrates excluded	(Kjellsson, 1985)
14.3–16.8% $week^{-1}$ during spring	*Poa annua*, *Stellaria media*, *Capsella bursa-pastoris*, and *Lamium amplexi-caule*	Cereal fields	Invertebrates (especially carabids), vertebrates excluded	(Saska et al., 2008)
10% day^{-1}, averaged over 4 year	*A. theophrasti*	Corn	Carabids, crickets (vertebrates excluded)	(Cardina et al., 1996)
> 50% 3 $week^{-1}$	Seven weed species; only three offered at a time	Oats and ley	All granivores	(Andersson, 1998)
2.9% day^{-1}; 82% $year^{-1}$	*Echinochloa crus-galli*	Corn	Carabids	(Cromar et al., 1999)
~ 4% day^{-1}	Four weed species	Corn	Vertebrates excluded	(Menalled et al., 2000)

58% day^{-1}	*S. faberi*	Wheat w/ red clover cover crop	Largely *G. pennsylvanicus*, ground beetles also captured	(Davis and Liebman, 2003)
57–70% year^{-1}	*Ambrosia trifida*	No-till corn	Invertebrates, *Harpalus pensylvanicus* (vertebrates excluded)	(Harrison et al., 2003)
0.1–3.54 day^{-1} (average among species)	*C. bursa-pastoris, Cirsium arvense, Descurainia Sophia, Lepidium ruderale, Sisymbrium loeselii, Taraxacum officinale*	Wheat, oilseed rape, soybean, millet, and corn	Carabids (vertebrates excluded)	(Honek et al., 2003)
38–70% year^{-1}	A community of up to seven species	Organic cereal fields	Carabids and mice	(Westerman et al., 2003b)
70% 2 week^{-1}	Four weed species	Spring barley	Carabids	(Mauschline et al., 2005)
1% day^{-1}	Seven weed species	Organic ley fields	Carabids and crickets (vertebrates excluded)	(Lundgren, 2005; Lundgren et al., 2006)

one study carabids and crickets focus on small weed seeds in North American cornfields, whereas rodents consume the larger seeds of *Datura stramonium* and *Triticum* (Brust and House, 1988b). Whether one argues that rodents or insects are the most important seed predators, entomophagous insects are invariably major components of post-dispersal, weed-seed-feeding guilds in cropland, and so merit attention when discussing the implementation of biological control of weed seeds.

The goal of weed management is a simple one: reduce or eliminate weeds that compete with the crop in a cost effective manner. For seed predation by granivorous entomophages to function a source of biological control, at least five prerequisites must be in place:

1. The target weed must be seed limited.
2. There must be minimal soil disturbance on the farm.
3. The habitat should be sufficiently diverse to support a rich and abundant granivore community.
4. Granivore foraging patterns cannot be inverse density dependent at economically important weed seed densities.
5. Granivores must be polyphagous, but display a preference for seeds, and must coincide with weed seeds temporally and spatially.

In the present chapter, these five criteria will be used as a framework for exploring the biological control of weed seed banks by granivorous entomophages. By the end of the chapter, a case will be made that under favorable circumstances weed biological control can be a reality for producers.

18.1 Are Weed Seeds Limited?

18.1.1 Seed Production in Agriculture

Agricultural practices select for a very distinct group of weeds that often mirror the crop with which they co-occur in a number of ways. First, the recurring nature of intense disturbance in most cropland selects for short-lived weeds (annuals and biennials) that predominate in disturbed habitats. Related to this, weeds often have a similar phenology to the crop with which they occur (Baker, 1989; Cavers and Benoit, 1989). As a result, fall-seeding species excel in many traditional agricultural systems (at least in North America), and harvesting processes favor weeds that produce seeds of shapes and sizes that ensure that they are transported with the grain through harvesting, threshing, and sowing (Cavers and Benoit, 1989; Harper et al., 1970). Additionally, farm management practices shape the distribution of weeds within a field; seeds are frequently dispersed by anthropogenic activities and thus can be aggregated along the tracks of tractors (Buhler et al., 1997; Cavers and Benoit, 1989; Forcella et al., 1996) or areas where manure is spread or livestock are fed (Buhler et al., 1997).

The seeds produced by weeds of cropland either reside on the soil surface or eventually accumulate into seed banks. Agronomic weeds are often extremely fecund, producing a staggering number of seeds per unit area (Buhler et al., 1997; Cavers and

Benoit, 1989; Forcella et al., 1996; Kegode et al., 1999). For example, five common agricultural weed species in Minnesota cornfields produce between 15,000 and 34,000 seeds m^{-2} (Forcella et al., 1996). However, only a fraction of these seeds are viable. Annual plants tend to produce some of the most persistent seeds known, and so arable lands frequently have some of the most long-lived seeds of a number of tested habitats (Fenner and Thompson, 2005). Many of the seeds die, but those that don't form a reservoir that fuels future generations of the weeds called the seed bank (Baker, 1989).

Agricultural practices are often successful in limiting seed production, so that weed problems in annual cropping systems frequently stem from the seed bank (Buhler et al., 1997). Thus, farm and weed management decisions drive the eventual importance of seed bank-derived weed problems (Buhler et al., 1997; Teasdale et al., 2004). The seed bank resides dormant beneath the soil surface, and disturbance greatly increases the number of germinating seedlings produced from the seedbank. However, seeds in the seed bank are extremely abundant, often numbering many 1,000 m^{-2} (Cavers and Benoit, 1989; Forcella et al., 1992). The magnitude of the seed bank can vary significantly within and among fields, leading to spatially aggregated proliferations of seedlings. Seed banks also diverge from the seed communities that are present on the soil surface, such that weeds absent from a given field for several years can become problematic after soil is disturbed, although the viability of the seed bank tends to peter out after around 4 years (Cavers and Benoit, 1989). While the percent of viable seeds within the seed bank decreases as the reservoir ages, a small component of the bank remains viable for extremely long periods, sometimes for several hundred years (Baker, 1989). But these viable seeds only germinate when conditions are appropriate, particularly the depth at which they are buried. The closer a seed comes to the surface, the greater the likelihood of its germination (Hughes and Westoby, 1992). Thus, disturbance of the soil is of paramount consideration in managing agricultural weeds.

18.1.2 Are Weed Seeds Limited?

A central concern when selecting a life stage to target with management is whether this life stage limits the population growth of the pest. Two factors that limit the number of seedlings are the number of seeds that are produced and the number of suitable germination sites in a habitat (Harper, 1977; Louda, 1989). This is to say that the availability of seeds may not limit the subsequent number of seedlings in habitats where there are only a few favorable sites for germination, or where other mortality factors remove the majority of viable seeds. In many natural habitats, it is presumed that plants are safe site-limited rather than seed-limited, and that plant populations can sustain intense levels of seed predation without incurring a reduction in seedling densities (Hughes and Westoby, 1992; Johnson and Fryer, 1996). Andersen (1989) presents an excellent example of this microsite-limited type of plant population structure. In this study, seed losses to insects by four plant species are considerable, but even in *Leptospermum juniperinum*, whose annual number of viable seeds exceeds 40 m^{-2}, the number of new plants recruited is similar to areas that are heavily depleted of seeds by predators. Although seeds in natural systems are frequently safe site-limited, a closer examination shows that seed limitation is probably more frequent than is generally appreciated. In a review on this topic, plants are seed-limited in 50% of published experiments on natural areas (Turnbull et al., 2000).

Current agricultural practices create a situation in which weeds are seed-rather than microsite-limited. Every spring, the land is laid bare to make way for the establishment of a new crop, and so microsites where weeds can germinate are extremely abundant. Under these circumstances, a large proportion of weed seeds find suitable safe sites (Andersen, 1989), and thus removing a portion of the seeds through predation could theoretically reduce weed competition. Indeed, annual and biennial plants (which stand out among cropland weeds) are known to be seed-limited over plants with other life-history characteristics (for instance, woody perennials) (Turnbull et al., 2000). However, annual plants are not always seed limited, since their seed banks contribute to the seed population, and they can be microsite limited just as perennials can (Crawley, 2000). In spite of persistent seed banks, targeting management at the seed stage remains a viable option for farmers. One model shows that removing the seed bank through repeated cultivations can be effective in reducing weed populations, demonstrating that this life stage is influential to weed populations (Jordan et al., 1995), and other models reinforce this notion that the seed stage is a vulnerable target in other weed species as well (Westerman et al., 2006, 2005).

18.2 Does Disturbance Associated with Crop Production Preclude Biological Control of Weed Seeds?

More stable habitats foster higher (or at least more even) levels of seed removal in natural and agricultural systems. Within natural ecosystems, disturbed systems with reduced vegetation cover such as those recently burned, plowed or logged, display lower levels of granivory by insects (Cote et al., 2005; Mittelbach and Gross, 1984; Nystrand and Granstrom, 2000; Reader, 1991). Not surprisingly, most natural systems tend to display higher levels of granivory than managed fields (Diaz, 1992; Mittelbach and Gross, 1984). For example, Mittlebach and Gross revealed that seed consumption increases in the following order: a plowed field has lowest seed removal, old-fields with reduced disturbances have intermediate levels of seed removal, and an undisturbed old-field has the highest level of granivory. Moreover, many ants prefer to establish their nests in stable habitats, often with plenty of vegetation cover (Diaz, 1991, 1992). Thus, when ants forage near their nests, more granivory is experienced in stable habitats.

The disturbance associated with the production of different rotations, crops, and tillage regimens are known to affect seed removal by and activity of beneficial insects. Since weeds are frequently adapted to the phenology of a specific crop, crop rotation can be a very effective method for reducing weed densities (Buhler et al., 1997). Some crop rotations can be made more effective when they are combined with reduced tillage systems, since constituents of the seedbank that are adapted to the following crop are not mobilized in no-till systems (Kegode et al., 1999). It isn't clear at this point how easily granivorous entomophages respond to annual changes in the crop, although perennial crops such as alfalfa fields tend to have higher (or at least more consistent) levels of seed removal than more ephemeral systems (Lundgren et al., 2006). Also, different crops (and the practices inherent in their production) are characteristically associated with different levels of seed predation (Anderson, 1998; Heggenstaller et al., 2007; Honek et al., 2003; O'Rourke et al., 2006; Westerman et al., 2005). A final disturbance associated with weed and farm management is tillage. Consistently, no- or reduced-tillage systems either favor or are benign to the establishment of granivorous entomophages and the biological control of weed seeds (Andersen, 1999; Brust and House, 1988b; Cromar et al., 1999; Menalled et al., 2005; Menalled et al., 2007). Also, the communities of granivores associated with no-till systems may be unique from tilled fields. Brust and House (1988b) found that *Selenophorus* and *Gryllus* are disproportionately more abundant in no-till cornfields than in conventionally tilled fields. Another consideration is that the weeds that colonize no-tillage systems tend to be perennials versus annuals, and may rely less on seedlings for plant regeneration. So there may be a trade off in no-till systems between the increased levels of seed predation and the relative importance of perennial weeds (which do not necessarily rely on seeds for their perpetuation) that are not as susceptible to this form of management.

18.3 Is Biological and Habitat Diversity on Farms Sufficient to Support Biological Control of Weed Seeds?

More diverse landscapes and vegetation supports more speciose and abundant granivore communities. Seeds produced by agricultural weeds differ in their sizes, shapes, nutrition, defensive properties, distributions, and densities. Responding to this variation in food quality, granivore communities partition seed communities such that many species can coexist within agricultural habitats. Because there is a level of specialization in each granivore species, it seems wise to conserve a rich community of granivores in and around farmland to ensure that the key granivores of a target weed are present and ready to work. Indeed, farms that foster a range of habitats tend to experience higher levels of seed predation and fewer weed problems.

18.3.1 The Effects of Landscape Diversity

Because many granivorous entomophages rely on non-crop habitats as overwintering sites and refuges from agricultural practices, the structure and diversity of habitats on a landscape scale have important implications for within field biodiversity and the ecological services this provides (Hendrickx et al., 2007). In general, landscape diversity improves granivory within farmland, although substantial field-to-field variability in granivory rates exists within a region (Menalled et al., 2000; Westerman et al., 2003a, b). Also, because weeds can immigrate to farmland from surrounding habitats, the diversity of weeds in surrounding habitats can have important implications for seed demography patterns (Crawley, 2000).

Organic farms tend to have a more heterogeneous landscape compared to conventional monocultures. Generally, organic farming systems have more diverse granivore communities (Doring and Kromp, 2003; Lundgren et al., 2006) and encourage greater seed predation rates (Gallandt et al., 2005) (but see Menalled et al. 2007). In part, this may be because organic farms have greater weed densities (Barberi et al., 1998). Given that organic producers have a reduced number of tools for managing weeds, and that this type of farming inherently fosters insect diversity, biological control of weed seed banks could be of particular importance in reducing weed pressure on organic farms.

18.3.2 Farm Practices that Promote Granivores

Improved habitat stability and structure in farmland often accompanies higher levels of granivory. Reducing or eliminating herbicide applications is one way to promote biodiversity within farmland. Although it is seldom realized by growers, there is a tolerable level of weeds within cropland, and maintaining crops under complete monoculture likely exacerbates pest problems more than it solves them. Allowing low levels of weeds to persist in farm fields can provide a number of valuable services to the crop habitat (Marshall et al., 2003) (see Chapter 16), including allowing biotic mortality sources of weeds to make a living.

Improved biological control of weed seeds is associated with various ground covers, especially cover crops, and degree of canopy cover is well correlated with seed removal rates in field crops (Heggenstaller et al., 2007). Allowing residue to persist in, or adding residue to a farm field in the form of a mulch may be one way to promote granivory by insects. A greenhouse study by House and Brust (1989) suggests that improved mortality of seeds resulting from mulches may be seen in agricultural systems. Adding wheat straw mulch to corn plots that contain weed seed and granivore densities reflective of field observations hampers the consumption of broadleaf weeds, but not grasses. But in the end, the straw mulch reduces the fitness of these broadleaf weeds such that there are more of them but they are significantly smaller than in the no-mulch treatment. Leaving crop residues in the field may be another way to diversify the habitat. Residues are not all created equally, and significantly different seed removal rates are associated with different crop residues (Cromar et al., 1999). Other studies either report no, or only very slight, positive trends in granivory associated with the presence of residues and mulches (Harrison et al., 2003; Jacob et al., 2006).

A more compelling practice that promotes granivore diversity and weed suppression is the use of cover crops. It appears that cover crops act synergistically with granivory to combat weed pressure, often resulting in higher levels of seedling suppression than by using either technique alone (Brust, 1994). In old-fields, the presence of ground cover (a variety of native forbs and grasses) amplifies the impact of ant predation on the establishment of three common weed species (Reader, 1991). The ground cover functions in reducing the establishment of larger-seeded weedy species, and the ants reduce the competitive ability of smaller seeded species, such that the entire seedling community is affected (Reader, 1993). In wheat fields, the use of a red clover cover crop increases the densities of *Gryllus pennsylvanicus* two-fold, and *Setaria faberi* densities are significantly lower in this treatment than those with no cover crops (Davis and Liebman, 2003) (Fig. 18.1). Again, weed density is likely lower because of direct competition from the cover crop and because granivorous entomophages are able to establish in the more diversified field. Not all cover crop-granivore interactions are equally beneficial for

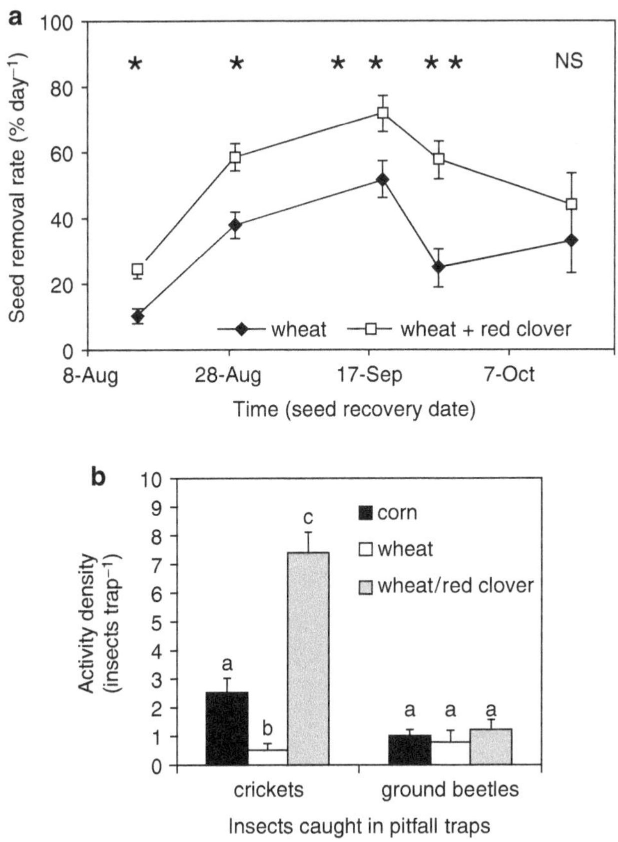

Fig. 18.1 Mean (±SD) activity density of granivorous insects in corn and wheat sole crops and a wheat-red clover crop mixture (*bottom*), and removal rates of *Setaria faberi* seeds in the wheat and wheat + red clover treatments (*top*) (Reproduced from Davis and Liebman, 2003. With permission of the Weed Science Society of America)

weed suppression. In one instance, red clover cover crops support significantly higher numbers of the granivorous carabid, *Harpalus rufipes*, although seed predation is similar among the different cover crops (Gallandt et al., 2005).

18.3.3 Community Interactions Among Granivores

An understanding of the behaviors and physiological requirements of vertebrate and invertebrate granivores and how they interact with individual weed species is essential to designing and implementing consistent and reliable biological control programs of weed seeds. Granivorous entomophages are only one component of most granivore communities, with vertebrates (rodents and birds) often being important competitors for seeds (Hulme, 1998). Generally, birds appear to play a minimal role in post-dispersal seed removal (Holmes and Froud-Williams, 2005), and rodents prefer to consume larger seeds than granivorous entomophages (although they will eat smaller seeds if larger ones become scarce) (Brust and House, 1988b; Davidson et al., 1985). Although the relative contributions of vertebrates and invertebrates to seed removal in agricultural systems are examined in several recent studies (Brust and House, 1988b; Marino et al., 1997; Menalled et al., 2000; Westerman et al., 2003a), the interactions between insects and vertebrates is best understood in desert systems. Direct competition between rodents and invertebrate granivores (especially ants) may partially explain why these two guilds are frequently reported to forage in different places and during non-overlapping seasonal and diel periods (at least in desert systems) (Abramsky, 1983; Brown et al., 1979; Davidson et al., 1985; Hansen, 1978; MacMahon et al., 2000). Furthermore, these guilds appear to exclude one another, at least to a degree, such that when either rodents or ants are barred from a habitat, the other guild becomes more abundant (Brown et al., 1979). In Australia, granivory by rodents and birds is minimal, and harvesting ants almost entirely encompass this ecological niche (Andersen and Ashton, 1985; Berg, 1979).

Within insect granivore communities, competition can lead to the partitioning of seed resources according to a number of seed traits (Hansen, 1978) (Table 18.2). Granivorous entomophage communities are particularly well-noted for partitioning seed resources based on seed size (Briese and Macauley, 1981; Kaspari, 1996).

Table 18.2 Percentage of total food items collected in each food class by and mean fresh weights for three granivorous ant species, indicating how this ant community partitions seed resources based on seed size (Reproduced from Hansen, 1978. With permission by Springer)

	Pogonomyrmex desertorum	*Pogonomyrmex maricopa*	*Pogonomyrmex rugosus*
Worker fresh weight (mean ± SD mg)	5.84 ± 0.11	8.86 ± 0.60	15.10 ± 0.33
Large seeds (1.2 g)	2.54	19.41	48.44
Medium seeds (0.6–1.2 g)	9.16	31.36	20.88
Small seeds (0.6 g)	85.22	30.09	25.71
Insects	3.00	19.13	4.40

Specializing on different seed densities may be another way for granivores to partition resources. Harvesting ants characteristically forage as groups or singly, depending on the species, and can thus exploit seed patches of different sizes. Thus, specializations on different seed densities can lead to several species coexisting in a habitat (Brown et al., 1979; Davidson, 1977; Reichman, 1979; Whitford, 1978). Finally, because conspecifics forage similarly, aggression between them may be more severe than interspecific interactions. Moggridge (1873) found that harvester ants are more aggressive toward other colonies of their own species than they are toward other ant species, and they frequently raid the granaries of conspecifics.

18.4 Do Granivores Respond Positively to Increasing Seed Densities?

A central question to biological control is whether or not granivorous entomophages exhibit a density dependent response to weed seed aggregations, such that the most abundant seeds (and likely the most problematic from a management standpoint) will suffer the highest levels of predation. Janzen (1970) and Connell (1971) independently hypothesize that seeds should suffer lower levels of predation the further they are from the parent plant. High densities should increase the likelihood that a granivore will encounter seeds, and dispersing even short distances from the parent should reduce the likelihood of predation for weeds (Fenner and Thompson, 2005). Harper (1977) considers aggregations of seeds as islands, and the closer they occur together and the larger they are, the more likely they will be colonized by predators. Implied in these ideas is that seed predation rates may be density dependent.

For some granivorous entomophages, the more seeds present, the greater the proportion that is consumed. Harvesting ants are particularly well-documented as displaying high fidelity and increasing ability to exhaust more dense sources of seeds (Brown et al., 1979; Holldobler, 1976; Platt, 1976). But carabids and crickets are not shown to exhibit density dependent foraging in agricultural systems.

More commonly, density independent foraging is seen in granivorous entomophages (Mittelbach and Gross, 1984). Marino et al. (2005) describe how seed aggregation patterns are more important to granivory rates than absolute seed density; aggregated seed patches are more fully exploited. So, even though aggregated seed sources are more attractive to granivores, these insects typically collect similar proportions regardless of the seed density in the patch. For instance, the proportion of weed seeds removed by carabids and crickets is similar across three seed-density treatments (Brust and House, 1988b). Cromar et al. (1999) present that the percentage of seed removal only increases with seed density at one site for one weed species, *Chenopodium album*. There is not density dependent consumption for *Echinochloa crus-galli*, nor for *C. album* at other sites. However, the authors claim that the absolute number of seeds consumed is a better measure of density dependence, since individual predators are easily satiated. Using this definition, both seed species would be consumed in a density-dependent manner. From a functional point of view, I think that the traditional definition of density dependence is the most applicable, since depleting sources of higher seed densities more quickly is desirable to biological control programs. However, density independent responses to seeds certainly do not preclude successful biological control, so long as the seeds are depleted below economically important levels.

Finally, satiation of granivore communities can and does occur, although reports of this in agricultural systems are scarce. After a certain density of seeds is exceeded, the granivores cannot consume additional seeds and inverse density dependence occurs (the bane of biological control…). In one instance, inverse density dependence at low seed densities turns to density independent foraging at higher seed densities. Cardina et al. (1996) describe how the greatest proportion of velvetleaf seeds removed from cornfields is at the lowest density of seeds tested (600 m^{-2}). Similar (but lower) proportions of seeds are removed in all of the treatments with higher seed densities (Fig. 18.2). Westerman et al. (2008) present data that inverse density dependence occurs in at least one large-scale field study where the densities of *Setaria faberi* are manipulated, and they suggest that this is likely more often the case than is currently reported in the literature. Put simply, granivores can easily overcome small caches of seeds commonly employed in removal studies, but under more realistic conditions that better represent seed rain, the granivore communities are overwhelmed. It would be revealing to discover the satiation point of granivores in additional cropping systems to see at what point biological control loses its function in arable land.

Ultimately, factors such as the mobility of predators, ambient densities of seeds and predators, the scale of the experiment, the range of seed densities employed, and relative availability of alternative foods are likely to influence the degree of density dependence on seeds, and so the relationships between seed density and granivory rates by insects are invariably quite plastic. For instance, similar proportions of single and grouped seeds of *Acacia linifolia* and *Bossiaea obcordata* are removed

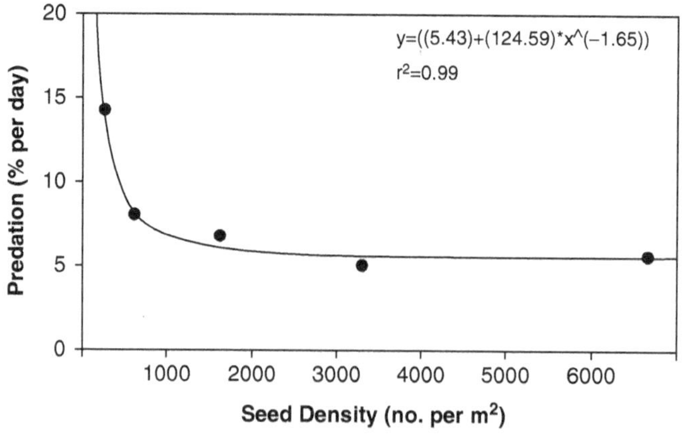

Fig. 18.2 Daily rate of predation on *Abutilon theophrasti* seeds as a function of seed density (Reproduced from Cardina et al., 1996. With permission by the Weed Science Society of America)

during the spring, but a greater proportion of the aggregated seeds is consumed during the fall (Hughes and Westoby, 1990). Another example of seasonal relationships with density dependence is observed with *Pogonomyrmex rugosus*, which prefers the numerically dominant seeds of *Chenopodium incanum* when it becomes available in August, versus *Eriogonum* seeds which are accessible over a longer interval (Whitford, 1978). Other *Pogonomyrmex* ant seed preferences track the seasonal availability of seed (Pulliam and Brand, 1975).

18.5 Characteristics of a Good Weed Seed Biological Control Agent

A question that remains is what characteristics are present in a good biological control agent of weed seeds. One intuitive criteria that must be present for effective biological control is that weed seeds and key granivores need to be present in the same place at the same time. Weed seeds occur patchily both temporally and spatially, and so it is no small matter for a granivore to track these populations. Most post-dispersal granivores are highly polyphagous and opportunistic feeders, which is important for granivores to maintain populations in cropland when seeds are scarce, but it is of little use for granivores to coincide with seeds if they don't consume them. Thus, for granivores to have the maximum impact on weed seed populations, they need to be able to quickly recognize seeds as a preferred food source.

18.5.1 Traits of Biological Control Agents

18.5.1.1 Temporal Overlap with Weed Seeds

The temporal synchrony of seed shedding by numerous weed species during a short window of time could have important adaptive significance in reducing seed predation. One of the evolutionary mechanisms for maintaining reproductive synchrony in plants is to satiate seed predators (Ims, 1990; Lalonde and Roitberg, 1992; Silvertown, 1980). Essentially, by producing seeds only during a brief period, a plant community can starve seed predators, and then overwhelm the remaining predator population with too many seeds for them to consume (Harper, 1977; Janzen, 1972; Silvertown, 1980). Silvertown (1980) points out that three factors must be present for the predator satiation theory to be functional. First, the plants must produce more seeds than the predators can consume. Second, the plants must have enough time between seed sheds that the predator populations will diminish. And third, there has to be reproductive synchrony among conspecific plants and even other plants within a habitat (a few individuals isn't enough). In natural systems, this reproductive strategy is termed masting, and often involves multiyear reproductive patterns (Kelly, 1994; Smith, 1970). As a case in point, in a manipulative experiment, granivorous ants regulate the populations of *Eucalyptus baxteri* seedlings when small numbers of seeds are dropped, but are not able to inflict similar levels of predation when seeds are at densities that mimic mast years (Andersen, 1987). There is evidence that the synchronization of reproduction in annual weed communities represents a form of masting and may functionally overwhelm the ability of resident granivore communities in conventionally managed cropland to reduce seed densities below acceptable levels (Westerman et al., 2008). This is not to say that seed predators are driving the timing of seed shed – clearly cropping practices are doing this. But the end result of synchronization within weed assemblages is the restriction in the types of granivores that will be effective as biological control agents.

As it turns out, granivorous entomophages are well synchronized to annual seed production patterns in agricultural systems. Post-dispersal granivorous entomophages abound during the late summer and early fall (O'Rourke et al., 2006; Harrison et al., 2003; Holmes and Froud-Williams, 2005; Menalled et al., 2005; Webster, 1903; Westerman et al., 2003a; Zhang et al., 1997), when weed seeds are commonly shed in many temperate field crops (Forcella et al., 1996; Leguizamon and Roberts, 1982). More importantly, invertebrate granivore communities track the patterns of weed seed shed in those crops that are not harvested in fall (e.g., some cereal crops), suggesting that these granivores and the level of seed consumption they inflict are clearly adapted to be present in or aggregate to fields when weed seeds are shed (Heggenstaller et al., 2007; Westerman et al., 2003b).

18.5.1.2 Spatial Relationships of Seeds and Granivores

Cropland is highly disturbed, and this simple fact has important implications for the dispersion patterns of both weeds and granivores alike. In part, making a living in cropland necessitates both that the granivore is physiologically adept at tolerating the sometimes harsh conditions imposed by this habitat, and also that the granivore's foraging patterns are such that discrete patches of weed seeds can be quickly found and exploited. Unfortunately, this is a lot to ask and predation does not always track seed dispersion patterns. Thus the impact of foraging granivores is often somewhat irregular within a field. In old-fields, Mittlebach and Gross (1984) found that more than 95% of seeds are removed from some seed sources, while other fields containing the same seed species have as few as 14% of seeds removed.

Research suggests that granivory by insects within agricultural fields tends to be evenly distributed within cropland. Although many granivorous populations that reside in cropland display an edge effect with higher densities near field margins, granivory by entomophagous insects is not consistently restricted to field margins (Saska et al., 2008; Marino et al., 1997; Westerman et al., 2003a). Indeed, seeds in one study were removed at higher rates in the middle of the fields than near the field margins, where carabids were captured most frequently (Saska et al., 2008). Ants may contribute an exception to this phenomenon when they create their nests in the margins and forage most intensively in nearby areas or maladapted to cropland (Azcarate and Peco, 2003; Jacob et al., 2006). But the literature suggests that regions that rely more on carabids and crickets probably do not experience edge effects with regard to seed removal rates. Of course, more research on this topic should be conducted before any hard and fast conclusions can be drawn.

18.5.1.3 Polyphagy with a Preference for Seeds

Polyphagous species are able to persist on other food sources (i.e., insect prey) when seeds are scarce. This allows the seed predator to persist in cropland and more quickly respond to local increases in the availability of seeds. Generally speaking, post-dispersal granivores are not specialists in the strictest sense of the word, but certain species within seed predator communities are known to specialize on specific weed seeds. One example is *Amara montivaga*, which loosely specializes on *Taraxacum officinale* seeds when they become available (Honek et al., 2005). This carabid tracks the seed populations of *T. officinale* both temporally and spatially, and inflicts heavy tolls on the number of seeds that ultimately develop into seedlings.

Both the frequency and degree of seed feeding in relation to animal and other vegetative material under field conditions give insight into the importance of seed-feeding to granivores. The nutritional importance of seeds relative to other foods is well illustrated by research of Johnson and Cameron (1969) with four species of granivorous carabids (*Anisodactylus discoideus*, *Notiobia terminata*, *Amara cupreolata*, *Harpalus affinis*, and *Stenolophus pallipes*) switching to grass seeds (*Poa annua*) from feeding on grass blades as seeds become available. All of these seed-feeding beetles also consume prey (*Hyperodes* dead or alive), but at least some preferred to eat grass seed over all other foods (Johnson and Cameron, 1969). Furthermore, these authors found these species consume 13–25% of their body weight in seeds in a single night (Johnson and Cameron, 1969). *Gryllus pennsylvanicus* and *Harpalus pensylvanicus* prefer to consume seeds of *Amaranthus retroflexus* over moribund aphids in the laboratory, whereas two predaceous carabid species (*Poecilus* spp.) prefer aphid prey over seeds, reinforcing the relative importance of seeds in the diets of these common North American agricultural species (O'Rourke et al., 2006). In another study, seeds are the predominant food of *H. affinis*, who consume seeds in 39% of their 'meals' (Sunderland et al., 1995). In yet another example, *Ophonus* and *Amara* species feed on other plant tissues during the summer and transition onto seeds during the autumn (Zhavoronkova, 1969). This same relationship is shown for more carnivorous carabids (e.g., *Poecilus cupreus*, *Dolichus halensis*, and *Synuchus vivalis*), which feed on insects throughout the season until seeds become available (Skruhavy, 1959; Zhavoronkova, 1969). Polyphagous species that preferentially consume seeds are ideal targets for conservation programs intended to inflict greater biological control on weeds.

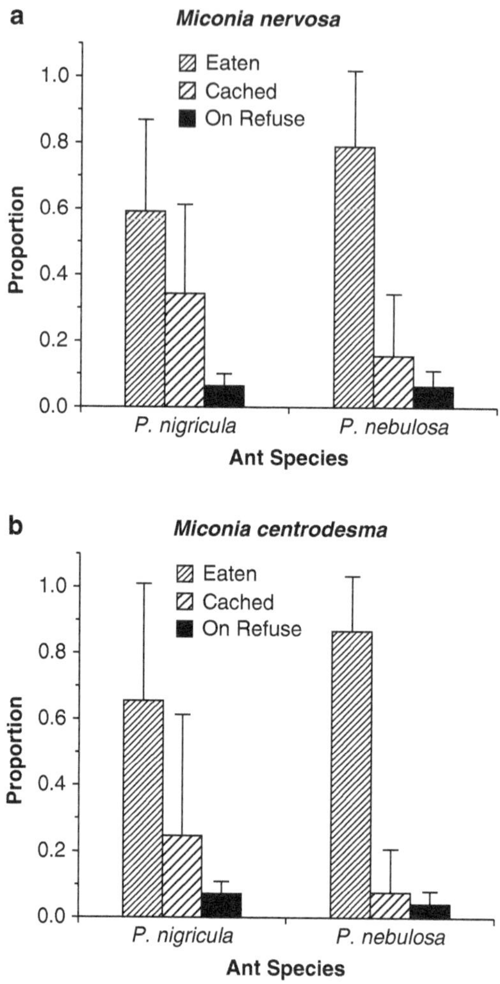

Fig. 18.3 Mean (SD) proportion of 100 seeds of (**a**) *Miconia nervosa* and (**b**) *Miconia centrodesma* eaten, cached, or discarded on refuse piles within 11 d by captive colonies of *Pheidole nigricula* and *P. nebulosa* (Reproduced from Levey and Byrne, 1993. With permission by the Ecological Society of America)

Finally, seed removal in and of itself is not sufficient for biological control – many granivorous entomophages transport seeds but do not actually kill them, particularly seeds with food bodies (vander Wall et al., 2005). Ants are particularly notorious for dropping seeds on the way back to their nests, placing viable seeds upon their middens, or leaving undamaged seeds in upper granaries of abandoned nests (Holldobler and Wilson, 1990; Levey and Byrne, 1993; MacMahon et al., 2000; Roberts and Heithaus, 1986) (Fig. 18.3). For example, 38% of seeds harvested by *Pheidole nebulosa* and 52% of seeds collected by *P. nigricula* are placed on the refuse pile of the nest unharmed. However, the seeds on the refuse piles have lower germination rates than those in the granaries. But, even though the seeds have lower germination on the refuse pile than in the nests, they still do better than they would in the surrounding leaf litter where they are collected. For some seed species, predation may actually promote seed germination through scarification of the seed coat (Hulme, 2002; Ollerton and Lack, 1996). Most granivorous species are messy eaters, and the seed coats are cracked and endosperm is severely damaged. This makes the seeds that survive susceptible to seed degrading microorganisms and other mortality factors. The role of non-lethal interactions between seeds and granivores in the population dynamics of weeds merits further research.

18.6 Seed Burial

A proportion of seeds that falls to the ground becomes buried or is dispersed to seed caches, sometimes with the help of granivorous insects, and the fate of these seeds and their contribution to the seed bank is a topic worthy of discussion (vander Wall et al., 2005). A number of factors affect the optimal depth of burial for a given seed species. Smaller seeded species tend to germinate best at shallower depths compared to larger seeds, and only a handful of species can germinate below 15 cm (Froud-Williams et al., 1984; Herr and Stroube, 1970). Most agricultural weeds germinate best between 1–4 cm beneath the soil surface (Froud-Williams et al., 1984; Herr and Stroube, 1970; Mestor and Buhler, 1990; Mohler, 1996). Burying seeds beneath the surface is not a death sentence; after all, many annual weeds create seed banks (discussed above). But carrying the seeds below ground removes the chances of them causing immediate problems, reduces their rate of germination, and decreases the fitness of those seedlings that are able to reach the surface (Mestor and Buhler, 1990). Of course, the higher level of germination success in shallow seeds is typically evaluated in the absence of predation, and so residing on the soil surface may not be an ideal situation for seeds even though seeds germinate well here (Hulme, 1998; Johnson and Fryer, 1996).

Because deeply burying seeds reduces germination and the likelihood of consumption by epigeal predators, the creation of seed caches by granivorous entomophages may have important implications for weed biological control. Ants and some carabid larvae are known to create seed caches in agricultural habitats. Seeds cached by animals tend to have poor germination (Hartke et al., 1998; Hulme, 1998). There are at least three explanations for this phenomenon. First, the animals may consume the weed seeds outright, as is the case for harvesting ants and certain carabid larvae. Second, the animal may be treating the seeds such that they will not germinate beneath the soil. For example, harvesting ants bite the radicles from the seeds to prevent germination, and then store the unviable seeds in granary chambers within the nest. Other harvesters appear to treat the seeds with a secretion from their abdomen that reduces their germination (Went et al., 1972). Third, simply placing the seed beneath its optimal germination depth will reduce its likelihood of germination (as was just discussed above). The burrows of *Harpalus* and *Ophonus* larvae can extend several meters into the ground, although seeds are typically cached between 8–23 cm deep (Hartke et al., 1998; Kirk, 1972; Luff, 1980), and granaries of harvesting ants can occur many meters under the ground.

At this point, it remains unclear whether those seeds that survive burial by granivorous entomophages will become problematic after tillage or whether they are permanently removed from the viable seed bank. One recent study shows that seeds stored in the seedbank can change in their suitability for granivores (Martinkova et al., 2006). Two carabid species consume fewer buried *Taraxacum officinale* seeds; but five other seed species are equally or more acceptable to the beetles after storage. The implications of storage on the palatability of different seeds to granivorous entomophages merits further attention.

18.7 Conclusions: How Can Biological Control of Weed Seeds Be Promoted?

> *A great many studies have been made of the magnitude of [seed] predation but very few help to determine whether predation is relevant either to the evolution or the population biology of plants* (Harper, 1977)

A number of criteria are present in agricultural systems to suggest that biological control of weed seeds can be an effective component of an integrated weed management program. First, annual weeds of cropland are often seed limited, although they create persistent seed banks that can give rise to weed problems after soil disturbance. Disturbance in agricultural systems can be reduced through the use of conservation tillage practices, tactics that also increase the diversity, abundance, and efficacy of granivorous entomophages. Habitat diversity is also central in improving granivory rates, and practices such as diversifying the landscape in ways that increase plant cover, and the use of mulches, residue, and cover crops typically improve biological control of weed seeds, in part by allowing a more diverse and abundant community of granivores to persist. Granivorous entomophages and weed seeds are well synchronized both in space and time, and a number of granivores may reduce seed densities by creating seed caches below the optimal burial depth of seeds. Although density dependence is seldom shown for granivorous entomophages, granivore communities are seldom reported as becoming satiated by seed availability, although more research on this topic is needed.

Within agricultural systems, granivorous entomophages inflict heavy losses upon weed seeds (Table 18.1). In fact, the majority of seeds that fall to the ground are predated by granivores within a short amount of time (Westerman et al., 2003b). Buhler et al. (1997) conclude that if seed rain from mature weeds is reduced or eliminated, then weed management tactics can drastically reduce the seed bank over a few years. Predation is as efficacious as other seed management tactics (Westerman et al., 2003b). A number of ecological models suggest that removing 50–86% of weed seeds from the seed bank can effectively stop population growth and lead to reduced weed densities for the tested species (Firbank and Watkinson, 1986; Jordan et al., 1995; Westerman et al., 2005, 2006). If most weeds follow this pattern, then biological control can have important effects on the demography of weeds. Still, biological control will be most effective in concert with management tactics that reduce seed rain.

The most sustainable systems involve an integrated approach to seed management that employs several synergistic tactics to inflict the maximum mortality to weeds (Hatcher and Melander, 2003). From the literature, a number of management tactics interact favorably with seed predation. Overall, no-till systems in diversified landscapes appear to be compatible with biological control. Within these systems, use of cover crops and other ground covers promote granivore diversity and ecosystem services. In general, encouraging habitat stability within our farmland is also key to reducing weed pressure.

A final and critical research area that needs to be established before seed predation can be sold as a viable option to farmers is how granivory affects the economics of farm management. The effects of seed predation on crop yield are seldom explored under field conditions. The economic impacts of weed infestations are extremely complicated (Buhler et al., 1997), and so it is not surprising that this issue is uninvestigated for weed seed biological control programs. Perhaps the question that needs to be posed is whether seed predation can reduce the impacts of weeds such that cost associated with yield reductions is smaller than the cost to apply herbicides. In any case, conventional farmers are going to have trouble adopting ecologically friendly approaches to weed management unless they are cost competitive with current herbicide applications and tillage regimens.

Chapter 19
Conclusions and the Relative Quality of Non-Prey Foods for Natural Enemies

Throughout this book, I discuss the nutritional ecology of four different classes of non-prey foods, the entomophagous species that consume these foods, their motivation for doing so, the physiological and behavioral adaptations that facilitate omnivory in entomophagous species, and the applied implications of consuming non-prey foods by entomophagous species within a pest management context. It is often presumed that non-prey foods are inferior to prey, and that entomophagous insects consume them only secondarily in their quest for prey. When the relative nutritional contents of prey and non-prey foods are examined empirically, it is very clear that this simply isn't so; in many ways non-prey foods are equal or superior to prey in terms of their energetic and nutritional contents. This is not to say that prey can be entirely substituted with non-prey foods (as discussed throughout, this is usually not the case). **Prey and non-prey foods are different entities, and both are critical components of a complete diet for entomophagous insects**. This fact is reinforced by the consistent and widespread capabilities of entomophagous species to recognize, collect and manipulate, ingest and digest non-prey nutritional resources. By focusing on the prey-based nutritional ecology of natural enemies and overlooking the influential non-prey components of their diets, biological control is weakened in many ways. Also, the omnivorous habits of traditionally regarded entomophagous species have important implications for IPM as new technologies are developed and the face of agriculture changes.

19.1 The Nutritional and Energetic Qualities of Prey and Non-Prey Foods

There is a widely held belief that non-prey foods are nutritionally inferior to prey, and for this reason it makes sense that entomophagy within arthropods is a successful strategy in terms of its prevalence at higher taxonomic levels and its longevity within evolutionary history (Southwood, 1973). Many of these suppositions are

J.G. Lundgren, *Relationships of Natural Enemies and Non-Prey Foods*, Progress in Biological Control 7,

Published by Springer Science+Business Media B.V.

based on the relative nutrient content of prey and non-prey foods, typically compared among foods within ecological studies on the basis of dry weights. When examining the diets of organisms, the water content of a food has important implications on the amount of food that is consumed over a unit of time. Thus, fresh weights (although admittedly a less stable measure of relative nutrition of different foods) give a better picture of the relative dietary quality of different food sources. When the fresh weights of prey and non-prey foods are examined, the notion of the nutritional supremacy of prey is rendered a fallacy.

The mean nutrient contents of fresh floral nectar, fungus, arthropod prey, pollen, and seeds from the literature were compared statistically in Figs. 19.1–19.4. Water contents of fungus, insects, pollen and seeds were typically greater than 85, greater than 50, 11 ± 0.54, and 10 ± 2.71% of fresh weights, respectively. Water contents were determined from the literature for fungi (Chang and Miles, 2004), arthropods

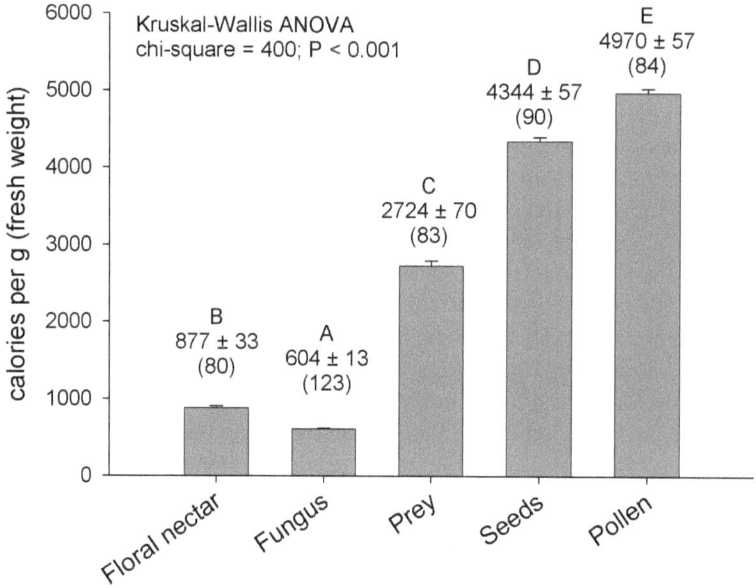

Fig. 19.1 The energy contents of prey and non-prey foods. Values presented represent mean ± SE calories, number of species in parentheses, and bars with different letters are significantly different from one another (LSD means separation, $\alpha = 0.05$)*

*Energy contents obtained from the following references for:

Floral nectar (Baker, 1975; Guerrant and Fiedler, 1981)

Fungus (Chang and Miles, 2004; Gray and Staff, 1967; La Guardia et al., 2005; Roe et al., 2002)

Pollen (Colin and Jones, 1980; Loper and Cohen, 1982; Petanidou and Vokou, 1990; Smith and Evenson, 1978)

Arthropod prey (Barclay et al., 1991; Barrentine, 1993; Caudell and Conover, 2006; Cummins and Wuycheck, 1971; French et al., 1957; Giles et al., 2002; Golley, 1961; Macon and Porter, 1995; McNeill, 1971; Schmolz et al., 1999; Singh and Yadava, 1973; Smalley, 1960; Southwood, 1973; Stepien and Rodriguez, 1982; van Hook, 1971; Weigert, 1965)

Seeds (Christian and Lederle, 1984; Kelrick et al., 1986; Kendeigh and West, 1965; O'Dowd and Hay, 1980; Reichman, 1976; Smith and Evenson, 1978)

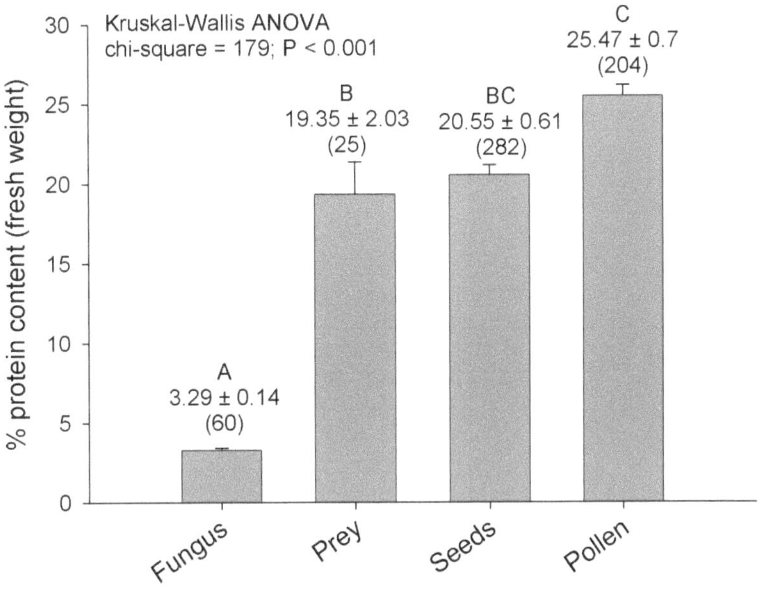

Fig. 19.2 The protein contents of prey and non-prey foods. Values presented represent mean ± SE percentages, number of species in parentheses, and bars with different letters are significantly different from one another (LSD means separation, $\alpha = 0.05$)*

*Protein contents were obtained from the following references for:

Fungus (Chang and Miles, 2004; Gray and Abou-el-Seoud, 1966; Gray and Karve, 1967; Gronwall and Pehrson, 1984; Mueller et al., 2001)

Pollen (Lundgren and Wiedenmann, 2004; Roulston et al., 2000; Todd and Bretherick, 1942)

Arthropod prey (Dunkel, 1996; Finke et al., 1989; French et al., 1957; Massieu et al., 1951; Oliveira et al., 1976; Shapiro and Legaspi, 2006)

Seeds (Barclay and Earle, 1974; Christian and Lederle, 1984; Crist and MacMahon, 1992; Kelrick et al., 1986)

(Chapman, 1998; Schmolz et al., 1999; Sinclair, 2000), pollen (Todd and Bretherick, 1942), and seeds (Christian and Lederle, 1984; Crist and MacMahon, 1992; Kelrick et al., 1986). In this analysis, when nutrient contents were reported for more than one life stage of an insect prey item in a single literature source, the life stage with the greatest fresh nutrient content was used in the analysis. If multiple sources reported a particular nutrient content for an organism, then a mean value from all sources was used in the analysis. Protein content expressed here was total N × 6.25, as determined by the Kjeldahl method. Of the sugar sources used by natural enemies, only floral nectars were included here and the caloric content of sucrose, 3,940 g^{-1} dry weight (Roe et al., 2002), was used to calculate their energy content. From a nutritional standpoint, nectars contained negligible amounts of lipids and proteins, and thus were not included in these comparisons here. Pollens were only analyzed if they were hand collected (no bee-collected samples were included). Dry weights for each food type were corrected by the relative water content to estimate nutrient and energy content in fresh foods. Comparisons were conducted using the non-parametric Kruskal-Wallis ANOVA, and means were separated by LSD comparisons.

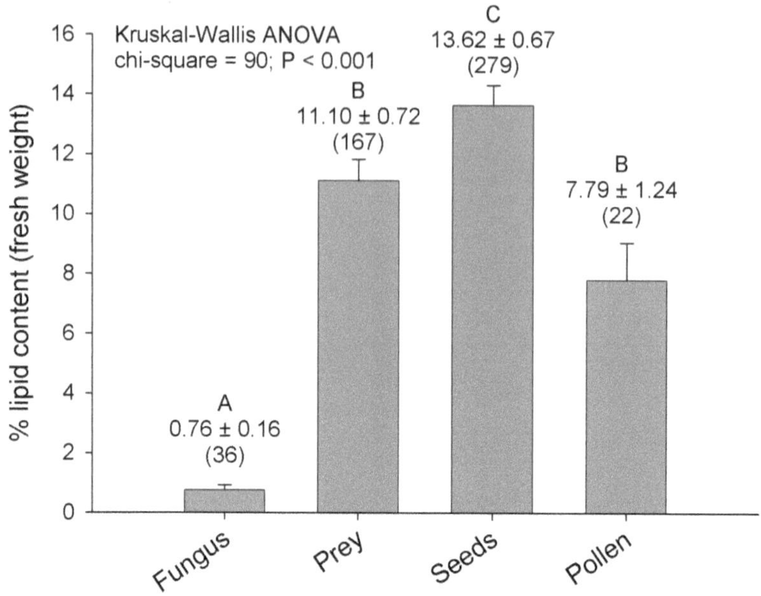

Fig. 19.3 The lipid contents of prey and non-prey foods. Values presented represent mean ± SE percentages, number of species in parentheses, and bars with different letters are significantly different from one another (LSD means separation, $\alpha = 0.05$)*

*Lipid contents were obtained from the following references for:

Fungus (Chang and Miles, 2004; Cochrane, 1958; Kurtzmann, 1997; Mueller et al., 2001; Roe et al., 2002)

Pollen (Human and Nicolson, 2006; Mandal et al., 1993; Piffanelli et al., 1997; Roulston and Cane, 2000; Stanley and Linskins, 1974; Todd and Bretherick, 1942)

Arthropod Prey (Dunkel, 1996; Fast, 1964; French et al., 1957; Shapiro and Legaspi, 2006)

Seeds (Barclay and Earle, 1974; Christian and Lederle, 1984; Crist and MacMahon, 1992; Kelrick et al., 1986)

The results of these analyses are that prey is a comparatively poor source of calories, proteins, lipids, and carbohydrates relative to some non-prey foods, especially pollen and seeds. In terms of energy content, prey has about half the calories of pollen or seeds per gram, but is superior to nectar and fungus. Prey has less protein than pollen, and has a similar percentage as seeds. For lipid content, prey is equivalent to pollen, and both contain significantly less than seeds. Prey and fungus are the poorest foods for carbohydrates, being out-ranked by nectar, pollen, and especially seeds. Fungus is the least nutritious food source examined in these comparisons. But it should be noted that yeasts, and fungal spores (both having lower water content) are more nutritious than the fungal tissues analyzed here (Nwokolo, 1986; Roe et al., 2002; Rumsey et al., 1991; Todd and Bretherick, 1942). Another fact worth noting is that even though nectar is nutritionally very limited, it is still on average a better source of carbohydrates than insect prey. EFN and honeydew may have a much higher sugar concentration than some floral nectars, and this will affect the relative values of these foods as sources of carbohydrates and energy

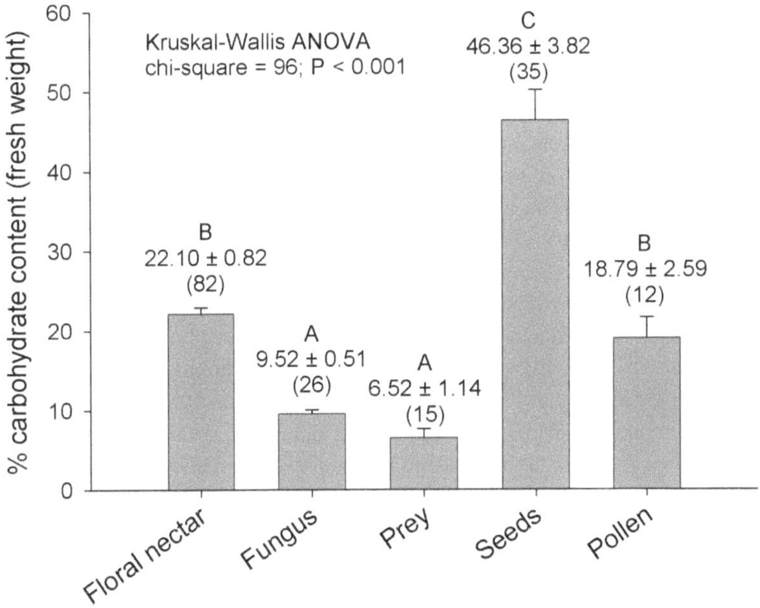

Fig. 19.4 The carbohydrate contents (reducing sugars and polysaccharides) of prey and non-prey foods. Values presented represent mean ± SE percentages, number of species in parentheses, and bars with different letters are significantly different from one another (LSD means separation, $\alpha = 0.05$)*
*Carbohydrate contents were obtained from the following references for:
Floral nectar (Baker, 1975; Guerrant and Fiedler, 1981)
Fungus (Chang and Miles, 2004; Mueller et al., 2001)
Pollen (Barbier, 1970; Human and Nicolson, 2006; Mandal et al., 1993; Piffanelli et al., 1997; Stanley and Linskins, 1974; Todd and Bretherick, 1942)
Arthropod prey (Dunkel, 1996; French et al., 1957; Oliveira et al., 1976; Schmidt and Mathur, 1967)
Seeds (Bewley and Black, 1994; Christian and Lederle, 1984; Crist and MacMahon, 1992; Kelrick et al., 1986; Pizo and Oliveira, 2001)

(Koptur, 2005; Wäckers et al., 2001). Perhaps as important is that these data are likely over-estimates of the relative quality of prey, since water content of insects often far exceeds 50% of fresh weight. Insect eggs are an exception to this, and are likely the most nutritious of prey life stages (McNeill, 1971). The conclusion from these analyses is that prey is not nutritionally superior to non-prey foods.

Given the high nutritional quality of non-prey foods, why is it often the case that natural enemies have higher fitness when fed prey compared to non-prey foods (Coll, 1998a)? The information presented in this book strongly suggests that entomophagous species that are reared on non-prey foods are not limited nutritionally in terms of these broad nutritional categories nor in terms of their energetic needs (e.g., proteins, lipids, and carbohydrates; an exception being nectar-fed species). It may be that non-prey foods are deficient in certain micronutrients that prey possess, but what these limiting micronutrients are remains to be substantiated. A strong argument as to why non-prey foods are often less suitable for omnivores may be made that these non-prey foods are frequently well defended, both chemically and structurally.

The role that the defenses of nectar, fungus, pollen, and seeds plays in limiting consumption by entomophagous species remains to be thoroughly explored.

Finally, it is often the case that prey does not support natural enemy fitness as well as non-prey foods, or that prey and non-prey foods are perfectly substitutable for one another in the diets of entomophagous species. Arthropod prey are often a consistently good source of protein, but vary widely in their other nutrient contents, especially lipid content (DeFoliart, 1992; Fast, 1964; Finke et al., 1989). Moreover, different prey species vary widely in their nutritional suitability for specific natural enemies (Eubanks and Denno, 1999; Giles et al., 2002; Shapiro and Legaspi, 2006, to note only a few). Consequently, certain non-prey foods are superior to prey in supporting the development and reproduction of a subset of natural enemies for at least a portion of their lives (Ahmad et al., 2006; Eubanks and Denno, 1999; Kamburov, 1971; Kiman and Yeargan, 1985; McMurtry and Croft, 1997; McMurtry et al., 1970; Tamaki and Weeks, 1972; Trichilo and Leigh, 1988; Zaher and Shehata, 1971). In other cases, it may be that non-prey foods are necessarily eaten in order to better consume prey. A case in point is with *Dicyphus hesperus*, which consumes plant material in order to obtain fluids that can then facilitate prey consumption in this lacerate-and-flush predator (Gillespie and McGregor, 2000).

19.2 Relative Conservation Benefits of Different Non-Prey Foods

A wide range of characteristics inherent to non-prey foods influence their utility as mechanisms for conserving natural enemies within managed land. A major factor influencing these traits is the ecological function that each non-prey food fulfills for the organism that produces it. Wäckers (2005) provides some discussion of the suitability of non-prey foods based on their availability, apparency, accessibility, chemical defenses, relative nutrition, use by herbivores, and the foraging risks associated with each food. As Jervis and Heimpel (2005) point out, the energetics of foraging entomophages depend on the amount of energy required to obtain a non-prey food, in addition to the energy obtained from it. So, the numerous facets of non-prey foods that influence the suitability of different non-prey foods become very important when considering the relative merits of these to the nutritional ecology of natural enemies. Heretofore these qualities have not been applied when systematically considering which non-prey foods would be wisest to target in conservation biological control programs.

19.2.1 Re-evaluating Flower-Bound Resources

Assessing the relative benefits of different sources of non-prey foods raises the question of why flowers are the focus of the vast majority of conservation biological control programs that focus on subsidizing non-prey foods for natural enemies in farmland. Of the non-prey foods discussed, flower-bound resources are the most restricted temporally (Lee and Heimpel, 2003), are some of the best defended of the non-prey foods in terms of their accessibility and chemical defenses, and are often the primary food source of lepidopteran pests and other herbivores. For these reasons, many natural enemies are simply unable to partake of flower-bound foods, and conservation programs that focus exclusively on these resources are likely going to differentially exclude potentially effective natural enemies.

Flower-bound resources are typically aimed at attracting pollinators, and many flowers make every effort to exclude potential nectar thieves. Thus, natural enemies that function as pollinators (such as syrphids and bombyliids) are likely to receive the most benefits from flower-bound resources. This being said, numerous parasitoid species that are of little use as pollinators visit flowers for nectar (Jervis et al., 1993; Kevan, 1973; Tooker and Hanks, 2000). Typically syrphids and parasitoids show a high fidelity to a very specific subset of the floral community (Bugg et al., 1987; Jervis and Heimpel, 2005; Jervis et al., 1993; Pontin et al., 2007; Tooker and Hanks, 2000), and a natural enemy's preference for a given flower species can change over the season (Cowgill et al., 1993). These facts make it challenging to develop conservation programs based on flower-bound resources that target a wide array of natural enemies. Moreover, the floral architecture that renders nectar accessible to the widest array of natural enemies (wide, open flowers such as in the Compositae) is also associated with the lowest nectar production rates (Chapter 3). Many of the predators found to consume floral nectar and pollen do so to pass the time while waiting on a flower for their next victim, and so flowers can be a hotspot for intraguild predation (Jervis, 1990). Finally, numerous parasitoids do not visit flowers at all (Elliott et al., 1987; Gardner, 1938, 1940; Hocking, 1967; Rogers and Potter, 2004), and so the nutritional ecology of each species needs to be considered before presuming that flower-bound resources will attract a key natural enemy. All of these reasons contribute to why flower-bound resources sometimes fail to contribute to biological control. Heimpel and Jervis (2005) review the literature pertaining to the contributions of floral resources to biological control by parasitoids. They estimate that flowers conclusively improve parasitism in only seven (35%) of 20 published studies. For flowers to be of use in conservation biological control, an array of species needs to be specifically chosen such that flowers are consistently available at key periods during the growing season and only those flowers that promote a specific suite of natural enemies of importance to pest management without aiding the pests themselves should be employed. All of this is not to say that flowers should be excluded from conservation programs (in fact, flowers can be quite beneficial to biological control under some circumstances), but it seems a better strategy that they should be incorporated with other more broadly available and suitable sources of non-prey foods.

19.2.2 Attributes of Alternative Non-Prey Foods

Analysis of the attributes of EFN and wind-pollinated plants (especially grasses) suggest that these sources of non-prey foods show great promise as nutritional subsidies for natural enemies. Honeydew also is particularly attractive to certain natural enemies, but is likely already widely available in cropland where sternorrhynchan pests are present. One review that may be in line with this hypothesis is that of Coll (1998b). In looking at how vegetational diversity affects parasitoid abundance and function, he describes a striking relationship of host order on the efficacy of intercropping on parasitoid populations. Intercropping improve the numbers and parasitism rates of sternorrhynchan parasitoids in only 25% of studies, but for parasitoids of lepidopteran and coleopteran hosts, intercropping improved their abundance or efficacy in 61% and 75% of studies, respectively. One possible explanation for this trend is that intercropping increases the nutritional subsidies of a habitat, but for natural enemies strongly associated with honeydew-producing insects, these augmented nutritional subsidies are less influential than for those of prey/hosts that do not produce non-prey foods. Still, localized honeydew resources can promote biological control under some circumstances. For example, predation by ants on insect eggs is greater near honeydew deposits on soybeans than near dry leaves, suggesting that the aggregation of natural enemies associated with honeydew has implications for pest management (Nickerson et al., 1977). Selecting plants known to host non-pest aphids may be useful in diversifying the honeydew resources available to natural enemies, not all of which are equally suitable as food.

Although less well-studied compared to flowers, a great diversity of natural enemies is known to visit EFN, and providing this resource within cropland increases natural enemy abundance substantially; this has been best studied within cotton systems (Naranjo and Gibson, 1996). In fact, the purpose of EFN is in part to attract natural enemies of herbivores, and plants bearing EFN tend to be better protected than non-EFN bearing plants (e.g., Beattie, 1985; Cuautle and Rico-Gray, 2003; Pemberton and Lee, 1996; Stephenson, 1982b). In addition to being widely exposed to foraging natural enemies, EFN is almost always available for a longer period of time than floral nectar over both diel and seasonal time scales (Bentley, 1977a; Heil et al., 2000; O'Dowd, 1979, but see Jakubska et al., 2005; Yokoyama, 1978), and so its benefits to biological control are likely going to be prolonged over those of flower-bound resources. Indeed, EFN production is closely tied to when plants need the most protection from herbivores, and so timing may be very appropriate in terms of encouraging top-down regulation of pests in cropland. Also, lepidopterans tend to have more trouble drinking concentrated sugar solutions, and so by containing higher sugar concentrations EFN may be less attractive to herbivores compared to floral nectar sources (Wäckers, 2005). A potentially major drawback to EFN is that this non-prey food is particularly attractive to ants, which often muscle out other natural enemies seeking a quick meal. Whether natural enemy exclusion by ants resulting from EFN is a benefit or a hindrance to biological control depends on the relative contributions of ants and the remaining natural enemy community to pest regulation (this is also true of sternorrhynchan-tending ants; Banks and Macauley, 1967; Beattie, 1985; McLain, 1980; Stary, 1966; Way, 1963). In many cases, ants are very efficacious as predators in the absence of other natural enemies (Tilman, 1978). Still, EFN is widely accessible to numerous natural enemies, and while ants may be disruptive to the function of other natural enemies, they are not empirically shown to entirely displace other entomophages from a habitat through dominating these nectaries.

Grass species also hold great promise as sources of pollen and seeds for natural enemies. The pollen of anemophilous plants is much more accessible to natural enemies than is flower-bound pollen, and it is often produced in much greater abundance. Also important, anemophilous pollen is widely dispersed within a habitat, and so may be more likely to be encountered by those species not prone to visiting flowers. Anthesis still occurs within a limited time period, and so pollen plants need to be selected based on their phenology just as traditionally flowering plants do (Bugg and Waddington, 1994; Fiedler and Landis, 2007). One study that tests the relative nutrition of grass and dicot pollens found that the grass pollens are as suitable as the best dicot pollens (apple and peach) for the development of the phytoseiid, *Euseius tularensis* (an exception being wheat pollen) (Ouyang et al., 1992). Other experiments reiterate the equivalency of anemophilous and entomophilous pollens for other predators (Smith, 1960, 1961; van Rijn and Tanigoshi, 1999b). Another benefit of grasses is that their seeds are some of the most preferred of granivores within cropland. For species associated with cropland, these grass seeds are most abundant late in the season, when granivorous arthropods thrive. Thus, if allowed to mature, grasses are able to provide two high quality non-prey foods in order to conserve both epigeal and foliar-dwelling predator communities.

Finally, understanding that myrmecochorous seeds rely on omnivorous insects to disperse them, and possess nutritional rewards that specifically target entomophagous species, more attention should be given to incorporating plants with this life-history strategy in or near cropland as a way of encouraging granivorous entomophage communities. Consciously and directly integrating EFN, honeydew diversity, myrmecochorous seeds, and grass species, in addition to flowering plants, with conservation biological control programs is likely to appeal to a broad suite of natural enemies and could feasibly offer a more predictable outcome of the programs.

19.3 Adaptations that Fuel Omnivory

Non-prey foods are chemically and structurally quite distinct from arthropod prey and many entomophagous insects have evolved the physiological, morphological and symbiotic prerequisites that allow them to exploit these abundant and rich sources of nutrients. Certain functional adaptations consistently occur across a phylogenetically wide swath of natural enemies that allow them to:

1. Recognize sugar sources, pollen grains, seeds, and fungi as food
2. Collect and manipulate non-prey foods
3. Access the nutrients housed in non-prey foods
4. Digest and assimilate the distinct nutrients frequently encountered within non-prey foods

Natural enemies may accomplish these sequential tasks in different ways, but each hurdle inevitably must be crossed and similar adaptations that have evolved across groups may be diagnostic of a convergence in their nutritional ecologies.

Natural enemies employ a variety of sensory adaptations that allow them to identify non-prey foods from the largely inedible world around them. Visual cues, especially color, are important for many flying entomophagous insects as they search for flowers bearing potential nectar and pollen meals. Chemoreceptors are also important in recognizing certain unique cues associated with non-prey foods. Some natural enemies are able to detect volatile signals, for instance from nectar sources, and it seems very common that taste receptors are employed upon contact with potential foods. For instance, pollen-feeding syrphids have tactile receptors that register the common pollen amino acid, proline, and carabids and ants are able to evaluate the fatty acid profiles present in the food bodies associated with seeds.

Once a food source has been identified, entomophagous arthropods need to be able to separate the food from contaminants, manipulate the food item and often transport it to a suitable dining locale. For floral nectar sources, the length of the labella dictates which flowers can be accessed by foraging flies and parasitoids (the concealed nectar extraction apparatus of parasitoid flies and wasps being an extreme example of this). Pollen grains and fungal spores are separated from the substrate using a series of combs and brushes located on various parts of the body (usually the mouthparts). Carrying large seeds that need to be transported back to the nests of ants is aided with the psamnophore in some harvester species, and specialized foraging castes in others.

The nutrients of many non-prey foods are structurally protected through rigid or waxy coverings, and the mouthparts of omnivores have evolved through strength or strategy to crack through these defenses. Granivorous carabids have comparatively stout and asymmetric mandibles, with increased musculature to the mouthparts that aid in cracking the seed coat. Pollinivorous mites have evolved spoon-like structures on their assymetrical mandibles that enable them to squash pollen grains and suck the contents into their oral cavities. Mycophagous staphylinids have evolved very distinct mouthpart structures according to the portion of the fungus that they specialize upon. All of these adaptations have evolved specifically to unlock the contents housed in non-prey foods.

Finally, non-prey foods are nutritionally distinct from prey, and require a unique set of digestive capabilities in order to maximize the extraction of energy from the food source. Cellulose, glucans, lignins, and pectins are all abundant structural components of plant-derived and fungal foods that purely entomophagous arthropods are ill-equipped to deal with. In some cases, entomophagous arthropods possess the digestive enzymes necessary for metabolizing these chemicals. But another source of digestive capabilities that biological control scientists are only beginning to explore is the role of microbial gut symbionts. The symbiotic relationships of yeasts and phytophagous green lacewings are long known to science. But recent evidence suggests that gut bacteria may play a role in the dietary breadth of granivorous carabid beetles and ants as well. Further investigations into these symbiotic relationships which improve the digestive capabilities of entomophagous species will likely reveal the pervasiveness of these interactions within biological control agents.

It should be reiterated that the degree of omnivory within an entomophagous species is not constant ontogenetically, and the importance of non-prey foods to an insect can even change within a life stage. In ants, the fourth instar of harvesting species is the only one which can digest solid seeds. Fourth instar *Coleomegilla maculata* are much more efficient at digesting pollen than any other instar. *Geocoris punctipes* can survive in the absence of prey on plant tissue, but only during the first two stadia. Later instars require insect prey in addition to plant material. The necessary adaptations to consuming and digesting non-prey foods will accompany the ontogenetic changes in diet within entomophagous species. Thus, a superficial glance at an insect may not tell the whole story of its nutritional ecology since the nutritional needs of an insect changes over its life and even within a life stage.

19.4 Applied Aspects of Omnivory – Complexity Within Multi-trophic Interactions

The omnivorous habits of most entomophagous species afford both opportunities and inflict constraints when integrating biological control into pest management systems. First, the impact of biological control is only maximized when the nutritional requirements of a natural enemy are recognized and addressed by the practitioner. Biological control is often improved when non-prey foods are incorporated into annual, monoculture cropping systems over those fields devoid of non-prey resources. But this is far from universally the case, and unintended outcomes are recorded frequently enough to say that each crop-pest-natural enemy scenario needs to be evaluated independently. Suffice it to say that incorporating non-prey foods into cropland invariably *changes* the predator community, but it does not always *improve* it from a functional standpoint.

Omnivory by entomophagous species takes on a different function when the non-prey food itself is considered a pest. Specifically, honeydew, fungus, and seeds are all non-prey foods that are regarded as detracting from crop production from time to time. In the case of weed seeds, entomophagous arthropods are viewed by many as a promising source of biological control, although there are several challenges that remain before farmers will deliberately incorporate weed seed predation into their pest management regimen. Although not discussed at length in this book, biological control of powdery mildews using mycophagous entomophages receives attention by researchers and may prove to be a valuable component of IPM for these pests under certain circumstances.

Finally, omnivory opens additional pathways for higher trophic levels to interact with other branches of pest management. Recognizing the importance of plant-based foods may allow breeders to select for greater quantities or higher quality of non-prey foods or synomones to be produced by the plant. At the very least, growers can take advantage of the natural variation already present in most commercial hybrids to maximize the effects of crop-based non-prey foods. But omnivory also has a downside in that it opens up a wide range of ecological pathways that plant-incorporated pesticidal properties can directly and adversely affect higher trophic levels. Thus, omnivory by natural enemies can challenge the integration of biological control with HPR, systemic insecticides, and insecticidal GM crops.

19.5 Concluding Remarks

A narrow view of the trophic designation of arthropods hampers our understanding of the ecological function of these organisms within food webs, and ultimately challenges our ability to understand the evolutionary history of feeding behavior itself. This book focuses on the complexity of feeding behavior in most species best appreciated for their ability to consume other arthropods. But the book also illustrates the intricate ecological functions of the non-prey foods themselves, particularly how nourishing, distasteful, or even deadly they can be as these foods perform the task assigned to them. Even more surprising are the important roles that entomophagous species play, for good or ill, in the lives of these non-prey foods and the organisms that produce them. The relationships between non-prey foods and entomophagous species are not simply passing or fortuitous. Whether they rely on or resist one another, the existences of non-prey foods and natural enemies have co-evolved together and are inextricably fused. Only when this is realized will we truly understand the complex ecology underway in even the most simplified food webs.

References

Abou-Awad, B.A. & Elsawi, S.A. (1992) Generations and reproduction in the predaceous mite *Amblysieus swirskii* Ath.-Hen. (Acari: Phytoseiidae) fed with artificial and natural diets. Anzeiger fuer Schaedlingskunde, Pflanzenschutz, Umweltschutz, 65, 115–117.

Abou-Setta, M.M. & Childers, C.C. (1989) Biology of *Euseius mesmbrinus* (Acari: Phytoseiidae): Life tables and feeding behavior on tetranychid mites on citrus. Environmental Entomology, 18, 665–669.

Abou-Setta, M.M., Fouly, A.H., & Childers, C.C. (1997) Biology of *Proprioseiopsis rotendus* (Acari: Phytoseiidae) reared on *Tetranychus urticae* (Acari: Tetranychidae) or pollen. Florida Entomologist, 80, 27–30.

Abramsky, Z. (1983) Experiments on seed predation by rodents and ants in the Israeli desert. Oecologia, 57, 328–332.

Acorn, J.H. & Ball, G.E. (1991) The mandibles of some adult ground beetles: Structure, function, and the evolution of herbivory (Coleoptera: Carabidae). Canadian Journal of Zoology, 69, 638–650.

Acosta, L.E. & Machado, G. (2007) Diet and foraging. In *Harvestmen: The Biology of Opiliones* (eds R. Pinto-da-Rocha, G. Machado & G. Giribet), pp. 309–338. Harvard University Press, Cambridge, MA.

Adams, P.A. & Garland, J.A. (1981) A new species of *Chrysopiella* Banks from western North America (Neuroptera: Chrysopidae). The Canadian Entomologist, 113, 1–4.

Adebooye, O.C. & Phillips, O.T. (2006) Studies on seed characteristics and chemical composition of three morphotypes of *Mucuna urens* (L.) Medikus – Fabaceae. Food Chemistry, 95, 658–663.

Adjei-Maafo, I.K. & Wilson, L.T. (1983) Factors affecting the relative abundance of arthropods on nectaried and nectariless cotton. Environmental Entomology, 12, 349–352.

Adkisson, P.L. & Dyck, V.A. (1980) Resistant varieties in pest management systems. In Breeding Plants Resistant to Insects, pp. 233–251. Wiley, New York.

Adler, L.S. (2000) The ecological significance of toxic nectar. Oikos, 91, 409–420.

Adler, L.S. & Irwin, R.E. (2005) Ecological costs and benefits of defenses in nectar. Ecology, 86, 2968–2978.

Adler, L.S., Wink, M., Distl, M., & Lentz, A.J. (2006) Leaf herbivory and nutrients increase nectar alkaloids. Ecology Letters, 9, 960–967.

Afifi, A.M., Potts, M.F., Patterson, C.G., & Rodriguez, J.G. (1988) Pollen diet of some predator mites. Transactions of the Kentucky Academy of Sciences, 49, 96–100.

Agarwal, A. (1976) A pollen analysis of spider-webs from the garden at Lucknow. Journal of Palynology, 12, 137–142.

Agelopoulos, N.G. & Keller, M.A. (1994) Plant-natural enemy association in the tritrophic system, *Cotesia rubecula-Pieris rapae*-Brassiceae (Cruciferae): I. Sources of infochemicals. Journal of Chemical Ecology, 20, 1725–1734.

Agnew, C.W., Sterling, W.L., & Dean, D.A. (1982) Influence of cotton nectar on red imported fire ants and other predators. Environmental Entomology, 11, 629–634.

Agrawal, A.A. & Rutter, M.T. (1998) Dynamic anti-herbivore defense in ant-plants: The role of induced responses. Oikos, 83, 227–236.

Ahmad, A., Wilde, G.E., Whitworth, R.J., & Zolnerowich, G. (2006) Effect of corn hybrids expressing the coleopteran-specific Cry3Bb1 protein for corn rootworm control on aboveground insect predators. Journal of Economic Entomology, 99, 1085–1095.

Ahmad, S., Govindarajan, S., Funk, C.R., & Johnson-Cicalese, J.M. (1985) Fatality of house crickets on perennial ryegrasses infected with a fungal endophyte. Entomologia Experimentalis et Applicata, 39, 183–190.

Ahmed, M.K., Newsom, L.D., Emerson, R.B., & Roussel, J.S. (1954) The effect of Systox on some common predators of the cotton aphid. Journal of Economic Entomology, 47, 445–449.

Akman, L., Yamashita, A., Watanabe, H., Oshima, K., Shiba, T., Hattori, M., & Aksoy, S. (2002) Genome sequence of the endocellular obligate symbiont of tse tse flies, *Wigglesworthia glossinidia*. Nature Genetics, 32, 402–407.

Albajes, R., Lopez, C., & Pons, X. (2003) Predatory fauna in cornfields and response to imidacloprid seed treatment. Journal of Economic Entomology, 96, 1805–1813.

Albuquerque, G.S., Tauber, C.A., & Tauber, M.J. (2001) *Chrysoperla externa* and *Ceraeochrysa* spp.: Potential for biological control in the New World tropics and subtropics. In Lacewings in the Crop Environment (eds P.K. McEwen, T.R. New, & A.E. Whittington), pp. 408–423. Cambridge University Press, Cambridge.

Alcock, J. (1976) The behaviour of the seed-collecting larvae of a carabid beetle (Coleoptera). Journal of Natural History, 10, 367–375.

Al-Deeb, M.A. & Wilde, G.E. (2003) Effect of Bt corn expressing the Cry3Bb1 toxin for corn rootworm control on aboveground nontarget arthropods. Environmental Entomology, 32, 1164–1170.

Al-Deeb, M.A., Wilde, G.E., & Zhu, K.Y. (2001) Effect of insecticides used in corn, sorghum, and alfalfa on the predator *Orius insidiosus* (Hemiptera: Anthocoridae). Journal of Economic Entomology, 94, 1353–1360.

Allawi, T.F. (1991) Biological and ecological studies on indigenous and introduced phytoseiid mites: 1. Mass rearing and study of the effect of different pollen and other plant derivatives on the fecundity and longevity of *Euseius scutalis* (A. H.). Dirasat, 18B, 57–67.

Allen, D.C., Kinght, F.B., & Foltz, J.L. (1970) Invertebrate predators of the jack-pine budworm, *Choristoneura pinus*, in Michigan. Annals of the Entomological Society of America, 63, 59–64.

Allen, H.W. (1929) An annotated list of the Tachinidae of Mississippi. Annals of the Entomological Society of America, 22, 676–690.

Allen, R.T. (1979) The occurrence and importance of ground beetles in agricultural and surrounding habitats. In Carabid Beetles: Their Evolution, Natural History, and Classification (eds T.L. Erwin, G.E. Ball, D.R. Whitehead, & A.L. Halpern), pp. 485–506. Dr. W. Junk, The Hague, The Netherlands.

Allen, W.R. & Hagley, E.A.C. (1982) Evaluation of immunoelectroosmophoresis on cellulose polyacetate for assessing predation of Lepidoptera (Tortricidae) by Coleoptera (Carabidae) species. Canadian Entomologist, 114, 1047–1054.

Altieri, M.A. & Letourneau, D.K. (1982) Vegetation management and biological control in agroecosystems. Crop Protection, 1, 405–430.

Altieri, M.A. & Whitcomb, W.H. (1979) The potential use of weeds in the manipulation of beneficial insects. HortScience, 14, 12–18.

Amylase Research Society of Japan, e. (1995) Enzyme Chemistry and Molecular Biology of Amylases and Related Enzymes. CRC Press, Boca Raton, FL.

Andersen, A. (1999) Plant protection in spring cereal production with reduced tillage. II. Pests and beneficial insects. Crop Protection, 18, 651–657.

Andersen, A.N. (1987) Effects of seed predation by ants on seedling densities at a woodland site in SE Australia. Oikos, 48, 171–174.

Andersen, A.N. (1988) Insect seed predators may cause far greater losses than they appear to. Oikos, 52, 337–340.

Andersen, A.N. (1989) How important is seed predation to recruitment in stable populations in long-lived perennials? Oecologia, 81, 310–315.

Andersen, A.N. (1990) Seed harvesting ant pests in Australia. In Applied Myrmecochory: A World Perspective (eds R.K. Vander Meer, K. Jaffee, & A. Cedeno), pp. 34–39. Westview, Boulder, CO.

Andersen, A.N. (1991) Seed harvesting by ants in Australia. In Ant-Plant Interactions (eds C.R. Huxley & D.F. Cutler), pp. 493–503, Oxford University Press, Oxford.

Andersen, A.N. & Ashton, D.H. (1985) Rates of seed removal by ants at heath and woodland sites in southeastern Australia. Australian Journal of Ecology, 10, 381–390.

Andersson, L. (1998) Post-dispersal seed removal in some agricultural weeds. Aspects of Applied Biology, 51, 159–164.

Anderson, J.M.E. (1981) Seasonal field analyses of fat content, live weight, dry weight and water content of the aphidophagous *Scymnodes lividigaster* (Mulsant) and mycophagous *Leptothea galbula* (Mulsant) (Coleoptera: Coccinellidae). Australian Journal of Zoology, 29, 679–689.

Anderson, J.M.E. (1982) Seasonal habitat utilization and food of the ladybirds *Scymnodes lividigaster* (Mulsant) and *Leptothea galbula* (Mulsant) (Coleoptera: Coccinellidae). Australian Journal of Zoology, 30, 59–70.

Anderson, J.M.E. & Hale, D.F. (1986) *Coccinella repanda*. Diapause? In Ecology of Aphidophaga (ed I. Hodek), pp. 233–238. Dr. W. Junk, Dordrecht, The Netherlands.

Anderson, L. (1998) Post-dispersal seed removal in some agricultural weeds. Aspects of Applied Biology, 51, 159–164.

Anderson, N.H. (1962a) Anthocoridae of the Pacific Northwest with notes on distributions, life histories, and habits (Heteroptera). The Canadian Entomologist, 94, 1325–1334.

Anderson, N.H. (1962b) Bionomics of six species of *Anthocoris* (Heteroptera: Anthocoridae) in England. Transactions of the Royal Entomological Society of London, 114, 67–95.

Andow, D.A. (1991) Vegetational diversity and arthropod population response. Annual Review of Entomology, 36, 561–586.

Andow, D.A. & Hilbeck, A. (2004) Science-based risk assessment for nontarget effects of transgenic crops. BioScience, 54, 637–649.

Andow, D.A. & Risch, S.J. (1985) Predation in diversified agroecosystems: Relations between a coccinellid predator *Coleomegilla maculata* and its food. Journal of Applied Ecology, 22, 357–372.

Anglade, P. (1971) Nouveaux degats de *Clivina fossor* L. sur mais dans les Pyrenees Atlantiques. Revue de zoologie Agricole et de Pathologie vegetate 70, 104–106.

Aparicio-Fernandez, X., Yousef, G.G., Loarca-Pina, G., de Mejia, E., & Lila, M.A. (2005) Characterization of polyphenolics in the seed coat of black Jamapa bean (*Phaseolus vulgaris* L.). Journal of Agricultural and Food Chemistry 53, 4615–4622.

Applebaum, S.W. & Birk, Y. (1979) Saponins. In Herbivores: Their Interactions with Secondary Plant Metabolites (eds G.A. Rosenthal & D.H. Janzen), pp. 539–566. Academic Press, New York.

Armer, C.A., Berry, R.E., & Kogan, M. (2000) Longevity of phytophagous heteropteran predators feeding on transgenic Btt-potato plants. Entomologia Experimentalis et Applicata, 95, 329–333.

Arndt, E. & Kirmse, S. (2002) Adaptation to seed-feeding in ground beetles (Coleoptera: Carabidae: Harpalini) of South Venezuela. *Studies on Neotropical Fauna and Environment*, 37(2), 151–157.

Ashe, J.S. (1984) Major features of the evolution of relationships between Gyrophaenine staphylinid beetles (Coleoptera: Staphylinidae: Aleocharinae) and fresh mushrooms. In Fungus-Insect Relationships (eds Q. Wheeler & M. Blackwell), pp. 227–255. Columbia University Press, New York.

Ashe, J.S. (1993) Mouthpart modifications correlated with fungivory among Aleocharine staphylinids (Coleoptera: Staphylinidae: Aleocharinae). In Functional Morphology of Insect Feeding (eds C.W. Schaefer & R.A.B. Leschen), pp. 105–130. Entomological Society of America, Lanham, MD.

Ashford, D.A., Smith, W.A., & Douglas, A.E. (2000) Living on a high sugar diet: The fate of sucrose ingested by a phloem-feeding insect, the pea aphid *Acyrthosiphon pisum*. Journal of Insect Physiology, 46, 335–341.

Ashley, T.R. & Gonzalez, D. (1974) Effect of various food substances on longevity and fecundity of *Trichogramma*. Environmental Entomology, 3, 169–171.

Atallah, Y.H. & Killebrew, R. (1967) Ecological and nutritional studies on Coleomegilla maculata (Coleoptera: Coccinellidae) IV. Amino acid requirements of the adults determined by the use of C^{14}-labelled acetate. Annals of the Entomological Society of America, 60, 186–188.

Atallah, Y.H. & Newsom, L.D. (1966) Ecological and nutritional studies on *Coleomegilla maculata* De Geer (Coleoptera: Coccinellidae). I. The development of an artificial diet and a laboratory rearing technique. Journal of Economic Entomology, 59, 1173–1178.

Auber, L. (1965) Atlas des Coleopteres de France, Belgique, Suisse.

Aubrook, E.W. (1949) *Amara familiaris* Duft. (Col., Carabidae) feeding in flower heads. Entomologist's Monthly Magazine, 85, 44.

Auclair, J.L. (1963) Aphid feeding and nutrition. Annual Review of Entomology, 8, 439–490.

Avidov, Z., Balshin, M., & Gerson, U. (1970) Studies on *Aphytis coheni*, a parasite of the California red scale, *Aonidiella auratnii*, in Israel. Entomophaga, 15, 191–207.

Azcarate, F.M., Arqueros, L., Sanchez, A.M., & Peco, B. (2005) Seed and fruit selection by harvester ants, *Messor barbarus*, in Mediterranean grassland and scrubland. Functional Ecology, 19, 273–283.

Azcarate, F.M. & Peco, B. (2003) Spatial patterns of seed predation by harvester ants (*Messor* Forel) in Mediterranean grassland and scrubland. Insectes Sociaux, 50, 120–126.

Azzouz, H., Giordenengo, P., Wäckers, F.L., & Kaiser, L. (2004) Effects of feeding frequency and sugar concentration on behavior and longevity of the adult aphid parasitoid: *Aphidius ervi* (Haliday) (Hymenoptera: Braconidae). Biological Control, 31, 445–452.

Ba, A.S. & Phillips, S.A., Jr. (1996) Yeast biota of the red imported fire ant. Mycological Research, 100, 740–746.

Bacheler, J.S. & Baranowski, R.M. (1975) *Paratriphleps laeviusculus*, a phytophagous anthocorid new to the United States (Hemiptera: Anthocoridae). Florida Entomologist, 58, 157–163.

Baggen, L.R., Gurr, G.M., & Meats, A. (1999) Flowers in tri-trophic systems: Mechanisms allowing selective exploitation by insect natural enemies for conservation biological control. Entomologia Experimentalis et Applicata, 91, 155–161.

Bai, Y.Y., Jiang, M.X., & Cheng, J.A. (2005) Effects of transgenic cry1Ab rice pollen on fitness of Propylea japonica (Thunberg). Journal of Pest Science, 78, 123–128.

Bailey, L. (1952) The action of the proventriculus of the worker honeybee, *Apis mellifera* L. Journal of Experimental Biology, 29, 310–327.

Baker, G.A. & O'Dowd, D.J. (1982) Effects of parent plant density on the production of achene types in the annual *Hypochoeris glabra*. Journal of Ecology, 70, 201–215.

Baker, H.G. (1972) Seed weight in relation to some environmental conditions in California. Ecology, 53, 997–1010.

Baker, H.G. (1975) Sugar concentrations in nectars from hummingbird flowers. Biotropica, 7, 37–41.

Baker, H.G. (1989) Some aspects of the natural history of seed banks. In Ecology of Soil Seed Banks (eds M.A. Leck, V.T. Parker, & R.L. Simpson), pp. 9–21. Academic Press, London.

Baker, H.G. & Baker, I. (1973) Amino-acids in nectar and their evolutionary significance. Nature, 241, 543–545.

Baker, H.G. & Baker, I. (1977) Intraspecific constancy of floral nectar amino acid complements. Botanical Gazette, 138, 183–191.

Baker, H.G. & Baker, I. (1978) Ants and flowers. Biotropica, 10, 80.

Baker, H.G. & Baker, I. (1979) Starch in angiosperm pollen grains and its evolutionary significance. American Journal of Botany, 66, 591–600.

Baker, H.G. & Baker, I. (1982) Starchy and starchless pollen in the Onagraceae. Annals of the Missouri Botanical Garden, 69, 748–754.

Baker, H.G. & Baker, I. (1983) A brief historical review of the chemistry of floral nectar. In The Biology of Nectaries (eds B.L. Bentley & T.S. Elias), pp. 126–152. Columbia University Press, New York.

Baker, H.G., Opler, P.A., & Baker, I. (1978) A comparison of the amino acid complements of floral and extrafloral nectars. Botanical Gazette, 139, 322–332.

Baker, I. & Baker, H.G. (1976) Analyses of amino acids in flower nectars of hybrids and their parents, with phylogenetic implications. New Phytologist, 76, 87–98.

Bakker, F.M. & Klein, M.E. (1992) Transtrophic interactions in cassava. Experimental and Applied Acarology, 14, 293–311.

Balachowsky, A.S. (1963) Entomologie appliquee a l'agriculture. In Coleopteres, Vol. 1 et 2, p. 1091, Paris. Fédération Français des sociétés de Sciences N

Balachowsky, A.S. & Mesnil, L. (1935) Les insectes nuisables aux plantes cultivees, Paris.

Balduf, W.V. (1935) The Bionomics of Entomophagous Coleoptera. John S. Swift, St. Louis, MO.

Balduf, W.V. (1939a) The bionomics of entomophagous insects. John S. Swift Company, St. Louis, MO.

Balduf, W.V. (1939b) Food habits of *Phymata pennsylvanica americana* Melin (Hemip.). Canadian Entomologist, 71, 66–74.

Balfour-Browne, F. (1944) The proventriculus of the Coleoptera (Adephaga) and other insects – a study in evolution. Journal of the Royal Microscopy Society, 6, 68–117.

Ball, G.E. (1960) A review of the taxonomy of the genus Eurydérus LeConte 1848. With notes on the North American Dapti (of authors). (Carabidae: Harpalini). Coleopterist's Bulletin, 14, 44–54.

Ball, O.J.-P., Barker, G.M., Prestidge, R.A., & Lauren, D.R. (1997) Distribution and accumulation of the alkaloid peramine in *Neotyphodium lolii*-infected perennial ryegrass. Journal of Chemical Ecology, 23, 1419–1434.

Baniameri, V., Soleiman-Nejadian, E., & Mohaghegh, J. (2005) Life table and age-dependent reproduction of the predatory bug *Orius niger* Wolff (Heteroptera: Anthocoridae) at three constant temperatures: A demographic analysis. Applied Entomology and Zoology, 40, 545–550.

Banks, C.J. (1957) The behaviour of individual coccinellid larvae on plants. Animal Behaviour, 5, 12–24.

Banks, C.J. & Macauley, E.D.M. (1967) Effects of *Aphis fabae* Scop. and of its attendant ants and insect predators on yields of field beans (*Vicia faba* L.). Annals of Applied Biology, 60, 445–453.

Barber, G.W. (1936) Orius insidiosus (Say), an important natural enemy of the corn earworm (ed U.S.D.A.T. Bulletin), Vol. 504, pp. 24.

Barberi, P., Cozzani, A., Macchia, M., & Bonari, E. (1998) Size and composition of the weed seedbank under different management systems for continuous maize cropping. Weed Research, 38, 319–334.

Barbier, M. (1970) Chemistry and biochemistry of pollens. Progress in Phytochemistry, 2, 1–33.

Barbosa, P. (1998) Agroecosystems and conservation biological control. In Conservation Biological Control (ed P. Barbosa), pp. 39–54. Academic Press, London.

Barbosa, P. & Benrey, B. (1998) The influence of plants on insect parasitoids: Implications for conservation biological control. In Conservation Biological Control (ed P. Barbosa), pp. 55–82. Academic Press, London.

Barbosa, P. & Wratten, S.D. (1998) Influence of plants on invertebrate predators: Implications to conservation biological control. In Conservation Biological Control (ed P. Barbosa), pp. 83–100. Academic Press, San Diego, CA.

Barclay, A.S. & Earle, F.R. (1974) Chemical analyses of seeds III: Oil and protein content of 1,253 species. Economic Botany, 28, 178–236.

Barclay, R.M.R., Dolan, M.-A., & Dyck, A. (1991) The digestive efficiency of insectivorous bats. Canadian Journal of Zoology, 69, 1853–1856.

Bargen, H., Saudhof, K., & Poehling, H.-M. (1998) Prey finding by larvae and adult females of *Episyrphus balteatus*. Entomologia Experimentalis et Applicata, 87, 245–254.

Barker, R.J., Lehner, Y., & Kunzmann, M.R. (1980) Pesticides and honey bees: Nectar and pollen contamination in alfalfa treated with dimethoate. Archives of Environmental Contamination and Toxicology, 9, 125–133.

Barlow, C.A. (1961) On the biology and reproductive capacity of *Syrphus corollae* Fab. (Syrphidae) in the laboratory. Entomologia Experimentalis et Applicata, 4, 91–100.

Barney, R.J. & Pass, B.C. (1986) Foraging behavior and feeding preference of ground beetles (Coleoptera: Carabidae) in Kentucky alfalfa. Journal of Economic Entomology, 79, 1334–1337.

Barrentine, C.D. (1993) Toads (*Bufo boreas* Baird & Girard) obtain no calories from ingested *Sphenophorous phoeniciensis* Chittenden (Coleoptera: Curculionidae). Pan-Pacific Entomologist, 69, 15–17.

Bartlett, B.R. (1962) The ingestion of dry sugars by adult entomophagous insects and the use of this feeding habit for measuring the moisture needs of parasites. Journal of Economic Entomology, 55, 749–753.

Barton, A.M. (1986) Spatial variation in the effect of ants on extrafloral nectary plant. Ecology, 67, 495–504.

Baskin, S.I. & Bliss, C.A. (1969) Sugar occurring in the extrafloral exudates of the Orchidaceae. Phytochemistry, 8, 1139–1145.

Bassett, P. (1978) Damage to winter cereals by *Zabrus tenebrioides* (Goeze) (Coleoptera: Carabidae). Plant Pathology, 27, 48.

Bates, R.B., Byrne, D.N., Kane, V.V., Miller, W.B., & Taylor, S.R. (1990) N.m.r. characterization of trehalulose from the excrement of the sweet potato whitefly, *Bemisia tabaci*. Carbohydrate Research, 201, 342–345.

Beach, J.P., Williams, L.I., Hendrix, D.L., & Price, L.D. (2003) Different food sources affect the gustatory response of *Anaphes iole*, an egg parasitoid of *Lygus* spp. Journal of Chemical Ecology, 29, 1203–1222.

Beattie, A.J. (1985) The Evolutionary Ecology of Ant-Plant Mutualisms. Cambridge University Press, Cambridge.

Beattie, A.J. & Culver, D.C. (1981) The guild of myrmecochores in the herbaceous flora of West Virginia forests. Ecology, 62, 107–115.

Beattie, A.J. & Hughes, L. (2002) Ant-plant interactions. In Plant-Animal Interactions: An Evolutionary Approach (eds C.M. Herrera & O. Pellmyr), pp. 211–235. Blackwell, Oxford.

Bebawi, F.F. & Campbell, S.D. (2004) Interactions between meat ants (*Iridomyrmex spadius*) and bellyache bush (*Jatropha gossypiifolia*). Australian Journal of Experimental Agriculture, 44, 1157–1164.

Becerra, J.X. & Venable, D.L. (1991) The role of ant-Homoptera mutualisms in the evolution of extrafloral nectaries. Oikos, 60, 105–106.

Beckman, N. & Hurd, L.E. (2003) Pollen feeding and fitness in praying mantids: The vegetarian side of a tritrophic predator. Environmental Entomology, 32, 881–885.

Beggs, J. (2001) The ecological consequences of social wasps (*Vespula* spp.) invading and ecosystem that has an abundant carbohydrate source. Biological Conservation, 99, 17–28.

Begum, M., Gurr, G.M., Wratten, S.D., Hedberg, P.R., & Nicol, H.I. (2006) Using selective food plants to maximize biological control of vineyard pests. Journal of Applied Ecology, 43, 547–554.

Beirne, B.P. (1962) Trends in applied biological control of insects. Annual Review of Entomology, 7, 387–400.

Beirne, B.P. (1975) Biological control attempts by introductions against pest insects in the field in Canada. Canadian Entomologist, 107, 225–236.

Bell, E.A. (1984) Toxic compounds in seeds. In Seed Physiology, Volume 1: Development (ed D.R. Murray), pp. 245–264. Academic Press, Orlando, FL.

Bell, E.A. & Janzen, D.H. (1971) Medical and ecological considerations of L-Dopa and 5-HTP in seeds. Nature, 229, 136–137.

Bell, E.A. & Tirimanna, A.S.L. (1965) Associations of amino acids and related compounds in the seeds of forty-seven species of *Vicia*: Their taxonomic and nutritional significance. Biochemistry Journal, 97, 104–111.

Bell, H.A., Fitches, E.C., Down, R.E., Marris, G.C., Edwards, J.P., Gatehouse, J.A., & Gatehouse, A.M.R. (1999) The effect of snowdrop lectin (GNA) delivered via artificial diet and transgenic plants on Eulophus pennicornis (Hymenoptera: Eulophidae), a parasitoid of the tomato moth Lacanobia oleracea (Lepidoptera: Noctuidae). Journal of Insect Physiology, 45, 983–991.

Bell, H.A., Fitches, E.C., Marris, G.C., Bell, J., Edwards, J.P., Gatehouse, J.A., & Gatehouse, A.M.R. (2001) Transgenic GNA expressing potato plants augment the beneficial biocontrol of *Lacanobia oleracea* (Lepidoptera; Noctuidae) by the parasitoid *Eulophus pennicornis* (Hymenoptera; Eulophidae). Transgenic Research, 10, 35–42.

Belliure, B. & Michaud, J.P. (2001) Biology and behavior of *Pseudodorus clavatus* (Diptera: Syrphidae), and important predator of citrus aphids. Annals of the Entomological Society of America, 94, 91–96.

Ben Saad, A.A. & Bishop, G.W. (1976a) Attraction of insect to potato plants through use of artificial honeydews and aphid juice. Entomophaga, 21, 49–57.

Ben Saad, A.A. & Bishop, G.W. (1976b) Effect of artificial honeydews on insect communities in potato fields. Environmental Entomology, 5, 453–457.

Bennett, R.N., Mellon, F.A., Rosa, E.A.S., Perkins, L., & Kroon, P.A. (2004) Profiling glucosinolates, flavonoids, alkaloids, and other secondary metabolites in tissues of *Azima tetracantha* (Salvadoraceae). Journal of Agricultural and Food Chemistry, 52, 5856–5862.

Bentley, B.L. (1977a) Extrafloral nectaries and protection by pugnacious bodyguards. Annual Review of Ecology and Systematics, 8, 407–427.

Bentley, B.L. (1977b) The protective function of ants visiting extrafloral nectaries of *Bixa orellana* (Bixaceae). Journal of Ecology, 65, 27–38.

Benton, A.H. & Crump, A.J. (1981) Observations on the spring and summer behavior of the 12-spotted ladybird beetle, *Coleomegilla maculata* (DeGeer) (Coleoptera: Coccinellidae). New York Entomological Society, 89, 102–108.

Bera, S.K., Trivedi, A., & Sharma, C. (2002) Trapped pollen and spores from spider webs of Lucknow environs. Current Science, 83, 1580–1585.

Berenbaum, M.R. (1988) Allelochemicals in insect-microbe-plant interactions; agents provacateurs in the coevolutionary arms race. In Novel Aspects of Insect-Plant Interactions (eds P. Barbosa & D.K. Letourneau), pp. 97–123. Wiley, New York.

Berg, R.Y. (1975) Myrmecochorous plants in Australia and their dispersal by ants. Australian Journal of Botany, 23, 475–508.

Berg, R.Y. (1979) Legume, seed, and myrmecochorous dispersal in Kennedia and Hardenbergia (Fabaceae), with a remark on the Durian theory. Norwegian Journal of Botany, 26, 229–254.

Bergman, J.M. & Tingey, W.M. (1979) Aspects of interaction between plant genotypes and biological control. Entomological Society of America Bulletin, 25, 275–278.

Bernal, C.C., Aguda, R.M., & Cohen, M.B. (2002a) Effect of rice lines transformed with *Bacillus thuringiensis* toxin genes on the brown planthopper and its predator *Cyrtorhinus lividipennis*. Entomologia Experimentalis et Applicata, 102, 21–28.

Bernal, J.S., Griset, J.G., & Gillogly, P.O. (2002b) Impacts of developing on *Bt* maize-intoxicated hosts on fitness parameters of a stem borer parasitoid. Journal of Entomological Science, 37, 27–40.

Berndt, L.A. & Wratten, S.D. (2005) Effects of alyssum flowers on the longevity, fecundity, and sex ratio of the leafroller parasitoid *Dolichogenidea tasmanica*. Biological Control, 32, 65–69.

Berndt, L.A., Wratten, S.D., & Scarratt, S.L. (2007) The influence of floral resource subsidies on parasitism rates of leafrollers (Lepidoptera: Tortricidae) in New Zealand vineyards. Biological Control, 37, 50–55.

Bersanetti, P.A., Park, H.-Y., Bae, K.S., Son, K.-H., Shin, D.-H., Hirata, I.Y., Juliano, M.A., Carmona, A.K., & Juliano, L. (2005) Characterization of arazyme, an exocellular metalloprotease isolated from *Serratia proteamaculans* culture medium. Enzyme and Microbial Technology, 37, 574–581.

Bertrand, J.A., Sudduth, T.Q., Condon, A., Jenkins, T.C., & Calhoun, M.C. (2005) Nutrient content of whole cottonseed. Journal of Dairy Science, 88, 1470–1477.

Bertrandi, F. & Zetto Brandmayr, T. (1991) Osservazioni sulla dieta e cenni sulla bionomia del genere *Harpalus* Latreille. [Diet preferences and bionomy of *Harpalus* Latr. s. str. (Coleoptera: Carabidae)]. Berichte des naturwissenschaft-lich-medizinischen Vereins in Innsbruck, 78, 145–155.

Best, R.L. & Beegle, C.C. (1977) Food preferences of five species of carabids commonly found in Iowa cornfields. Environmental Entomology, 6, 9–12.

Betts, A.D. (1920) Nectar yeasts. Bee World, 1, 132.

Beveridge, T.H.J., Girard, B., Kopp, T., & Drover, J.C.G. (2005) Yield and composition of grape seed oils extracted by supercritical carbon dioxide and petroleum ether: Varietal effects. Journal of Agricultural and Food Chemistry, 53, 1799–1804.

Bewley, J.D. & Black, M. (1978) Physiology and Biochemistry of Seeds. Springer, New York.

Bewley, J.D. & Black, M. (1994) Seeds: Physiology of Development and Germination, 2nd edn. Plenum Press, New York.

Bewley, J.D., Hempel, F.D., McCormick, S., & Zambryski, P. (2000) Reproductive development. In Biochemistry & Molecular Biology of Plants (eds B. Buchanan, W. Gruissem, & R. Jones), pp. 988–1043. American Society of Plant Physiologists, Waldorf, MD.

Bhatti, M.A., Duan, J.J., Head, G., Jiang, C., McKee, M.J., Nickson, T.E., Pilcher, C.L., & Pilcher, C.D. (2005a) Field evaluation of the impact of corn rootworm (Coleoptera: Chrysomelidae)-protected Bt corn on ground-dwelling predators. Environmental Entomology, 34, 1325–1335.

Bhatti, M.A., Duan, J.J., Head, G.P., Jiang, C., McKee, M.J., Nickson, T.E., Pilcher, C.L., & Pilcher, C.D. (2005b) Field evaluation of the impact of corn rootworm (Coleoptera: Chrysomelidae)-protected Bt corn on foliage-dwelling arthropods. Environmental Entomology, 34, 1336–1345.

Bigger, J.H. & Blanchard, R.A. (1959) Insecticidal control of underground insects of corn. University of Illinois Agricultured Experiment station Bulletin 641, 1–28.

Bignell, D.E. (1984) The arthropod gut as an environment for microorganisms. In Invertebrate-Microbial Interactions (eds J.M. Anderson, A.D.M. Raynor, & D.W.H. Walton), pp. 205–228. Cambridge University Press, Cambridge.

Birch, A.N.E., Geoghegan, I.E., Majerus, M.E.N., McNicol, J.W., Hackett, C.A., Gatehouse, A.M.R., & Gatehouse, J.A. (1999) Tri-trophic interactions involving pest aphids, predatory 2-spot ladybirds, and transgenic potatoes expressing snowdrop lectin for aphid resistance. Molecular Breeding, 5, 75–83.

Birch, A.N.E., Geoghegan, I.E., Griffiths, D.W., & McNicol, J.W. (2002) The effect of genetic transformations for pest resistance on foliar solanidine-based glycoalkaloids of potato (*Solanum tuberosum*). Annals of Applied Biology, 140, 143–149.

Birken, E.M. & Cloyd, R.A. (2007) Food preference of the rove beetle, Atheta coriaria Kraatz (Coleoptera: Staphylinidae) under laboratory conditions. Insect Science, 14, 53–56.

Blaszyk, P. (1963) Beobachtungen bei einem Massequftreten des Laufkafers (*Amara convexiuscula* Mrsch.) in Osfriesland. Gesunde Pflanzen, 15, 225–229.

Blatchley, W.S. (1910) Coleoptera or Beetles Known to Occur in Indiana. The Nature Publishing Company, Indianapolis, IN.

Blunck, H. (1925) Syllabus der insektenbiologie. Coleoptera Parcey, Berlin.

Bluthgen, N. & Fiedler, K. (2004) Preferences for sugars and amino acids and their conditionality in a diverse nectar-feeding ant community. Journal of Animal Ecology, 73, 155–166.

Bonadona, P. (1971) Catalogue des Colepteres Carabiques de France. Nouvelle Revue d'Entomologie, Suppl. 1, 1–177.

Bonmatin, J.M., Moineau, I., Charvet, R., Fleche, C., Colin, M.E., & Bengsch, E.R. (2003) A LC/APCI-MS/MS method for analysis of imidacloprid in soils, in plants, and in pollens. Analytical Chemistry, 75, 2027–2033.

Bonmatin, J.M., Marchand, P.A., Charvet, R., Moineau, I., Bengsch, E.R., & Colin, M.E. (2005) Quantification of imidacloprid uptake in maize crops. Journal of Agricultural and Food Chemistry, 53, 5336–5341.

Bono, J.M. & Heithaus, E.R. (2002) Sex ratios and the distribution of elaiosomes in colonies of the ant *Aphaenogaster rudis*. Insectes Sociaux, 49, 320–325.

Bose, S.R. (1947) Hereditary (seed-borne) symbiosis in *Casuarina equisetifolia* forst. Nature, 159, 512–514.

Bottrell, D.G., Barbosa, P., & Gould, F. (1998) Manipulating natural enemies by plant variety selection and modification: A realistic strategy? Annual Review of Entomology, 43, 347–367.

Bouchard, Y. & Cloutier, C. (1984) Honeydew as a source of host-searching kairomones for the aphid parasitoid *Aphidius nigripes* (Hymenoptera: Aphidiidae). Canadian Journal of Zoology, 62, 1513–1520.

Bouchard, Y. & Cloutier, C. (1985) Role of olfaction in host finding by aphid parasitoid *Aphidius nigripes* (Hymenoptera: Aphidiidae). Journal of Chemical Ecology, 11, 801–808.

Boucher, D.H., James, S., & Keeler, K.H. (1982) The ecology of mutualism. Annual Review of Ecology and Systematics, 13, 315–347.

Boulay, R., Fedriani, J.M., Manzaneda, A.J., & Cerda, X. (2005) Indirect effects of alternative food resources in an ant-plant interaction. Oecologia, 144, 72–79.

Boush, G.M. & Matsumura, F. (1967) Insecticidal degradation by *Pseudomonas melophthora*, the bacterial symbiote of the apple maggot. Journal of Economic Entomology, 60, 918–920.

Bowden, B.N. (1970) The sugars in the extrafloral nectar of *Andropogon gayanus* var. *bisquamulatus*. Phytochemistry, 9, 2315–2318.

Box, H.E. (1927) The introduction of braconid parasites of Diatraea saccharalis Fabr., into certain of the West Indian islands. Bulletin of Entomological Research, 28, 365–370.

Boyd, D.W., Jr. & Cohen, A.C. (2002) Digestive enzymes and stylet morphology of *Deraeocoris nebulosus* (Hemiptera: Miridae), a predacious plant bug. Annals of the Entomological Society of America, 95, 395–401.

Bozsik, A. (1992) Natural adult food of some important *Chrysopa* species (Planipennia: Chrysopidae). Acta Phytopathologica et Entomologica Hungarica, 27, 141–146.

Bozsik, A. (2000) Nahrungsanalytische Untersuchungen an einigen mitteleuropaischen Chrysopiden-imagines. Beitrage Zur Entomologie, 50, 237–246.

Bozsik, A. (2006) Susceptibility of adult Coccinella septempunctata (Coleoptera: Coccinellidae) to insecticides with different modes of action. Pest Management Science, 62, 651–654.

Bracken, G.K. (1966) Role of ten dietary vitamins on fecundity of the parasitoid *Exeristes comstockii* (Cresson) (Hyemenoptera: Ichneumonidae). Canadian Entomologist, 98, 918–922.

Braman, S.K., Pendley, A.F., & Corley, W. (2002) Influence of commercially available wildflower mixes on beneficial arthropod abundance and predation in turfgrass. Environmental Entomology, 31, 564–572.

Brandmayr, P. (1972) Studio ecologico sui Coleotteri Carabid di due formaxioni vegetali dell-Istria Montana. Bolletino d: Zoologia 39, 523–529.

Brandmayr, P. (1973/1974) La cure parentali di *Carterus (Sabienus) calydonius* Rossi (Coleoptera: Carabidae). Atti Accademia delle Scienze d: Torino, 108, 811–818.

Brandmayr, P. & Brandmayr, T.Z. (1987) The problem of presocial behaviour in Ditomine ground beetles. *Publicazioni dell'Istituto di Entomologia dell'Universita di Pavia*, 36, 15–18.

Brandmayr, P., Ferrero, E., & Zetto-Brandmayr, T. (1980) Larval versus imaginal taxonomy and the systematic status of the ground beetle taxa *Harpalus* and *Ophonus* (Coleoptera: Carabidae: Harpalini). Entomologia Generalis, 6, 335–353.

Brandmayr, T. (1990) Spermophagous (seed-eating) ground beetles: First comparison of the diet and ecology of the Harpaline genera *Harpalus* and *Ophonus* (Col., Carabidae). In The Role of Ground Beetles in Ecological and Environmental Studies (ed N.E. Stork), pp. 307–316. Intercept, Andover.

Brandmayr, T.Z. (1976a) Nutrizione e allevamento di Carabidi exclusivamente fitofagi: spermofagia larvale di *Ophonus ardosiacus* Lutsh. Redia, 54, 197–206.

Brandmayr, T.Z. (1976b) Studi sulla fitofagia nei Carabidi: spermofagia larvale di *Ophonus ardosiacus* Lutsh. (Coleoptera: Carabidae). In *XI Congresso Nazionale Italiano di Entomologia*, pp. 59–64, Portici-Sorrento.

Brandmayr, T.Z. (1983) Life cycle, control of propagation rhythem and fecundity of *Ophonus rotundicollis* Fairm. et Lab. (Coleoptera: Carabidae: Harpalini) as an adaptation to the main feeding plant *Daucus carota* L. (Umbelliferae). *Report of the 4th Symposium on Carabidology*, 81, 93–103.

Brandmayr, T.Z. & Brandmayr, P. (1978) Sur la biologie d'un carabique tres interessant: *Tetraplatypus ganglbaueri* Apfb. Cordulia, 4(4), 156–159.

Brandmayr, T.Z., Giglio, A., Marano, I. & Brandmayr, P. (1998) Morphofunctional and ecological features in carabid (Coleoptera) larvae. Mus. reg. Sci. nat. Torino, 1998, 449–490.

Brem, D. & Leuchtmann, A. (2001) *Epichloe* grass endophytes increase herbivore resistance in the woodland grass *Brachypodium sylvaticum*. Oecologia, 126, 522–530.

Bresinsky, A. (1963) Bau, Entwicklungsgeschichte, und Inhaltsstoffe der Eliaosomen. Studien zur myrmekochoren Verbreitung von Samen und Fruchten. Bibliotheca Botanica, 126, 1–54.

Brew, C.R., O'Dowd, D.J., & Rae, I.D. (1989) Seed dispersal by ants: Behaviour-releasing compounds in eliaosomes. Oecologia, 80, 490–497.

Breznak, J.A. & Brune, A. (1994) Role of microorganisms in the digestion of lignocellulose by termites. Annual Review of Entomology, 39, 453–487.

Brian, M.V. (1973) Feeding and growth in the ant *Myrmica*. Journal of Animal Ecology, 42, 37–53.

Briel, J. (1964) Les *Ophonus* de France (Coleopteres: Carabidae). Thiebault et Cie, Chateau-Salines.

Briese, D.T. & Macauley, B.J. (1981) Food collection within an ant community in semi-arid Australia, with special reference to seed harvesters. Australian Journal of Ecology, 6, 1–19.

Briggs, C.L., Morris, E.C., & Ashford, A.E. (2005) Investigations into seed dormancy in *Grevillea linearifolia*, *G. buxifolia*, and *G. sericea*: Anatomy and histochemistry of the seed coat. Annals of Botany, 96, 965–980.

Briggs, J.B. (1965) Biology of some ground beetles (Col., Carabidae) injurious to strawberries. Bulletin of Entomological Research, 56, 79–93.

Briggs, J.B. & Tew, R.P. (1965) Control of ground beetles (Col., Carabidae) attacking strawberries. Bulletin of Entomological Research, 54, 83–92.

Brink, D. & deWet, J.M.J. (1980) Interpopulation variation in nectar production in *Aconitum columbianum* (Ranunculaceae). Oecologia, 47, 160–163.

Bristow, C.M. (1991) Are ant-aphid associations a tritrophic interaction? Oleander aphids and Argentine ants. Oecologia, 87, 514–521.

Broadhead, E.C. (1984) Adaptations for fungal grazing in Lauxaniid flies. Journal of Natural History, 18, 639–649.

Brooks, S.J. & Barnard, P.C. (1990) The green lacewings of the world: A generic review (Neuroptera: Chrysopidae). Bulletin of the British Museum of Natural History (Entomology), 59, 117–286.

Broufas, G.D. & Koveos, D.S. (2000) Effect of different pollens on development, survivorship, and reproduction of *Euseius finlandicus* (Acari: Phytoseiidae). Environmental Entomology, 29, 743–749.

Brown, J.L., Reichman, O.J., & Davidson, D.W. (1979) Granivory in desert systems. Annual Review of Ecology and Systematics, 10, 201–227.

Bruce-Oliver, S.J., Hoy, M.A., & Yaninek, J.S. (1996) Effect of some food sources associated with cassava in Africa on the development, fecundity and longevity of *Euseius fustis* (Pritchard and Baker) (Acari: Phytoseiidae). Experimental and Applied Acarology, 20, 73–85.

Brust, G.E. (1994) Seed-predators reduce broadleaf weed growth and competitive ability. Agriculture, Ecosystems and Environment, 48, 27–34.

Brust, G.E. & House, G.J. (1988a) A study of *Tyrophagus putrescentiae* (Acari: Acaridae) as a factultative predator of southern corn rootworm eggs. Experimental and Applied Acarology, 4, 335–344.

Brust, G.E. & House, G.J. (1988b) Weed seed destruction by arthropods and rodents in low-input soybean agroecosystems. American Journal of Alternative Agriculture, 3, 19–25.

Brust, G.E., Stinner, B.R., & McCartney, D.A. (1986) Predator activity and predation in corn agroecosystems. Environmental Entomology, 15, 1017–1021.

Buchner, P. (1965) Endosymbiosis of Animals with Plant Microorganisms. Wiley, New York.

Budenberg, W.J. (1990) Honeydew as a contact kairomone for aphid parasitoids. Entomologia Experimentalis et Applicata, 55, 139–148.

Budenberg, W.J. & Powell, W. (1992) The role of honeydew as an ovipositional stimulant for two species of syrphids. Entomologia Experimentalis et Applicata, 64, 57–61.

Bugbee, R.E. (1942) Damage to kernels of wheat by *Harpalus amputatus* Say, in western Texas. Journal of the Kansas Entomological Society, 15, 108.

Bugg, R.L. (1987) Observations on insects associated with a nectar-bearing Chilean tree, *Quillaja saponaria* Molina (Rosaceae). Pan-Pacific Entomologist, 63, 60–64.

Bugg, R.L. & Pickett, C.H. (1998) Introduction: Enhancing biological control – habitat management to promote natural enemies of agricultural pests. In Enhancing Biological Control: Habitat Management to Promote Natural Enemies of Agricultural Pests (eds C.H. Pickett & R.L. Bugg), pp. 1–24. University of California Press, Berkeley, CA.

Bugg, R.L. & Waddington, C. (1994) Using cover crops to manage arthropod pests of orchards: A review. Agriculture, Ecosystems and Environment, 50, 11–28.

Bugg, R.L., Ehler, L.E., & Wilson, L.T. (1987) Effect of common knotweed (*Polygonum aviculare*) on abundance and efficiency of insect predators of crop pests. Hilgardia, 55, 1–51.

Buhler, D.D., Hartzler, R.G., & Forcella, F. (1997) Implications of weed seedbank dynamics to weed management. Weed Science, 45, 329–336.

Burakowski, B. (1967) Biology, ecology, and distribution of *Amara pseudocommunis* Burak. (Coleoptera, Carabidae). Annales Zoologici, 24, 485–526.

Burger, H.C., de Goffau, L.J.W., & Ulenberg, S.A. (1984) Unusual damage by insects in 1983. Entomologische Berichten, 44, 145–150.

Burger, J.M.S., Reinjen, T.M., van Lenteren, J.C., & Vet, L.E.M. (2004) Host feeding in insect parasitoids: Why destructively feed upon a host that excretes an alternative? Entomologia Experimentalis et Applicata, 112, 207–215.

Burger, J.M.S., Kormany, A., van Lenteren, J.C., & Vet, L.E.M. (2005) Importance of host feeding for parasitoids that attack honeydew-producing hosts. Entomologia Experimentalis et Applicata, 117, 147–154.

Burgess, L., Dueck, J., & McKenzie, D.L. (1983) Insect vectors of the yeast *Nematospora coryli* in mustard, *Brassica juncea*, crops in southern Saskatchewan. Canadian Entomologist, 115, 25–30.

Burgin, S.G. & Hunter, F.F. (1997) Nectar versus honeydew as sources of sugar for male and female black flies (Diptera: Simuliidae). Journal of Medical Entomology, 34, 605–608.

Burgio, G., Lanzoni, A., Accinelli, G., Dinelli, G., Bonetti, A., Marotti, I., & Ramilli, F. (2007) Evaluation of Bt-toxin uptake by the non-target herbivore, Myzus persicae (Hemiptera: Aphididae), feeding on transgenic oilseed rape. Bulletin of Entomological Research, 97, 211–215.

Burmeister, F. (1939) Biologie, Okologie und Verbreitung der europaischer Kafer. H. Goecke, Krefeld.

Butler, G.D., Jr. & Ritchie, P.L., Jr. (1971) Feed Wheast and the abundance and fecundity of Chrysopa carnea. Journal of Economic Entomology, 64, 933–934.

Butler, G.D., Jr., Loper, G.M., McGregor, S.E., Webster, J.L., & Margolis, H. (1972) Amounts and kinds of sugars in the nectar of cotton (*Gossypium* spp.) and the time of their secretion. Agronomy Journal, 64, 364–368.

Byrne, D.N., Hendrix, D.L., & Williams, L.H.I. (2003) Presence of trehalulose and other oligosaccharides in hemipteran honeydew, particularly Aleyrodidae. Physiological Entomology, 28, 144–149.

Byzov, B.A., Thanh, V.N., & Babjeva, I.P. (1993) Interrelationships between yeasts and soil diplopods. Soil Biology and Biochemistry, 25, 1119–1126.

Cadogan, B.L. & Laing, J.E. (1977) A technique for rearing the predaceous mite *Balaustium putnami* (Acarina: Erythraeidae), with notes on its biology and life history. Canadian Entomologist, 109, 1535–1544.

Caldwell, D.L. & Gerhardt, K.O. (1986) Chemical analysis of peach extrafloral nectary exudate. Phytochemistry, 25, 411–413.

Callaway, J.C. (2004) Hempseed as a nutritional resource: An overview. Euphytica, 140, 65–72.

Campbell, B.C. (1989) On the role of microbial symbiotes in herbivorous insects. In Insect-Plant Interactions (ed E.A. Bernays), Vol. I, pp. 1–44. CRC Press, Boca Raton, FL.

Canard, M. (2001) Natural food and feeding habits of lacewings. In Lacewings in the Crop Environment (eds P.K. McEwen, T.R. New, & A.E. Whittington), pp. 116–129. Cambridge University Press, Cambridge.

Canard, M., H., Kokubu, H., & Duelli, P. (1990) Tracheal trunks supplying air to the foregut and feeding habits in adults of European green lacewing species. In Advances in Neuropterology, Proceedings of the Third International Symposium on Neuropterology (eds M.W. Masell & H. Aspock), pp. 277–286. Berg en Dal, Kruger National Park, R. S. A.

Canas, L.A. & O'Neil, R.J. (1998) Applications of sugar solutions to maize, and the impact of natural enemies on fall armyworm. International Journal of Pest Management, 44, 59–64.

Canedo, T.D.V. & Lizarraga, T.A. (1988) Artificial diets for laboratory rearing of *Chrysoperla externa* Hagen (Neuroptera: Chrysopidae). Revista Peruana de Entomologia, 31, 83–85.

Capra, F. (1947) Note sui Coccinellidi (Col.) III. La larva ed il regime pollinivoro di *Bulaea lichtschovi* Hummel. Memorie della Societa Entomologica Italiana Suppl. 26, 80–86.

Carayon, J. & Steffan, J.R. (1959) Observations sur le regime alimentaire des *Orius* particulierement d' *Orius pallidicornis* Reuter (Heteroptera: Anthocoridae). Cahiers des Naturalistes, Bulletin N. P., 15, 53–63.

Cardina, J., Norquay, H.M., Stinner, B.R., & McCartney, D.A. (1996) Postdispersal predation of velvetleaf (*Abutilon theophrasti*) seeds. Weed Science, 44, 534–539.

Carlson, R.E. & Chiang, H.C. (1973) Reduction of an *Ostrinia nubilalis* population by predatory insects attracted by sucrose sprays. Entomophaga, 18, 205–211.

Carmona, D.M., Menalled, F.D., & Landis, D.A. (1999) *Gryllus pennsylvanicus* (Orthoptera: Gryllidae): Laboratory weed seed predation and within field activity-density. Journal of Economic Entomology, 92, 825–829.

Carrel, J.E., Burgess, H.K., & Shoemaker, D.M. (2000) A test of pollen feeding by a lyniphiid spider. Journal of Acarology, 28, 243–244.

Carroll, C.R. & Janzen, D.H. (1973) Ecology of foraging by ants. Annual Review of Ecology and Systematics, 4, 231–257.

Carroll, C.R. & Risch, S.J. (1984) The dynamics of seed harvesting in early successional communities by a tropical ant, *Solenopsis geminata*. Oecologia, 61, 388–392.

Carroll, G. (1988) Fungal endophytes in stems and leaves: From latent pathogen to mutualistic symbiont. Ecology, 69, 2–9.

Carter, C., Shafir, S., Yehonatan, L., Palmer, R.G., & Thornburg, R. (2006) A novel role for proline in plant floral nectars. Naturwissenschaften, 93, 72–79.

Carter, M.C. & Dixon, A.F.G. (1984) Honeydew: An arrestant stimulus for coccinellids. Ecological Entomology, 9, 383–387.

Casper, B.B. (1987) Spatial patterns of seed dispersal and postdispersal seed predation of *Cryptantha flava* (Boraginaceae). American Journal of Botany, 74, 1646–1655.

Cassill, D. (2003) Rules of supply and demand regulate recruitment to food in an ant society. Behavioral Ecology and Sociobiology, 54, 441–450.

Cate, J.R., Jr., Ridgway, R.L., & Lingren, P.D. (1972) Effects of systemic insecticides applied to cotton on adults of an ichneumonid parasite, *Campoletis perdistinctus*. Journal of Economic Entomology, 65, 484–488.

Cate, J.R., Jr., Bottrell, D.G., & Teetes, G.L. (1973) Management of the greenbug on grain sorghum. 2. Testing seed and soil treatments for greenbut and corn leaf aphid control. Journal of Economic Entomology, 66, 953–959.

Caudell, J.N. & Conover, M.R. (2006) Energy content and digestibility of brine shrimp (*Artemia franciscana*) and other prey items of eared grebes (*Podiceps nigricollis*) on the Great Salt Lake, Utah. Biological Conservation, 130, 251–254.

Cavers, P.B. (1983) Seed demography. Canadian Entomologist, 61, 3578–3590.

Cavers, P.B. & Benoit, D.L. (1989) Seed banks in arable land. In Ecology of Soil Seed Banks (eds M.A. Leck, V.T. Parker, & R.L. Simpson), pp. 309–328. Academic Press, London.

Chandler, A.E.F. (1968a) Some factors influencing the occurrence and site of oviposition by aphidophagous Syrphidae (Diptera). Annals of Applied Biology, 61, 435–446.

Chandler, A.E.F. (1968b) Some host-plant factors affecting oviposition by aphidophagous Syrphidae (Diptera). Annals of Applied Biology, 61, 415–423.

Chang, S.-T. & Miles, P.G. (2004) Mushrooms: Cultivation, Nutritional Value, Medicinal Effect, and Environmental Impact, 2nd edn. CRC Press, Boca Raton, FL.

Chant, D.A. (1959) Phytoseiid mites: Part 1. Bionomics of seven species in Southeastern England. Canadian Entomologist, Suppl. 12, 1–45.

Chant, D.A. & Fleschner, C.A. (1960) Some observations on the ecology of phytoseiid mites (Acarina: Phytoseiidae) in California. Entomophaga, 5, 131–139.

Chapman, R.F. (1998) The Insects: Structure and Function, 4th edn. Cambridge University Press, Cambridge.

Charlet, L.D., Aiken, R.M., Meyer, R.F., & Gebre-Amlak, A. (2007) Impact of combining planting date and chemical control to reduce larval densities of stem-infesting pests of sunflower in the central plains. Journal of Economic Entomology, 100, 1248–1257.

Cheeseman, M.T. & Pritchard, G. (1984) Proventricular trituration in adult carabid beetles (Coleoptera: Carabidae). Journal of Insect Physiology, 30, 203–209.

Chen, T.-Y., Chu, C.-C., Hu, C., Mu, J.-Y., & Henneberry, T.J. (2006) Observations on midgut structure and content of *Chrysoperla carnea* (Neuroptera: Chrysopidae). Annals of the Entomological Society of America, 99, 917–919.

Cherrett, J.M., Powell, R.J., & Stradling, D.J. (1989) The mutualism between leaf-cutting ants and their fungus. In Insect-Fungus Interactions (eds N. Wilding, N.M. Collins, P.M. Hammond, & J.F. Webber), pp. 93–120. Academic Press, London.

Childers, C.C. & Rock, G.C. (1980) Observations on the occurrence and feeding habits of *Balaustium putnami* (Acari: Erythraeidae) in North Carolina apple orchards. International Journal of Acarology, 7, 63–67.

Chiverton, P.A. & Sotherton, N.W. (1991) The effects on beneficial arthropods of the exclusion of herbicides from cereal crop edges. Journal of Applied Ecology, 28, 1027–1039.

Choe, D.-H. & Rust, M.K. (2006) Homopteran chemical signatures reduce aggression of tending ants. Chemoecology, 16, 175–178.

Choh, Y. & Takabayashi, J. (2006) Herbivore-induced extrafloral nectar production in lima bean plants enhanced by previous exposure to volatiles from infested conspecifics. Journal of Chemical Ecology, 32, 2073–2077.

Choh, Y., Kugimiya, S., & Takabayashi, J. (2006) Induced production of extra-floral nectar in intact lima bean plants in response to volatiles from spider mite-infested conspecific plants as a possible indirect defense against spider mites. Oecologia, 147, 455–460.

Chreighton, W.S. (1952) Studies on Arizona ants (3) The habits of *Pogonomyrmex huachucanus* wheeler and a description of the sexual castes. Psyche, 59, 71–81.

Christian, D.P. & Lederle, P.E. (1984) Seed properties and water balance in desert granivores. Southwestern Entomologist, 29, 181–188.

Chu, Y.-i. (1969) On the bionomics of *Lyctocoris beneficus* (Hiura) and *Xylocoris galactinus* (Fieber) (Anthocoridae: Heteroptera). Journal of the Faculty of Agriculture, Kyushu University, 15, 1–136.

Churchill, D.M. & Christensen, P. (1970) Observations on pollen harvesting by brush-tongued lorikeets. Australian Journal of Zoology, 18, 427–437.

Clark, A. (1998) Managing Cover Crops Profitably, 2nd edn. USDA, Beltsville, MD.

Clark, E.W. & Lukefahr, M.J. (1956) A partial analysis of cotton extrafloral nectar and its approximation as a nutritional medium for adult pink bollworms (*Pectinophora gossypiella*). Journal of Economic Entomology, 49, 875–876.

Clay, K. (1988) Fungal endophytes of grasses: A defensive mutualism between plants and fungi. Ecology, 69, 10–16.

Clay, K. (1992) Fungal endophytes of plants: Biological and chemical diversity. Natural Toxins, 1, 147–149.

Clay, K. & Schardl, C. (2002) Evolutionary origins and ecological consequences of endophyte symbioses with grasses. The American Naturalist, 160, S99–S127.

Clay, K., Holah, J., & Rudgers, J.A. (2005) Herbivores cause a rapid increase in hereditary symbiosis and alter plant community composition. Proceedings of the National Academy of Sciences, 102, 12465–12470.

Clement, S.L. & Quisenberry, S.S. (1999) Global Plant Genetic Resources for Insect-Resistant Crops, p. 295. CRC Press, Boca Raton, FL.

Clinch, P.G., Palmer-Jones, T., & Forster, I.W. (1972) Effect on honey bees of nectar from the yellow kowhai (*Sophora microphylla* Ait.). New Zealand Journal of Agricultural Research, 15, 194–201.

Cochrane, V.W. (1958) Physiology of Fungi, Wiley, New York.

Cocuzza, G.E., De Clerq, P., Van de Veire, M., De Cock, A., Degheele, D., & Vacante, V. (1997) Reproduction of *Orius laevigatus* and *Orius albidipennis* on pollen and *Ephestia kuehniella* eggs. Entomologia Experimentalis et Applicata, 82, 101–104.

Cohen, A.C. (1990) Feeding adaptations of some predaceous Hemiptera. Annals of the Entomological Society of America, 83, 1215–1223.

Cohen, A.C. (1996) Plant feeding by predatory Heteroptera: Evolutionary and adaptational aspects of trophic switching. In Zoophytophagous Heteroptera: Implications for Life History and Integrated Pest Management (eds O. Alomar & R.N. Wiedenmann), pp. 1–17. Entomological Society of America, Lanham, MD.

Cohen, A.C. (2004) Insect Diets: Science and Technology. CRC Press, Boca Raton, FL.

Colby, D.M., Windham, M.T., & Grant, J.F. (1995) *Hippodamia convergens* (Coleoptera: Coccinellidae) dissemination of dogwood anthracnose fungus (Melanconiales: Melanconeacae). Environmental Entomology, 24, 1075–1079.

Colby, D.M., Windham, M.T., & Grant, J.F. (1996) Transportation and viability of conidia of *Discula destructiva* on *Hippodamia convergens*. Plant Disease, 80, 804–805.

Colin, L.J. & Jones, C.E. (1980) Pollen energetics and pollination modes. American Journal of Botany, 67, 210–215.

Coll, M. (1998a). Living and feeding on plants in predatory Heteroptera. In Predatory Heteroptera: Their Ecology and Use in Biological Control (eds M. Coll & J.R. Ruberson), pp. 89–129. Entomological Society of America, Lanham, MD.

Coll, M. (1998b). Parasitoid activity and plant species composition in intercropped systems. In Enhancing Biological Control: Habitat Management to Promote Natural Enemies of Agricultural Pests (eds C.H. Pickett & R.L. Bugg), pp. 85–120. University of California Press, Berkeley, CA.

Coll, M. & Bottrell, D.G. (1991) Microhabitat and resource selection of the European corn borer (Lepidoptera: Pyralidae) and its natural enemies in Maryland field corn. Environmental Entomology, 20, 526–533.

Coll, M. & Guershon, M. (2002) Omnivory in terrestrial arthropods: Mixing plant and prey diets. Annual Review of Entomology, 47, 267–297.

Colley, M.R. & Luna, J.M. (2000) Relative attractiveness of potential beneficial insectary plants to aphidophagous hoverflies (Diptera: Syrphidae). Environmental Entomology, 29, 1054–1059.

Compton, S.G. & Robertson, H.G. (1988) Complex interactions between mutualisms: Ants tending homopterans protect fig seeds and pollinators. Ecology, 69, 1302–1305.

Comstock, J.H. (1913) The Spider Book. Doubleday, Page & Company, Garden City, New York.

Conn, E.E. (1979) Cyanide and Cyanogenic Glycosides. In Herbivores: Their Interactions with Secondary Plant Metabolites (eds G.A. Rosenthal & D.H. Janzen), pp. 387–412. Academic Press, New York.

Connell, J.H. (1971) One the role of natural enemies in preventing competitive exclusion in some marine animals and in rain forest trees. In Dynamics of Populations (eds P.J. den Boer & G.R. Gradwell), pp. 298–310. Centre for Agricultural Publishing and Documentation, Wageningen, The Netherlands.

Corbet, S.A., Unwin, D.M., & Prys-Jones, O.E. (1979) Humidity, nectar and insect visits to flowers, with special reference to *Crataegus*, *Tilia* and *Echium*. Ecological Entomology, 4, 9–22.

Corbett, A. (1998) The importance of movement in the response of natural enemies to habitat manipulation. In Enhancing Biological Control: Habitat Management to Promote Natural Enemies of Agricultural Pests (eds C.H. Pickett & R.L. Bugg), pp. 25–48. University of California Press, Berkeley, CA.

Corey, D., Kamhampati, S., & Wilde, G.E. (1998) Electrophoretic analysis of *Orius insidiosus* (Hemiptera: Anthocoridae) feeding habits in field corn. Journal of the Kansas Entomological Society, 71, 11–17.

Cornic, J.F. (1973) Etude du régime alimentaire de trios espèces de Carabiques et de ses variations en vergers de pommiers. Annales de la Société Entomologique de France, N. S., 9, 69–87.

Cortesero, A.M., Stapel, J.O., & Lewis, W.J. (2000) Understanding and manipulating plant attributes to enhance biological control. Biological Control, 17, 35–49.

Costa, H.S., Toscano, N.C., Hendrix, D.L., & Henneberry, T.J. (1999) Patterns of honeydew droplet production by nymphal stages of *Bemisia argentifolii* (Homoptera: Aleyrodidae) and relative composition of honeydew sugars. Journal of Entomological Science, 34, 305–313.

Cote, M., Ferron, J., & Gagnon, R. (2005) Invertebrate predation of postdispersal seeds and juvenile seedlings of black spruce (*Picea mariana*) in the boreal forest of eastern Canada. Canadian Journal of Forest Research, 35, 674–681.

Cottrell, T.E. & Yeargan, K.V. (1998) Effect of pollen on *Coleomegilla maculata* (Coleoptera: Coccinellidae) population density, predation, and cannibalism in sweet corn. Environmental Entomology, 27, 1402–1410.

Couty, A., Clark, S.J., & Poppy, G.M. (2001a) Are fecundity and longevity of *Aphelinus abdominalis* affected by development in GNA-dosed *Macrosiphum euphorbiae*? Physiological Entomology, 26, 287–293.

Couty, A., de la Vina, G., Clark, S.J., Kaiser, L., Pham-Delegue, M.-H., & Poppy, G.M. (2001b) Direct and indirect sublethal effects of *Galanthus nivalis* agglutinin (GNA) on the development of a potato-aphid parasitoid, *Aphelinus abdominalis* (Hymenoptera: Aphelinidae). Journal of Insect Physiology, 47, 553–561.

Cowdry, E.V. (1923) The distribution of *Rickettsia* in the tissues of insects and arachnids. Journal of Experimental Medicine, 37, 431–456.

Cowgill, S.E., Wratten, S.D., & Sotherton, N.W. (1993) The selective use of floral resources by the hoverfly *Episyrphus balteatus* (Diptera: Syrphidae) on farmland. Annals of Applied Biology, 122, 223–231.

Crane, E. (1990) Bees and Beekeeping: Science, Practice, and World Resources. Cornell University Press, Ithaca, NY.

Cranshaw, W., Sclar, D.C., Spriggs, A., & Bishop, J. (1996) Flea beetle control in broccoli, Ft. Collins, CO, 1995. Arthropod Management Tests, 21, 87–88.

Crawley, M.J. (2000) Seed predators and plant population dynamics. In Seeds: The Ecology of Regeneration in Plant Communities (ed M. Fenner), pp. 167–182. CABI, Oxon.

Criddle, N. (1925) Field crickets in Manitoba. Canadian Entomologist, 57, 79–84.

Crist, T.O. & Friese, C.F. (1993) The impact of fungi on soil seeds: Implications for plants and granivores in a semiarid shrub-steppe. Ecology, 74, 2231–2239.

Crist, T.O. & MacMahon, J.A. (1992) Harvester ant foraging and shrub-steppe seeds: Interactions of seed resources and seed use. Ecology, 73, 1768–1779.

Crocker, R.L. & Whitcomb, W.H. (1980) Feeding niches of the big-eyed bugs *Geocoris bullatus*, *G. punctipes*, and *G. uliginosus* (Hemiptera: Lygaeidae: Geocorinae). Environmental Entomology, 9, 508–513.

Cromar, H.E., Murphy, S.D., & Swanton, C.J. (1999) Influence of tillage and crop residue on postdispersal predation of weed seeds. Weed Science, 47, 184–194.

Cruden, R.W., Hermann, S.M., & Peterson, S. (1983) Patterns of nectar production and plant-pollinator coevolution. In The Biology of Nectaries (eds B.L. Bentley & T.S. Elias), pp. 80–125. Columbia University Press, New York.

Cuautle, M. & Rico-Gray, V. (2003) The effect of wasps and ants on the reproductive success of the extrafloral nectaried plant *Turnera ulmifolia* (Turneraceae). Functional Ecology, 17, 417–423.

Cuautle, M., Rico-Gray, V., & Diaz-Castelazo, C. (2005) Effects of ant behaviour and presence of extra-floral nectaries on seed dispersal of the neotropical *Turnera ulmifolia* L. (Turneraceae). Biological Journal of the Linnean Society, 86, 67–77.
Culver, D.C. & Beattie, A.J. (1978) Myrmecochory in *Viola*: Dynamics of seed-ant interactions in some West Virginia species. Journal of Ecology, 66, 53–72.
Culver, D.C. & Beattie, A.J. (1980) The fate of *Viola* seeds dispersed by ants. American Journal of Botany, 67, 710–714.
Culver, D.C. & Beattie, A.J. (1983) Effects of ant mounds on soil chemistry and vegetation patterns in a Colorado montane meadow. Ecology, 64, 485–492.
Cummins, K.V. & Wuycheck, J.C. (1971) Caloric equivalents for investigations in ecological energetics. Internationale Vereinigung fur Theoretische und Angewandte Limnologie, 18, 1–158.
Dadd, R.H. (1985) Nutrition: Organisms. In Comprehensive Insect Physiology, Biochemistry, and Pharmacology (eds G.A. Kerkut & L.I. Gilbert), Vol. 4, pp. 313–390. Permagon Press, Oxford.
Darling, C.D. (1988) Comparative morphology of the labrum in Hymenoptera: The digitate labrum of Perilampidae and Eucharitidae (Chalcidoidea). Canadian Journal of Zoology, 66, 2811–2835.
Davidson, D.W. (1977) Foraging ecology and community organization in desert seed-eating ants. Ecology, 58, 725–737.
Davidson, D.W. (1993) The effects of herbivory and granivory on terrestrial plant succession. Oikos, 68, 23–35.
Davidson, D.W. & Morton, S.R. (1981) Competition for dispersal in ant-dispersed plants. Science, 213, 1259–1261.
Davidson, D.W., Samson, D.A., & Inouye, R.S. (1985) Granivory in the Chihuahuan desert: Interactions within and between trophic levels. Ecology, 66, 486–502.
Davidson, E.W., Segura, B.J., Steele, T., & Hendrix, D.L. (1994) Microorganisms influence the composition of honeydew produced by the silverleaf whitefly, *Bemisia argentifolii*. Journal of Insect Physiology, 40, 1069–1076.
Davidson, W.M. (1921) Observations on *Psyllobora taedata* LeConte, a coccinellid attacking mildews. Entomological News, 32, 83–89.
Davies, M. (1959) A contribution to the ecology of species of *Notiophilus* and allied genera (Col., Carabidae). Entomologist's Monthly Magazine, 95, 25–28.
Davies, M.J. (1953) The contents of the crops of some British carabid beetles. Entomologist's Monthly Magazine, 89, 18–23.
Davis, A.R., Shuel, R.W., & Peterson, R.L. (1988) Distribution of carbofuran and dimethoate in flowers and their secretion in nectar as related to nectary vascular supply. Canadian Journal of Botany, 66, 1248–1255.
Davis, A.S. & Liebman, M. (2003) Cropping system effects on giant foxtail (*Setaria faberi*) demography: I. Green manure and tillage timing. Weed Science, 51, 919–929.
Dawson, N. (1965) A comparative study of the ecology of eight species of Fenland Carabidae (Coleoptera). Journal of Animal Ecology, 34, 299–314.
de Charmoy, D.E. (1917) An attempt to introduce scoliid wasps from Madagascar to Mauritius. Bulletin of Entomological Research, 13, 245–254.
De Clercq, P., Bonte, M., Van Speybroeck, K., Bolckmans, K., & Deforce, K. (2006) Development and reproduction of *Adalia bipunctulata* (Coleoptera: Coccinellidae) on eggs of *Ephestia kuehniella* (Lepidoptera: Phycitidae) and pollen. Pest Management Science, 61, 1129–1132.
de Costa Lima, A. (1952) Insetos do Brasil. In 7 tomo Coleopteros, Vol. I, pp. 1–372. Escala Nacional de Agronomia, Sao Paulo.
de la Poza, M., Pons, X., Farinos, G.P., Lopez, C., Ortego, F., Eizaguirre, M., Castanera, P., & Albajes, R. (2005) Impact of farm-scale Bt maize on abundance of predatory arthropods in Spain. Crop Protection, 24, 677–684.
De Lima, J.O.G. & Leigh, T.F. (1984) Effect of cotton genotypes on the western bigeyed bug (Heteroptera: Miridae). Journal of Economic Entomology, 77, 898–902.

De Luca, V. & St. Pierre, B. (2000) The cell and developmental biology of alkaloid biosynthesis. Trends in Plant Science, 5, 1360–1385.
De Tullio, M.C. & Arrigoni, O. (2003) The ascorbic acid system in seeds: To protect and serve. Seed Science Research, 13, 249–260.
Dean, G.J. & Satasook, C. (1983) Response of Chrysoperla carnea (Stephens) (Neuroptera: Chrysopidae) to some potential attractants. Bulletin of Entomological Research, 73, 619–624.
DeFoliart, G. (1992) Insects as human food. Crop Protection, 11, 395–399.
Deinzer, M.L., Thomson, P.A., Burgett, D.M., & Isaacson, D.L. (1977) Pyrrolizidine alkaloids: Their occurrence in honey from tansy ragwort (*Senecio jacobaea* L.). Science, 195, 497–499.
Del-Claro, K. & Oliveira, P.S. (1993) Ant-homoptera interaction: Do alternative sugar sources distract tending ants? Oikos, 68, 202–206.
DellaPenna, D. & Pogson, B. (2006) Vitamin synthesis in plants: Tocopherols and carotenoids. Annual Review of Plant Biology, 57, 711–738.
DeLoach, C.J. & Peters, J.C. (1972) Effect of strip-planting vs.solid-planting on predators of cotton insects in southeastern Missouri, 1969. Environmental Entomology, 1, 94–102.
Dennis, P., Wratten, S.D., & Sotherton, N.W. (1991) Mycophagy as a factor limiting predation of aphids (Hemiptera: Aphididae) by staphylinid beetles (Coleoptera: Staphylinidae) in cereals. Bulletin of Entomological Research, 81, 25–31.
Dennis, P.A. & Sotherton, N.W. (1994) Behavioural aspects of staphylinid beetles that limit their aphid feeding potential in cereal crops. Pedobiologia, 38, 222–237.
Derenne, E. (1957) Catalogue des Coleopteres de Belgique. Fasc. I. Carabiodea: Carabidae. Societe Royale Entomologique de Belgique, Bruxelles.
DeVries, P.J. & Baker, I. (1989) Butterfly exploitation of an ant-plant mutualism: Adding insult to herbivory. Journal of the New York Entomological Society, 97, 332–340.
Deyrup, M.A. (1988) Pollen-feeding by Poecilognathus punctipennis (Diptera: Bombyliidae). Florida Entomologist, 71, 597–605.
Diaz, M. (1991) Spatial patterns of granivorous ant nest abundance and nest site selection in agricultural landscapes of central Spain. Insectes Sociaux, 38, 351–363.
Diaz, M. (1992) Spatial and temporal patterns of granivorous ant seed predation in patchy cereal crop areas of central Spain. Oecologia, 91, 561–568.
Dick, J. & Johnson, N.E. (1958) Carabid beetles damage Douglas-fir seed. Journal of Economic Entomology, 51, 542–544.
Dicke, F.F. & Jarvis, J.L. (1962) The habits and seasonal abundance of *Orius insidiosus* (Say) (Hemiptera: Heteroptera: Anthocoridae) on corn. Journal of Kansas Entomological Society, 35, 339–344.
Dicke, M. (1988a). Microbial allelochemicals affecting the behavior of insects, mites, nematodes, and protozoa in different trophic levels. In Novel Aspects of Insect-Plant Interactions (eds P. Barbosa & D.K. Letourneau), pp. 125–163. Wiley, New York.
Dicke, M. (1988b) Prey preference of the phytoseiid mite *Typhlodromus pyri* 1. Response to volatile kairomones. Experimental and Applied Acarology, 4, 1–13.
Dicke, M., Sabelis, M.W., & Groeneveld, A. (1986) Vitamin A deficiency modifies response of predatory mite *Amblyseius potentillae* to volatile kairomone of two-spotted spider mite, *Tetranychus urticae*. Journal of Chemical Ecology, 12, 1389–1396.
Dicke, M., Sabelis, M.W., Takabayashi, J., Bruin, J., & Posthumus, M.A. (1990) Plant strategies of manipulating predator/prey interactions through allelochemicals: Prospects for application in pest control. Journal of Chemical Ecology, 16, 3091–3118.
Ding, H., Lamb, R.J., & Ames, N. (2000) Inducible production of phenolic acids in wheat and antibiotic resistance to *Sitodiplosis mosellana*. Journal of Chemical Ecology, 26, 969–985.
Dively, G.P., Follett, P.A., Linduska, J.J., & Roderick, G.K. (1998) Use of imidacloprid-treated row mixtures for Colorado potato beetle (Coleoptera: Chrysomelidae) management. Journal of Economic Entomology, 91, 376–387.
Do Carmo-Sousa, L. (1969) Distribution of yeasts in nature. In The Yeasts (eds A.H. Rose & J.S. Harrison), Vol. I, pp. 79–105. Academic Press, London.

Dogan, E.B., Berry, R.E., Reed, G.L., & Rossignol, P.A. (1996) Biological parameters of convergent lady beetle (Coleoptera: Coccinellidae) feeding on aphids (Homoptera: Aphididae) on transgenic potato. Journal of Economic Entomology, 89, 1105–1108.

Dominguez, C.A., Dirzo, R., & Bullock, S.H. (1989) On the function of floral nectar in *Croton suberosus* (Euphorbiaceae). Oikos, 56, 109–114.

Doring, T.F. & Kromp, B. (2003) Which carabid beetles benefit from organic agriculture? – a review of comparative studies in winter cereals from Germany and Switzerland. Agriculture, Ecosystems and Environment, 98, 153–161.

Dosse, G. (1961) Uber die bedeutung der pollennahrung fur *Typhlodromus* (*T.*) *pyri* Scheuten (= *Tiliae* Oud.) (Acari, Phytoseiidae). Entomologia Experimentalis et Applicata, 4, 191–195.

Douglas, A.E. (1998) Nutritional interactions in insect-microbial symbioses: Aphids and their symbiotic bacteria *Buchnera*. Annual Review of Entomology, 43, 17–37.

Down, R.E., Ford, L., Woodhouse, S.D., Raemaekers, R.J.M., Leitch, B., Gatehouse, J.A., & Gatehouse, A.M.R. (2000) Snowdrop lectin (GNA) has no acute toxic effects on a beneficial insect predator, the 2-spot ladybeetle (*Adalia bipunctata* L.). Journal of Insect Physiology, 46, 379–391.

Downes, J.A. (1974) Sugar feeding by the larva of *Chrysopa* (Neuroptera). Canadian Entomologist, 106, 121–125.

Downes, W.L., Jr. & Dahlem, G.A. (1987) Keys to the evolution of Diptera: Role of Homoptera. Environmental Entomology, 16, 847–854.

Dreyer, B.S., Neuenschwander, P., Baumgartner, J., & Dorn, S. (1997) Trophic influences on survival, development and reproduction of *Hyperaspis notata* (Col., Coccinellidae). Journal of Applied Entomology, 121, 249–256.

Duan, J.J., Head, G., McKee, M.J., Nickson, T.E., Martin, J.H., & Sayegh, F.S. (2002) Evaluation of dietary effects of transgenic corn pollen expressing Cry3Bb1 protein on a non-target ladybird beetle, *Coleomegilla maculata*. Entomologia Experimentalis et Applicata, 104, 271–280.

Duan, J.J., Head, G., Jensen, A., & Reed, G. (2004) Effects of Bacillus thuringiensis potato and conventional insecticides for Colorado potato beetle (Coleoptera: Chrysomelidae) management on the abundance of ground-dwelling arthropods in Oregon potato ecosystems. Environmental Entomology, 33, 275–281.

Duan, J.J., Paradise, M.S., Lundgren, J.G., Bookout, J., Jiang, C., & Wiedenmann, R.N. (2005) Assessing non-target impacts of Bt corn resistant to corn rootworms: Tier-1 testing with larvae of *Poecilus chalcites* (Coleoptera: Carabidae). Environmental Entomology, 35, 135–142.

Duelli, P. (1987) The influence of food on the oviposition-site selection in a predatory and a honeydew-feeding lacewing species (Planipennia: Chrysopidae). Neuroptera International, 4, 205–210.

Dunkel, F.V. (1996) Nutritional value of various insects. In The Food Insects Newsletter, Vol. 9. www. food-insects.com accessed 8/2/08.

Durkee, L.T. (1983) The ultrastructure of floral and extrafloral nectaries. In The Biology of Nectaries (eds B.L. Bentley & T.S. Elias), pp. 1–29. Columbia University Press, New York.

Duso, C. & Camporese, P. (1991) Developmental times and oviposition rates of predatory mites *Typhlodromus pyri* and *Amblyseius andersoni* (Acari: Phytoseiidae) reared on different foods. Experimental and Applied Acarology, 13, 117–128.

Duso, C., Pozzebon, A., Capuzzo, C., Bisol, P.M., & Otto, S. (2003) Grape downy mildew spread and mite seasonal abundance in vineyards: Evidence for the predatory mites *Amblyseius andersoni* and *Typhlodromus pyri*. Biological Control, 27, 229–241.

Duso, C., Pozzebon, A., Capuzzo, C., Malagnini, V., Otto, S., & Borgo, M. (2005) Grape downy mildew spread and mite seasonal abundance in vineyards: Effects on *Tydeus caudatus* and its predators. Biological Control, 32, 143–154.

Dutton, A., Klein, H., Romeis, J., & Bigler, F. (2002) Uptake of Bt-toxin by herbivores feeding on transgenic maize and consequences for the predator *Chrysoperla carnea*. Ecological Entomology, 27, 441–447.

Earle, F.R. & Jones, Q. (1962) Analysis of seed samples from 113 plant families. Economic Botany, 16, 221–250.

Ecker, J.E. (1933) Buckeye poisoning of the honeybe – a progress report. Journal of Economic Entomology, 26, 181–187.

Edmunds, M. (1978) On the association between *Myrmarachne* spp. (Salticidae) and ants. Bulletin of the British Arachnological Society, 4, 149–160.

Ehlers, B.K. & Olesen, J.M. (1997) The fruit-wasp route to toxic nectar in *Epipactis* orchids? Flora, 192, 223–229.

Eijs, I.E., Ellers, J., & van Duinen, G.-J. (1998) Feeding strategies in drosophilid parasitoids: The impact of natural food resources on energy reserves in females. Ecological Entomology, 23, 133–138.

Eisikowitch, D., Kevan, P.G., & Lachance, M.-A. (1990a) The nectar-inhabiting yeasts and their effect on pollen germination in common milkweed, *Asclepias syriaca* L. Israel Journal of Botany, 39, 217–225.

Eisikowitch, D., Lachance, M.-A., Kevan, P.G., Willis, S., & Collins-Thompson, D.L. (1990b) The effect of the natural assemblage of microorganisms and selected strains of the yeast *Metschnikowia reukaufii* in controlling the germination of pollen of the common milkweed *Asclepias syriaca*. Canadian Journal of Botany, 68, 1163–1165.

Eisner, T. (1957) A comparative morphological study of the proventriculus of ants (Hymenoptera: Formicidae). Bulletin of the Museum of Comparative Zoology at Harvard University, 116, 439–490.

Eisner, T. & Happ, G.M. (1962) The infrabuccal pocket of a formicine ant: A social filtration device. Psyche, 69, 107–116.

Elbadry, E.A. (1968) Biological studies on *Amblyseius aleyrodis* a predator of the cotton whitefly (Acarina: Phytoseiidae). Entomophaga, 13, 323–329.

Elbadry, E.A. & Elbenhawy, E.M. (1968a) The effect of non-prey food, mainly pollen, on the development, survival, and fecundity of *Amblyseius gossipi* (Acarina: Phytoseiidae). Entomologia Experimentalis et Applicata, 11, 269–272.

Elbadry, E.A. & Elbenhawy, E.M. (1968b) The effects of pollen feeding on the predatory efficiency of *Amblyseius gossipi* (Acarina: Phytoseiidae). Entomologia Experimentalis et Applicata, 11, 273–276.

Elden, T.C. & McCaslin, M. (1997) Potato leafhopper (Homoptera: Cicadellidae) resistance in perennial glandular-haried alfalfa clones. Journal of Economic Entomology, 90, 842–847.

Elias, T.S. (1983) Extrafloral nectaries: Their structure and distribution. In The Biology of Nectaries (eds B.L. Bentley & T.S. Elias), pp. 174–203. Columbia University Press, New York.

Elias, T.S. & Gelband, H. (1975) Nectar: Its production and functions in trumpet creeper. Science, 189, 289–291.

Elliott, N.C., Simmons, G.A., & Sapio, F.J. (1987) Honeydew and wildflowers as food for the parasites *Glypta fumiferanae* (Hymenoptera: Ichneumonidae) and *Apantales fumiferanae* (Hymenoptera: Braconidae). Journal of the Kansas Entomological Society, 60, 25–29.

Elzen, G.W., Williams, H.J., & Vinson, S.B. (1986) Wind tunnel flight responses by hymenopterous parasitoid *Campoletis sonorensis* to cotton cultivars and lines. Entomologia Experimentalis et Applicata, 42, 285–289.

England, S. & Evans, E.W. (1997) Effects of pea aphid (Homoptera: Aphididae) honeydew on longevity and fecundity of the alfalfa weevil (Coleoptera: Curculionidae) parasitoid *Bathyplectes curculionis* (Hymenoptera: Ichneumonidae). Environmental Entomology, 26, 1437–1441.

English-Loeb, G., Norton, A.P., Gadoury, D., Seem, R.C., & Wilcox, W.F. (1999) Control of powdery mildew in wild and cultivated grapes by a tydeid mite. Biological Control, 14, 97–103.

Erhardt, A. & Baker, I. (1990) Pollen amino acids – an additional diet for a nectar feeding butterfly? Plant Systematics and Evolution, 169, 111–121.

Erwin, T.L. & Erwin, L.J.M. (1976) Relationships of predaceous beetles to tropical forest wood decay. Part II. The natural history of neotropical Eurycoleus macularis Chevrolat (Carabidae: Lebiini) and its implications in the evolution of ectoparasitoidism. Biotropica, 8, 215–224.

Esau, K. (1977) The Anatomy of Seeds, 2nd edn. Wiley, New York.

Ettershank, G. (1966) A generic revision of the world Myrmicinae related to *Solenopsis* and *Pheidologeton* (Hymenoptera: Formicidae). Australian Journal of Zoology, 14, 73–171.

Eubanks, M.D. & Denno, R.F. (1999) The ecological consequences of variation in plants and prey for an omnivorous insect. Ecology, 80, 1253–1266.

Eubanks, M.D. & Denno, R.F. (2000) Host plants mediate omnivore-herbivore interactions and influence prey suppression. Ecology, 81, 936–947.

Eubanks, M.D. & Styrsky, J.D. (2005) Effects of plant feeding on the performance of omnivorous "predators". In Plant Provided Foods for Carnivorous Insects (eds F.L. Wäckers, P.C.J. van Rijn, & J. Bruin), pp. 148–177. Cambridge University Press, Cambridge.

Evans, E.W. (2000) Egg production in response to combined alternative foods by the predator *Coccinella transversalis*. Entomologia Experimentalis et Applicata, 94, 141–147.

Evans, E.W. & Dixon, A.F.G. (1986) Cues for oviposition by ladybird beetles (Coccinellidae): Response to aphids. Journal of Animal Ecology, 55, 1027–1034.

Evans, E.W. & England, S. (1996) Indirect interactions in biological control of insects: Pests and natural enemies in alfalfa. Ecological Applications, 6, 920–930.

Evans, E.W. & Richards, D.R. (1997) Managing the dispersal of ladybird beetles (Coleoptera: Coccinellidae): Use of artificial honeydew to manipulate spatial distributions. Entomophaga, 42, 93–102.

Evans, E.W. & Swallow, J.G. (1993) Numerical responses of natural enemies to artificial honeydew in Utah alfalfa. Environmental Entomology, 22, 1392–1401.

Evans, M.E.G. (1965) A comparative account of the feeding methods of the beetles *Nebria brevicollis* (F.) (Carabidae) and *Philonthus decorus* (Grav.) (Staphylinidae). Transactions of the Royal Society of Edinburgh, 66, 91–109.

Eveleigh, E.S. & Chant, D.A. (1982) Expeimental studies on acarine predator-prey interactions: The effects of predator density on immature survival, adult fecundity and emigration rates, and the numerical response to prey density (Acarina: Phytoseiidae). Canadian Journal of Zoology, 60, 630–638.

Ewart, W.H. & Metcalf, R.L. (1956) Preliminary studies of sugars and amino acids in the honeydew of five species of coccids feeding on citrus in California. Annals of the Entomological Society of America, 49, 441–447.

Ewert, M.A. & Chiang, H.C. (1966) Dispersal of three species of coccinellids in corn fields. Canadian Entomologist, 98, 999–1003.

Ewing, H.E. (1913) Notes on Oregon Coccinellidae. Journal of Economic Entomology, 6, 404–407.

Fadamiro, H.Y. & Chen, L. (2005) Utilization of aphid honeydew and floral nectar by *Pseudacteon tricuspis* (Diptera: Phoridae), a parasitoid of imported fire ants, *Solenopsis* spp. (Hymenoptera: Formicidae). Biological Control, 34, 73–82.

Faegri, K. & van der Pijl, L. (1966) The Principles of Pollination Ecology. Pergamon Press, London.

Faeth, S.H. (2002) Are endophytic fungi defensive plant mutualists? Oikos, 98, 25–36.

Fahn, A. (1949) Studies in the ecology of nectar secretion. Palestine Journal of Botany, Jerusalem Series 4, 207–224.

Fahn, A. (1988) Tansley review no 14 secretory tissues in vascular plants. New Phytologist, 108, 229–257.

Fahn, A. (2000) Structure and function of secretory cells. In Plant Trichomes (eds D.L. Hallahan, J.C. Gray, & J.A. Callow), Vol. 31, pp. 37–66. Academic Press, London.

Falade, M.S., Owoyomi, O., Harwood, C.E., & Adewusi, S.R.A. (2005) Chemical composition and starch hydrolysis of *Acacia colei* and *Acacia tumida* seeds. Cereal Chemistry, 82, 479–484.

Faria, C.A., Wäckers, F.L., Pritchard, J., Barrett, D.A., & Turlings, T.C.J. (2007) High susceptibility of Bt maize to aphids enhances the performance of parasitoids of lepidopteran pests. PLoS One, 2(7), e600.

Fassati, M. (1957) Revision der Arten *Amara tricuspidata* Dej. und *Amara pseudostrenua* Kult (Coleoptera: Carabidae). Acta Faunistica Entomologica Musei Nationalis Pragae 2, 35–50.

Fast, P.G. (1964) Insect lipids: A review. Memoirs of the Entomological Society of Canada, 37, 1–50.

Fauvel, G. (1974) Sur l'alimentation pollinique d'un anthocoride predateur *Orius* (*Heterorius*) *vicinus* Rib. (Hemiptere). Annales de Zoologie Ecologie Animale, 6, 245–258.

Fawki, S. & Toft, S. (2005) Food preferences and the value of animal food for the carabid beetle *Amara similata* (Gyll.) (Col., Carabidae). Journal of Applied Entomology, 129(9/10), 551–556.

Fearing, P.L., Brown, D., Vlachos, D., Meghji, M., & Privalle, L. (1997) Quantitative analysis of CryIA(b) expression in Bt maize plants, tissues, and silage and stability of expression over successive generations. Molecular Breeding, 3, 167–176.

Feese, H. & Harlan, G. (1975) Planting time applications of systemic insecticides on grain sorghum for greenbug control: Interactions with herbicides and effect on predators (Hemiptera (Homoptera): Aphididae). Journal of the Kansas Entomological Society, 48, 396–402.

Feinsinger, P. & Swarm, L.A. (1978) How common are ant-repellent nectars? Biotropica, 10, 238–239.

Feldhaar, H., Straka, J., Kirschke, M., Berthold, K., Stoll, S., Mueller, M.J., & Gross, R. (2007) Nutritional upgrading for omnivorous carpenter ants by the endosymbiont *Blochmannia*. BMC Biology, 5(48).

Fenner, M. (1983) Relationships between seed weight, ash content and seedling growth in twenty-four species of Compositae. New Phytologist, 95, 697–706.

Fenner, M. (1992) Environmental influences on seed size and composition. Horticultural Reviews, 13, 183–213.

Fenner, M. (2004) Seed size and chemical composition: The allocation of minerals to seeds and their use in early seedling growth. Botanical Journal of Scotland, 56, 163–173.

Fenner, M. & Thompson, K. (2005) The Ecology of Seeds. Cambridge University Press, Cambridge.

Ferragut, F., Garcia-Mari, F., Costa-Comelles, J., & Laborda, R. (1987) Influence of food and temperature on development and oviposition of *Euseius stipulatus* and *Typhlodromus phialatus* (Acari: Phytoseiidae). Experimental and Applied Acarology, 3, 317–329.

Ferry, N., Mulligan, E.A., Majerus, M.E.N., & Gatehouse, A.M.R. (2007) Bitrophic and tritrophic effects of Bt Cry3A transgenic potato on beneficial, non-target, beetles. Transgenic Research, 16, 795–812.

Fiedler, A.K. & Landis, D.A. (2007a) Attractiveness of Michigan native plants to arthropod natural enemies and herbivores. Environmental Entomology, 36, 751–765.

Fiedler, A.K. & Landis, D.A. (2007b) Plant characteristics associated with natural enemy abundance at Michigan native plants. Environmental Entomology, 36(4), 878–886.

Finke, M.D., DeFoliart, G.D., & Benevenga, N.J. (1989) Use of a four-parameter logistic model to evaluate the quality of the protein from three insect species when fed to rats. Journal of Nutrition, 119, 864–871.

Finlayson, L.R. & Finlayson, T. (1957) Influence of adult food on viability of early stages of Aptesis basizonia (Grav.) (Hymenoptera: Ichneumonidae), a parastie of pine sawflies (Diprionidae). Canadian Entomologist, 89, 507–509.

Finney, G.L. (1948) Culturing *Chrysopa californica* and obtaining eggs for field distribution. Journal of Economic Entomology, 41, 719–721.

Firbank, L.G. & Watkinson, A.R. (1986) Modelling the population dynamics of an arable weed and its effects upon crop yield. Journal of Applied Ecology, 23, 147–159.

Fischer, M.K. & Shingleton, A.W. (2001) Host plant and ants influence the honeydew sugar composition of aphids. Functional Ecology, 15, 544–550.

Flaherty, D., Lynn, C., Jensen, F., & Hoy, M. (1971) Influence of environment and cultural practices on spider mite abundance in southern San Joaquin Thompson seedless vineyards. California Agriculture, 25, 6–8.

Flechtmann, C.H.W. & McMurtry, J.A. (1992a) Studies of cheliceral and deutosternal morphology of some Phytoseiidae (Acari: Mesostigmata) by scanning electron microscopy. International Journal of Acarology, 18, 163–169.

Flechtmann, C.H.W. & McMurtry, J.A. (1992b) Studies on how phytoseiid mites feed on spider mites and pollen. International Journal of Acarology, 18, 157–162.

Fleschner, C.A. (1950) Studies on searching capacity of the larvae of three predators of the citrus red mite. Hilgardia, 20, 233–265.

Foissac, X., Loc, N.T., Christou, P., Gatehouse, A.M.R., & Gatehouse, J.A. (2000) Resistance to green leafhopper (*Nephotettix virescens*) and brown plant hopper (*Nilaparvata lugens*) in transgenic rice expressing snowdrop lectin (*Galanthus nivalis* agglutinin; GNA). Journal of Insect Physiology, 46, 573–583.

Forbes, S.A. (1881) Notes on insectivorous Coleoptera. Bulletin of the Illinois State Laboratory of Natural History, 1, 153–160.

Forbes, S.A. (1883) The food relations of the Carabidae and Coccinellidae. Bulletin of the Illinois State Laboratory of Natural History, 3, 33–64.

Forbes, S.A. (1892) Bacteria normal to digestive organs of Hemiptera. Bulletin of the Illinois State Laboratory of Natural History, IV, 1–7.

Forcella, F., Wilson, R.G., Renner, K.A., Dekker, J., Harvey, R.G., Alm, D.A., Buhler, D.D., & Cardina, J. (1992) Weed seedbanks of the U. S. corn belt: Magnitude, variation, emergence, and application. Weed Science, 40, 636–644.

Forcella, F., Peterson, D.H., & Barbour, J.C. (1996) Timing and measurement of weed seed shed in corn (*Zea mays*). Weed Technology, 10, 535–543.

Forrest, J.M.S. & Knights, B.A. (1972) Presence of phytosterols in the food of the aphid, *Myzus persicae*. Journal of Insect Physiology, 18, 723–728.

Forsse, E., Smith, S.M., & Bourchier, R.S. (1992) Flight initiation in the egg parasitoid *Trichogramma minutum*: Effects of ambient temperature, mates, food, and host eggs. Entomologia Experimentalis et Applicata, 62, 147–154.

Forsythe, T.G. (1982a) Feeding mechanisms of certain ground beetles (Coleoptera: Carabidae). Coleopterist's Bulletin, 36, 26–73.

Forsythe, T.G. (1982b) Qualitative analyses of certain enzymes of the oral defence fluids of *Pterostichus madidus* (F.), including a list of other carabid beetles which produce oral defence fluids. Entomologist's Monthly Magazine, 118, 1–5.

Forsythe, T.G. (1983) Mouthparts and feeding of certain ground beetles (Coleoptera: Carabidae). Zoological Journal of the Linnean Society, 79, 319–376.

Foster, M.A. & Ruesink, W.G. (1984) Influence of flowering weeds associated with reduced tillage in corn on a black cutworm (Lepidoptera: Noctuidae) parasitoid, *Meteorus rubens* (Nees von Esenbeck). Environmental Entomology, 13, 664–668.

Fouly, A.H., Abou-Setta, M.M., & Childers, C.C. (1995) Effects of diet on the biology and life tables of *Typhlodromalus peregrinus* (Acari: Phytoseiidae). Environmental Entomology, 24, 870–874.

Fowden, L. (1964) The chemistry and metabolism of recently isolated amino acids. Annual Review of Biochemistry, 33, 173–204.

Franceschi, V.R. & Nakata, P.A. (2005) Calcium oxalate in plants: Formation and function. Annual Review of Plant Biology, 56, 41–71.

Franklin, R.T. & Holdaway, F.G. (1966) A relationship of the plant to parasitism of European corn borer by the tachinid parasite *Lydella grisescens*. Journal of Economic Entomology, 59, 440–441.

Freitag, R. (1969) A revision of the species of the genus *Evarthrus* Leconte. Quaestiones Entomologicae, 5, 89–212.

French, C.E., Liscinsky, S.A., & Miller, D.R. (1957) Nutrient composition of earthworms. Journal of Wildlife Management, 21, 348.

Fritzsche Hoballah, M.E., Tamo, C., & Turlings, T.C.J. (2002) Differential attractiveness of induced odors emitted by eight maize varieties for the parasitoid *Cotesia marginiventris*: Is quality or quantity important? Journal of Chemical Ecology, 28, 951–968.

Froud-Williams, R.J., Chancellor, R.J., & Drennan, D.S.H. (1984) The effects of seed burial and soil disturbance on emergence and survival of arable weeds in relation to minimal cultivation. Journal of Applied Ecology, 21, 629–641.

Fuchsberg, J.R., Yong, T.-H., Losey, J.E., Carter, M.E., & Hoffman, M.P. (2007) Evaluation of corn leaf aphid (*Rhopalosiphum maidis*; Homoptera: Aphididae) honeydew as a food source for the egg parasitoid *Trichogramma ostriniae* (Hymenoptera: Trichogrammatidae). Biological Control, 40, 230–236.

Fuller, P.J. & Hay, M.E. (1983) Is glue production by seeds of *Salvia columbariae* a deterrent to desert granivores? Ecology, 64, 960–963.

Funau, T. & Yoshiyasu, Y. (1995) Development and fecundity of *Orius sauteri* (Hemiptera: Anthocoridae) reared on *Aphis gossypii* Glover and corn pollen. Japanese Journal of Applied Entomology and Zoology, 39, 84–85.

Gallandt, E.R., Molloy, T., Lynch, R.P., & Drummond, F.A. (2005) Effect of cover-cropping systems on invertebrate seed predation. Weed Science, 53, 69–76.

Gallego, V.C., Baltazar, C.R., Cadapan, E.P., & Abad, R.G. (1983) Some ecological studies on the coconut leafminer, *Promecotheca cumingii* Baly (Coleoptera: Hispidae) and its hymenopterous parasitoids in the Phillipines. Phillipine Entomologist, 6, 417–494.

Galletto, L. & Bernardello, G. (2004) Floral nectaries, nectar production dynamics and chemical composition in six Ipomoea species (Convolvulaceae) in relation to pollinators. Annals of Botany, 94, 269–280.

Gardener, M.C. & Gillman, M.P. (2001a) Analyzing variability in nectar amino acids: Composition is less variable than concentration. Journal of Chemical Ecology, 27, 2545–2558.

Gardener, M.C. & Gillman, M.P. (2001b) The effects of soil fertilizer on amino acids in the floral nectar of corncockle, Agrostemma githago (Caryophyllaceae). Oikos, 92, 101–106.

Gardener, M.C. & Gillman, M.P. (2002) The taste of nectar – a neglected area of pollination ecology. Oikos, 98, 552–557.

Gardner, T.R. (1938) Influence of feeding habits of *Tiphia vernalis* on the parasitization of the Japanese beetle. Journal of Economic Entomology, 31, 204–207.

Gardner, T.R. (1940) Investigations of the Parasites of Popillia japonica and Related Scarabaeidae in the Far East from 1929 to 1933, Inclusive. United States Department of Agriculture Agricultural Research Service, Technical Bulletin No. 738, pp. 1–36.

Garraway, M.O. & Evans, R.C. (1984) Fungal Nutrition and Physiology. Wiley, New York.

Garrido, J.L., Rey, P.J., Cerda, X., & Herrera, C.M. (2002) Geographical variation in diaspore traits of an ant-dispersed plant (*Helleborus foetidus*): Are ant community composition and diaspore traits correlated? Journal of Ecology, 90, 446–455.

Gatehouse, A.M.R., Down, R.E., Powell, K.S., Sauvion, N., Rhabe, Y., Newell, C.A., Merryweather, A., Hamilton, W.D.O., & Gatehouse, J.A. (1996) Transgenic potato plants with enhanced resistance to the peach-potato aphid *Myzus persicae*. Entomologia Experimentalis et Applicata, 79, 295–307.

Gebrehiwot, L. & Beuselink, P.R. (2001) Seasonal variations in hydrogen cyanide concentration of three *Lotus* species. Agronomy Journal, 93, 603–608.

Gegear, R.J., Manson, J.S., & Thomson, J.D. (2007) Ecological contex influences pollinator deterrence by alkaloids in floral nectar. Ecology Letters, 10, 375–382.

Geiler, H. (1956/1957) Zum Ekologie und Phanologie der auf mitteldeutschen Feldern lebenden Carabiden. Wissenschaftliche Zeitschrift der Karl-Marx Universität 6, 35–61.

Geng, J.-H., Shen, Z.-R., Song, K., & Zheng, L. (2006) Effect of pollen of regular cotton and transgenic Bt + CpTI cotton on the survival and reproduction of the parasitoid wasp *Trichogramma chilonis* (Hymenoptera: Trichogrammatidae) in the laboratory. Environmental Entomology, 35, 1661–1668.

George, D.A. & Rincker, C.M. (1982) Residues of commercially used-insecticides in the environment of *Megachile rotundata*. Journal of Economic Entomology, 75, 319–323.

Gersdorf, E. (1937) Okolgisch-faunistische Untersuchungun über die Carabiden der mecklenburgischen Landschaft. Zool. Jb. Syst., 70, 17–86.

Geyer, J.W.C. (1947) A study of the biology and ecology of *Exochomus flavipes* Thunb. (Coccinellidae, Coleoptera). Journal of the Entomological Society of South Africa, 9, 219–234.

Gibson, C.M. & Hunter, M.S. (2005) Reconsideration of the role of yeasts associated with *Chrysoperla* green lacewings. Biological Control, 32, 57–64.

Gil, R., Silva, F.J., Zientz, E., Delmotte, F., Gonzalez-Candelas, F., Latorre, A., Rausell, C., Kamerbeek, J., Gadau, J., Holldobler, B., van Ham, R.C.H.J., Gross, R., & Moya, A. (2003) The genome sequence of *Blochmannia floridanus*: Comparative analysis of reduced genomes. Proceedings of the National Academy of Sciences, 100, 9388–9393.

Gilbert, F. & Jervis, M. (1998) Functional, evolutionary and ecological aspects of feeding-related mouthpart specializations in parasitoid flies. Biological Journal of the Linnean Society, 63, 495–535.

Gilbert, F.S. (1981) Foraging ecology of hoverflies: Morphology of the mouthparts in relation to feeding on nectar and pollen in some common urban species. Ecological Entomology, 6, 245–262.

Gilbert, F.S. (1985a) Diurnal activity patterns in hoverflies (Diptera, Syrphidae). Ecological Entomology, 10, 385–392.

Gilbert, F.S. (1985b) Ecomorphological relationships in hoverflies (Diptera, Syrphidae). Proceedings of the Royal Society of London, Series B, 224, 91–165.

Gilbert, F.S. (1986a). Ecomorphology of Syrphidae. In Ecology of Aphidophaga (ed I. Hodek), pp. 23–28. Dr. W. Junk, Dordrecht, The Netherlands.

Gilbert, F.S. (1986b) Hoverflies. Cambridge University Press, Cambridge.

Gilbert, L.E. (1972) Pollen feeding and reproductive biology of *Heliconius* butterflies. Proceedings of the National Academy of Sciences, 69, 1403–1407.

Giles, K.L., Madden, R.D., Stockland, R., Payton, M.E., & Dillwith, J.W. (2002) Host plants affect predator fitness via the nutritional value of herbivore prey: Investigation of a plant-aphid-ladybeetle system. BioControl, 47, 1–21.

Gillespie, D.R. & McGregor, R.R. (2000) The functions of plant feeding in the omnivorous predator *Dicyphus hesperus*: Water places limits on predation. Ecological Entomology, 25, 380–386.

Gilliam, M., Moffett, J.O., & Kauffeld, N.M. (1983) Examination of floral nectar of citrus, cotton, and Arizona desert plants for microbes. Apidologie, 14, 299–302.

Girling, R.D., Hassall, M., Turner, J.G., & Poppy, G.M. (2006) Behavioural responses of the aphid parasitoid *Diaeretiella rapae* to volatiles from Arabidopsis thaliana induced by *Myzus persicae*. Entomologia Experimentalis et Applicata, 120, 1–9.

Giroux, S., Coderre, D., Vincent, C., & Cote, J.-C. (1994) Effects of *Bacillus thuringiensis* var. *San Diego* on predation effectiveness, development and mortality of *Coleomegilla maculata legni* (Col.: Coccinellidae) larvae. Entomophaga, 39, 61–69.

Glancey, B.M., Vander Meer, R.K., Glover, A., Lofgren, C.S., & Vinson, S.B. (1981) Filtration of microparticles from liquids ingested by the red imported fire ant *Solenopsis invicta* Buren. Insectes Sociaux, 28, 395–401.

Glascow, H. (1914) The gastric caeca and the caecal bacteria of Heteroptera. Biological Bulletin, 26, 101–170.

Glynne Jones, G.D. & Thomas, W.D.E. (1953) Experiments on the possible contamination of honey with Schradan. Annals of Applied Biology, 40, 546–555.

Goidanich, A. (1947) Due Coccinelle igrofile e pollinivore sul riso (*Hippodama* 13-*punctata* L., *Anisosticta* 19-*punctata* L.). Risicoltura, 33, 145–156, 169–177.

Goldschmidt, H. & Toft, S. (1997) Variable degrees of granivory and phytophagy in insectivorous carabid beetles. Pedobiologia, 41, 521–525.

Golley, F.B. (1961) Energy values of ecological materials. Ecology, 42, 581–584.

Gomez, L.D. & Espadaler, X. (1998) Myrmecochorous dispersal distances: A world survey. Journal of Biogeography, 25, 573–580.

Gooding, R.H. & Huang, C.-T. (1969) Trypsin and chymotrypsin from the beetle *Pterostichus melanarius*. Journal of Insect Physiology, 15, 325–339.

Goolsbey, A. (1998) Poison pollen from death camas. American Bee Journal, 138, 459–460.

Gorb, S.N. & Gorb, E.V. (1999) Effects of ant species composition on seed removal in deciduous forest in eastern Europe. Oikos, 84, 110–118.

Gottsberger, G., Schrauwen, J., & Linskins, H.F. (1984) Amino acids and sugars in nectar, and their putative evolutionary significance. Plant Systematics and Evolution, 145, 55–77.

Gould, F. (1998) Sustainability of transgenic insecticidal cultivars: Integrating pest genetics and ecology. Annual Review of Entomology, 43, 701–726.

Gould, F., Kennedy, G.G., & Johnson, M.T. (1991) Effects of natural enemies on the rate of herbivore adaptation to resistant host plants. Entomologia Experimentalis et Applicata, 58, 1–14.

Goulson, D. & Wright, N.P. (1998) Flower constancy in the hoverflies *Episyrphus balteatus* (DeGeer) and *Syrphus ribesii* (L.) (Syrphidae). Behavioral Ecology, 9, 213–219.

Grafton-Cardwell, E.E., Ouyang, Y., & Bugg, R.L. (1999) Leguminous cover crops to enhance population development of *Euseius tularensis* (Acari: Phytoseiidae) in citrus. Biological Control, 16, 73–80.

Grandjean, F. (1959) Les stases du developpement ontogenetique chez *Balaustium florale* (acarien erythroide) deuxieme partie. Annales de la Societe Entomologique de France, 128, 159–177.

Gray, H.E. & Fraenkel, G. (1954) The carbohydrate components of honeydew. Physiological Zoology, 27, 56–65.

Gray, R.A. (1952) Composition of honeydew excreted by pineapple mealybugs. Science, 115, 129–133.

Gray, W.D. & Abou-el-Seoud, M.O. (1966) Fungal protein for food and feeds. III. Manioc as a potential crude raw material for tropical areas. Economic Botany, 20, 251–255.

Gray, W.D. & Karve, M.D. (1967) Fungal proteins for food and feeds. V. Rice as a source of carbohydrate for the production of fungal protein. Economic Botany, 21, 110–114.

Gray, W.D. & Staff, I.A. (1967) Fungal protein for food and feeds. VII. Caloric values of fungus mycelium. Economic Botany, 21, 341–344.

Grayum, G.H. (1985) Evolutionary and ecological significance of starch storage in pollen of the Araceae. American Journal of Botany, 72, 1565–1577.

Greenstone, M.H. (1979) Spider feeding behavior optimizes dietary essential amino acid composition. Nature, 282, 502–503.

Griffen, M.L. & Yeargan, K.V. (2002a) Factors potentially affecting oviposition site selection by the lady beetle *Coleomegilla maculata* (Coleoptera: Coccinellidae). Environmental Entomology, 31, 112–119.

Griffen, M.L. & Yeargan, K.V. (2002b) Oviposition site selection by the spotted ladybeetle *Coleomegilla maculata* (Coleoptera: Coccinellidae): Choices among plant species. Environmental Entomology, 31, 107–111.

Griffeths, J.T. & Swanton, C.J. (1999) Predispersal weed seed predation in corn (*Zea mays*). Abstracts of the Weed Society of America, 39, 76.

Griffin, D.H. (1994) Fungal Physiology, 2nd edn. Wiley-Liss, New York.

Grogan, D.E. & Hunt, J.H. (1979) Pollen proteases: Their potential role in insect digestion. Insect Biochemistry, 9, 309–313.

Gronwall, O. & Pehrson, A. (1984) Nutrient content in fungi as a primary food of the red squirrel *Sciurus vulgaris* L. Oecologia, 64, 230–231.

Groot, A.T. & Dicke, M. (2002) Insect-resistant transgenic plants in a multi-trophic context. The Plant Journal, 31, 387–406.

Grossi-de-Sa, M.F., Lucena, W., Souza, M.L., Nepomuceno, A.L., Osir, E.O., Amugune, N., Hoa, T.T.C., Hai, T.N.H., Somers, D.A., & Romano, E. (2006) Transgene expression and locus structure of Bt cotton. In Environmental Risk Assessment of Genetically Modified Organisms. Methodologies for Assessing Bt cotton in Brazil (eds A. Hilbeck, D.A. Andow, & E.M.G. Fontes), Vol. 2, pp. 93–107. CABI, Wallinford.

Grout, T.G. & Richards, R.I. (1990) The influence of windbreak species on citrus thrips (Thysanoptera: Thripidae) populations and their damage to South African citrus orchards. Journal of the Entomological Society of South Africa, 53, 151–157.

Grout, T.G. & Richards, R.I. (1992) The dietary effect of windbreak pollens on longevity and fecundity of a predaceous mite *Euseius addoensis addoensis* (Acari: Phtyoseiidae) found in citrus orchards in South Africa. Bulletin of Entomological Research, 82, 317–320.

Grusak, M.A. & DellaPenna, D. (1999) Improving the nutrient composition of plants to enhance human nutrition and health. Annual Review of Plant Physiology and Plant Molecular Biology, 50, 133–161.

Guerrant, E.O., Jr. & Fiedler, P.L. (1981) Flower defenses against nectar-pilferage by ants. Biotropica, 13, 25–33.

Guimaraes, P.R., Jr. & Cogni, R. (2002) Seed cleaning of *Cupania vernalis* (Sapindaceae) by ants: Edge effect in a highland forest in south-east Brazil. Journal of Tropical Ecology, 18, 303–307.

Gupta, G. & Krischik, V.A. (2007) Professional and consumer insecticides for management of adult Japanese beetle on hybrid tea rose. Journal of Economic Entomology, 100, 830–837.
Gurr, G.M. & Wratten, S.D. (1999) 'Integrated biological control': A proposal for enhancing success in biological control. International Journal of Pest Management, 45, 81–84.
Gurr, G.M., van Emden, H.F., & Wratten, S.D. (1998) Habitat manipulation and natural enemy efficiency: Implications for the control of pests. In Conservation Biological Control (ed P. Barbosa), pp. 155–185. Academic Press, London.
Gurr, G.M., Wratten, S.D., & Luna, J.M. (2003) Multi-function agricultural biodiversity: Pest management and other benefits. Basic and Applied Ecology, 4, 107–116.
Gurr, G.M., Wratten, S.D., Tylianakis, J.M., Kean, J., & Keller, M.A. (2005) Providing plant foods for natural enemies in farming systems: Balancing practicalities and theory. In Plant Provided Food for Carnivorous Insects: A Protective Mutualism and Practical Applications (eds F. Wäckers, P. van Rijn, & J. Bruin), pp. 341–347. Cambridge University Press, Cambridge.
Gutterman, Y. & Shem-Tov, S. (1997) The efficiency of the strategy of mucilaginous seeds of some common annuals of the Negev adhering to the soil crust to delay collection by ants. Israel Journal of Plant Sciences, 45, 317–327.
Györfi, J. (1945) Beobachtungen über die Ernahrung der Schlupwespenimagos. Erdészeti kisérletek, 45, 100–112.
Haas, F. & Konig, H. (1988) *Coriobacterium glomerans* gen. nov., sp. nov. from the intestinal tract of the red soldier bug. International Journal of Systematic Bacteriology, 38, 382–384.
Haber, W.A., Frankie, G.W., Baker, H.G., Baker, I., & Koptur, S. (1981) Ants like flower nectar. Biotropica, 13, 211–214.
Hackman, W. & Meinander, M. (1979) Diptera feeding as larvae on macrofungi in Finland. Annales Zoologici Fennici, 16, 50–83.
Hagen, K.S. (1962) Biology and ecology of predaceous Coccinellidae. Annual Review of Entomology, 7, 289–326.
Hagen, K.S. (1986) Ecosystem analysis: Plant cultivars (HPR), entomophagous species and food supplements. In Interactions of Plant Resistance and Parasitoids and Predators of Insects (eds D.J. Boethel & R.D. Eikenbary), pp. 151–197. Ellis Horwood, Chichester.
Hagen, K.S. (1987) Nutritional ecology of terrestrial insect predators. In Nutritional Ecology of Insects, Mites, Spiders, and Related Invertebrates (eds F.J. Slansky & J.G. Rodriguez), pp. 533–577. Wiley, New York.
Hagen, K.S. & Hale, R. (1974) Increasing natural enemies through use of supplementary feeding and non-target prey. In Summer Institute on Biological Control of Plant Insects and Diseases (eds F.G. Maxwell & F.A. Harris), pp. 170–181. The University Press of Mississippi, Jackson, MS.
Hagen, K.S. & Tassan, R.L. (1966) The influence of protein hydrolysates of yeasts and chemically defined diets upon the fecundity of *Chrysopa carnea* Stephens (Neuroptera). Vestnik Ceskoslovenske Spolecnosti Zoologicke, 30, 219–227.
Hagen, K.S. & Tassan, R.L. (1970) The influence of food wheast and related *Saccharomyces fragilis* yeast products on the fecundity of *Chrysopa carnea* (Neuroptera: Chrysopidae). Canadian Entomologist, 102, 806–811.
Hagen, K.S. & Tassan, R.L. (1972) Exploring nutritional roles of extracellular symbiotes on the reproduction of honeydew feeding adult chrysopids and tephritids. In Insect and Mite Nutrition (ed J.G. Rodriguez), pp. 323–351. North Holland, Amsterdam, The Netherlands.
Hagen, K.S., Tassan, R.L., & Sawall, E.F.J. (1970) Some ecophysiological relationships between certain *Chrysopa*, honeydews and yeasts. Bolletino dell'Istituto di Entomologia della Universita di Bologna, 28, 113–134.
Hagen, K.S., Sawall, E.F.J., & Tassan, R.L. (1971) The use of food sprays to increase effectiveness of entomophagous insects. In Proceedings of the Tall Timbers Conference on Ecological Animal Control by Habitat Management (ed E.V. Komarek, Sr.), Vol. 3, pp. 59–81. Tall Timbers Research Station, Tallahassee, FL.
Hagen, K.S., Greany, P., Sawall, E.F.J., & Tassan, R.L. (1976) Tryptophan in artificial honeydews as a source of an attractant for adult *Chrysopa carnea*. Environmental Entomology, 5, 458–468.
Hagler, J.R. & Buchmann, S.L. (1993) Honey bee (Hymenoptera: Apidae) foraging responses to phenolic-rich nectars. Journal of the Kansas Entomological Society, 66, 223–230.

Hagley, E.A.C. & Barber, D.R. (1992) Effect of food sources on the longevity and fecundity of *Pholetesor ornigis* (Weed) (Hymenoptera: Braconidae). Canadian Entomologist, 124, 341–346.
Hagley, E.A.C. & Simpson, C.M. (1981) Effect of food sprays on numbers of predators in an apple orchard. Canadian Entomologist, 113, 75–77.
Hagley, E.A.C., Holliday, N.J., & Barber, D.R. (1982) Laboratory studies of the food preferences of some orchard carabids (Coleoptera: Carabidae). Canadian Entomologist, 114, 431–437.
Hagvar, E.B. & Hofsvang, T. (1989) Effect of honeydew and hosts on plant colonization by the aphid parasitoid *Ephedrus cerasicola*. Entomophaga, 34, 495–501.
Halitschke, R., Keßler, A., Kahl, J., Lorenz, A., & Baldwin, I.T. (2000) Ecophysiological comparison of direct and indirect defenses in *Nicotiana attenuata*. Oecologia, 124, 408–417.
Halkier, B.A. & Gershenzon, J. (2006) Biology and biochemistry of glucosinolates. Annual Review of Plant Biology, 57, 303–333.
Hall, R.W. & Ehler, L.E. (1979) Rate of establishment of natural enemies in classical biological control. Entomological Society of America Bulletin, 25, 280–282.
Halm, M.-P., Rortais, A., Arnold, G., Tasei, J.N., & Rault, S. (2006) New risk assessment approach for systemic insecticides: The case of honeybees and imidacloprid (Gaucho). Environmental Science and Technology, 40, 2448–2454.
Hammond, P.M. & Lawrence, J.F. (1989) Mycophagy in insects: A summary. In Insect-Fungus Interactions (eds N. Wilding, N.M. Collins, P.M. Hammond, & J.F. Webber), pp. 275–324. Academic Press, London.
Han, B. & Chen, Z. (2002) Behavioral and electrophysiological responses of natural enemies to synomones from tea shoots and kairomones from tea aphids, *Toxoptera aurantii*. Journal of Chemical Ecology, 28, 2203–2219.
Handel, S.N. (1978) The competitive relationship of three woodland sedges and its bearing on the evolution of ant-dispersal of *Carex pedunculata*. Evolution, 32, 151–163.
Handel, S.N. & Beattie, A.J. (1990) Seed dispersal by ants. Scientific American, 263, 76–83.
Hanney, B.W. & Elmore, C.D. (1974) Amino acid composition of cotton nectar. Journal of Agricultural and Food Chemistry, 22, 476–478.
Hansen, S.R. (1978) Resource utilization and coexistence of three species of *Pogonomyrmex* ants in an upper Sonoran grassland community. Oecologia, 35, 109–117.
Hansen, V. & Larsson, S.G. (1968) Biller. XXIV. Sandspringere og Lobebiller (Cicindelidae og Carabidae). Danmarks Fauna, 76, 1–316.
Hanski, I. (1989) Fungivory: Fungi, insects, and ecology. In Insect-Fungus Interactions (eds N. Wilding, N.M. Collins, P.M. Hammond, & J.F. Webber), pp. 25–68. Academic Press, London.
Haq, S.K., Atif, S.M., & Khan, R.H. (2004) Protein proteinase inhibitor genes in combat against insects, pests, and pathogens: Natural and engineered phytoprotection. Archives of Biochemistry and Biophysics, 431, 145–159.
Haque, M.R. & Bradbury, J.H. (2002) Total cyanide determination of plants and foods using the picrate and acid hydrolysis methods. Food Chemistry, 77, 107–114.
Harbourne, J.B. (1979) Flavonoid pigments. In Herbivores: Their Interactions with Secondary Plant Metabolites (eds G.A. Rosenthal & D.H. Janzen), pp. 619–656. Academic Press, New York.
Harder, L.D. & Barrett, S.C.H. (1992) The energy cost of bee pollination for *Pontederia cordata* (Pontederiaceae). Functional Ecology, 6, 226–233.
Hare, J.D. (1992) Effects of plant variation on herbivore-natural enemy interactions. In Plant Resistance to Herbivores and Pathogens (eds R.S. Fritz & E.L. Simms), pp. 278–298. University of Chicago Press, Chicago, IL.
Harmon, J.P., Ives, A.R., Losey, J.E., Olson, A.C., & Rauwald, K.S. (2000) *Coleomegilla maculata* (Coleoptera: Coccinellidae) predation on pea aphids promoted by proximity to dandelions. Oecologia, 125, 543–548.
Harper, J.L. (1977) Population Biology of Plants. Academic Press, London.
Harper, J.L., Lovell, P.H., & Moore, K.G. (1970) The shapes and sizes of seeds. Annual Review of Ecology and Systematics, 1, 327–356.
Harris, D.L. (1969) An evaluation of pollen from corn and cotton as a diet for *Coleomegilla maculata* (DeGeer). University of Arkansas, Fayetteville.

Harrison, S.J. & McEwen, P.K. (1998) Acid hydrolysed L-tryptophan and its role in the attraction of the green lacewing *Chrysoperla carnea* (Stephens) (Neuroptera: Chrysopidae). Journal of Applied Entomology, 122, 343–344.

Harrison, S.K., Regnier, E.E., & Schmoll, J.T. (2003) Postdispersal predation of giant ragweed (*Ambrosia trifida*) seed in no-till corn. Weed Science, 51, 955–964.

Hartke, A., Drummond, F.A., & Liebman, M. (1998) Seed feeding, seed caching, and burrowing behaviors of *Harpalus rufipes* De Geer larvae (Coleoptera: Carabidae) in the Maine potato agroecosystem. Biological Control, 13, 91–100.

Hartzell, A. (1967) Insect endosymbiosis. In Symbiosis (ed S.M. Henry), Vol. II, pp. 142–206. Academic Press, New York.

Harwood, J.D., Wallin, H., & Obrycki, J.J. (2005) Uptake of Bt-endotoxins by non-target herbivores and higher order arthropod predators: Molecular evidence from a transgenic corn agroecosystem. Molecular Ecology, 14, 2815–2823.

Harwood, J.D., Samson, R.A., & Obrycki, J.J. (2007) Temporal detection of Cry1Ab-endotoxins in coccinellid predators from fields of *Bacillus thuringiensis* corn. Bulletin of Entomological Research, 97, 643–648.

Harwood, J.L. & Russell, N.J. (1984) Lipids in Plants and Microbes. George Allen & Unwin, London.

Haslett, J.R. (1983) A photographic account of pollen digestion by adult hoverflies. Physiological Entomology, 8, 167–171.

Haslett, J.R. (1989) Adult feeding by holometabolous insects: Pollen and nectar as complementary nutrient sources for *Rhingia campestris* (Diptera: Syrphidae). Oecologia, 81, 361–363.

Haslett, J.R. & Entwistle, P.F. (1980) Further notes on *Eriozona syrphoides* (Fall.) (Dipt., Syrphidae) in Hafren Forest, mid-Wales. Entomologist's Monthly Magazine, 116, 36.

Hassan, V.E. (1967) Untersuchungen uber die Bedeutung der Kraut- und Strauchschicht als Nahrungsquelle fur Imagines entomophager Hymenopteren. Journal of Applied Entomology, 60, 238–265.

Hatcher, P.E. & Melander, B. (2003) Combining physical, cultural and biological methods: Prospects for integrated non-chemical weed management strategies. Weed Research, 43, 303–322.

Hausammann, A. (1996) Strip-management in rape crop: Is winter rape endangered by negative impacts of sown weed strips? Journal of Applied Entomology, 120, 505–512.

Hausmann, C., Wäckers, F.L., & Dorn, S. (2005) Sugar convertability in the parasitoid *Cotesia glomerata* (Hymenoptera: Braconidae). Archives of Insect Biochemistry and Physiology, 60, 223–229.

Hawkeswood, T.J. & Turner, J.R. (2002) Observations on the adults of *Micraspis frenata* (Erichson 1942) (Coleoptera: Coccinellidae) feeding on the pollen of native and non-native grasses (Poaceae) in eastern South Wales, Australia. Journal of the Entomological Research Society, 4, 21–29.

Hed, B.E., Windham, M.T., & Grant, J.F. (1999) Survival of conidia of *Discula destructiva* in frass of the convergent lady beetle. Plant Disease, 83, 806–809.

Heike, F. (1970) Die palaarktischen-Arten des Subgenus *Zezea* Csiki (Carabidae: Coleoptera). Deutsche entomologische Zeitschrift 17, 119–214.

Heil, M. (2004) Induction of two indirect defences benefits lima bean (*Phaseolus lunatus*, Fabaceae) in nature. Journal of Ecology, 92, 527–536.

Heil, M., Fiala, B., Baumann, B., & Linsenmair, K.E. (2000) Temporal, spatial and biotic variations in extrafloral nectar secretion by *Macaranga tanarius*. Functional Ecology, 14, 749–757.

Heil, M., Koch, T., Hilpert, A., Fiala, B., Boland, W., & Linsenmair, K.E. (2001) Extrafloral nectar production of the ant-associated plant, Macaranga tanarius, is an induced, indirect, defensive response elicited by jasmonic acid. Proceedings of the National Academy of Sciences, 98, 1083–1088.

Heil, M., Hilpert, A., Kruger, R., & Linsenmair, K.E. (2004) Competition among visitors to extrafloral nectaries as a source of ecological costs of an indirect defence. Journal of Tropical Ecology, 20.

Heimpel, G.E. & Collier, T.R. (1996) The evolution of host-feeding behaviour in insect parasitoids. Biological Reviews, 71, 373–400.

Heimpel, G.E. & Jervis, M.A. (2005) Does floral nectar improve biological control by parasitoids? In Plant-Provided Food for Carnivorous Insects (eds F.L. Wäckers, P.C.J. van Rijn, & J. Bruin), pp. 267–304. Cambridge University Press, Cambridge.

Heimpel, G.E. & Lundgren, J.G. (2000) Sex ratios of commercially reared biological control agents. Biological Control, 19, 77–93.

Heimpel, G.E., Rosenheim, J.A., & Kattari, D. (1997) Adult feeding and lifetime reproductive success in the parasitoid *Aphytis melinus*. Entomologia Experimentalis et Applicata, 83, 305–315.

Heimpel, G.E., Lee, J.C., Wu, Z., Weiser, L., Wäckers, F., & Jervis, M.A. (2004) Gut sugar analysis in field-caught parasitoids: Adapting methods originally developed for biting flies. International Journal of Pest Management, 50, 193–198.

Heinrich, B. (1975) Energetics of pollination. Annual Review of Ecology and Systematics, 6, 139–170.

Heitmans, W.R.B., Overmeer, W.P.J., & van der Geest, L.P.S. (1986) The role of *Orius vicinus* Ribaut (Heteroptera; Anthocoridae) as a predator of phytophagous and predacious mites in a Dutch orchard. Journal of Applied Entomology, 102, 391–402.

Hemptinne, J.L. & Desprets, A. (1986) Pollen as spring food for *Adalia bipunctata*. In Ecology of Aphidophaga (ed I. Hodek), pp. 29–35. Academia, Prague & Dr. W. Junk, Dordrecht, The Netherlands.

Hemptinne, J.L., Naisse, J., & Os, S. (1988) Glimps of the life history of *Propylea quatuordecimpunctata* (L.) (Coleoptera: Coccinellidae). Mededelingen Faculteit Landbouwwetenschappen Rijks universiteit Gent 53, 1175–1182.

Hendrickx, F., Maelfait, J.-P., van Wingerden, W., Schweiger, O., Speelmans, M., Aviron, S., Augenstein, I., Billeter, R., Bailey, D., Bukacek, R., Burel, F., Diekotter, T., Dirksen, J., Herzog, F., Liira, J., Roubalova, M., Vandomme, V., & Bugter, R. (2007) How landscape structure, land-use intensity and habitat diversity affect components of total arthropod diversity in agricultural landscapes. Journal of Applied Ecology, 44, 340–351.

Hendrix, D.L. & Wei, Y. (1994) Bemisiose: An unusual trisaccharide in *Bemisia* honeydew. Carbohydrate Research, 253, 329–334.

Hendrix, D.L., Wei, Y., & Leggett, F.L. (1992) Homopteran honeydew sugar composition is determined by both the insect and plant species. Comparative Biochemistry and Physiology, 101B, 1–2.

Hengeveld, R. (1980a) Food specialization in ground beetles: An ecological or a phylogenetic process? (Coleoptera: Carabidae). Netherlands Journal of Zoology, 30, 585–594.

Hengeveld, R. (1980b) Polyphagy, oligophagy and food specialization in ground beetles (Coleoptera: Carabidae). Netherlands Journal of Zoology, 30, 564–584.

Hengeveld, R. (1980c) Qualitative and quantitative aspects of the food of ground beetles (Coleoptera: Carabidae): A review. Netherlands Journal of Zoology, 30, 555–563.

Henneberry, T.J., Bariola, L.A., & Kittock, D.L. (1977) Nectariless cotton: Effect on cotton leafperforator and other cotton insects in Arizona. Journal of Economic Entomology, 70, 797–199.

Herr, D.E. & Stroube, E.W. (1970) Velvetleaf control as influenced by herbicide placement and seed depth. Weed Science, 18, 459–461.

Herrebout, W.M. (1967) Habitat selection in *Eucarcelia rutilla* Vill. (Diptera: Tachinidae). Zeitschrift für Angewandt Entomologie, 60, 219–229.

Herrebout, W.M. & van der Veer, J. (1969) Habitat selection in Eucarcelia rutilla Vill. (Diptera: Tachinidae). III. Preliminary results of olfactometer experiments with females of known age. Zeitschrift für Angewandt Entomologie, 64, 55–61.

Herting, B. (1960) Biologie der wespaläarktischen Raupenfliegen (Diptera: Tachinidae). Monographien zur Angewandte Entomologie, 16, 1–202.

Hespenheide, H.A. (1985) Insect visitors to extrafloral nectaries of *Byttneria aculeata* (Sterculiaceae): Relative importance and roles. Ecological Entomology, 10, 191–204.

Hickman, J.C. (1974) Pollination by ants: A low-energy system. Science, 184, 1290–1292.

Higashi, S. & Ito, F. (1991) Ground beetles and seed dispersal of the myrmeochorous plant *Trillium tschonoskii* (Trilliaceae). In Ant-Plant Interactions (eds C.R. Huxley & D.F. Cutler), pp. 486–517. Oxford University Press, Oxford.

Hilbeck, A., Baumgartner, M., Fried, P.M., & Bigler, F. (1998a) Effects of transgenic *Bacillus thuringiensis* corn-fed prey on mortality and development time of immature *Chrysoperla carnea* (Neuroptera: Chrysopidae). Environmental Entomology, 27, 480–487.

Hilbeck, A., Moar, W.J., Pusztai-Carey, M., Filippini, A., & Bigler, F. (1998b) Toxicity of Bacillus thuringiensis Cry1Ab toxin to the predator *Chrysoperla carnea* (Neuroptera: Chrysopidae). Environmental Entomology, 27, 1255–1263.

Hilbeck, A., Moar, W.J., Pusztai-Carey, M., Filippini, A., & Bigler, F. (1999) Prey-mediated effects of Cry1Ab toxin and protoxin and Cry2A protoxin on the predator *Chrysoperla carnea.* Entomologia Experimentalis et Applicata, 91, 305–316.

Hilder, V.A., Gatehouse, A.M.R., Sheerman, S.E., Barker, R.F., & Boulter, D. (1987) A mechanism of insect resistance engineered into tobacco. Nature, 330, 160–163.

Hilder, V.A., Powell, K.S., Gatehouse, A.M.R., Gatehouse, J.A., Gatehouse, L.N., Shi, Y., Hamilton, W.D.O., Merryweather, A., Newell, C.A., Timans, J.C., Peumans, W.J., van Damme, E., & Boulter, D. (1995) Expression of snowdrop lectin in transgenic tobacco plants results in added protection against aphids. Transgenic Research, 4, 18–25.

Hill, N.S., Bouton, J.H., Hiatt, E.E., III, & Kittle, B. (2005) Seed maturity, germination, and endophyte relationships in tall fescue. Crop Science, 45, 859–863.

Hitchcock, J.D. (1959) Poisoning of honey bees by death camas blossoms. American Bee Journal, October, 418–419.

Hocking, H. (1967) The influence of food on longevity and oviposition in *Rhyssa persuasoria* (L.) (Hymenoptera: Ichneumonidae). Journal of the Australian Entomological Society, 6, 83–88.

Hodek, I. (1967) Bionomics and ecology of predaceous Coccinellidae. Annual Review of Entomology, 12, 79–104.

Hodek, I. & Honěk, A. (1996) Ecology of Coccinellidae. Kluwer, Dordrecht, The Netherlands.

Hogan, M.E., Slaytor, M., & O'Brien, R.W. (1985) Transport of volatile fatty acids across the hindgut of the cockroach *Panesthia cribrata* Saussure and the termite, *Mastotermes darwiniensis* Froggatt. Journal of Insect Physiology, 31, 587–591.

Hogervorst, P.A.M., Wäckers, F.L., & Romeis, J. (2007) Effects of honeydew sugar composition on the longevity of *Aphidius ervi.* Entomologia Experimentalis et Applicata, 1–10.

Holldobler, B. (1976) Recruitment behavior, home range orientation and territoriality in harvester ants, *Pogonomyrmex.* Behavioral Ecology and Sociobiology, 1, 3–44.

Holldobler, B. & Wilson, E.O. (1990) The Ants. Belknap Press, Cambridge, MA.

Holloway, B.A. (1976) Pollen-feeding in hover-flies (Diptera: Syrphidae). New Zealand Journal of Zoology, 3, 339–350.

Holmes, R.J. & Froud-Williams, R.J. (2005) Post-dispersal weed seed predation by avian and non-avian predators. Agriculture, Ecosystems and Environment, 105, 23–27.

Holser, R.A., Bost, G., & Van Boven, M. (2004) Phytosterol composition of hybrid *Hybiscus* seed oils. Journal of Agricultural and Food Chemistry, 52, 2546–2548.

Honek, A. & Martinkova, Z. (2001) Aggregation of ground beetles (Carabidae, Coleoptera) on winter rape seeds dispersed on the ground. Plant Protection Science, 37, 97–102.

Honek, A. & Martinkova, Z. (2003) Seed consumption by ground beetles. In The BCPC International Congress: Crop Science & Technology, Vol. 1, pp. 451–456. British Crop Protection Council, Hamptonshire/Glasgow, Scotland.

Honek, A., Martinkova, Z., & Jarosik, V. (2003) Ground beetles (Carabidae) as seed predators. European Journal of Entomology, 100, 531–544.

Honek, A., Martinkova, Z., & Saska, P., (2005) Post-dispersal predation of *Taraxacum officinale* (dandelion) seed. Journal of Ecology, 93, 345–352.

Honek, A., Saska, P., & Martinkova, Z. (2006) Seasonal variation in seed predation by adult carabid beetles. Entomologia Experimentalis et Applicata, 118, 157–162.

Honek, A., Martinkova, Z., Saska, P., & Pekar, S. (2007) Size and taxonomic constraints determine the seed preferences of Carabidae (Coleoptera). Basic and Applied Ecology, 8, 343–353.

Hood Henderson, D.E. (1982) Fine structure and neurophysiology of a gustatory sensillum on the ovipositors of *Metasyrphus venablesi* and *Eupeodes volucris* (Diptera: Syrphidae). Canadian Journal of Zoology, 60, 3187–3195.

Hood Henderson, D.E. & Wellington, W.G. (1982) Antennal sensilla of some aphidophagous Syrphidae (Diptera): Fine structure and electroantennogramme study. Canadian Journal of Zoology, 60, 3172–1386.
Hoogendoorn, M. & Heimpel, G.E. (2004) Competitive interactions between an exotic and native ladybeetle: A field cage study. Entomologia Experimentalis et Applicata, 111, 19–28.
Horvitz, C.C. (1981) Analysis of how ant behaviors affect germination in a tropical myrmecochore *Calathea microcephala* (P. & E.) Koernicke (Marantaceae): Microsite selection and aril removal by neotropical ants, *Odontomachus*, *Pachycondyla*, and *Solenopsis* (Formicidae). Oecologia, 51, 47–52.
Horvitz, C.C. & Beattie, A.J. (1980) Ant dispersal of *Calathea* (Marantaceae) seeds by carnivorous ponerines (Formicidae) in a tropical rain forest. American Journal of Botany, 67, 321–326.
Horvitz, C.C. & Schemske, D.W. (1986) Seed dispersal of a neotropical myrmecochore: Variation in removal rates and dispersal distance. Biotropica, 18, 319–323.
Hoshikawa, K. (1995) Bionomics of *Coccinula crotchi* in Hokkaido (Coccinellidae: Coleoptera). Japanese Journal of Entomology, 63, 419–424.
Hosokawa, T., Kikuchi, Y., Shimada, M., & Fukatsu, T. (2007) Obligate symbiont involved in pest status of host insect. Proceedings of the Royal Society B, 274, 1979–1984.
House, G.J. & Brust, G.E. (1989) Ecology of low-input, no-tillage systems. Agriculture, Ecosystems and Environment, 27, 331–345.
Houssain, Z., Gurr, G.M., Wratten, S.D., & Raman, A. (2002) Habitat manipulation in lucerne *Medicago sativa*: Arthropod population dynamics in harvested and 'refuge' crop strips. Journal of Applied Ecology, 39, 445–454.
Howard, D.F. & Tschinkel, W.R. (1981) The flow of food in colonies of the fire ant, *Solenopsis invicta*: A multifactorial study. Physiological Entomology, 6, 297–306.
Howard, D.J., Bush, G.L., & Breznak, J.A. (1985) The evolutionary significance of bacteria associated with *Rhagoletis*. Evolution, 39, 405–417.
Hoy, M.A. & Jeyaprakash, A. (2005) Microbial diversity in the predatory mite, *Metaseiulus occidentalis* (Acari: Phytoseiidae) and its prey, *Tetranychus urticae* (Acari: Tetranychidae). Biological Control, 32, 427–441.
Hubbell, S.P., Wiemer, D.F., & Adajari, A. (1983) An antifungal terpenoid defends a neotropical tree (*Hymenaea*) against attack by fungus-growing ants (*Atta*). Oecologia, 60, 321–327.
Huffaker, C.B. & Flaherty, D.L. (1966) Potential of biological control of two-spotted spider mites on strawberries in California. Journal of Economic Entomology, 59, 836–892.
Huffaker, C.B., van de Vrie, M., & McMurtry, J.A. (1970) II. Tetranychid populations and their possible control by predators: An evaluation. Hilgardia, 40, 391–458.
Hughes, L. & Westoby, M. (1990) Removal rates of seeds adapted for dispersal by ants. Ecology, 71, 138–148.
Hughes, L. & Westoby, M. (1992) Fate of seeds adapted for dispersal by ants in Australian sclerophyll vegetation. Ecology, 73, 1285–1299.
Hughes, L., Westoby, M., & Johnson, A.D. (1993) Nutrient costs of vertebrate- and ant-dispersed fruits. Functional Ecology, 7, 54–62.
Hughes, L., Westoby, M., & Jurado, E. (1994) Convergence of eliasomes and insect prey: Evidence from ant foraging behaviour and fatty acid composition. Functional Ecology, 8, 358–365.
Hughes, S.J. (1976) Sooty moulds. Mycologia, 68, 693–820.
Hukusima, S. & Itoh, K. (1976) Pollen and fungus as food for some coccinellid beetles. Research Bulletin of the Faculty of Agriculture at Gifu University, 29, 53–66.
Hulme, P.E. (1996) Herbivory, plant regeneration, and species coexistence. Journal of Ecology, 84, 609–615.
Hulme, P.E. (1998) Post-dispersal seed predation: Consequences for plant demography and evolution. Perspectives in Plant Ecology, Evolution, and Systematics, 1, 32–46.
Hulme, P.E. (2002) Seed-eaters: Seed dispersal, destruction and demography. In Seed Dispersal and Frugivory: Ecology, Evolution, and Conservation (eds D.J. Levey, W.R. Silva, & M. Galetti), pp. 257–274. CABI, Oxon.

Hulshof, J. & Jurchenko, O. (2000) *Orius laevigatus* in a choice situation: Thrips or pollen? Mededelingen Faculteit Landbouwkindige en Toegepaste Biologische Wetenschappen Universiteit Gent, 65, 351–358.

Hulspas-Jordaan, P.M. & van Lenteren, J.C. (1978) The relationship between host-plant leaf structure and parasitization efficiency of the parasitic wasp *Encarsia formosa* Gahan (Hymenoptera: Aphelinidae). *Mededelingen Faculteit Landbouwwetenschappen Rijksuniversiteit Gent*, 43, 431–440.

Human, H. & Nicolson, S.W. (2006) Nutritional content of fresh, bee-collected and stored pollen of *Aloe greatheadii* var. *davyana* (Asphodelaceae). Phytochemistry, 67, 1486–1492.

Hungate, R.E. (1938) Studies on the nutrition of *Zootermes* II. The relative importance of the termite and protozoa in wood digestion. Ecology, 19, 1–25.

Hunt, J.H., Brown, P.A., Sago, K.M., & Kerker, J.A. (1991) Vespid wasps eat pollen (Hymenoptera: Vespidae). Journal of the Kansas Entomological Society, 64, 127–130.

Hunter, F.F. & Ossowski, A.M. (1999) Honeydew sugars in wild-caught female horse flies (Diptera: Tabanidae). Journal of Medical Entomology, 36, 896–899.

Hurd, L.E. (1989) The importance of late season flowers to the fitness of an insect predator, *Tenordera sinensis* Saussure (Orthoptera: Mantidae) in an old-field community. Entomologist, 108, 223–228.

Hulspas-Jordaan, P.M. & van Lenteren, J.C. (1978) The relationship between host-plant leaf structure and parasitization efficiency of the parasitic wasp *Encarsia formosa* Gahan (Hymenoptera: Aphelinidae). *Mededelingen Faculteit Landbouwwetenschappen Rijksuniversiteit Gent*, 43, 431–440.

Hůrka, K. & Jarošík, V. (2003) Larval omnivory in *Amara aenea* (Coleoptera: Carabidae). European Journal of Entomology, 100, 329–335.

Ibrahim, M.M. (1955) Studies on *Coccinella undecimpunctata aegyptiaca* Reiche. 2. Biology and life-history. Bulletin of the Entomological Society of Egypt, 39, 395–423.

Idris, A.B. & Grafius, E.J. (1995) Wildflowers as nectar sources for Diadegma insulare (Hymenoptera: Ichneumonidae), a parasitoid of diamondback moth (Lepidoptera: Yponomeutidae). Environmental Entomology, 24, 1726–1735.

Ims, R.A. (1990) The ecology and evolution of reproductive synchrony. Trends in Ecology and Evolution, 5, 135–140.

Inouye, D.W. (1980) The terminology of floral larceny. Ecology, 61, 1251–1253.

Irvin, N.A., Hoddle, M.S., & Castle, S.J. (2007) The effect of resource provisioning and sugar composition of foods on longevity of three *Gonatocerus* spp., egg parasitoids of *Homalodisca vitripennis*. Biological Control, 40, 69–79.

Isenhour, D.J. & Marston, N.L. (1981) Seasonal cycles of *Orius insidiosus* (Hemiptera: Anthocoridae) in Missouri soybeans. Journal of the Kansas Entomological Society, 54, 129–142.

Islam, F.M.A., Rengifo, J., Redden, R.J., Basford, K.E., & Beebe, S.E. (2003) Association between seed coat polyphenolics (tannins) and disease resistance in common bean. Plant Foods for Human Nutrition, 58, 285–297.

Iwanami, Y., Sasakuma, T., & Yamada, Y. (1988) Pollen: Illustrations and Scanning Electronmicrographs. Springer, Heidelberg, Germany.

Jacob, H.S. & Evans, E.W. (1998) Effects of sugar spray and aphid honeydew on field populations of the parasitoid *Bathyplectes curculionis* (Hymenoptera: Ichneumonidae). Environmental Entomology, 27, 1563–1568.

Jacob, H.S., Minkey, D.M., Gallagher, R.S., & Borger, C.P. (2006) Variation in postdispersal weed seed predation in a crop field. Weed Science, 54, 148–155.

Jakubska, A., Przado, D., Steininger, M., Aniol-Kwiatkowska, J., & Kadej, M. (2005) Why do pollinators become "sluggish"? Nectar chemical constituents from *Epipactis helleborine* (L.) Crantz (Orchidaceae). Applied Ecology and Environmental Research, 3, 29–38.

James, D.G. (1989) Influence of diet on development, survival and oviposition in an Australian phytoseiid, *Amblyseius victoriensis* (Acari: Phytoseiidae). Experimental and Applied Acarology, 6, 1–10.

James, D.G. (1993) Pollen, mould mites and fungi: Improvements to mass rearing of *Typhlodromus doreenae* and *Amblyseius victoriensis*. Experimental and Applied Acarology, 17, 271–276.
Jänner, G. (1905) Die Thüringer Laufkäfer. Ent. Jahrb., Leipzig, 14.
Janzen, D.H. (1969) Seed-eaters versus seed size, number, toxicity and dispersal. Evolution, 23, 1–27.
Janzen, D.H. (1970) Herbivores and the number of tree species in tropical forests. American Naturalist, 104, 501–528.
Janzen, D.H. (1971) Seed predation by animals. Annual Review of Ecology and Systematics, 2, 465–492.
Janzen, D.H. (1972) Association of a rainforest palm and seed-eating beetles in Puerto Rico. Ecology, 53, 258–261.
Janzen, D.H. (1976) Reduction of *Mucuna andreana* (Leguminosae) seedling fitness by artificial seed damage. Ecology, 57, 826–828.
Janzen, D.H. (1977) Why don't ants visit flowers? Biotropica, 9, 252.
Janzen, D.H. (1979) New horizons in the biology of plant defenses. In Herbivores: Their Interactions with Secondary Plant Metabolites (eds G.A. Rosenthal & D.H. Janzen), pp. 331–350. Academic Press, New York.
Janzen, D.H., Juster, H.B., & Liener, I.E. (1976) Insecticidal action of phytohemagglutinin in black beans on a bruchid beetle. Science, 192, 795–796.
Janzen, D.H., Juster, H.B., & Bell, E.A. (1977) Toxicity of secondary compounds to the seed-eating larvae of the bruchid beetle *Callosobruchus maculatus*. Phytochemistry, 16, 223–227.
Jaycox, E.R. (1964) Effect on honey bees of nectar from systemic insecticide-treated plants. Journal of Economic Entomology, 57, 31–35.
Jeannel, R. (1941) Coleopteres Carabiques. Faune de France, Suppl. 39, 51.
Jeannel, R. (1942) Coleopteres Carabiques. Faune de France, Suppl. 40, 1–1173.
Jensen, A.M.D. (2005) Endophyte persistence and toxin (lolitrem b) production in Danish seed crop of perennial ryegrass. European Journal of Agronomy, 23, 68–78.
Jervis, M.A. & Kidd, N.A.C. (1996). Phytophagy. In *Natural Enemies*, pp. 375–394.
Jervis, M. (1998) Functional and evolutionary aspects of mouthpart structure in parasitoid wasps. Biological Journal of the Linnean Society, 63, 461–493.
Jervis, M.A. (1990) Predation of Lissonota coracinus (Gmelin) (Hymenoptera: Ichneumonidae) by Dolichonabis limbatus (Dahlbom) (Hemiptera: Nabidae). Entomologist's Gazette, 41, 231–233.
Jervis, M.A. & Boggs, C.L. (2005) Linking nectar amino acids to fitness in female butterflies. Trends in Ecology and Evolution, 20, 585–587.
Jervis, M.A. & Heimpel, G.E. (2005) Phytophagy. In Insects as Natural Enemies (ed M.A. Jervis), pp. 525–550. Springer, Dordrecht, The Netherlands.
Jervis, M.A., Kidd, N.A.C., Fitton, M.G., Huddleston, T., & Dawah, H.A. (1993) Flower-visiting by hymenopteran parasitoids. Journal of Natural History, 27, 67–105.
Jervis, M.A., Hawkins, B.A., & Kidd, N.A.C. (1996a) The usefulness of destructive host feeding parasitoids in classical biological control: Theory and observation conflict. Ecological Entomology, 21, 41–46.
Jervis, M.A., Kidd, N.A.C., & Heimpel, G.E. (1996b) Parasitoid adult feeding behaviour and biocontrol – a review. Biocontrol News and Information, 17, 11N–26N.
Johanson, C.A., Mayer, D.F., Eves, J.D., & Kious, C.W. (1983) Pesticides and bees. Environmental Entomology, 12, 1513–1518.
Johnson, D.R. (1949) Chemical treatment of seed corn for control of *Agonoderus comma*. Journal of Economic Entomology, 42, 801–805.
Johnson, E.A. & Fryer, G.I. (1996) Why Engelmann spruce does not have a persistent seed bank. Canadian Journal of Forest Research, 26, 872–878.
Johnson, N.E. & Cameron, R.S. (1969) Phytophagous ground beetles. Annals of the Entomological Society of America, 62, 909–914.
Johnson, N.E., Lawrence, W.H., & Ellis, I.D. (1966) Seasonal occurrence of ground beetles (Coleoptera: Carabidae) in three habitats in southwestern Washington. Annals of the Entomological Society of America, 59, 1055–1059.

Jones, G.D. (1984) Microorganisms as mediators of plant resource exploitation by insect herbivores. In A New Ecology: Novel Approaches to Interactive Systems (eds P.W. Price, C.N. Slobodchikoff, & W.S. Gaud), pp. 53–99. Wiley, New York.

Jones, P.R., Andersen, M.D., Nielsen, J.S., Hoj, P.B., & Moller, B.L. (2000) The biosynthesis, degradation, transport and possible function of cyanogenic glucosides. In Evolution of Metabolic Pathways (eds J.T. Romeo, R. Ibrahim, L. Varin, & V. DeLuca), Vol. 34, pp. 191–248. Pergamon Press, Amsterdam, The Netherlands.

Jones, Q. & Earle, F.R. (1966) Chemical analyses of seeds II: Oil and protein content of 759 species. Economic Botany, 20, 127–155.

Jordan, N., Mortensen, D.A., Prenzlow, D.M., & Cox, K.C. (1995) Simulation analysis of crop rotation effects on weed seedbanks. American Journal of Botany, 82, 390–398.

Jorgensen, H.B. & Toft, S. (1997a) Food preference, diet dependent fecundity and larval development in *Harpalus rufipes* (Coleoptera: Carabidae). Pedobiologia, 41, 307–315.

Jorgensen, H.B. & Toft, S. (1997b) Role of granivory and insectivory in the life cycle of the carabid beetle *Amara similata*. Ecological Entomology, 22, 7–15.

Judd, W.W. (1947) The proventriculus of a ground beetle, *Chlaenius* sp. (Coleoptera: Carabidae). Canadian Entomologist, 79, 58–60.

Jules, E.S. (1996) Yellow jackets (*Vespula vulgaris*) as a second seed disperser for the myrmecochorous plant, *Trillium ovatum*. American Midland Naturalist, 135, 367–369.

Justus, M., Witte, L., & Hartmann, T. (1997) Levels and tissue distribution of loline alkaloids in endophyte-infected *Festuca pratensis*. Phytochemistry, 44, 51–57.

Kaczorowski, R.L., Gardener, M.C., & Holtsford, T.P. (2005) Nectar traits of Nicotiana section Alatae (Solanaceae) in relation to floral traits, pollinators, and mating systems. American Journal of Botany, 92, 1270–1283.

Kalberer, P.P. (1990) Influence of the water potential of the casing soil on crop yield and on dry-matter content, osmotic potential and mannitol content of the fruit bodies of *Agaricus bisporus*. Journal of Horticultural Science, 65, 573–581.

Kamburov, S.S. (1971) Feeding, development, and reproduction of *Amblyseius largoensis* on various food substances. Journal of Economic Entomology, 64, 643–648.

Kammer, A.E. & Heinrich, B. (1978) Insect flight metabolism. Advances in Insect Physiology, 8, 133–228.

Kanecka, P. (1993) Ichneumon flies on larvae of false spruce webworm sawfly (*Cephalcia abietis* (L.)) and information on their breedings. Lesnictvi Prague, 39, 297–303.

Kanrar, S., Venkateswari, J., Kirti, P.B., & Chopra, V.L. (2002) Transgenic Indian mustard (*Brassica juncea*) with resistance to the mustard aphid (*Lipaphis erysimi* Kalt.). Plant Cell Reports, 20, 976–981.

Karban, R. & Lowenberg, G. (1992) Feeding by seed bugs and weevils enhances germination of wild *Gossypium* species. Oecologia, 92, 196–200.

Karl, E. & Hartleb, H. (1963) Beobachtungen zur Schedigung von Erfeerfruchten durch Laufkafer (Carabidae). NachrBl. dtsch. PflSchDienst (N. F.), 17, 79–83.

Karley, A.J., Ashford, D.A., Minto, L.M., Pritchard, J., & Douglas, A.E. (2005) The significance of gut sucrase activity for osmoregulation in the pea aphid, *Acyrthosiphon pisum*. Journal of Insect Physiology, 51, 1313–1319.

Karise, R., Mand, M., Ivask, M., Koskor, E., & Bender, A. (2006) The effect of pollen amount on and its caloric value in hybrid lucerne (*Medicago x varia*) on its attractiveness to bumble bees (*Bombus terrestris*). Agronomy Research, 4, 211–216.

Kartohardjono, A. & Heinrichs, E.A. (1984) Populations of the brown planthopper, Nilaparvata lugens (Stal) (Homoptera: Delphacidae), and its predators on rice varieties with different levels of resistance. Environmental Entomology, 13, 359–365.

Kaspari, M. (1996) Worker size and seed size selection by harvester ants in a neotropical forest. Oecologia, 105, 397–404.

Kauffman, W.C. & Flanders, R.V. (1985) Effects of variably resistant soybean and lima bean cultivars on *Pediobius foveolatus* (Hymenoptera: Eulophidae), a parasitoid of the Mexican bean beetle, *Epilachna varivestis* (Coleoptera: Coccinellidae). Environmental Entomology, 14, 678–682.

Kaufman, M.G. & Klug, M.J. (1991) The contribution of hindgut bacteria to dietary carbohydrate utilization by crickets (Orthoptera: Gryllidae). Comparative Biochemistry and Physiology, 98A, 117–123.

Kaufman, M.G., Klug, M.J., & Merritt, R.W. (1989) Growth and food utilization parameters of germ-free house crickets, *Acheta domesticus*. Journal of Insect Physiology, 35, 957–967.

Keeler, K.H. (1977) The extrafloral nectaries of Ipomoea carnea (Convulvulaceae). American Journal of Botany, 64, 1182–1188.

Keeler, K.H. (1978) Insects feeding at extrafloral nectaries of *Ipomoea carnea* (Convolvulaceae). Entomological News, 89, 163–168.

Keeler, K.H. (1980) The extrafloral nectaries of Ipomoea leptophylla (Convolvulaceae). American Journal of Botany, 67, 216–222.

Keeler, K.H. (1981) Function of Mentzelia nuda (Loasaceae) postfloral nectaries in seed defense. American Journal of Botany, 68, 295–299.

Kegode, G.O., Forcella, F., & Clay, S. (1999) Influence of crop rotation, tillage, and management inputs on weed seed production. Weed Science, 47, 175–183.

Kelly, D. (1994) The evolutionary ecology of mast seeding. Tree, 9, 465–470.

Kelrick, M.I. & MacMahon, J.A. (1985) Nutritional and physical attributes of seeds of some common sagebrush-steppe plants: Some implications for ecological theory and management. Journal of Range Management, 38, 65–69.

Kelrick, M.I., MacMahon, J.A., Parmenter, R.R., & Sisson, D.V. (1986) Native seed preferences of shrub-steppe rodents, birds, and ants: The relationships of seed attributes and seed use. Oecologia, 68, 327–337.

Kemp, J.C. & Barrett, G.W. (1989) Spatial patterning: Impact of uncultivated corridors on arthropod populations within soybean agroecosystems. Ecology, 70, 114–128.

Kendeigh, S.C. & West, G.C. (1965) Caloric values of plant seeds eaten by birds. Ecology, 46, 553–555.

Kennedy, G.G., Kishaba, A.N., & Bohn, G.W. (1975) Response of several pest species to *Cucumis melo* L. lines resistant to *Aphis gossypii* Glover. Environmental Entomology, 4, 653–657.

Kennett, C.E. & Hamai, J. (1980) Oviposition and development in predaceous mites fed with artificial and natural diets (Acari: Phytoseiidae). Entomologia Experimentalis et Applicata, 28, 116–122.

Kennett, C.E., Flaherty, D.L., & Hoffmann, R.W. (1979) Effect of wind-borne pollens on the population dynamics of *Amblysieus hibisci* [*Acarina: Phytoseiidae*]. Entomophaga, 34, 83–98.

Kessler, A. & Baldwin, I.T. (2002) Plant responses to insect herbivory: The emergin molecular analysis. Annual Review of Plant Biology, 53, 299–328.

Kevan, P.G. (1973) Parasitoid wasps as flower visitors in the Canadian high arctic. Anzeiger Schadlungskunde, 46, 3–7.

Kevan, P.G., Eisikowitch, D., Fowle, S., & Thomas, K. (1988) Yeast-contaminated nectar and its effects on bee foraging. Journal of Apicultural Research, 27, 26–29.

Kikuchi, Y., Hosokawa, T., & Fukatsu, T. (2007) Insect-microbe mutualism without vertical transmission: A stinkbug acquires a beneficial gut symbiont from the environment every generation. Applied and Environmental Microbiology, 73, 4308–4316.

Kiman, Z.B. & Yeargan, K.V. (1985) Development and reproduction of the predator *Orius insidiosus* (Hemiptera: Anthocoridae) reared on diets of selected plant material and arthropod prey. Annals of the Entomological Society of America, 78, 464–467.

Kingsolver, J.G. & Daniel, T.L. (1995) Mechanics of food handling by fluid-feeding insects. In Regulatory Mechanisms in Insect Feeding (eds R.F. Chapman & G. de Boer), pp. 32–73. Chapman & Hall, New York.

Kirchner, H.A. (1939) Laufkaferschaden an Erdbeeren. Zeitschrift für Pflanzen krankheiten, 49, 267–271.

Kirk, V.M. (1972) Seed-caching by larvae of two ground beetles, *Harpalus pensylvanicus* and *H. erraticus*. Annals of the Entomological Society of America, 65, 1426–1428.

Kirk, V.M. (1973) Biology of a ground beetle, *Harpalus pensylvanicus*. Annals of the Entomological Society of America, 66, 513–518.

Kittel, G. (1873/1874) Systematische Ubersicht der Kafer, welche in Bayern und der nachsten Umbegung vorkommen. Corr.-Blatt. zool.-miner., 27, 28.

Kjellsson, G. (1985) Seed fate in a population of *Carex pilulifera* L. II. Seed predation and its consequences for dispersal and seed bank. Oecologia, 67, 424–429.

Klimaszewski, J. & Peck, S.B. (1987) Succession and phenology of beetle faunas (Coleoptera) in the fungus Polyporellus squamosus (Huds.: Fr) Karst. (Polporaceae) in Silesia, Poland. Canadian Journal of Zoology, 65, 542–550.

Klingenberg, C.P., Leigh, R.H.B., Keddie, B.A., & Spence, J.R. (1997) Influence of gut parasites on growth performance in the water strider *Gerris buenoi* (Hemiptera: Gerridae). Ecography, 20, 29–36.

Knoch, T.R., Faeth, S.H., & Arnott, D.L. (1993) Endophytic fungi alter foraging and dispersal by desert seed-harvesting ants. Oecologia, 95, 470–473.

Koca, I. & Koca, A.F. (2007) Poisoning by mad honey: A brief review. Food and Chemical Toxicology, 45, 1315–1318.

Koch, A. (1960) Intracellular symbiosis in insects. Annual Review of Microbiology, 14, 121–140.

Koch, R.L., Burkness, E.C., Wold-Burkness, S.J., & Hutchison, W.D. (2004) Phytophagous preferences of the multicolored Asian ladybeetle (Coleoptera: Coccinellidae) for autumn-ripening fruit. Journal of Economic Entomology, 97, 539–544.

Koilpillai, R.R. & Haniffa, M.A. (1996) Substrate specific and age related digestive enzyme changes in the house cricket, *Gryllodes sigillatus* (Walker) (Orthroptera: Gryllidae). Journal of Entomological Research, 20, 325–334.

Koiwa, H., Bressan, R.A., & Hasegawa, P.M. (1997) Regulation of protease inhibitors and plant defense. Trends in Plant Science, 2, 379–384.

Kollmann, J., Coomes, D.A., & White, S.M. (1998) Consistencies in post-dispersal seed predation of temperate fleshy-fruited species among season, years and sites. Functional Ecology, 12, 683–690.

Koptev, V. (1948) The reasons for the summer deaths of bees in Siberia (in Russian). Pchelovodstvo, 10, 45–47.

Koptur, S. (1994) Floral and extrafloral nectars of Costa Rican Inga trees: A comparison of their constituents and composition. Biotropica, 26, 276–284.

Koptur, S. (2005) Nectar as fuel for plant protectors. In Plant-Provided Food for Carnivorous Insects (eds F.L. Wäckers, P.C.J. van Rijn, & J. Bruin), pp. 75–108. Cambridge University Press, Cambridge.

Koptur, S. & Truong, N. (1998) Facultative ant-plant interactions: Nectar sugar preferences of introduced pest ant species in south Florida. Biotropica, 30, 179–189.

Kost, C. & Heil, M. (2005) Increased availability of extrafloral nectar reduces herbivory in lima bean plants (*Phaseolus lunatus*, Fabaceae). Basic and Applied Ecology, 6, 237–248.

Kost, C. & Heil, M. (2006) Herbivore-induced plant volatiles induce an indirect defence in neighbouring plants. Journal of Ecology, 94, 619–628.

Kovach, J. (2004) Impact of multicolored Asian lady beetles as a pest of fruit and people. American Entomologist, 50, 159–161.

Kovar, I. (1996) Morphology and anatomy. In Ecology of Coccinellidae (eds I. Hodek & A. Honek), pp. 1–18. Kluwer, Dordrecht, The Netherlands.

Koveos, D.S. & Broufas, G.D. (1999) Feeding history affects the response of the predatory mite *Typhlodromus kerkirae* (Acari: Phytoseiidae) to volatiles of plants infested with spider mites. Experimental and Applied Acarology, 23, 429–436.

Krantz, G.W. & Lindquist, E.E. (1979) Evolution of phytophagous mites (Acari). Annual Review of Entomology, 24, 121–158.

Krischik, V.A., Landmark, A.L., & Heimpel, G.E. (2007) Soil-applied imidacloprid is translocated to nectar and kills nectar-feeding *Anagyrus pseudococci* (Girault) (Hymenoptera: Encyrtidae). Environmental Entomology, 36, 1238–1245.

Krishnamoorthy, A. (1984) Influence of adult diet on the fecundity of the predator, *Chrysopa scelestes* [Neur.: Chrysopidae]. Entomophaga, 29, 445–450.

Kritchevsky, D. (1997) Phytosterols. In Advances in Experimental Medicine and Biology (eds D. Kritchevsky & C.T. Bonfield), pp. 235–243. Plenum Press, New York.

Krochmal, C. (1994) Poison honeys. American Bee Journal, 134, 549–550.

Krombein, K.V. (1951) Wasp visitors of tulip-tree honeydew at Dunn Loring, Virginia. Annals of the Entomological Society of America, 44, 141–143.

Kugler, C. & Hincapie, M.C. (1983) Ecology of the ant *Pogonomyrmex mayri*: Distribution, ant abundance, nest structure, and diet. Biotropica, 15, 190–198.

Kukor, J.J. & Martin, M.M. (1987) Nutritional ecology of fungus-feeding arthropods. In Nutritional Ecology of Insects, Mites, Spiders and Related Invertebrates (eds F.J. Slansky & J.G. Rodriguez), pp. 791–814. Wiley, New York.

Kuo, H.-L. (1986) Resistance of oats to cereal aphids: Effects on parasitism by *Aphelinus asychis* (Walker). In Interactions of Plant Resistance and Parasitoids and Predators of Insects (eds D.J. Boethel & R.D. Eikenbary), pp. 125–137. Ellis Horwood, Chichester.

Kurtzmann, R.H., Jr. (1997) Nutrition from mushrooms, understanding and reconciling available data. Mycoscience, 38, 247–253.

Kusmenoglu, S., Rockwood, L.L., & Gretz, M.R. (1989) Fatty acids and diacylglycerols from elaiosomes of some ant-dispersed seeds. Phytochemistry, 28, 2601–2602.

La Guardia, M., Venturella, G., & Venturella, F. (2005) On the chemical composition and nutritional value of *Pleurotus* taxa growing on umbelliferous plants (Apiaceae). Agricultural and Food Chemistry, 53, 5997–6002.

Lachance, M.-A., Bowles, J.M., Chavarria Diaz, M.M., & Janzen, D.H. (2001) *Candida cleridarum*, *Candida tilneyi* and *Candida powellii*, three new yeast species isolated from insects associated with flowers. International Journal of Systematic and Evolutionary Microbiology, 51, 1201–1207.

Lalonde, R.G. & Roitberg, B.D. (1992) On the evolution of masting behavior in trees: Predation or weather? American Naturalist, 139, 1293–1304.

LaMana, M.L. & Miller, J.C. (1996) Field observations on *Harmonia axyridis* Pallas (Coleoptera: Coccinellidae) in Oregon. Biological Control, 6, 232–237.

Lamb, K.P. (1959) Composition of the honeydew of the aphid *Brevicoryne brassicae* (L.) feeding on swedes (*Brassica napobrassica* DC.). Journal of Insect Physiology, 3, 1–13.

Landholt, P.J. & Lenczewski, B. (1993) Lack of evidence for the toxic nectar hypothesis: A plant alkaloid did not deter nectar feeding by Lepidoptera. Florida Entomologist, 76, 556–566.

Landis, D.A. & Menalled, F.D. (1998) Ecological considerations in the conservation of effective parasitoid communities in agricultural systems. In Conservation Biological Control (ed P. Barbosa), pp. 101–123. Academic Press, London.

Landis, D.A., Wratten, S.D., & Gurr, G.M. (2000) Habitat management to conserve natural enemies of arthropod pests in agriculture. Annual Review of Entomology, 45, 175–201.

Landis, D.A., Menalled, F.D., Costamagna, A.C., & Wilkinson, T.K. (2005) Manipulating plant resources to enhance beneficial arthropods in agricultural landscapes. Weed Science, 53, 902–908.

Lanza, J. (1988) Ant preferences for *Passiflora* nectar mimics that contain amino acids. Biotropica, 20, 341–344.

Lanza, J. & Krauss, B.R. (1984) Detection of amino acids in artificial nectars by two tropical ants, *Leptothorax* and *Monomorium*. Oecologia, 63, 423–425.

Lanza, J., Shmitt, M.A., & Awad, A.B. (1992) Comparative chemistry of elaiosomes of three species of *Trillium*. Journal of Chemical Ecology, 18, 209–221.

Lanza, J., Smith, G.C., Sack, S., & Cash, A. (1995) Variation in nectar volume and composition of Impatiens capensis at the individual, plant, and population levels. Oecologia, 102, 113–119.

Larochelle, A. (1974) A world list of prey of *Chlaenius* (Coleoptera: Carabidae). Great Lakes Entomologist, 7, 137–142.

Larochelle, A. (1977) The food of the genus *Anisodactylus* Dejean (Coleoptera: Carabidae). Cordulia, 3, 53–55.

Larochelle, A. (1990) The food of carabid beetles (Coleoptera: Carabidae, including Cicindelinae). Fabreries, Suppl. 5, 1–132.

Last, F.T. & Price, D. (1969) Yeasts associated with living plants and their environs. In The Yeasts (eds A.H. Rose & J.S. Harrison), Vol. I, pp. 183–218. Academic Press, London.

Lattanzio, V., Terzano, R., Cicco, N., Cardinali, A., Di Venere, D., & Linsalata, V. (2005) Seed coat tannins and bruchid resistance in stored cowpea seeds. Journal of the Science of Food and Agriculture, 85, 839–845.

Lattin, J.D. (1989) Bionomics of the Nabidae. Annual Review of Entomology, 34, 383–400.

Lattin, J.D., Asquith, A., & Booth, S. (1989) *Orius minutus* (Linneaus) in North America (Hemiptera: Heteroptera: Anthocoridae). Journal of the New York Entomological Society, 97, 409–416.

Laurema, S., Husberg, G.B., & Markkula, M. (1986). Composition and functions of the salivary gland of the larvae of the aphid midge *Aphidoletes aphidimyza*. In *Ecology of Aphidophaga* (ed I. Hodek), pp. 113–118. Dr. W. Junk N. V., Dordrecht, The Netherlands.

Laurent, F.M. & Rathahao, E. (2003) Distribution of [^{14}C]imidacloprid in sunflowers (Helianthus annuus L.) following seed treatment. Journal of Agricultural and Food Chemistry, 51, 8005–8010.

Lavendero, B., Wratten, S.D., Shishehbor, P., & Worner, S.P. (2005) Enhancing the effectiveness of the parasitoid Diadegma semiclausum (Helen): Movement after use of nectar in the field. Biological Control, 34, 152–158.

Lavigne, R. (1977) Checklist of the Carabidae of Wyoming, U. S. A. Cordulia, 3, 43–48.

Lavigne, R.J. (1976) Rangeland insect-plant associations on the Pawnee site. Annals of the Entomological Society of America, 69, 753–763.

Lawrence, J.F. (1989) Mycophagy in the Coleoptera: Feeding strategies and morphological adaptations. In Insect-Fungus Interactions (eds N. Wilding, N.M. Collins, P.M. Hammond, & J.F. Webber), pp. 2–24. Academic Press, London.

Lawrence, J.R. & Newton, A.F., Jr. (1980) Coleoptera associated with the fruiting bodies of slime molds (Myxomycetes). Coleopterist's Bulletin, 34, 129–143.

Leatemia, J.A., Laing, J.E., & Corrigan, J.E. (1995) Effects of adult nutrition on longevity, fecundity, and offspring sex ratio of *Trichogramma minutum* Riley (Hymenoptera: Trichogrammatidae). Canadian Entomologist, 127, 245–254.

Lebiedzinska, A. & Szefer, P. (2006) Vitamin B in grain and cereal-grain food, soy-products, and seeds. Food Chemistry, 95, 116–122.

Lee, J.C., Andow, D.A., & Heimpel, G.E. (2006) Influence of floral resources on sugar feeding and nutrient dynamics of a parasitoid in the field. Ecological Entomology, 31, 470–480.

Lee, J.C. & Heimpel, G.E. (2003) Nectar availability and parasitoid sugar feeding. In International Symposium on Biological Control of Arthropods (ed R.G. Van Driesche), pp. 220–225. United States Department of Agriculture, Forest Service, Morgantown, WV/Honolulu, HA.

Lee, J.C., Heimpel, G.E., & Leibee, G.L. (2004) Comparing floral nectar and aphid honeydew diets on the longevity and nutrient levels of a parasitoid wasp. Entomologia Experimentalis et Applicata, 111, 189–199.

Leemann, A.C. (1933) A short summary on our botanical knowledge of Lolium temulentum L. Onderstepoort Journal of Veterinary Science and Animal Industry, I, 213–218.

Leereveld, H. (1982) Anthecological relations between reputedly anemophilous flowers and syrphid flies. III. Worldwide survey of crop and intestine contents of certain anthophilous syrphid flies. Tijdschrift voor Entomologie, 125, 25–35.

Leguizamon, E.S. & Roberts, H.A. (1982) Seed production by an arable weed community. Weed Research, 22, 35–39.

Lehman, R.M., Lundgren, J.G., & Petzke, L.M. (in press) Bacterial communities associated with the digestive tract of the predatory ground beetle, *Poecilus chalcites*, and their response to laboratory rearing and antibiotic treatment. Microbial Ecology.

Lehninger, A.L. (1982) Principles of Biochemistry. Worth, New York.

Leishman, M.R., Westoby, M., & Jurado, E. (1995) Correlates of seed size variation: A comparison among five temperate floras. Journal of Ecology, 83, 517–530.

Leius, K. (1961a) Influence of food on fecundity and longevity of adults of *Itoplectis conquisitor* (Say) (Hymenoptera: Ichneumonidae). Canadian Entomologist, 93, 771–780.

Leius, K. (1961b) Influence of various foods on fecundity and longevity of adults of *Scambus buolianae* (Htg.) (Hymenoptera: Ichneumonidae). Canadian Entomologist, 93, 1079–1084.

Leius, K. (1963) Effects of pollens on fecundity and longevity of adult *Scambus buolianae* (Htg.) (Hymenoptera: Ichneumonidae). Canadian Entomologist, 95, 202–207.

Leius, K. (1967) Food sources and preferences of adults of a parasite, *Scambus buolianae* (Hym.: Ichn.), and their consequences. Canadian Entomologist, 99, 865–871.

Lengkerken, H.v. (1924) Extra-intestinale Verdauung. Biologischen Zentralblat, 44, 273–295.

Leschen, R.A.B. (1993) Evolutionary patterns of feeding in selected Staphylinoidea (Coleoptera): Shifts among food textures. In Functional Morphology of Insect Feeding (eds C.W. Schaefer & R.A.B. Leschen), pp. 59–104. Entomological Society of America, Lanham, MD.

Leschen, R.A.B. (2000) Beetles feeding on bugs (Coleoptera, Hemiptera): Repeated shifts from mycophagous ancestors. Invertebrate Taxonomy, 14, 917–929.

Leschen, R.A.B. & Beutel, R.G. (2001) Pseudotracheal tubes, larval head, and mycophagy in *Sepedophilus* (Coleoptera: Staphylinidae: Tachyporinae). Journal of Zoological Systematics and Evolutionary Research, 39, 25–35.

Leuchtmann, A., Schmidt, D., & Bush, L.P. (2000) Different levels of protective alkaloids in grasses with stroma-forming and seed-transmitted *Epichloe/Neotyphodium* endophytes. Journal of Chemical Ecology, 26, 1025–1036.

Levey, D.J. & Byrne, M.M. (1993) Compex ant-plant interactions: Rain-forest ants as secondary dispersers and post-dispersal seed predators. Ecology, 74, 1802–1812.

Levin, D.A. (1974) The oil content of seeds: An ecological perspective. American Naturalist, 108, 193–206.

Lewis, G.P. & Worthen, W.B. (1992) Effects of ant predation and mushroom dessication on the survival of mycophagous *Drosophila tripunctata* larvae. Oikos, 64, 553–559.

Lewis, W.J. & Takasu, K. (1990) Use of learned odours by a parasitic wasp in accordance with host and food needs. Nature, 348, 635–636.

Lewis, W.J., Stapel, J.O., Cortesero, A.M., & Takasu, K. (1998) Understanding how parasitoids balance food and host needs: Importance to biological control. Biological Control, 11, 175–183.

Liber, H. & Niccoli, A. (1988) Observations on the effectiveness of an attractant food spray in increasing chrysopid predation on Prays oleae (Bern.) eggs. Redia, 71, 467–482.

Liener, I.E. (1979) Phytohemagglutinins. In Herbivores: Their Interaction with Secondary Plant Metabolites (eds G.A. Rosenthal & D.H. Janzen), pp. 567–598. Academic Press, New York.

Limburg, D.D. & Rosenheim, J.A. (2001) Extrafloral nectar consumption and its influence on survival and development of an omnivorous predator, larval *Chrysoperla plorabunda* (Neuroptera: Chrysopidae). Environmental Entomology, 30, 595–604.

Lindroth, C.H. (1961–1969) The ground beetles (Carabidae, excl. Cicindellinae) of Canada and Alaska. Opuscula Entomologica, Suppl. XX, XXIV, XXIX, XXXIII, XXXIV, XXXV, 1–1192.

Lingren, P.D. & Lukefahr, M.J. (1977) Effects of nectariless cotton on caged populations of *Campoletis sonorensis*. Environmental Entomology, 6, 586–588.

Linskins, H.F. & Pfahler, P.L. (1973) Biochemical composition of maize (*Zea mays* L.) pollen. Theoretical and Applied Genetics, 43, 49–53.

Linskins, H.F. & Schrauwen, J. (1969) The release of free amino acids from germinating pollen. Acta Botanica Neerlandica, 18, 605–614.

Linskins, H.F., Ciampolini, F., & Cresti, M. (1993) Spider webs as pollen traps. Proc. Kon. Ned. Akad. v. Wetensch., 96, 415–425.

Loper, G.M. & Cohen, A.C. (1982) The caloric content of bee-gathered pollen. American Bee Journal, 65, 709–710.

Lopes, A.R., Juliano, M.A., Juliano, L., & Terra, W.R. (2004) Coevolution of insect trypsins and inhibitors. Archives of Insect Biochemistry and Physiology, 55, 140–152.

Lord, K.A., May, M.A., & Stevenson, J.H. (1968) The secretion of the systemic insecticides dimethoate and phorate into nectar. Annals of Applied Biology, 61, 19–27.

Lott, J.N.A. (1984) Accumulation of seed reserves of phosphorus and other minerals. In Seed Physiology, Volume 1: Development (ed D.R. Murray), pp. 139–166. Academic Press, Orlando, FL.

Louda, S.M. (1989) Predation in the dynamics of seed regeneration. In Ecology of Seed Banks (eds M.A. Leck, V.T. Parker, & R.L. Simpson), pp. 25–51. Academic Press, London.

Louda, S.M., Potvin, M.A., & Collinge, S.K. (1990) Predispersal seed predation, postdispersal seed predation and competition in the recruitment of seedlings of the native thistle in sandhills prairie. American Naturalist, 124, 105–113.

Loughrin, J.H., Manukian, A., Heath, R.R., & Tumlinson, J.H. (1995) Volatiles emitted by different cotton varieties damaged by feeding beet armyworm larvae. Journal of Chemical Ecology, 21, 1217–1227.

Lubin, Y.D. (1978) Seasonal abundance and diversity of web-building spiders in relation to habitat structure on Barro Colorado Island, Panama. Journal of Arachnology, 6, 31–51.

Ludy, C. & Lang, A. (2006a) A 3-year field-scale monitoring of foliage-dwelling spiders (Aranae) in transgenic Bt maize fields and adjacent field margins. Biological Control, 38, 314–324.

Ludy, C. & Lang, A. (2006b) Bt maize pollen exposure and impact on the garden spider, *Araneus diadematus*. Entomologia Experimentalis et Applicata, 118, 145–156.

Luff, M.L. (1974) Adult and larval feeding habits of *Pterostichus madidis* (F.) (Coleoptera: Carabidae). Journal of Natural History, 8, 403–409.

Luff, M.L. (1980) The biology of the ground beetle *Harpalus rufipes* in a strawberry field in Northumberland. Annals of Applied Biology, 94, 153–164.

Lunau, K. & Wacht, S. (1994) Optical releasers of the innate proboscis extension in the hoverfly *Eristalis tenax* L. (Syrphidae, Diptera). Journal of Comparative Physiology A, 174, 575–579.

Lund, R.D. (1975) The feeding of *Harpalus pensylvanicus* DeGeer (Coleoptera: Carabidae) on weed seeds commonly occurring in corn fields. Proceeding of the North Central Branch – E. S. A., 30, 104.

Lund, R.D. & Turpin, F.T. (1977) Carabid damage to weed seeds found in Indiana cornfields. Environmental Entomology, 6, 695–698.

Lundblad, O. (1927) Noxious animals in Sweden for the years 1922–1926 [in Swedish]. Medd. Centralanst. forsoksvas. jordbr., Stockholm.

Lundgren, J.G. (2005) Ground beetles as weed control agents: The influence of farm management on granivory. American Entomologist, 51, 224–226.

Lundgren, J.G. & Heimpel, G.E. (2003) Quality assessment of three species of commercially produced *Trichogramma* and the first report of thelytoky in commercially produced *Trichogramma*. Biological Control, 26, 68–73.

Lundgren, J.G. & Rosentrater, K.A. (2007) The strength of seeds and their destruction by granivorous insects. Arthropod-Plant Interactions, 1, 93–99.

Lundgren, J.G. & Wiedenmann, R.N. (2002) Coleopteran-specific Cry3Bb toxin from transgenic corn pollen does not affect the fitness of a non-target species, *Coleomegilla maculata* DeGeer (Coleoptera: Coccinellidae). Environmental Entomology, 31, 1213–1218.

Lundgren, J.G. & Wiedenmann, R.N. (2004) Nutritional suitability of corn pollen for the predator *Coleomegilla maculata* (Coleoptera: Coccinellidae). Journal of Insect Physiology, 50, 567–575.

Lundgren, J.G. & Wiedenmann, R.N. (2005) Tritrophic interactions among Bt (Cry3Bb1) corn, aphid prey, and the predator *Coleomegilla maculata* (Coleoptera: Coccinellidae). Environmental Entomology, 34, 1621–1625.

Lundgren, J.G., Heimpel, G.E., & Bomgren, S.A. (2002) Comparison of Trichogramma brassicae (Hymenoptera: Trichogrammatidae) augmentation with organic and synthetic pesticides for control of cruciferous Lepidoptera. Environmental Entomology, 31, 1231–1239.

Lundgren, J.G., Razzak, A.A., & Wiedenmann, R.N. (2004) Population responses and food consumption by predators *Coleomegilla maculata* and *Harmonia axyridis* (Coleoptera: Coccinellidae) during anthesis in an Illinois cornfield. Environmental Entomology, 33, 958–963.

Lundgren, J.G., Huber, A., & Wiedenmann, R.N. (2005) Quantification of consumption of corn pollen by the predator *Coleomegilla maculata* (Coleoptera: Coccinellidae) during anthesis in an Illinois cornfield. Agricultural and Forest Entomology, 7, 53–60.

Lundgren, J.G., Shaw, J.T., Zaborski, E.R., & Eastman, C.E. (2006) The influence of organic transition systems on beneficial ground-dwelling arthropods and predation of insects and weed seeds. Renewable Agriculture and Food Systems, 21, 227–237.

Lundgren, J.G., Lehman, R.M., & Chee-Sanford, J. (2007) Bacterial communities within digestive tracts of ground beetles (Coleoptera: Carabidae). Annals of the Entomological Society of America, 100, 275–282.

Lundgren, J.G., Fergen, J.K., & Riedell, W.E. (2008a) The influence of plant anatomy on oviposition and reproductive success of the omnivorous bug, *Orius insidiosus*. Animal Behaviour, 75, 1495–1502.

Lundgren, J.G., Wyckhuys, K.A.G., & Desneux, N. (2008b) Population responses by *Orius insidiosus* to vegetational diversity. BioControl, in press.

Mabry, T.J. & Gill, J.E. (1979) Sesquiterpene lactones and other terpenoids. In Herbivores: Their Interactions with Secondary Plant Metabolites (eds G.A. Rosenthal & D.H. Janzen), pp. 502–538. Academic Press, New York.

MacGregor, A.W. (1983) Cereal α-amylases: Synthesis and action pattern. In Seed Proteins (eds J. Daussant, J. Mosse, & J. Vaughan), pp. 1–34. Academic Press, London.

MacKay, W.P. (1990). The biology and economic impact of *Pogonomyrmex* harvester ants. In *Applied Myrmecology: A World Perspective.* (eds R.K. Vander Meer, K. Jaffee & A. Cedeno), pp. 533–541. Westview Press, Oxford, UK.

MacLeod, A., Wratten, S.D., Sotherton, N.W., & Thomas, M.B. (2004) 'Beetle banks' as refuges for beneficial arthropods in farmland: Long-term changes in predator communities and habitat. Agricultural and Forest Entomology, 6, 147–154.

Macleod-Carey, F., Lewis, J.J., MacGregor, J.L., & Martin-Smith, M. (1959) Pharmacological and chemical observations on some toxic nectars. Journal of Pharmacy and Pharmacology, Suppl. 11, 269T–274T.

MacMahon, J.A., Mull, J.F., & Crist, T.O. (2000) Harvester ants (*Pogonomyrmex* spp.): Their community and ecosystem influences. Annual Review of Ecology and Systematics, 31, 265–291.

Macon, T.E. & Porter, S.D. (1995) Food and energy requirements of laboratory fire ant colonies (Hymenoptera: Formicidae). Environmental Entomology, 24, 387–391.

Magalhaes, S. & Bakker, F.M. (2002) Plant feeding by a predatory mite inhabiting cassava. Experimental and Applied Acarology, 27, 27–37.

Maier, C.T. (1978) The immature stages and biology of *Mallota posticata* (Fabricius) (Diptera: Syrphidae). Proceedings of the Entomological Society of Washington, 80, 424–440.

Maier, C.T. & Waldbauer, G.P. (1979) Diurnal activity patterns of flower flies (Diptera: Syrphidae) in an Illinois sand area. Annals of the Entomological Society of America, 72, 237–245.

Majerus, M.E.N., Sloggett, J.J., Godeau, J.-F., & Hemptinne, J.-L. (2007) Interactions between ants and aphidophagous and coccidophagous ladybirds. Population Ecology, 49, 15–27.

Malcolm, S.B. (1990) Chemical defence in chewing and sucking insect herbivores: Plant-derived cardenolides in the monarch butterfly and oleander aphid. Chemoecology, 1, 12–21.

Mali, S. & Borges, R.M. (2003) Phenolics, fibre, alkaloids, saponins, and cyanogenic glycosides in a seasonal cloud forest in India. Biochemical Systematics and Ecology, 31, 1221–1246.

Malone, L.A. & Pham-Delegue, M.-H. (2001) Effects of transgene products on honey bees (*Apis mellifera*) and bumblebees (*Bombus* sp.). Apidologie, 32, 287–304.

Maltais, J.B. & Auclair, J.L. (1952) Occurrence of amino acids in the honeydew of the crescent-marked lily aphid, *Myzus circumflexus* (Buck.). Canadian Journal of Zoology, 30, 191–193.

Mandal, S., Datta, B.K., & Bhattakaryya, G.N. (1993) Biochemical analysis of the allergenic pollen of *Madhuca indica* Gmel. and *Spathodea campanulata* Beauv. Aerobiologia, 9, 197–200.

Manley, G.V. (1971) A seed-cacheing carabid (Coleoptera). Annals of the Entomological Society of America, 64, 1474–1475.

Manzaneda, A.J., Fedriani, J.M., & Rey, P.J. (2005) Adaptive advantages of myrmecochory: The predator-avoidance hypothesis tested over a wide geographic range. Ecography, 28, 583–592.

Marcone, M.F., Kakuda, Y., & Yada, R.Y. (2004) Amaranth as a rich dietary source of B-sitosterol and other phytosterols. Plant Foods for Human Nutrition, 58, 207–211.

Marino, P.C., Gross, K.L., & Landis, D.A. (1997) Weed seed loss due to predation in Michigan maize fields. Agriculture, Ecosystems and Environment, 66, 189–196.

Marino, P.C., Westerman, P.R., Pinkert, C., & van der Werf, W. (2005) Influence of seed density and aggregation on post-dispersal weed seed predation in cereal fields. Agriculture, Ecosystems and Environment, 106, 17–25.

Marisa, C. & Sauro, S. (1990) Biological observations and life table parameters of *Amblyseius cucumeris* (Oud.) (Acarina: Phytoseiidae) reared on different diets. Redia, 73, 569–583.

Mark, S. & Olesen, J.M. (1996) Importance of elaiosome size to removal of ant-dispersed seeds. Oecologia, 107, 95–101.

Marshall, D.L., Beattie, A.J., & Bollenbacher, W.E. (1979) Evidence for diglycerides as attractants in an ant-seed interaction. Journal of Chemical Ecology, 5, 335–344.

Marshall, E.J.P., Brown, V.K., Boatman, N.D., Lutman, P.J.W., Squire, G.R., & Ward, L.K. (2003) The role of weeds in supporting biological diversity within crop fields. Weed Research, 43, 77–89.

Martin, M.M. (1984) The role of ingested enzymes in the digestive processes of insects. In Invertebrate-Microbial Interactions (eds J.M. Anderson, A.D.M. Raynor, & D.W.H. Walton), pp. 155–172. Cambridge University Press, Cambridge.

Martin, M.M. (1992) The evolution of insect-fungus associations: From contact to stable symbiosis. American Zoologist, 32, 593–605.

Martin, M.M., Martin, J.S., Kukor, J.J., & Merritt, R.W. (1980) The digestion of protein and carbohydrate by the stream detritivore, *Tipula abdominalis* (Diptera, Tipulidae). Oecologia, 46, 360–364.

Martin, M.M., Kukor, J.J., Martin, J.S., O'Toole, T.E., & Johnson, M.W. (1981) Digestive enzymes of fungus-feeding beetles. Physiological Zoology, 54, 137–145.

Martin, W.R.J., Grisham, M.P., Kenerley, C.M., Sterling, W.L., & Morgan, P.W. (1987) Microorganisms associated with cotton fleahopper, *Pseudatomoscelis seriatus* (Heteroptera: Miridae). Annals of the Entomological Society of America, 80, 251–255.

Martinkova, Z., Saska, P., & Honek, A. (2006) Consumption of fresh and buried seed by ground beetles (Coleoptera: Carabidae). European Journal of Entomology, 103, 361–364.

Marvier, M., McCreedy, C., Regetz, J., & Kareiva, P. (2007) A meta-analysis of effects of Bt cotton and maize on nontarget invertebrates. Science, 316, 1475–1477.

Massieu, G.H., Guzman, J.G., Cravioto, R.O., & Calvo, J.d.l.T. (1951) Nutritive value of some primitive Mexican foods. Journal of the American Dietetic Association, 27, 212–214.

Matheson, N.K. (1984) The synthesis of reserve oligosaccharides and polysaccharides in seeds. In Seed Physiology, Volume 1: Development (ed D.R. Murray), pp. 167–208. Academic Press, Orlando, FL.

Matteoni, J.A. (2003) Economics of banker plant systems in Canadian greenhouse crops. In Proceedings of the International Symposium on Biological Control of Arthropods (ed R.G. VanDriesche), pp. 154–157, Honolulu, HA.

Matthewman, W.G. & Pielou, D.P. (1971) Arthropods inhabiting the sporophores of *Fomes fomentarius* (Polysporaceae) in Gatineau Park, Quebec. Canadian Entomologist, 103, 775–847.

Matthews, C.R., Brown, M.W., & Bottrell, D.G. (2007) Leaf extrafloral nectaries enhance biological control of a key economic pest, *Grapholita molesta* (Lepidoptera: Tortricidae), in peach (Rosales: Rosaceae). Environmental Entomology, 36, 383–389.

Mauchline, A.L., Watson, S.J., Brown, V.K., & Froud-Williams, R.J. (2005) Post-dispersal seed predation of non-target weeds in arable crops. Weed Research, 45, 157–164.

Maurizio, A. (1941) Über ein massensterben von bienen, verursacht durch pollen von *Ranunculus puberulus* Koch. Schweiz. Naturfors. Ges., 121, 149–150.

Maurizio, A. (1945) Trachtkrankheiten der bienen. I. Vergiftungen bei einseitiger tracht von rosskastanien. Schweiz. Bienen-Ztg., 1, 430–440.

Maxwell, F.G. & Jennings, P.R., eds. (1980) Breeding Plants Resistant to Insects, p. 683. Wiley, New York.

McCaffrey, J.P. & Horsburgh, R.L. (1986) Biology of *Orius insidiosus* (Heteroptera: Anthocoridae) a predator in Virginia apple orchards. Environmental Entomology, 15, 984–988.

McClanahan, R.J. (1967) Food-chain toxicity of systemic acaricides to predaceous mites. Nature, 215, 1001.

McCoy, C.W., Selhime, A.G., & Kanavel, R.F. (1969) The feeding behavior and biology of *Parapronematus acaciae* (Acarina: Tydaeidae). Florida Entomologist, 51, 13–20.

McDougall, S.J. & Mills, N.J. (1997) The influence of hosts, temperature and food sources on the longevity of *Trichogramma platneri*. Entomologia Experimentalis et Applicata, 83, 195–203.

McEwen, P.K., Clow, S., Jervis, M.A., & Kidd, N.A.C. (1993a) Alteration in searching behaviour of adult female green lacewings *Chrysoperla carnea* (Neur.: Chrysopidae) following contact with honeydew of the black scale *Saissetia oleae* (Hom.: Coccidae) and solutions containing acidhydrolysed L-tryptophan. Entomophaga, 38, 347–354.

McEwen, P.K., Jervis, M.A., & Kidd, N.A.C. (1993b) Influence of artificial honeydew on larval development and survival in *Chrysoperla carnea* [Neur., Chrysopidae]. Entomophaga, 38, 241–244.

McEwen, P.K., Jervis, M.A., & Kidd, N.A.C. (1994) Use of a sprayed L-tryptophan solution to concentrate numbers of the green lacewing *Chrysoperla carnea* in olive tree canopy. Entomologia Experimentalis et Applicata, 70, 97–99.

McEwen, P.K., Jervis, M.A., & Kidd, N.A.C. (1996) The influence of an artificial food supplement on larval and adult performance in the green lacewing *Chrysoperla carnea* (Stephens). International Journal of Pest Management, 42, 25–27.

McFayden, R.E.C. (1998) Biological control of weeds. Annual Review of Entomology, 43, 369–393.

McGregor, S.E. & Todd, F.E. (1952) Cantaloup production with honeybees. Journal of Economic Entomology, 45, 43–47.

McKey, D. (1979) The distribution of secondary compounds within plants. In Herbivores: The Interactions with Secondary Plant Metabolites (eds G.A. Rosenthal & D.H. Janzen), pp. 56–122. Academic Press, New York.

McLain, D.K. (1980) Relationships among ants, aphids, and coccinellids on wild lettuce. Journal of the Georgia Entomological Society, 15, 413–417.

McMurtry, J.A. (1985) Chapter 3.2.7 Citrus. In Spider Mites: Their Biology, Natural Enemies, and Control (eds W. Helle & M.W. Sabelis), Vol. 1B, pp. 339–347. Elsevier, Amsterdam, The Netherlands.

McMurtry, J.A. & Croft, B.A. (1997) Life-styles of phytoseiids mites and their role in biological control. Annual Review of Entomology, 42, 291–231.

McMurtry, J.A., de Moraes, G.J., & Johnson, H.G. (1991) Arrestment responses of some phytoseiid mites to extracts of *Oligonychus punicae*, *Tetranychus urticae* and pollen. Isreal Journal of Entomology, 25–26, 29–34.

McMurtry, J.A., Huffaker, C.B., & van de Vrie, M. (1970) I. Tetranychid enemies: Their biological characters and the impact of spray practices. Hilgardia, 40, 331–390.

McMurtry, J.A. & Johnson, H.G. (1965) Some factors influencing the abundance of the predaceous mite *Amblyseius hibisci* in southern California (Acarina: Phytoseiidae). Annals of the Entomological Society of America, 58, 49–56.

McMurtry, J.A. & Rodriguez, J.G. (1987) Nutritional ecology of phytoseiid mites. In Nutritional Ecology of Insects, Mites, Spiders, and Related Invertebrates (eds F.J. Slansky & J.G. Rodriguez), pp. 609–644. Wiley, New York.

McMurtry, J.A. & Scriven, G.T. (1964a) Biology of the predaceous mite *Typhlodromus rickeri* (Acarina: Phytoseiidae). Annals of the Entomological Society of America, 57, 362–367.

McMurtry, J.A. & Scriven, G.T. (1964b) Studies on the feeding, reproduction, and development of *Amblyseius hibisci* (Acarina: Phytoseiidae) on various food substances. Annals of the Entomological Society of America, 57, 649–655.

McMurtry, J.A. & Scriven, G.T. (1965) Life-history studies of *Amblyseius limonicus*, with comparative observations on *Amblyseius hibisci* (Acarina: Phytoseiidae). Annals of the Entomological Society of America, 58, 106–111.

McMurtry, J.A. & Scriven, G.T. (1966) The influence of pollen and prey density on the number of prey consumed by *Amblyseius hibisci* (Acarina: Phytoseiidae). Annals of the Entomological Society of America, 59, 147–149.

McNeill, S. (1971) The energetics of a population of *Leptopterna dolabrata* (Heteroptera: Miridae). Journal of Animal Ecology, 40, 127–140.

Meier, M.S. & Hilbeck, A. (2001) Influence of transgenic *Bacillus thuringiensis* corn-fed prey on prey preference of immature *Chrysoperla carnea* (Neuroptera: Chrysopidae). Basic and Applied Ecology, 2, 35–44.

Meissle, M. & Lang, A. (2005) Comparing methods to evaluate the effects of *Bt* maize and insecticide on spider assemblages. Agriculture, Ecosystems and Environment, 107, 359–370.

Menalled, F.D., Marino, P.C., Renner, K.A., & Landis, D.A. (2000) Post-dispersal weed seed predation in Michigan corn fields as a function of agricultural landscape structure. Agriculture, Ecosystems and Environment, 77, 193–202.

Menalled, F.D., Dauer, J., Fox, T.B., & Renner, K.A. (2001) Managing Your Farm to Increase Weed Seed Predation, Vol. Bulletin E-2749. Michigan State University Extension, East Lansing, Michigan.

Menalled, F.D., Liebman, M., & Renner, K.A. (2005) The ecology of weed seed predation in herbaceous crop systems. In Handbook of Sustainable Weed Management (ed D. Batish), pp. 297–327. Hawthorn Press, Binghamton, NY.

Menalled, F.D., Smith, R.G., Dauer, J.T., & Fox, T.B. (2007) Impact of agricultural management on carabid communities and weed seed predation. Agriculture, Ecosystems and Environment, 118, 49–54.

Mensah, R.K. (1996) Suppression of *Helicoverpa* spp. (Lepidoptera: Noctuidae) oviposition by use of the natural enemy food supplement Envirofeast®. Australian Journal of Entomology, 35, 323–329.

Mensah, R.K. (1997) Local density responses of predatory insects of *Helicoverpa* spp. to a newly developed food supplement 'Envirofeast' in commercial cotton in Australia. International Journal of Pest Management, 43, 221–225.

Mensah, R.K. & Singleton, A. (2003) Optimum timing and placement of a supplementary food spray Envirofeast® for the establishment of predatory insects of *Helicoverpa* spp. in cotton systems in Australia. International Journal of Pest Management, 49, 163–168.

Mensah, R.K., Verneau, S., & Frerot, B. (2000) Deterrence of oviposition of adult *Ostrinia nubilalis* Hubner (Lepidoptera: Pyralidae) by a natural enemy food supplement Envirofeast® on maize in France. International Journal of Pest Management, 46, 49–53.

Mestor, T.C. & Buhler, D.D. (1990) Effect of planting depth on velvetleaf (*Abutilon theophrasti*) seedling development and response to cyanazine. Weed Science, 38, 34–38.

Metcalf, R.L. (1955) Organic Insecticides. Interscience, New York.

Mevi-Schutz, J. & Erhardt, A. (2005) Amino acids in nectar enhance butterfly fecundity: A long-awaited link. The American Naturalist, 165, 411–419.

Meyling, N.V. & Pell, J.K. (2006) Detection and avoidance of an entomopathogenic fungus by a generalist insect predator. Ecological Entomology, 31, 162–171.

Michaud, J.P. (2000) Development and reproduction of ladybeetles (Coleoptera: Coccinellidae) on the citrus aphids *Aphis spiraecola* Patch and *Toxoptera citricida* (Kirkaldy) (Homoptera: Aphididae). Biological Control, 18, 287–297.

Michaud, J.P. & Grant, A.K. (2005) Suitability of pollen sources for the development and reproduction of *Coleomegilla maculata* (Coleoptera: Coccinellidae) under simulated drought conditions. Biological Control, 32, 363–370.

Michaud, J.P. & Qureshi, J.A. (2006) Reproductive diapause in Hippodamia convergens (Coleoptera: Coccinellidae) and its life history consequences. Biological Control, 39, 193–200.

Middledorf, J. & Ruthmann, A. (1984) Yeast-like endosymbionts in an ichneumonid wasp. Zeitschrift für Naturforschung, 39C, 322–326.

Midgley, J.J. & Bond, W.J. (1995) Relative attractiveness of seeds of myrmecochorous Australian and South African plants to ants, and the chemical basis for this attraction. South African Journal of Botany, 61, 230–232.
Mikola, J. (1983) Proteinases, peptidases, and inhibitors of endogenous proteinases in germinating seeds. In Seed Proteins (eds J. Daussant, J. Mosse, & J. Vaughan), pp. 35–52. Academic Press, London.
Miles, P.W. (1972) The saliva of Hemiptera. Advances in Insect Physiology, 9, 183–255.
Miles, P.W. (1987) Plant-sucking bugs can remove the contents of cells without mechanical damage. Experientia, 43, 937–939.
Miller, N.C.E. (1971) The Biology of the Heteroptera. E.W. Casey, Hampton, Middlesex, England.
Mitchell, B. (1963) Ecology of two carabid beetles, *Bembidion lampros* (Herbst) and *Trechus quadristriatus* (Schrank). Journal of Animal Ecology, 32, 289–299.
Mittelbach, G.G. & Gross, K.L. (1984) Experimental studies of seed predation in old-fields. Oecologia, 65, 7–13.
Mittler, T.E. (1957) Studies on the feeding and nutrition of Tuberolachnus salignus (Gmelin) (Homptera: Aphididae). I. Uptake of phloem sap. Journal of Experimental Biology, 34, 334–341.
Mittler, T.E. (1958a) Studies on the feeding and nutrition of Tuberolachnus salignus (Gmelin) (Homoptera: Aphididae). II. The nitrogen and sugar composition of ingested phloem sap and excreted honeydew. Journal of Experimental Biology, 35, 74–84.
Mittler, T.E. (1958b) Studies on the feeding and nutrition of Tuberolachnus salignus (Gmelin) (Homoptera: Aphididae). III. The nitrogen economy. Journal of Experimental Biology, 35, 626–638.
Mizuno, T. (1958) Nippon Kagaku Zasschi. Japan Journal of Pure Chemistry, 79, 192.
Moggridge, J.T. (1873) Harvesting Ants and Trap-Door Spiders. L. Reeve, London.
Mohler, C.L. (1996) Ecological bases for cultural control of annual weeds. Journal of Production Agriculture, 9, 468–474.
Mohr, N.A. & Jay, S.C. (1990) Nectar production of selected cultivars of *Brassica campestris* L. and *Brassica napus* L. Journal of Apicultural Research, 29, 95–100.
Molyneux, R.J., Campbell, B.C., & Dreyer, D.L. (1990) Honeydew analysis for detecting phloem transport of plant natural products. Implications for host-plant resistance to sap-sucking insects. Journal of Chemical Ecology, 16, 1899–1909.
Mommsen, T.P. (1978) Digestive enzymes of a spider (*Tegenaria atrica* Koch) – II. Carbohydrases. Comparative Biochemistry and Physiology, 60A, 371–375.
Mondor, E.B. & Addicott, J.F. (2003) Conspicuous extra-floral nectaries are inducible in *Vicia faba*. Ecology Letters, 6, 495–497.
Monsrud, C. & Toft, S. (1999) The aggregative numerical response of polyphagous predators to aphids in cereal fields: Attraction to what? Annals of Applied Biology, 134, 265–270.
Morales, M.A. & Heithaus, E.R. (1998) Food from seed-dispersal mutualism shifts sex ratios in colonies of the ant *Aphaenogaster rudis*. Ecology, 79, 734–739.
Morales-Ramos, J.A., Rojas, M.G., Sittertz-Bhaktar, H., & Saldana, G. (2000) Symbiotic relationship between *Hypothenemus hampei* (Coleoptera: Scolytidae) and *Fusarium solani* (Monilales: Tuberculariaceae). Annals of the Entomological Society of America, 93, 541–547.
Moreau, R.A., Singh, V., & Hicks, K.B. (2001) Comparison of oil and phytosterol levels in germplasm accessions of corn, teosinte, and Job's tears. Journal of Agricultural and Food Chemistry, 49, 3793–3795.
Mori, H. (1977) Notes on the biology of phytoseiid mites (Acarina: Phytoseiidae) as predators of tetranychid mites in Japan. Review of Plant Protection Research, 10, 91–101.
Morrison, D.E., Bradley, J.R., Jr., & Van Duyn, J.W. (1979) Populations of corn earworm and associated predators after applications of certain soil-applied pesticides to soybeans. Journal of Economic Entomology, 72, 97–100.
Morrison, J.E.J., Williams, D.F., Oi, D.H., & Potter, K.N. (1997) Damage to dry crop seed by red imported fire ant (Hymenoptera: Formicidae). Journal of Economic Entomology, 90, 218–222.

Morrison, J.E.J., Williams, D.F., & Oi, D.H. (1999) Effect of crop seed water content on the rate of seed damage by red imported fire ants (Hymenoptera: Formicidae). Journal of Economic Entomology, 92, 215–219.

Morse, D.H. (1986) Foraging behavior of crab spiders (*Misumena vatia*) hunting on inflorescences of different quality. American Midland Naturalist, 116, 341–347.

Morse, D.H. & Fritz, R.S. (1982) Experimental and observational studies of patch choice at different scales by the crab spider, *Misumena vatia*. Ecology, 63, 172–182.

Mubarak, A.E. (2005) Nutritional composition and antinutritional factors of mung bean seeds (*Phaseolus aureus*) as affected by some home traditional processes. Food Chemistry, 89, 489–495.

Muchuweti, M., Ndhlala, A.R., & Kasiamhuru, A. (2006) Analysis of phenolic compounds including tannins, gallotannins and flavanols of *Uapaca kirkiana* fruit. Food Chemistry, 94, 415–419.

Muck, O. (1939) Bericht der amtlichen untersuchungsstelle für ansteckende bienenkrankheiten ander tierärztlichen hochschule über dar jahr. Wien. Tieräerztl. Monatsschr., 23, 168–173.

Mueller, U.G., Schultz, T.R., Currie, C.R., Adams, R.M.M., & Malloch, D. (2001) The origin of the attine ant-fungus mutualism. Quarterly Review of Biology, 76, 169–197.

Muggeridge, J. (1939) A maize and strawberry pest, (*Clivina rugithorax* Putz.). New Zealand Journal of Science and Technology, 21A, 184–185.

Muller, C.L. (1948) Die trachtkrankheiten der honigbiene. Tierzucht, 2, 117–120.

Müller, H. (1883) The Fertilisation of Flowers. MacMillan, London.

Mullin, C.A., Saunders, M.C., II, Leslie, T.W., Biddinger, D.J., & Fleischer, S.J. (2005) Toxic and behavioral effects to Carabidae of seed treatments used on Cry3Bb1- and Cry1Ab/c-protected corn. Environmental Entomology, 34, 1626–1636.

Murray, D.R. (1984a). Accumulation of seed reserves of nitrogen. In Seed Physiology, Volume 1: Development (ed D.R. Murray), pp. 84–137. Academic Press, Orlando, FL.

Murray, D.R. (1984b). The seed and survival. In Seed Physiology, Volume 1: Development (ed D.R. Murray), pp. 1–37. Academic Press, Orlando, FL.

Nalepa, C.A., Bambara, S.B., & Burroughs, A.M. (1992) Pollen and nectar feeding by *Chilocorus kuwanae* (Silvestri) (Coleoptera: Coccinellidae). Proceedings of the Entomological Society of Washington, 94, 596–597.

Naranjo, S.E. & Gibson, R.L. (1996) Phytophagy in predaceous Heteroptera: Effects on life history and population dynamics. In Zoophytophagous Heteroptera: Implications for Life History and Integrated Pest Management (eds O. Alomar & R.N. Wiedenmann), pp. 57–93. Entomological Society of America, Lanham, MD.

Naranjo, S.E. & Stimac, J.L. (1985) Development, survival, and reproduction of *Geocoris punctipes* (Hemiptera: Lygaeidae): Effects of plant feeding on soybean and associated weeds. Environmental Entomology, 14, 523–530.

Navasero, R.C. & Elzen, G.W. (1989) Responses of *Microplitis croceipes* to host and nonhost plants of *Heliothis virescens* in a wind tunnel. Entomologia Experimentalis et Applicata, 53, 57–63.

Nayar, J.L. (1965) Digestion and excretion in *Syrphus balteatus* DeGeer (Syrphidae: Diptera). Agra University Journal of Research, 14, 111–125.

Nedved, O., Ceryngier, P., Hodkova, M., & Hodek, I. (2001) Flight potential and oxygen uptake during early dormancy in *Coccinella septempunctata*. Entomologia Experimentalis et Applicata, 99, 371–380.

Nentwig, W., Frank, T., & Lethmayer, C. (1998) Sown weed strips: Artificial ecological compensation areas as an important tool in conservation biological control. In Conservation Biological Control (ed P. Barbosa), pp. 133–154. Academic Press, London.

Ness, J.H. & Bronstein, J.L. (2004) The effects of invasive ants on prospective ant mutualists. Biological Invasions, 6, 445–461.

Neuenschwander, P. & Hagen, K.S. (1980) Role of the predator *Hemerobius pacificus* in a non-insecticide treated artichoke field. Environmental Entomology, 9, 492–495.

Newton, A.F., Jr. (1984) Mycophagy in Staphylinoidea (Coleoptera). In Fungus-Insect Interactions (eds Q. Wheeler & M. Blackwell), pp. 302–353. Columbia University Press, New York.

Nguyen, N.H., Suh, S.-O., Erbil, C.K., & Blackwell, M. (2006) *Metschnikowia noctiluminum* sp. nov., *Metschnikowia corniflorae* sp. nov., and *Candida chrysomelidarum* sp. nov., isolated from green lacewings and beetles. Mycological Research, 110, 346–356.

Nichols, P.R. & Neel, W.W. (1977) The use of food wheast as a supplemental food for *Coleomegilla maculata* (DeGeer) (Coleoptera: Coccinellidae) in the field. Southwestern Entomologist, 2, 102–105.

Nickerson, J.C., Rolph Kay, C.A., Buschmann, L.L., & Whitcomb, W.H. (1977) The presence of *Spissistilus festinus* as a factor affecting egg predation by ants in soybeans. Florida Entomologist, 60, 193–199.

Nickle, D.A. & Neal, T.M. (1972) Observations on the foraging behavior of the southern harvester ant, *Pogonomyrmex badius*. Florida Entomologist, 55, 65–66.

Nicolson, S.W. (1994) Pollen feeding in the eucalypt nectar fly, *Drosophila flavohirta*. Physiological Entomology, 19, 58–60.

Nicolson, S.W. & Van Wyck, B.-E. (1998) Nectar sugars in Proteaceae: Patterns and processes. Australian Journal of Botany, 46, 489–504.

Nomikou, M., Janssen, A., & Sabelis, M.W. (2003a) Phytoseiid predator of whitefly feeds on plant tissue. Experimental and Applied Acarology, 31, 27–36.

Nomikou, M., Janssen, A., & Sabelis, M.W. (2003b) Phytoseiid predators of whiteflies feed and reproduce on non-prey food sources. Entomologia Experimentalis et Applicata, 31, 15–26.

Nomikou, M., Janssen, A., Schraag, R., & Sabelis, M.W. (2004) Vulnerability of *Bemisia tabaci* immatures to phytoseiid predators: Consequences for oviposition and influence of alternative food. Entomologia Experimentalis et Applicata, 110, 95–102.

Norton, A.P., English-Loeb, G., Gadoury, D., & Seem, R.C. (2000) Mycophagous mites and foliar pathogens: Leaf domatia mediate tritrophic interactions in grapes. Ecology, 81, 490–499.

Nusslin, O. & Rhumbler, L. (1922) Fortinsektenkunde, 3rd edn. Verlagsbuchhandlung, Paul Parey, Berlin.

Nwokolo, E. (1986) A comparison of metabolizable energy content of eight common feed ingredients determined with young guinea fowls (keets) and pullet chicks. Animal Feed Science and Technology, 15, 1–6.

Nystrand, O. & Granstrom, A. (2000) Predation on *Pinus sylvestris* seeds and juvenile seedlings in a Swedish boreal forest in relation to stand disturbance by loggin. Journal of Applied Ecology, 37, 449–463.

O'Connor, B.M. (1984) Acarine-fungal relationships: The evolution of symbiotic associations. In Fungus-Insect Relationships (eds Q. Wheeler & M. Blackwell), pp. 354–381. Columbia University Press, New York.

O'Dowd, D.J. (1979) Foliar nectar production and ant activity on a neotropical tree, *Ochroma pyramidale*. Oecologia, 43, 233–248.

O'Dowd, D.J. & Hay, M.E. (1980) Mutualism between harvester ants and a desert ephemeral: Seed escape from rodents. Ecology, 61, 531–540.

O'Neal, M.E., Zontek, E.L., Szendrei, Z., Landis, D.A., & Isaacs, R. (2005) Ground predator abundance affects prey removal in highbush blueberry (*Vaccinium corymbosum*) fields and can be altered by aisle ground covers. BioControl, 50, 205–222.

O'Neal, R.J. & Waller, G.D. (1984) On the pollen harvest by the honey bee (*Apis mellifera* L.) near Tucson, Arizona (1976–1981) Desert Plants, 6, 81–109.

O'Rourke, M.E., Heggenstaller, A.H., Liebman, M., & Rice, M.E. (2006) Post-dispersal weed seed predation by invertebrates in conventional and low-external-input crop rotation systems. Agriculture, Ecosystems and Environment, 116, 280–288.

Obrist, L.B., Dutton, A., Albajes, R., & Bigler, F. (2006a) Exposure of arthropod predators to Cry1Ab toxin in Bt maize fields. Ecological Entomology, 31, 143–154.

Obrist, L.B., Klein, H., Dutton, A., & Bigler, F. (2006b) Assessing the effects of Bt maize on the predatory mite *Neoseiulus cucumeris*. Experimental and Applied Acarology, 38, 125–139.

Obrycki, J.J., Ruberson, J.R., & Losey, J.E. (2004) Interactions between natural enemies and transgenic insecticidal crops. In Genetics, Evolution and Biological Control (eds L.E. Ehler, R. Sforza, & T. Mateille), pp. 183–206. CABI, Wallingford.

Ohara, M. & Higashi, S. (1987) Interference by ground beetles with the dispersal by ants of seeds of *Trillium* species (Liliaceae). Journal of Ecology, 75, 1091–1098.

Ohkawara, K. & Akino, T. (2005) Seed cleaning behavior by tropical ants and its anti-fungal effect. Journal of Ethology, 23, 93–98.

Olesen, J.M. & Warncke, E. (1989) Predation and potential transfer of pollen in a population of *Saxifraga hirculus*. Holarctic Ecology, 12, 87–95.

Oliveira, J.F.S., de Carvalho, J.P., de Sousa, R.F.X.B., & Simao, M.M. (1976) The nutritional value of four species of insects consumed in Angola. Ecology of Food and Nutrition, 5, 91–97.

Oliveira, P.S. & Leito-Filho, H.F. (1987) Extrafloral nectaries: Their taxonomic distribution and abundance in the woody flora of cerrado vegetation in southeast Brazil. Biotropica, 19, 140–148.

Ollerton, J. & Lack, A. (1996) Partial pre-dispersal seed predation in *Lotus corniculatus* L. (Fabaceae). Seed Science Research, 6, 65–69.

Olson, D.M. & Andow, D.A. (1998) Larval crowding and adult nutrition effects on longevity and fecundity of female *Trichogramma nubilale* Ertle & Davis (Hymenoptera: Trichogrammatidae). Environmental Entomology, 27, 508–514.

Olson, D.M., Fadamiro, H.Y., Lundgren, J.G., & Heimpel, G.E. (2000) Effects of sugar feeding on carbohydrate and lipid metabolism in a parasitoid wasp. Physiological Entomology, 25, 17–26.

Olson, D.M., Rains, G.C., Meiners, T., Takasu, K., Tertuliano, M., Tumlinson, J.H., Wäckers, F.L., & Lewis, W.J. (2003) Parasitic wasps learn and report diverse chemicals with unique conditional behaviors. Chemical Senses, 28, 545–549.

Olson, D.M., Takasu, K., & Lewis, W.J. (2005) Food needs of adult parasitoids: Behavioral adaptations and consequences. In Plant-Provided Foods for Carnivorous Insects: A Protective Mutualism and Its Applications (eds F.L. Wäckers, J. Bruin, & P.C.J. van Rijn), pp. 137–147. Cambridge University Press, Cambridge, MA.

Omkar (2006) Suitability of different foods for a generalist ladybird, Micraspis discolor (Coleoptera: Coccinellidae). International Journal of Tropical Insect Science, 26, 35–40.

Opler, P.A. (1983) Nectar production in a tropical ecosystem. In The Biology of Nectaries (eds B.L. Bentley & T.S. Elias), pp. 30–79. Columbia University Press, New York.

Orr, D.B. & Boethel, D.J. (1986) Influence of plant antibiosis through four trophic levels. Oecologia, 70, 242–249.

Orr, D.B. & Landis, D.A. (1997) Oviposition of European corn borer (Lepidoptera: Pyralidae) and impact of natural enemy populations in transgenic versus isogenic corn. Journal of Economic Entomology, 90, 905–909.

Osakabe, M. (1988) Relationships between food substances and developmental success in *Amblyseius sojaensis* Ehara (Acarina: Phytoseiidae). Applied Entomology and Zoology, 23, 45–51.

Osakabe, M., Inoue, K., & Ashihara, W. (1986) Feeding, reproduction and development of *Amblyseius sojaensis* Ehara (Acarina: Phytoseiidae) on two species of spider mites and on tea pollen. Applied Entomology and Zoology, 21, 322–327.

Osbourn, A.E., Qi, X., Townsend, B., & Qin, B. (2003) Dissecting plant secondary metabolism – constitutive chemical defences in cereals. New Phytologist, 159, 101–108.

Osbourn, A.E., Wubben, J.P., Melton, R.E., Carter, J.P., & Daniels, M.J. (1998) Saponins and plant defense. In Chemical Signals and Plant-Microbe Interactions (eds J.T. Romeo, K.R. Downum, & R. Verpoorte), Vol. 32, pp. 1–16. Plenum Press, New York.

Ott, J. (1998) The delphic bee: Bees and toxic honeys as pointers to psychoactive and other medicinal plants. Economic Botany, 52, 260–266.

Ouyang, Y., Grafton-Cardwell, E.E., & Bugg, R.L. (1992) Effects of various pollens on development, survivorship, and reproduction of *Euseius tularensis* (Acari: Phytoseiidae). Environmental Entomology, 21, 1371–1376.

Overmeer, W.P.J. (1981) Notes on breeding phytoseiid mites from orchards (Acarina: Phytoseiidae) in the laboratory. Mededelingen Faculteit Landbouwwetenschappen Rijks universiteit Gent 46, 503–509.

Overmeer, W.P.J. (1985) Chapter 2.1.3.2, Alternative prey and other food resources. In Spider Mites: Their Biology, Natural Enemies, and Control (eds W. Helle & M.W. Sabelis), Vol. 1B, pp. 131–140. Elsevier, Amsterdam, The Netherlands.

Overmeer, W.P.J. & van Zon, A.Q. (1983) The effect of different kinds of food on the induction of diapause in the predaceous mite *Amblyseius potentillae*. Entomologia Experimentalis et Applicata, 33, 27–30.

Pacini, E., Nepi, M., & Vesprini, J.L. (2003) Nectar biodiversity: A short review. Plant Systematics and Evolution, 238, 7–21.

Paarmann, W. (2002a) Determination of food quality for successful development of seed-feeding ground beetles (Col., Carabidae) from tropical lowland rainforests. In *How to Protect or What We Know About Carabid Beetles.* (ed J.e.a. Szyszko), Warsaw Agricultural University Press, Warsaw, Poland. Pages 45–57

Paarmann, W., Gutzmann, B., Stumpe, P., Bolte, H., Kuppers, S., Holzkamp, K., & Niers, C. (2002b) The structure of ground beetle assemblages (Coleoptera: Carabidae) at fruit falls of Melastomataceae trees in a Brazilian terre firme rain forest. Biotropica, 34(3), 368–375.

Paarmann, W., Adis, J., Stork, N.E., Gutzmann, B., Stumpe, P., Staritz, B., Bolte, H., Kuppers, S., Holzkamp, K., Niers, C., & Da Fonseca, C.R.V. (2001) The structure of ground beetles assemblages (Coleoptera: Carabidae) at fig fruit falls (Maraceae) in a terra firme rain forest near Manaus (Brazil). Journal of Tropical Ecology, 17, 549–561.

Paarmann, W., Adis, J., Stork, N.E., Stumpe, P., Gutzmann, B., & Holzkamp, K. (2003) Finding the host tree species of *Notiobia nebrioides* Perty (Coleoptera: Carabidae), a member of the seed-feeding guild at fruit falls in Amazonian non-inundated lowland rainforest. Journal of Natural History, 37, 839–844.

Paarmann, W., Faust, N., Arndt, E., Luchtrath, I., & Rohe, W. (2006) Constant seed size and mandible growth- a fundamental problem for granivorous ground beetle larvae (Coleoptera: Carabidae). Entomologica Fennica, 17, 334–339.

Painter, R.H. (1951) Insect Resistance in Crop Plants. Macmillan, New York.

Pair, S.D., Wiseman, B.R., & Sparks, A.N. (1986) Influence of four corn cultivars on fall armyworm (Lepidoptera: Noctuidae) establishment and parasitization. Florida Entomologist, 69, 566–570.

Palmer-Jones, T. & Line, L.J.S. (1962) Poisoning of honey bees by nectar from the karaka tree (Corynocarpus laevigata J. R. et G. Forst.). New Zealand Journal of Agricultural Research, 5, 433–436.

Parmenter, L. (1953a) The hover-flies (Syrphidae). Entomologist's Record, 65, 154–159.

Parmenter, L. (1953b) The hover-flies (Syrphidae). Entomologist's Record, 65, 185–190.

Patt, J.M., Hamilton, G.C., & Lashomb, J.H. (1997) Foraging success of parasitoid wasps on flowers: Interplay of insect morphology, floral architecture and searching behavior. Entomologia Experimentalis et Applicata, 83, 21–30.

Patt, J.M., Hamilton, G.C., & Lashomb, J.H. (1999) Response of two parasitoid wasps to nectar odors as a function of experience. Entomologia Experimentalis et Applicata, 90, 1–8.

Patt, J.M., Wainwright, S.C., Hamilton, G.C., Whittinghill, D., Bosley, K., Dietrick, J., & Lashomb, J.H. (2003) Assimilation of carbon and nitrogen from pollen and nectar by a predaceous larva and its effects on growth and development. Ecological Entomology, 28, 717–728.

Pausch, R.D. (1979) Observations on the biology of the seed corn beetles, *Stenolophus comma* and *Stenolophus lecontei*. Annals of the Entomological Society of America, 72, 24–28.

Pausch, R.D. & Pausch, L.M. (1980) Observations on the biology of the slender seedcorn beetle, *Clivina impressifrons* (Coleoptera: Carabidae). Great Lakes Entomologist, 13, 189–194.

Payne, P.I. (1983) Breeding for protein quantity and protein quality in seed crops. In Seed Proteins (eds J. Daussant, J. Mosse, & J. Vaughan), pp. 223–253. Academic Press, London.

Peakall, R., Handel, S.N., & Beattie, A.J. (1991) The evidence for, and importance of, ant pollination. In Ant-Plant Interactions (eds C.R. Huxley & D.F. Cutler), pp. 421–429. Oxford University Press, Oxford.

Pellmyr, O. (1985) Yellow jackets disperse *Vancouveria* seeds (Berberidaceae). Madroño, 32, 56.

Pemberton, R.W. (1992) Fossil extrafloral nectaries, evidence for the ant-guard antiherbivore defense in an Oligocene *Populus*. American Journal of Botany, 79, 1242–1246.
Pemberton, R.W. (1993) Observations of extrafloral nectary feeding by predaceous and fungivorous mites. Proceedings of the Entomological Society of Washington, 95, 642–643.
Pemberton, R.W. (1998) The occurrence and abundance of plants with extrafloral nectaries, the basis for antiherbivore defensive mutualisms, along with a latitudinal gradient in east Asia. Journal of Biogeography, 25, 661–668.
Pemberton, R.W. & Lee, J.-H. (1996) The influence of extrafloral nectaries on parasitism of an insect herbivore. American Journal of Botany, 83, 1187–1194.
Pemberton, R.W. & Vandenberg, N.J. (1993) Extrafloral nectar feeding by ladybird beetles (Coleoptera: Coccinellidae). Proceedings of the Entomological Society of Washington, 95, 139–151.
Percival, M.S. (1961) Types of nectar in angiosperms. New Phytologist, 60, 235–281.
Pereira, J.F. & Splittstoesser, W.E. (1987) Exudate from cassava leaves. Agriculture, Ecosystems and Environment, 18, 191–194.
Pernollet, J.C. & Mosse, J. (1983) Structure and location of legume and cereal seed storage proteins. In Seed Proteins (eds J. Daussant, J. Mosse, & J. Vaughan), pp. 155–191. Academic Press, London.
Perrin, R.M. (1975) The role of the perennial stinging nettle, *Urtica dioica*, as a reservoir of beneficial natural enemies. Annals of Applied Biology, 81, 289–297.
Petanidou, T. (2005) Sugars in Mediterranean floral nectars: An ecological and evolutionary approach. Journal of Chemical Ecology, 31, 1065–1088.
Petanidou, T. & Vokou, D. (1990) Pollination and pollen energetics in Mediterranean ecosystems. American Journal of Botany, 77, 986–992.
Petanidou, T., Van Laere, A.J., & Smets, E. (1996) Change in floral nectar components from fresh to senescent flower of *Capparis spinosa* (Capparidaceae), a nocturnally flowering Mediterranean shrub. Plant Systematics and Evolution, 199, 79–92.
Petanidou, T., Goethals, V., & Smets, E. (1999) The effect of nutrient and water availability on nectar secretion and nectary structure of the dominant Labiatae species of Phrygana. Systematics and Geography of Plants, 68, 233–244.
Petanidou, T., Goethals, V., & Smets, E. (2000) Nectary structure of Labiatae in relation to their nectar secretion and characteristics in a Mediterranean shrub community – does flowering time matter? Plant Systematics and Evolution, 225, 103–118.
Petanidou, T., Van Laere, A., Ellis, W.N., & Smets, E. (2006) What shapes amino acid and sugar composition in Mediterranean floral nectars? Oikos, 115, 155–169.
Petralia, R.S. & Vinson, S.B. (1978) Feeding in the larvae of the imported fire ant, *Solenopsis invicta*: Behavior and morphological adaptations. Annals of the Entomological Society of America, 71, 643–648.
Phaff, H.J., Miller, M.W., & Mrak, E.M. (1966) The Life of Yeasts. Harvard University Press, Cambridge, MA.
Phillips, K.M., Ruggio, D.M., & Ashraf-Khorassani, M. (2005) Phytosterol composition of nuts and seeds commonly consumed in the United States. Journal of Agricultural and Food Chemistry, 53, 9436–9445.
Picard-Nizou, A.L., Pham-Delegue, M.-H., Kerguelen, V., Douault, P., Marillaeu, R., Olsen, L., Grison, R., Toppan, A., & Masson, C. (1995) Foraging behaviour of honey bees (*Apis mellifera* L.) on transgenic oilseed rape (*Brassica napus* L. var. *oleifera*). Transgenic Research, 4, 270–276.
Pickett, C.H. & Clark, W.D. (1979) The function of extrafloral nectaries in *Opuntia acanthocarpa* (Cactaceae). American Journal of Botany, 66, 618–625.
Pickett, C.H., Schoenig, S.E., & Hoffmann, M.P. (1996) Establishment of the squash bug parasitoid, *Trichopoda pennipes* Fabr. (Diptera: Tachinidae), in northern California. Pan-Pacific Entomologist, 72, 220–226.
Pielou, D.P. & Verma, A.N. (1968) The arthropod fauna associated with the birch bracket fungus, *Polyporus betulinus*, in eastern Canada. Canadian Entomologist, 100, 1179–1199.
Pierce, W.D. (1917) A Manual of Dangerous Insects. U.S. Department of Agriculture, Government Printing Office, Washington, DC.

Pierre, J., Marsault, D., Genecque, E., Renard, M., Champolivier, J., & Pham-Delegue, M.-H. (2003) Effects of herbicide-tolerant transgenic oilseed rape genotypes on honey bees and other pollinating insects under field conditions. Entomologia Experimentalis et Applicata, 108, 159–168.

Piffanelli, P., Ross, J.H.E., & Murphy, D.J. (1997) Intra- and extracellular lipid composition and associated gene expression patterns during pollen development in *Brassica napus*. The Plant Journal, 11, 549–562.

Pilcher, C.D., Obrycki, J.J., Rice, M.E., & Lewis, L.C. (1997) Preimaginal development, survival, and field abundance of insect predators on transgenic *Bacillus thuringiensis* corn. Environmental Entomology, 26, 446–454.

Pimentel de Carvalho, A.C. & Message, D. (2004) A scientific note on the toxic pollen of *Stryphnodendron polyphyllum* (Fabaceae, Mimosoideae) which causes sacbrood-like symptoms. Apidologie, 35, 89–90.

Pimentel, D. & Wheeler, A.G., Jr. (1973) Influence of alfalfa resistance on a pea aphid population and its associated parasites, predators, and competitors. Environmental Entomology, 2, 1–11.

Pizo, M.A. & Oliveira, P.S. (1998) Interaction between ants and seeds of a nonmyrmecochorous neotropical tree, *Cabralea canjerana* (Meliaceae), in the Atlantic forest of southeast Brazil. American Journal of Botany, 85, 669–674.

Pizo, M.A. & Oliveira, P.S. (2001) Size and lipid content of nonmyrmecochorous diaspores: Effects on the interaction with litter-foraging ants in the Atlantic rainforest of Brazil. Plant Ecology, 157, 37–52.

Platt, W.J. (1976) The natural history of a fugitive prairie plant (*Mirabilis hirsuta* (Pursh) MacM.). Oecologia, 22, 399–409.

Pollard, S.D., Beck, M.W., & Dodson, G.N. (1995) Why do male crab spiders drink nectar? Animal Behaviour, 49, 1443–1448.

Poltev, V.I. (1956) Food toxicosis of bees and their diagnosis. (In Russian).

Ponsard, S., Gutierrez, A.P., & Mills, N.J. (2002) Effect of *Bt*-toxin (Cry1Ac) in transgenic cotton on the adult longevity of four heteropteran predators. Environmental Entomology, 31, 1197–1205.

Pontin, D.R., Wade, M.R., Kehrli, P., & Wratten, S.D. (2007) Attractiveness of single and multiple species flower patches to beneficial insects in agroecosystems. Annals of Applied Biology, 148, 39–47.

Poppy, G.M. & Powell, W. (2004) Genetic manipulation of natural enemies: Can we improve biological control by manipulating the parasitoid and/or the plant? In Genetics, Evolution and Biological Control (eds L.E. Ehler, R. Sforza, & T. Mateille), pp. 219–233. CABI, Wallingford.

Potrikus, C.J.B., J. A. (1981) Gut bacteria recycle uric acid nitrogen in termites: A strategy for nutrient conservation. Proceedings of the National Academy of Sciences, 78, 4601–4605.

Poutaraud, A. & Girardin, P. (2005) Influence of chemical characteristics of soil on mineral and alkaloid seed contents of *Colchicum autumnale*. Environmental and Experimental Botany, 54, 101–108.

Povey, F.D., Smith, H., & Watt, T.A. (1993) Predation of annual grass weed seeds in arable field margins. Annals of Applied Biology, 122, 323–328.

Powell, W. & Zhi-Li, Z. (1983) The reactions of two cereal aphid parasitoids, *Aphidius uzbekistanicus* and *A. ervi* to host aphids and their food plants. Physiological Entomology, 8, 439–443.

Prado, S.S., Rubinoff, D., & Almeida, R.P.P. (2006) Vertical transmission of a pentatomid caeca-associated symbiont. Annals of the Entomological Society of America, 99, 577–585.

Prasifka, J.R., Schmidt, N.P., Kohler, K.A., O'Neal, M.E., Hellmich, R.L., & Singer, J.W. (2006) Effects of living mulches on predator abundance and sentinel prey in a corn–soybean–forage rotation. Environmental Entomology, 35, 1423–1431.

Prosser, W.A. & Douglas, A.E. (1991) The aposymbiotic aphid: An analysis of chlortetracycline-treated pea aphid, *Acyrthosiphon pisum*. Journal of Insect Physiology, 37, 713–719.

Pryce-Jones, J. (1942) Some problems associated with nectar, pollen, and honey. Proceedings of the Linnean Society, 155, 129–171.

Pryce-Jones, J. (1944) Some problems associated with nectar, pollen, and honey. Proceedings of the Linnean Society, 155, 129–171.

Prys-Jones, O.E. & Willmer, P.G. (1992) The biology of alkaline nectar in the purple toothwort (*Lathraea clandestina*): Ground level defences. Biological Journal of the Linnean Society, 45, 373–388.

Pulliam, H.R. & Brand, M.R. (1975) The production and utilization of seeds in plains grassland of Southeastern Arizona. Ecology, 56, 1158–1166.

Pusztai, A., Croy, R.R.D., Grant, G., & Stewart, J.C. (1983) Seed lectins: Distribution, location and biological role. In Seed Proteins (eds J. Daussant, J. Mosse, & J. Vaughan), pp. 53–82. Academic Press, London.

Putnam, W.L. (1955) Bionomics of *Stethorus punctillum* Weise (Coleoptera: Coccinellidae) in Ontario. Canadian Entomologist, 87, 9–33.

Putnam, W.L. (1962) Life-history and behaviour of the predacious mite *Typhlodromus* (T.) *caudiglans* Schuster (Acarina: Phytoseiidae) in Ontario, with notes on prey of related species. Canadian Entomologist, 94, 163–177.

Putnam, W.L. (1963) Nectar of peach leaf glands as insect food. Canadian Entomologist, 95, 108–109.

Putnam, W.L. (1964) Occurrence and food of some coccinellids (Coleoptera) in Ontario peach orchards. Canadian Entomologist, 96, 1149–1155.

Pyke, G.H. (1991) What does it cost a plant to produce floral nectar? Nature, 350, 58–59.

Quicke, D.L.J. (1997) Parasitic Wasps. Chapman & Hall, London.

Quinlan, R.J. & Cherrett, J.M. (1978) Studies on the role of the infrabuccal pocket of the leaf-cutting ant *Acromyrmex octospinosus* (Reich) (Hym., Formicidae). Insectes Sociaux, 25, 237–245.

Ragsdale, D.W., Larson, A.D., & Newsom, L.D. (1979) Microorganisms associated with feeding and from various organs of *Nezara viridula*. Journal of Economic Entomology, 72, 725–727.

Ragusa, S. & Swirski, E. (1977) Feeding habits, post-embryonic and adult survival, mating virility and fecundity of the predacious mite *Amblyseius swirskii* [Acarina: Phytoseiidae] on some coccids and mealybugs. Entomophaga, 22, 383–392.

Rankin, M.A. & Rankin, S. (1980) Some factors affecting presumed migratory flight activity of the convergent ladybeetle, *Hippodamia convergens* (Coccinellidae: Coleoptera). Biological Bulletin, 158, 356–369.

Rao, K.V., Rathore, K.S., Hodges, T.K., Fu, X., Stoger, E., Sudhaker, D., Williams, S., Christou, P., Bharathi, M., Brown, D.P., Powell, K.S., Spence, J., Gatehouse, A.M.R., & Gatehouse, J.A. (1998) Expression of snowdrop lectin (GNA) in transgenic rice plants confers resistance to rice brown planthopper. Plant Journal, 15, 469–477.

Raps, A., Kehr, J., Gugerli, P., Moar, W.J., Bigler, F., & Hilbeck, A. (2001) Immunological analysis of phloem sap of *Bacillus thuringiensis* corn and of the nontarget herbivore *Rhopalosiphum padi* (Homoptera: Aphididae) for the presence of Cry1Ab. Molecular Ecology, 10, 525–533.

Rapusas, H.R., Bottrell, D.G., & Coll, M. (1996) Intraspecific variation in chemical attraction of rice to insect predators. Biological Control, 6, 394–400.

Rashbrook, V.K., Compton, S.G., & Lawton, J.H. (1992) Ant-herbivore interactions: Reasons for the absence of benefits to a fern with foliar nectaries. Ecology, 73, 2167–2174.

Rassam, M. & Laing, W. (2005) Variation in ascorbic acid and oxalate levels in the fruit of *Actinidia chinensis* tissues and genotypes. Journal of Agricultural and Food Chemistry, 53, 2322–2326.

Raven, P.H., Evert, R.F. & Eichhorn, S.E. (1992) *Biology of Plants*, 5th edn. Worth Publishers, New York, NY.

Read, D.P., Feeny, P.P., & Root, R.B. (1970) Habitat selection by the aphid parasite *Diaeretiella rapae* (Hymenoptera: Braconidae) and hyperparasite *Charips brassicae* (Hymenoptera: Cynipidae). Canadian Entomologist, 102, 1567–1578.

Reader, R.J. (1991) Control of seedling emergence using ground cover: A potential mechanism involving seed predation. Canadian Journal of Botany, 69, 2084–2087.

Reader, R.J. (1993) Control of seedling emergence by ground cover and seed predation in relation to seed size for some old field species. Journal of Ecology, 81, 169–175.

Reader, T., MacCleod, I., Elliott, P.T., Robinson, O.J., & Manica, A. (2005) Inter-order interactions between flower-visiting insects: Foraging bees avoid flowers previously visited by hoverflies. Journal of Insect Behavior, 18, 51–57.

Ready, C.C. & Vinson, S.B. (1995) Seed selection by the red imported fire ant (Hymenoptera: Formicidae) in the laboratory. Environmental Entomology, 24, 1422–1431.

Rebek, E.J. & Sadof, C.S. (2003) Effects of pesticide applications on the euonymous scale (Homoptera: Diaspididae) and its parasitoid, *Encarsia citrina* (Hymenoptera: Aphelinidae). Journal of Economic Entomology, 96, 446–452.

Reed, G.L., Jenson, A.S., Riebe, J., Head, G., & Duan, J.J. (2001) Transgenic Bt potato and conventional insecticides for Colorado potato beetle management: Comparative efficacy and non-target impacts. Entomologia Experimentalis et Applicata, 100, 89–100.

Reichman, O.J. (1976) Relationships between dimensions, weights, volumes, and calories of some Sonoran desert seeds. Southwestern Entomologist, 20, 573–586.

Reichman, O.J. (1979) Desert granivore foraging and its impact on seed densities and distributions. Ecology, 60, 1085–1092.

Reid, C.D. & Lampman, R.L. (1989) Olfactory response of *Orius insidiosus* (Hemiptera: Anthocoridae) to volatiles of corn silks. Journal of Chemical Ecology, 15, 1109–1115.

Rhoades, D.F. & Bergdahl, J.C. (1981) Adaptive significance of toxic nectar. The American Naturalist, 117, 798–803.

Ricci, C. (1982) Sulla costituzione e funyione delle mandibole delle larve di *Tytthaspis sedecimpunctata* (L.) e *Tytthaspis trilineata* (Weise). Frustula Entomologica, III, 205–212.

Ricci, C. (1986a). Food strategy of *Tytthaspis sedecimpunctata* in different habitats. In Ecology of Aphidophaga (ed I. Hodek), pp. 311–316. Dr. W. Junk, Dordrecht, The Netherlands.

Ricci, C. (1986b). Seasonal food preferences and behavior of *Rhyzobius litura* (F.). In Ecology of Aphidophaga (ed I. Hodek), pp. 119–123. Dr. W. Junk, Dordrecht, The Netherlands.

Ricci, C. & Ponti, L. (2005) Seasonal food of *Ceratomegilla notata* (Coleoptera: Coccinellidae) in mountain environments of the Northern Italian Alps. European Journal of Entomology, 102, 527–530.

Ricci, C., Fiori, G., & Colazza, S. (1983) Regime alimentare dell'adulto di *Tytthaspsis sedecimpunctata* (L.) (Coleoptera: Coccinellidae) in ambiente a influenza antropica primaria: Prato polifita. Atti XIII Congresso Nazionalte Italiana de Entomologia Sestriere-Torino, 691–698.

Ricci, C., Ponti, L., & Pires, A. (2005) Migratory flights and pre-diapause feeding of Coccinella septempunctata (Coleoptera: Coccinellidae) adults in agricultural and mountain ecosystems of Central Italy. European Journal of Entomology, 102, 531–538.

Richards, D.R. & Evans, E.W. (1998) Reproductive responses of aphidophagous lady beetles (Coleoptera: Coccinellidae) to nonaphid diets: An example from alfalfa. Annals of the Entomological Society of America, 91, 632–640.

Richards, P.C. & Schmidt, J.M. (1996) The suitability of some natural and artificial substrates as oviposition sites for the insidious flower bug, *Orius insidiosus*. Entomologia Experimentalis et Applicata, 80, 325–333.

Richmond, D.S., Grewel, P.S., & Cardina, J. (2003) Competition between *Lolium perenne* and *Digitaria sanguinalis*: Ecological consequences for habouring an endosymbiotic fungus. Journal of Vegetation Science, 14, 835–840.

Riddick, E.W. & Barbosa, P. (1998) Impact of Cry3A-intoxicated *Leptinotarsa decemlineata* (Coleoptera: Chrysomelidae) and pollen on consumption, development, and fecundity of *Coleomegilla maculata* (Coleoptera: Coccinellidae). Annals of the Entomological Society of America, 91, 303–307.

Ridgway, R.L. & Jones, S.L. (1968) Plant feeding by *Geocoris pallens* and *Nabis americoferus*. Annals of the Entomological Society of America, 61, 232–233.

Ridgway, R.L. & Lindquist, D.A. (1966) Systemic activity of Shell SD-9129 in cotton plants. Journal of Economic Entomology, 59, 961–964.

Ridgway, R.L., Lingren, P.D., Cowan, C.B., Jr., & Davis, J.W. (1967) Populations of arthropod predators and *Heliothis* spp. after applications of systemic insecticides to cotton. Journal of Economic Entomology, 60, 1012–1016.

Ridley, H.N. (1930) The Dispersal of Plants Throughout the World. L. Reeve, Ashford, Kent.

Ridley, P., Howse, P.E., & Jackson, C.W. (1996) Control of the behaviour of leaf-cutting ants by their 'symbiotic' fungus. Experientia, 52, 631–635.

Riggin, T.M., Wiseman, B.R., Isenhour, D.J., & Espelie, K.E. (1992) Incidence of fall armyworm (Lepidoptera: Noctuidae) parasitoids on resistant and susceptible corn genotypes. Environmental Entomology, 21, 888–895.

Risch, S.J., Andow, D.A., & Altieri, M.A. (1983) Agroecosystem diversity and pest control: Data, tentative conclusions, and new research directions. Environmental Entomology, 12, 625–629.

Rissing, S.W. (1981) Foraging specializations in individual seed-harvester ants. Behavioral Ecology and Sociobiology, 9, 149–152.

Rissing, S.W. (1986) Indirect effects of granivory by harvester ants: Plant species composition and reproductive increase near nest sites. Oecologia, 68, 231–234.

Rissing, S.W. (1987) Annual cycles in worker size of the seed-harvester ant, *Veromessor pergandei* (Hymenoptera: Formicidae). Behavioral Ecology and Sociobiology, 20, 117–124.

Rissing, S.W. & Wheeler, J. (1976) Foraging responses of *Veromessor pergandei* to changes in seed production. Pan-Pacific Entomologist, 52, 63–72.

Roberts, J.T. & Heithaus, E.R. (1986) Ants rearrange the vertebrate-generated seed shadow of a neotropical fig tree. Ecology, 67, 1046–1051.

Robinson, K.A., Jonsson, M., Wratten, S.D., Wade, M.R., & Buckley, H.L. (2008) Implications of floral resources for predation by an omnivorous lacewing. Basic and Applied Ecology, 9, 172–181.

Robinson, T. (1979) The evolutionary ecology of alkaloids. In Herbivores: The Interaction with Secondary Plant Metabolites (eds G.A. Rosenthal & D.H. Janzen), pp. 419–448. Academic Press, New York.

Rockwood, L.P. (1952) Notes on coccinellids in the Pacific northwest. Pan-Pacific Entomologist, 28, 139–147.

Rodgerson, L. (1998) Mechanical defense in seeds adapted for ant dispersal. Ecology, 79, 1669–1677.

Rodriguez, D.B., Lee, T.-C., & Chichester, C.O. (1975) Comparative study of the carotenoid composition of seeds of ripening *Momordica charantia* and tomatoes. Plant Physiology, 56, 626–629.

Roe, M.A., Finglas, P.M., & Church, S.M. (2002) McCance and Widdowson's The Composition of Foods, 6th edn. Royal Society of Chemistry, London.

Rogers, D.J. & Sullivan, J. (1987) Growth of *Geocoris punctipes* (Hemiptera: Lygaeidae) on attached and detached leaves of pest-resistant soybeans. Journal of Entomological Science, 22, 282–285.

Rogers, D.J. & Sullivan, M.J. (1986) Nymphal performance of *Geocoris punctipes* (Hemiptera: Lygaeidae) on pest-resistant soybeans. Environmental Entomology, 15, 1032–1036.

Rogers, M.A., Krischik, V.A., & Martin, L.A. (2007) Effect of soil application of imidacloprid on survival of adult green lacewing, *Chrysoperla carnea* (Neuroptera: Chrysopidae), used for biological control in greenhouse. Biological Control, 42, 172–177.

Rogers, M.E. & Potter, D.A. (2004) Potential for sugar sprays and flowering plants to increase parasitism of white grubs (Coleoptera: Scarabaeidae) by tiphiid wasps (Hymenoptera: Tiphiidae). Environmental Entomology, 33, 619–626.

Rohi, L., Bourarach, K., Chemseddine, M., & Pintureau, B. (2002) Effect of food on the longevity of two Hymenopteran egg parasitoids. Bulletin Mensuel de la Societe Linneenne de Lyon, 71, 118–122.

Rohlfs, M., Albert, M., Keller, N.P., & Kempken, F. (2007) Secondary chemicals protect mould from fungivory. Biology Letters, 223, 523–525.

Romeis, J. & Zebitz, C.P.W. (1997) Searching behaviour of *Encarsia formosa* as mediated by colour and honeydew. Entomologia Experimentalis et Applicata, 82, 299–309.

Romeis, J., Babendreier, D., & Wäckers, F.L. (2003) Consumption of snowdrop lectin (*Galanthus nivalis* agglutinin) causes direct effects on adult parasitic wasps. Oecologia, 134, 528–536.

Romeis, J., Dutton, A., & Bigler, F. (2004) *Bacillus thuringiensis* toxin (Cry1Ab) has no direct effect on larvae of the green lacewing Chrysoperla carnea (Stephens) (Neuroptera: Chrysopidae). Journal of Insect Physiology, 50, 175–183.

Romeis, J., Meissle, M., & Bigler, F. (2006) Transgenic crops expressing *Bacillus thuringiensis* toxins and biological control. Nature Biotechnology, 24, 63–71.

Rondon, S.I., Price, J.F., & Cantliffe, D.J. (2006) Developmental time, reproduction, and feeding of two subspecies of *Coleomegilla maculata* (Coleoptera: Coccinellidae) in the laboratory. Florida Entomologist, 89, 85–88.

Rose, U.S.R., Lewis, J., & Tumlinson, J.H. (2006) Extrafloral nectar from cotton (*Gossypium hirsuitum*) as food source for parasitic wasps. Functional Ecology, 20, 67–74.

Rosenthal, G.A. & Bell, E.A. (1979) Naturally occurring, toxic nonprotein amino acids. In Herbivores: Their Interactions with Secondary Plant Metabolites (eds G.A. Rosenthal & D.H. Janzen), pp. 353–386. Academic Press, New York.

Rotheray, G.E. (1990) A new species of *Bioblapsis* (Hymenoptera: Ichneumonidae) from Scotland parasitising a mycophagous hoverfly, *Cheilosia longula* (Diptera: Syrphidae). Entomologica Scandinavica, 21, 277–280.

Rotheray, G.E. & Gilbert, F. (1999) Phylogeny of Palaearctic Syrphidae (Diptera): Evidence from larval stages. Zoological Journal of the Linnean Society, 127, 1–112.

Rothschild, M., von Euw, J., & Reichstein, T. (1970) Cardiac glycosides in the oleander aphid, *Aphis nerii*. Journal of Insect Physiology, 16, 1141–1145.

Roulston, T.A. & Buchmann, S.L. (2000) A phylogenetic reconsideration of the pollen starch-pollination correlation. Evolutionary Ecology Research, 2, 627–643.

Roulston, T.A. & Cane, J.H. (2000) Pollen nutritional content and digestibility for animals. Plant Systematics and Evolution, 222, 187–209.

Roulston, T.A., Cane, J.H., & Buchmann, S.L. (2000) What governs protein content of pollen: Pollinator preferences, pollen-pistil interactions, or phylogeny? Ecological Monographs, 70, 617–643.

Ruffner, G.A. & Clark, W.D. (1986) Extrafloral nectar of Ferocactus acanthodes (Cactaceae): Composition and its importance to ants. American Journal of Botany, 73, 185–189.

Ruiz, M.A. & Sotelo, A. (2001) Chemical composition, nutritive value, and toxicology evaluation of Mexican wild lupins. Journal of Agricultural and Food Chemistry, 49, 5336–5339.

Rumsey, G.L., Hughes, S.G., Smith, R.R., Kinsella, J.E., & J., S.K. (1991) Digestibility and energy values of intact, disrupted and extracts from brewer's dried yeast fed to rainbow trout (*Oncorhynchus mykiss*). Animal Feed Science and Technology, 33, 185–193.

Rurhen, S. & Handel, S.N. (1999) Jumping spiders (Salticidae) enhance the seed production of a plant with extrafloral nectaries. Oecologia, 119, 227–230.

Russell, E.P. (1989) Enemies hypothesis: A review of the effect of vegetational diversity on predatory insects and parasitoids. Environmental Entomology, 18, 590–599.

Rusterholz, H.P. & Erhardt, A. (1998) Effects of elevated CO_2 on flowering phenology and nectar production of nectar plants important for butterflies of calcareous grasslands. Oecologia, 113, 341–349.

Ryan, C.A. (1973) Proteolytic enzymes and their inhibitors in plants. Annual Review of Plant Physiology, 24, 173–196.

Ryan, C.A. (1979) Proteinase inhibitors. In Herbivores: Their Interactions with Plant Secondary Metabolites (eds G.A. Rosenthal & D.H. Janzen), pp. 599–618. Academic Press, New York.

Sadasivam, K.V. (1970) On the composition of leaf exudate and leaf leachate of tapioca (*Manihot utilissima* Phol.) foliage. Science and Culture, 36, 608–609.

Sahayaraj, K. (2002) Small scale laboratory rearing of a reduviid predator, *Rhynocoris marginatus* Fab. (Hemiptera: Reduviidae) on *Corcyra cephalonica* Stainton larvae by larval card method. Journal of Central European Agriculture, 3, 137–147.

Saikkonen, K., Helander, M., Faeth, S.H., Schulthess, F., & Wilson, D. (1999) Endophyte-grass-herbivore interactions: The case of *Neotyphodium* endophytes in Arizona fescue populations. Oecologia, 121, 411–420.

Sakurai, H. (1968) Physiological studies on the digestion of coccinellid beetles (Coleoptera: Coccinellidae), with special reference to their food habits. Applied Entomology and Zoology, 3, 130–138.

Salas-Aguilar, J. & Ehler, L.E. (1977) Feeding habits of *Orius tristicolor*. Annals of the Entomological Society of America, 70, 60–62.

Samways, M.J., Osborn, R., & Saunders, T.L. (1997) Mandible form relative to the main food type in ladybirds (Coleoptera: Coccinellidae). Biocontrol Science and Technology, 7, 275–286.

Sanchez-Pena, S.R. (2005) New view on origin of attine ant-fungus mutualism: Exploitation of preexisting insect-fungus symbiosis (Hymenoptera: Formicidae). Annals of the Entomological Society of America, 98, 151–164.
Sanchez-Tinoco, M.Y. & Engleman, E.M. (2004) Seed coat anatomy of *Ceratozamia mexicana* (Cycadales). The Botanical Review, 70, 24–38.
Sandhu, D.K. & Waraich, M.K. (1985) Yeasts associated with pollinating bees and flower nectar. Microbial Ecology, 11, 51–58.
Saska, P. (2004) Carabid larvae as predators of weed seeds: granivory in larvae of *Amara eurynota* (Coleoptera: Carabidae). *Community and Applied Biological Sciences, Ghent University*, 69(3), 27–31.
Saska, P., Van der Werf, W., de Vries, E. & Westerman, P. (2008) Spatial and temporal patterns of carabid activity-density in cereals do not explain levels of predation on weed seeds. Bulletin of Entomological Research, 98, 179–191.
Saska, P. (2005) Contrary food requirements of the larvae of two *Curtonotus* (Coleoptera: Carabidae: *Amara*) species. Annals of Applied Biology, 147, 139–144.
Saska, P. & Jarosik, V. (2001) Laboratory study of larval food requirements in nine species of *Amara* (Coleoptera: Carabidae). Plant Protection Science, 37, 103–110.
Savile, D.B.O. (1976) Evolution of the rust fungi (Uredinales) as reflected by their ecological problems. Evolutionary Biology, 9, 137–207.
Savoiskaya, G.I. (1970) Introduction and acclimatization of some coccinellids in the Alma-Ata reserve. Trudy Alma-Atin. Gos. Zapov., 9, 138–162.
Savoiskaya, G.I. (1983) Kokcinellidy. In Izdetelstvo Nauka Kazachskoi SSR, Alma-Ata., 246 pp.
Saxena, D. & Stotzky, G. (2001) *Bt* corn has a higher lignin content than non-*Bt* corn. American Journal of Botany, 88, 1704–1706.
Schaefer, C.W. & Panizzi, A.R. (2000) Heteroptera of Economic Importance. CRC Press, Boca Raton, FL.
Schausberger, P. (1992) Comparative investigations on the effect of different foods on development and reproduction of *Amblyseius aberrans* Oud. and *Amblyseius finlandicus* Oud. (Acarina: Phytoseiidae). Journal of Applied Entomology, 113, 476–486.
Schemske, D.W. (1980) The evolutionary significance of extrafloral nectar production by *Costus woodsonii* (Zingiberaceae): An experimental analysis of ant protection. Journal of Ecology, 68, 959–967.
Schiefelbein, J.W. & Chiang, H.C. (1966) Effects of spray of sucrose solution in a corn field on the populations of predatory insects and their prey. Entomophaga, 11, 333–339.
Schmale, I., Wäckers, F.L., Cardona, C., & Dorn, S. (2001) Control potential of three hymenopteran parasitoid species against the bean weevil in stored beans: The effect of adult parasitoid nutrition on longevity and progeny production. Biological Control, 21, 134–139.
Schmalhofer, V.R. (2001) Tritrophic interactions in a pollination system: Impacts of species composition and size of flower patches on the hunting success of a flower-dwelling spider. Oecologia, 129, 292–303.
Schmid, R. (1988) Reproductive versus extra-reproductive nectaries: Histological perspective and terminological recommendations. Botanical Review, 54, 179–227.
Schmidt, G.H. & Mathur, M. (1967) Changes in the carbohydrate content during the metamorphosis of different castes of *Formica polyctena* (Hymenoptera: Formicidae). Entomologia Experimentalis et Applicata, 10, 421–437.
Schmolz, E., Drutschmann, S., Schricker, B., & Lamprecht, I. (1999) Calorimetric measurements of energy contents and heat production rates during development of the wax moth *Galleria mellonella*. Thermochimica Acta, 337, 83–88.
Schmuck, R., Schoning, R., Stork, A., & Schramel, O. (2001) Risk posed to honeybees (Apis mellifera L, Hymenoptera) by an imidacloprid seed dressing if sunflowers. Pest Management Science, 57, 225–238.
Schneider, F. (1948) Beitrag zur Kenntnis der Generationsverhältnisse und Diapause räuberischer Schwebfliegen. Mitteilungen der Schweizertschen Entomologischen Gesellschaft 21, 249–285.

Schneider, F. (1969) Bionomics and physiology of aphidophagous Syrphidae. Annual Review of Entomology, 14, 103–124.

Schoener, T.W. (1971) Theory of feeding strategies. Annual Review of Ecology and Systematics, 2, 369–404.

Scholz, D. & Poehling, H.-M. (2000) Oviposition site selection of *Episyrphus balteatus*. Entomologia Experimentalis et Applicata, 94, 149–158.

Schoning, C., Espadaler, X., Hensen, I., & Roces, F. (2004) Seed predation of the tussock-grass *Stipa tenacissima* L. by ants (*Messor* spp.) in south-eastern Spain: The adaptive value of trypanocarpy. Journal of Arid Environments, 56, 43–61.

Schremmer, F. (1960) Beitrag zur Biologie von *Ditomus clypeatus* Rossi, eines kornersammelinden Carabiden. Zeitschrift der Arbeitsgemeinschaft österreichischer Entomologen 3, 140–146.

Schubart, H.O.B. & Anderson, A.B. (1978) Why don't ants visit flowers? A reply to D. H. Janzen. Biotropica, 10, 310–311.

Schuhmacher, H. & Hoffmann, H. (1982) Zur Funktion der Mundwerkzeuge von Schwebfliegen bei der Nahrungsaufnahme (Diptera: Syrphidae). Entomologia Generalis, 7, 327–342.

Schuler, T.H., Pottings, R.P.J., Denholm, I., & Poppy, G.M. (1999) Parasitoid behaviour and Bt plants. Nature, 400, 825–826.

Schuster, D.J. & Starks, K.J. (1975) Preference of *Lysiphlebus testaceipes* for greenbug resistant and susceptible small grain species. Environmental Entomology, 4, 887–888.

Schuster, M.F. & Calderon, M. (1986) Interactions of host plant resistant genotypes and beneficial insects in cotton ecosystems. In Interactions of Plant Resistance and Parasitoids and Predators of Insects (eds D.J. Boethel & R.D. Eikenbary), pp. 84–97. Ellis Horwood, Chichester.

Schuster, M.F., Lukefahr, M.J., & Maxwell, F.G. (1976) Impact of nectariless cotton on plant bugs and natural enemies. Journal of Economic Entomology, 69, 400–402.

Sclar, D.C., Gerace, D., & Cranshaw, W.S. (1998) Observations of population increases and injury by spider mites (Acari: Tetranychidae) on ornamental plants treated with Imidacloprid. Journal of Economic Entomology, 91, 350–355.

Scott, W.P., Snodgrass, G.L., & Smith, J.W. (1988) Tarnished plant bug (Hemiptera: Miridae) and predaceous arthropod populations in commercially produced selected nectaried and nectariless cultivars of cotton. Journal of Entomological Science, 23, 280–286.

Seagraves, M.P. & Yeargan, K.V. (2006) Selection and evaluation of a companion plant to indirectly augment densities of *Coleomegilla maculata* (Coleoptera: Coccinellidae) in sweet corn. Environmental Entomology, 35, 1334–1341.

Seigler, D.S. (2003) Phytochemistry of *Acacia – sensu lato*. Biochemical Systematics and Ecology, 31, 845–873.

Sernander, R. (1906) Entwurf einer Monographie der europaischen Myrmekochoren. Kunglica Svenska Vetenskapsakademinens Handlingar 41, 1–410.

Setamou, M., Bernal, J.S., Legaspi, J.C., & Mirkov, T.E. (2002a) Effects of snowdrop lectin (*Galanthus nivalis* agglutinin) expressed in transgenic sugarcane on fitness of *Cotesia flavipes* (Hymenoptera: Braconidae), a parasitoid of the nontarget pest *Diatraea saccharalis* (Lepidoptera: Crambidae). Annals of the Entomological Society of America, 95, 75–83.

Setamou, M., Bernal, J.S., Legaspi, J.C., & Mirkov, T.E. (2002b) Parasitism and location of sugarcane borer (Lepidoptera: Pyralidae) by *Cotesia flavipes* (Hymenoptera: Braconidae) on transgenic and conventional sugarcane. Environmental Entomology, 31, 1219–1225.

Shaginyan, E.G. (1956) Pchelovodstvo, 33, 45.

Shahjahn, M. (1974) *Erigeron* flowers as a food and attractive odor source for *Peristenus pseudopallipes*, a braconid parasitoid of the tarnished plant bug. Environmental Entomology, 3(1), 69–72.

Shaltiel, L. & Ayal, Y. (1998) The use of kairomones for foraging decisions by an aphid parasitoid in small host aggregations. Ecological Entomology, 23, 319–329.

Shaltiel, L. & Coll, M. (2004) Reduction of pear psylla damage by the predatory bug *Anthocoris nemoralis* (Hemiptera: Anthocoridae): The importance of orchard colonization time and neighboring vegetation. Biocontrol Science and Technology, 14, 811–821.

Shands, W.A., Simpson, G.W., & Brunson, M.H. (1972) Insect predators for controlling aphids on potatoes. 2. In small plots with two kinds of barriers, in small fields, or in large cages. Journal of Economic Entomology, 65, 514–518.

Shapiro, J.P. & Legaspi, J.C. (2006) Assessing biochemical fitness of predator *Podisus maculiventris* (Heteroptera: Pentatomidae) in relation to food quality: Effects of five species of prey. Annals of the Entomological Society of America, 99, 321–326.

Sheldon, J.K. & MacLeod, E.G. (1971) Studies on the biology of the Chrysopidae II. The feeding behavior of the adult of *Chrysopa carnea* (Neuroptera). Psyche, March–June, 107–121.

Shi, Y., Wang, W.B., Powell, K.S., Van Damme, E., Hilder, V.A., Gatehouse, A.M.R., Boulter, D., & Gatehouse, J.A. (1994) Use of the rice sucrose synthase-1 promoter to direct phloem-specific expression of B-glucoronidase and snowdrop lectin genes in transgenic tobacco plants. Journal of Experimental Botany, 45, 623–631.

Shieh, J.-N., Berry, R.E., Reed, G.L., & Rossignol, P.A. (1994) Feeding activity of green peach aphid (Homoptera: Aphididae) on transgenic potato expressing a *Bacillus thuringiensis* ssp. *tenebrionis* δ-endotoxin gene. Journal of Economic Entomology, 87, 618–622.

Shigenobu, S., Watanabe, H., Hattori, M., Sakaki, Y., & Ishikawa, H. (2000) Genome sequence of the endocellular bacterial symbiont of aphids *Buchnera* sp. APS. Nature, 407, 81–86.

Shih, C.T., Chang, H.Y., Hsu, P.H., & Hwang, Y.F. (1993) Responses of *Amblysieus ovalis* (Evans) (Acarina: Phytoseiidae) to natural food resources and two artificial diets. Experimental and Applied Acarology, 17, 503–519.

Shiraishi, A. & Kuwabara, M. (1970) The effects of amino acids on the labellar hair chemosensory cells of the fly. Journal of General Physiology, 56, 768–782.

Shou-Jian, C. & Fen-Wei, C. (1982) Bionomics and utilization of *Amblyseius deleoni* Muma et Demark (Acarina, Phytoseiidae). Acta Entomologica Sinica, 25, 49–55.

Shuel, R.W. (1955) Nectar secretion. American Bee Journal, 95, 229–234.

Shuel, R.W. (1957) Some aspects of the relation between nectar secretion and nitrogen, phosphorus, and potassium nutrition. Canadian Journal of Plant Science, 37, 220–236.

Siegler, D.S. (1979) Toxic seed lipids. In Herbivores: Their Interaction with Secondary Plant Metabolites (eds G.A. Rosenthal & D.H. Janzen), pp. 449–470. Academic Press, New York.

Siekmann, G., Keller, M.A., & Tenhumberg, B. (2004) The sweet tooth of adult parasitoid Cotesia rubecula: Ignoring hosts for nectar? Journal of Insect Behavior, 17, 459–476.

Silvertown, J.W. (1980) The evolutionary ecology of mast seeding in trees. Biological Journal of the Linnean Society, 14, 235–250.

Simmonds, F.J. (1948) The influence of maternal physiology on the incidence of diapause. Philosophical Transactions of the Royal Society of London Series B, 233, 385–414.

Sims, S.R. (1995) *Bacillus thuringiensis* var. *kurstaki* [CryIA9C)] protein expressed in transgenic cotton: Effects on beneficial and other non-target insects. Southwestern Entomologist, 20, 493–500.

Sinclair, B.J. (2000) Water relations of the freeze-tolerant New Zealand alpine cockroach *Celatoblatta quinquemaculata* (Dictyoptera: Blattidae). Journal of Insect Physiology, 46, 869–876.

Singaravelan, N., Inbar, M., Ne'eman, G., Distl, M., Wink, M., & Izhaki, I. (2006) The effects of nectar-nicotine on colony fitness of caged honeybees. Journal of Chemical Ecology, 31, 49–58.

Singh, D.P. & Singh, A. (2005) Disease and Insect Resistance in Plants. Science Publishers, Enfield, NH.

Singh, J.S. & Yadava, P.S. (1973) Caloric values of plant and insect species of a tropical grassland. Oikos, 24, 186–194.

Sisterton, M.S., Biggs, R.W., Olson, C., Carriere, Y., Dennehy, T.J., & Tabashnik, B.E. (2004) Arthropod abundance and diversity in Bt and non-Bt cotton fields. Environmental Entomology, 33, 921–929.

Skidmore, B.A. & Heithaus, E.R. (1988) Lipid cues for seed-carrying by ants in *Hepatica americana*. Journal of Chemical Ecology, 14, 2185–2196.

Skirvin, D.J., Kravar-Garde, L., Reynolds, K., Jones, J., Mead, A., & Fenlon, J. (2007) Supplemental food affects thrips predation and movement of *Orius laevigatus* (Hemiptera: Anthocoridae) and *Neoseiulus cucumeris* (Acari: Phytoseiidae). Bulletin of Entomological Research, 97, 309–315.

Skruhavy, V. (1959) Potrava polnich strevlikovitych – Die Nahrung der Feldcarabiden. Acta Societatis Entomologicae Cechosloveniae., 56, 1–18.

Slack, C.R. & Browse, J.A. (1984) Synthesis of storage lipids in developing seeds. In Seed Physiology, Volume 1: Development (ed D.R. Murray), pp. 209–244. Academic Press, Orlando, FL.

Slosser, J.E., Parajulee, M.N., & Bordovsky, D.G. (2000) Evaluation of food sprays and relay strip crops for enhancing biological control of bollworms and cotton aphids in cotton. International Journal of Pest Management, 46, 267–275.

Sluss, R.R. (1967) Population dynamics of the walnut aphid, *Chromaphis juglandicola* (Kalt.) in northern California. Ecology, 48, 41–58.

Smalley, A.E. (1960) Energy flow of a salt marsh grasshopper population. Ecology, 41, 672–677.

Smallwood, P.D., Steele, M.A., & Faeth, S.H. (2001) The ultimate basis of the caching preferences of rodents, and the oak-dispersal syndrome: Tannins, insects, and seed germination. American Zoology, 41, 840–851.

Smith, B.C. (1960) A technique for rearing coccinellid beetles on dry foods, and influence of various pollens on the development of *Coleomegilla maculata* Legni Timb. (Coleoptera: Coccinellidae). Canadian Journal of Zoology, 38, 1047–1049.

Smith, B.C. (1961) Results of rearing some coccinellid (Coleoptera: Coccinellidae) larvae on various pollens. Proceedings of the Entomological Society of Ontario, 91, 270–271.

Smith, C.A. & Evenson, W.E. (1978) Energy distribution in reproductive structures of *Amaryllis*. American Journal of Botany, 65, 714–716.

Smith, C.C. (1970) The coevolution of pine squirrels (*Tamiasciurus*) and conifers. Ecological Monographs, 40, 350–371.

Smith, C.M. (2005) Plant Resistance to Arthropods. Springer, Dordrecht, The Netherlands.

Smith, L.L., Lanza, J., & Smith, G.C. (1990) Amino acid concentrations in extrafloral nectar of *Impatiens sultani* increase after simulated herbivory. Ecology, 71, 107–115.

Smith, R.B. & Mommsen, T.P. (1984) Pollen feeding in an orb-weaving spider. Science, 226, 1330–1332.

Smith, S.F. & Krischik, V.A. (1999) Effects of systemic imidacloprid on Coleomegilla maculata (Coleoptera: Coccinellidae). Environmental Entomology, 28, 1189–1195.

Smith, S.M., Hubbes, M., & Carrow, J.R. (1986) Factors affecting inundative releases of *Trichogramma minutum* Rol. against the spruce budworm. Journal of Applied Entomology, 101, 29–39.

Solbreck, C. (1974) Maturation of post-hibernation flight behaviour in the coccinellid *Coleomegilla maculata* (DeGeer). Oecologia, 17, 265–275.

Sorensen, A.A., Kamas, R.S., & Vinson, S.B. (1983) The influence of oral secretions from larvae on levels of proteinases in colony members of *Solenopsis invicta* Buren (Hymenoptera: Formicidae). Journal of Insect Physiology, 29, 163–168.

Sotherton, N.W. (1995) Beetle banks – helping nature to control pests. Pesticide Outlook, 6, 13–17.

Soukup, V.G. & Holman, R.T. (1987) Fatty acids of seeds of North American pedicillate *Trillium* species. Phytochemistry, 26, 1015–1018.

Souquet, J.-M., Labarbe, B., Le Guerneve, C., Cheynier, V., & Moutounet, M. (2000) Phenolic composition of grape stems. Journal of Agricultural and Food Chemistry, 48, 1076–1080.

Southwick, E.E. (1984) Photosynthate allocation to floral nectar: A neglected energy investment. Ecology, 65, 1775–1779.

Southwood, T.R.E. (1973) The insect/plant relationship – an evolutionary perspective. Symposia of the Royal Entomological Society of London, 6, 3–30.

Speight, M.R. (1983) The potential of ecosystem management for pest control. Agriculture, Ecosystems and Environment, 10, 183–199.

Spellman, B., Brown, M.W., & Matthews, C.R. (2006) Effect of floral and extrafloral resources on predation of *Aphis spiraecola* by *Harmonia axyridis* on apple. BioControl, 51, 715–724.

Standifer, L.N. (1966) Some lipid constituents of pollens collected by honeybees. Journal of Apicultural Research, 5, 93–98.

Standifer, L.N. (1967) A comparison of the protein quality of pollens for growth-stimulation of the hypopharyngeal glands and longevity of honey bees, *Apis mellifera* L. (Hymenoptera: Apidae). Insectes Sociaux, Paris, 14, 415–426.

Standifer, L.N., Devys, M., & Barbier, M. (1968) Pollen sterols – a mass spectrographic survey. Phytochemistry, 7, 1361–1365.

Stang, M., Klinkhamer, P.G.L., & van der Meijden, E. (2006) Size constraints and flower abundance determine the number of interactions in a plant-flower visitor web. Oikos, 112, 111–121.

Stanley, R.G. & Linskins, H.F. (1965) Protein diffusion from germinating pollen. Physiologia Plantarum, 18, 47–53.

Stanley, R.G. & Linskins, H.F. (1974) Pollen: Biology, Biochemistry, Management. Springer, New York.

Stapel, J.O., Cortesero, A.M., De Moraes, C.M., Tumlinson, J.H., & Lewis, W.J. (1997) Extrafloral nectar, honeydew, and sucrose effects on searching behavior and efficiency of *Microplitis croceipes* (Hymenoptera: Braconidae) in cotton. Environmental Entomology, 26, 617–623.

Stapel, J.O., Cortesero, A.M., & Lewis, W.J. (2000) Disruptive sublethal effects of insecticides on biological control: Altered foraging ability and life span of a parasitoid after feeding on extrafloral nectar of cotton treated with systemic insecticides. Biological Control, 17, 243–249.

Starks, K.J. & Lilly, J.H. (1955) Some effects of insecticide treatment on dent corn. Journal of Economic Entomology, 48, 549–555.

Starks, K.J., Munippan, R., & Eikenbary, R.D. (1972) Interaction between plant resistance and parasitism against the greenbug on barley and sorghum. Annals of the Entomological Society of America, 65, 650–655.

Stary, P. (1966) Aphid parasites (Hym., Aphidiidae) and their relationships to aphid attending ants, with respect to biological control. Insectes Sociaux, 13, 185–202.

Steele, M.A., Knowles, T., Bridle, K., & Simms, E.L. (1993) Tannins and partial consumption of acorns: Implications for dispersal of oaks by seed predators. American Midland Naturalist, 130, 229–238.

Steinhaus, E.A. (1941) A study of bacteria associated with thirty species of insects. Journal of Bacteriology, 42, 757–790.

Stelzl, M. (1990) Nahrungsanalystische Untersuchungen an Hemerobiiden-Imagines (Insecta: Planipennia). Mitteilungen der deutschen Gesellschaft fur allgemeine und angewande Entomologie, 7, 670–676.

Stelzl, M. (1991) Untersuchungen zu Nahrungsspektren mitteleuropäischer Neuropteren-Imagines (Neuropteroidea, Insecta). Mit einer diskussion über deren Nützlichkeit als Opponentenbon Pflanzenschädlingen. Journal of Applied Entomology, 111, 469–477.

Stelzl, M. & Hassan, S.A. (1992) Culturing of *Micromus angulatus* Steph. (Neuropteroidea: Hemerobiidae) a new candidate beneficial insect for the biological control of insect pests in greenhouses. Journal of Applied Entomology, 114, 32–37.

Stephens, M.J., France, C.M., Wratten, S.D., & Frampton, C. (1998) Enhancing biological control of leafrollers (Lepidoptera: Tortricidae) by sowing buckwheat (*Fagopyrum esculentum*) in an orchard. Biocontrol Science and Technology, 8, 547–558.

Stephenson, A.G. (1982a) Iridoid glycosides in the nectar of *Catalpa speciosa* are unpalatable to nectar thieves. Journal of Chemical Ecology, 8, 1025–1034.

Stephenson, A.G. (1982b) The role of extrafloral nectaries of *Catalpa speciosa* in limiting herbivory and increasing fruit production. Ecology, 63, 663–669.

Stepien, Z.A. & Rodriguez, J.G. (1982) Caloric values of mites for bioenergetical studies. International Journal of Acarology, 8, 189–192.

Stern, V.M., Smith, R.F., van den Bosch, R., & Hagen, K.S. (1959) The integrated control concept. Hilgardia, 29, 81–101.

Stiles, F.G. & Freeman, C.E. (1993) Patterns in floral nectar characteristics of some bird-visited plant species from Costa Rica. Biotropica, 25, 191–205.

Stoffolano, J.G., Jr. (1995) Regulation of a carbohydrate meal in adult Diptera, Lepidoptera, and Hymenoptera. In Regulatory Mechanisms in Insect Feeding (eds R.F. Chapman & G. de Boer), pp. 210–248. Chapman & Hall, New York.

Stoll, S., Gadau, J., Gross, R., & Feldhaar, H. (2007) Bacterial microbiota associated with ants of the genus *Tetraponera*. Biological Journal of the Linnean Society, 90, 399–412.

Stone, T.B., Pitre, H.N., & Thompson, A.C. (1984) Relationships of cotton phenology, leaf soluble protein, extrafloral nectar carbohydrate and fatty acid concentrations with populations of five predator species. Journal of the Georgia Entomological Society, 19, 204–212.

Stoner, A. (1970) Plant feeding by a predaceous insect, *Geocoris punctipes*. Journal of Economic Entomology, 63, 1911–1915.

Stoner, A. (1972) Plant feeding by *Nabis*, a predaceous genus. Environmental Entomology, 1, 557–558.

Stoner, A., Metcalfe, A.M., & Weeks, R.E. (1974) Plant feeding by a predaceous insect, *Podisus acutissimus*. Environmental Entomology, 3, 187–189.

Stoner, A., Metcalfe, A.M., & Weeks, R.E. (1975) Plant feeding by Reduviidae, a predaceous family (Hemiptera). Journal of the Kansas Entomological Society, 48, 185–188.

Stradling, D.J. (1987) Nutritional ecology of ants. In Nutritional Ecology of Insects, Mites, Spiders and Related Invertebrates (eds F.J. Slansky & J.G. Rodriguez), pp. 927–969. Wiley, New York.

Strickland, A.H. (1947) Coccids attacking cacao (*Theobroma cacao,* L.) in West Africa, with descriptions of five new species. Bulletin of Entomological Research, 38, 497–523.

Sudjaroen, Y., Haubner, R., Wurtele, G., Hull, W.E., Erben, G., Spiegelhalder, B., Changbumrung, S., Bartsch, H., & Owen, R.W. (2005) Isolation and structure elucidation of phenolic antioxidants from Tamarind (*Tamarindus indica* L.) seeds and pericarp. Food and Chemical Toxicology, 43, 1673–1682.

Suh, S.-O. & Blackwell, M. (2005). The beetle gut as a habitat for new species of yeasts. In *Insect-Fungal Associations: Ecology and Evolution* (eds F.E. Vega & M. Blackwell), pp. 244–256. Oxford University Press, Oxford, UK.

Suh, S.-O., Gibson, C.M., & Blackwell, M. (2004) *Metschnikowia chrysoperlae* sp. nov., *Candida picachoensis* sp. nov. and *Candida pimensis* sp. nov., isolated from the green lacewings *Chrysoperla comanche* and *Chrysoperla carnea* (Neuroptera: Chrysopidae). International Journal of Systematic and Evolutionary Microbiology, 54, 1883–1890.

Summers, C.G. (1976) Population fluctuations of selected arthropods in alfalfa: Influence of two harvesting practices. Environmental Entomology, 5, 103–109.

Sundby, R.A. (1967) Influence of food on the fecundity of *Chrysopa carnea* Stephens [Neuroptera: Chrysopidae]. Entomophaga, 12, 475–479.

Sunderland, K.D. (1975) The diet of some predatory arthropods in cereal crops. Journal of Applied Ecology, 12, 507–515.

Sunderland, K.D., Lovei, G.L., & Fenlon, J.S. (1995) Diets and reproductive phenologies of the introduced ground beetles *Harpalus affinis* and *Clivina australasiae* (Coleoptera: Carabidae) in New Zealand. Australian Journal of Zoology, 43, 39–50.

Sutherland, J.P., Sullivan, M.S., & Poppy, G.M. (2001) Oviposition behaviour and host colony size discrimination in *Episyrphus balteatus* (Diptera: Syrphidae). Bulletin of Entomological Research, 91, 411–417.

Suzuki, Y., Tsuji, H., & Sasakawa, M. (1984) Sex allocation and effects of superparasitism on secondary sex ratios in the gregarious parasitoid, *Trichogramma chilonis* (Hymenoptera: Trichogrammatidae). Animal Behavior, 32, 478–484.

Svensson, B.G. & Janzon, L.-A. (1984) Why does the hoverfly *Metasyrphus corollae* migrate? Ecological Entomology, 9, 329–335.

Svoboda, J.A., Thompson, M.J., Robbins, W.E., & Kaplanis, J.N. (1978) Insect steroid metabolism. Lipids, 13, 742–753.

Swain, T. (1979) Tannins and lignins. In Herbivores: Their Interactions with Secondary Plant Metabolites (eds G.A. Rosenthal & D.H. Janzen), pp. 657–682. Academic Press, New York.

Swift, M.J., Heal, O.W., & Anderson, J.M. (1979) Decomposition in Terrestrial Ecosystems. University of California Press, Berkeley, CA.

Swirski, E., Amitai, S., & Dorzia, N. (1967) Laboratory studies on the feeding, development, and oviposition of the predaceous mite *Typhlodromus athiasae* P. & S. [*Acarina: Phytoseiidae*] on various kinds of food substances. Israel Journal of Agricultural Research, 17, 213–218.

Swirski, E., Amitai, S., & Dorzia, N. (1970) Laboratory studies on the feeding habits, post-embryonic survival and oviposition of the predaceous mites *Amblyseius chilensis* Dosse and

Amblyseius hibisci Chant [Acarina: Phytoseiidae] on various kinds of food substances. Entomophaga, 15, 93–106.

Szentkiralyi, F. (2001) Ecology and habitat relationships. In Lacewings in the Crop Environment (eds P.K. McEwen, T.R. New, & A.E. Whittington), pp. 82–115. Cambridge University Press, New York.

Takasu, K. & Lewis, W.J. (1993) Host- and food-foraging of the parasitoid *Microplitis croceipes*: Learning and physiological state effects. Biological Control, 3, 70–74.

Takasu, K. & Lewis, W.J. (1995) Importance of adult food sources to host searching of the larval parasitoid *Microplitis croceipes*. Biological Control, 5, 25–30.

Takeuchi, M., Sasaki, Y., Sato, C., Iwakuma, S., Isozaki, A., & Tamura, M. (2000) Seasonal host utilization of mycophagous ladybird *Ileis koebelei* (Coccinellidae: Coleoptera). Japanese Journal of Applied Entomology and Zoology, 44, 89–94.

Tallamy, D.W. & Wood, T.K. (1986) Convergence patterns in subsocial insects. Annual Review of Entomology, 31, 369–390.

Tamaki, G. & Weeks, R.E. (1972) Biology and Ecology of Two Predators, *Geocoris pallens* Stal and *G. bullatus* (Say) (United States Department of Agriculture. Agricultural Research Service), Technical Bulletin No. 1446, pp. 1–46.

Tanada, Y. & Kaya, H.K. (1993) Insect Pathology. Academic Press, San Diego, CA.

Tanigoshi, L.K., Megevand, B., & Yaninek, J.S. (1993) Non-prey food for subsistence of *Amblyseius idaeus* (Acari: Phytoseiidae) on cassava in Africa. Experimental and Applied Acarology, 17, 91–96.

Tarczynski, M.C., Byrne, D.N., & Miller, W.B. (1992) High performance liquid chromatography analysis of carbohydrates of cotton-phloem sap and of honeydew produced by *Bemisia tabaci* feeding on cotton. Plant Physiology, 98, 753–756.

Tasei, J.-N., Sabik, H., Pirastru, L., Lungiu, E., Blanche, J.M., Fournier, J., & Taglioni, J.P. (1994) Effects of sublethal doses of delatmethrin (Decis Ce) on *Bombus terrestris*. Journal of Apicultural Research, 33, 129–135.

Tassan, R.L., Hagen, K.S., & Sawall, E.F.J. (1979) The influence of field food sprays on the egg production rate of *Chrysopa carnea*. Environmental Entomology, 8, 81–85.

Taylor, R.M. & Foster, W.A. (1996) Spider nectarivory. American Entomologist, 42, 82–86.

Taylor, R.M. & Pfannenstiel, R.S. (in press) Plant nectar increases a spider's survival, growth, and fecundity.

Teasdale, J.R., Mangum, R.W., Radhakrishnan, J., & Cavigelli, M.A. (2004) Weed seedbank dynamics in three organic farming crop rotations. Agronomy Journal, 96, 1429–1435.

Tennant, L.E. & Porter, S.D. (1991) Comparison of diets of two fire ant species (Hymenoptera: Formicidae): Solid and liquid components. Journal of Entomological Science, 26, 450–465.

Teo, L.H. & Woodring, J.P. (1994) Comparative total activities of digestive enzymes in different gut regions of the house cricket, *Acheta domesticus* L. (Orthoptera: Gryllidae). Annals of the Entomological Society of America, 87, 886–890.

Tertuliano, M., Olson, D.M., Rains, G.C., & Lewis, W.J. (2004) Influence of handling and conditioning protocol on learning and memory of *Microplitis croceipes*. Entomologia Experimentalis et Applicata, 110, 165–172.

Tevis, L.J. (1958) Interrelations between the harvestor ant *Veromessor pergandei* (Mayr) and some desert ephemerals. Ecology, 39, 695–704.

Theile, H.-U. (1977) Carabid Beetles in Their Environments. Springer, Berlin.

Therond, J. (1975) Catalogue des Coleopteres de la Camargue et du Gard. Sociétè des Sciences Naturelle de Nimes, Mémoire 10, 1–410.

Thomas, K.K. & Nation, J.L. (1984a) Absorption of glucose, glycine and palmitic acid by isolated midgut and hindgut from crickets. Comparative Biochemistry and Physiology, 79A, 289–295.

Thomas, K.K. & Nation, J.L. (1984b) Protease, amylase and lipase activities in the midgut and hindgut of the cricket, *Gryllus rubens* and the mole cricket, *Scapteriscus acletus*. Comparative Biochemistry and Physiology, 79A, 297–304.

Thomas, W.D.E. & Glynne Jones, G.D. (1955) The systematic properties of diethyl-S-2(ethylthioethyl) phosphorothiolate (Demeton-S) with reference to the contamination of nectar. Annals of Applied Biology, 43, 182–191.

Thompson, S.N. (1973) A review and comparative characterization of the fatty acid compositions of seven insect orders. Comparative Biochemistry and Physiology B, 45, 467–482.

Tilman, D. (1978) Cherries, ants, and tent caterpillars: Timing of nectar production in relation to susceptibility of caterpillars to ant predation. Ecology, 59, 686–692.

Tillman, P.G. (2006) Susceptibility of pest *Nezera viridula* (Heteroptera: Pentatomidae) and parasitoid *Trichopoda pennipes* (Diptera: Tachinidae) to selected insecticides. Journal of Economic Entomology, 99, 648–657.

Timmons, F.D., Brook, T.S., & Harris, F.A. (1973) Effects of aldicarb applied side-dress to cotton on some arthropods in the Monroe County, Mississippi, boll weevil diapause-control area in 1969. Journal of Economic Entomology, 66, 151–153.

Tjeder, B. (1966) Neuroptera-Planipennia. The lace-wings of southern Africa. 5. Family Chrysopidae. In South African Animal Life: Results of the Lund University Expeditions in 1950–1951 (eds B. Hanström, P. Brinck, & G. Rudebeck), Vol. 12, pp. 228–534. Swedish National Science Research Council, Stockholm, Sweden.

Todd, F.E. & Bretherick, O. (1942) The composition of pollens. Journal of Economic Entomology, 35, 312–317.

Todd, F.E. & Vansell, G.H. (1942) Pollen grains in nectar and honey. Journal of Economic Entomology, 35, 728–731.

Todd, R. (1967) Coleridge and Paracelsus; honeydew and LSD. London Magazine, 6, 52–62.

Toft, C.A. (1983) Community patterns of nectarivorous adult parasitoids (Diptera, Bombyliidae) on their resources. Oecologia, 57, 200–215.

Toft, C.A. (1984) Resource shifts in bee flies (Bombyliidae): Interactions among species determine choice of resources. Oikos, 43, 104–112.

Tooker, J.F. & Hanks, L.M. (2000) Flowering plant hosts of adult hymenopteran parasitoids of Central Illinois. Annals of the Entomological Society of America, 93, 580–588.

Tooley, J.A. & Brust, G.E. (2002) Weed seed predation by carabid beetles. In The Agroecology of Carabid Beetles (ed J.M. Holland), pp. 215–230. Intercept, Andover.

Torres, C. & Galetto, L. (2002) Are nectar sugar composition and corolla tube length related to the diversity of insects that visit Asteraceae flowers? Plant Biology, 4, 360–366.

Torres, J.A. (1984) Niches and coexistence of ant communities in Puerto Rico: Repeated patterns. Biotropica, 16, 284–295.

Torres, J.B. & Ruberson, J.R. (2004) Toxicity of thiamethoxam and imidacloprid to *Podisus nigrispinus* (Dallas) (Heteroptera: Pentatomidae) nymphs associated to aphid and whitefly control in cotton. Neotropical Entomology, 33, 99–106.

Torres, J.B., Ruberson, J.R., & Adang, M.J. (2006) Expression of Bacillus thuringiensis Cry1Ac protein in cotton plants, acquisition by pests and predators: A tritrophic analysis. Agricultural and Forest Entomology, 8, 191–202.

Townes, H. (1958) Some biological characteristics of the Ichneumonidae (Hymenoptera) in relation to biological control. Journal of Economic Entomology, 51, 650–652.

Traveset, A. (1993) Deceptive fruits reduce seed predation by insects in *Pistacia terebinthus* L. (Anacardiaceae). Evolutionary Ecology, 7, 357–361.

Treacy, M.F., Benedict, J.H., Walmsley, M.H., Lopez, J.D., & Morrison, R.K. (1987) Parasitism of bollworm (Lepidoptera: Noctuidae) eggs on nectaried and nectariless cotton. Environmental Entomology, 16, 420–423.

Trelease, W. (1881) The foliar nectar glands of *Populus*. Botanical Gazette, 6, 284–290.

Trichilo, P.J. & Leigh, T.F. (1988) Influence of resource quality on the reproductive fitness of flower thrips (Thysanoptera: Thripidae). Annals of the Entomological Society of America, 81, 64–70.

Triltsch, H. (1997) Gut contents in field sampled adults of *Coccinella septempunctata* (Col.: Cocinellidae). Entomophaga, 42, 125–131.

Triltsch, H. (1999) Food remains in the guts of *Cocinella septempunctata* (Coleoptera: Coccinellidae) adults and larvae. European Journal of Entomology, 96, 355–364.

Tschinkel, W.R. (2006) The Fire Ants. The Belknap Press, Cambridge, MA.

Turnbull, L.A., Crawley, M.J., & Rees, M. (2000) Are plant populations seed-limited? A review of seed sowing experiments. Oikos, 88, 225–238.

Tylianakis, J.M., Didham, R.K., & Wratten, S.D. (2004) Improved fitness of aphid parasitoids receiving resource subsidies. Ecology, 85, 658–666.

Umoren, U.E., Essien, A.I., Ukorebi, B.A., & Essien, E.B. (2005) Chemical evaluation of the seeds of *Milletia obanensis*. Food Chemistry, 91, 195–201.

Vacante, V., Cocuzza, G.E., de Clercq, P., Van de Veire, M., & Tirry, L. (1997) Development and survival of *Orius albidipennis* and *O. laevigatus* (Het.: Anthocoridae) on various diets. Entomophaga, 42, 493–498.

Vadivel, V. & Janardhanan, K. (2005) Nutritional and antinutritional characteristics of seven South Indian wild legumes. Plant Foods for Human Nutrition, 60, 69–75.

van den Meiracker, R.A.F., Hammond, W.N.O., & van Alphen, J.J.M. (1990) The role of kairomones in prey finding by *Diomus* sp. and *Exochomus* sp., two coccinellid predators of the cassava mealybug, *Phenacoccus manihoti*. Entomologia Experimentalis et Applicata, 56, 209–217.

van der Goot, V.S. & Grabandt, R.A.J. (1970) Some species of the genera Melanostoma, Platycheirus and Pyrophaena (Diptera, Syrphidae) and their relation to flowers. Entomologische Berichten, 30, 135–143.

van Emden, H.F. (1965) The effect of uncultivated land on the distribution of cabbage aphid (*Brevicoryne brassicae*) on an adjacent crop. Journal of Applied Ecology, 2, 171–196.

van Emden, H.F. (1966) Plant insect relationships and pest control. World Review of Pest Control, 5, 115–123.

van Emden, H.F. (1986) The interaction of plant resistance and natural enemies: Effects on populations of sucking insects. In Interactions of Plant Resistance and Parasitoids and Predators of Insects (eds D.J. Boethel & R.D. Eikenbary), pp. 138–150. Ellis Horwood, Chichester.

van Emden, H.F. (1995) Host plant-aphidophaga interactions. Agriculture, Ecosystems and Environment, 52, 3–11.

van Emden, H.F. (1999) Transgenic host plant resistance to insects – some reservations. Annals of the Entomological Society of America, 92, 788–797.

van Emden, H.F. (2003) Conservation biological control: From theory to practice. In International Symposium on Biological Control of Arthropods (ed R.G. Van Driesche), pp. 199–208. United States Department of Agriculture, Forest Service, Morgantown, WV/Honolulu, HA.

van Emden, H.F. & Hagen, K.S. (1976) Olfactory reactions of the green lacewing, *Chrysopa carnea*, to tryptophan and certain breakdown products. Environmental Entomology, 5, 469–473.

Van Etten, C.H. & Tookey, H.L. (1979) Chemistry and biological effects of glucosinolates. In Herbivores: Their Interactions with Secondary Plant Metabolites (eds G.A. Rosenthal & D.H. Janzen), pp. 471–501. Academic Press, New York.

van Hook, R.I., Jr. (1971) Energy and nutrient dynamics of spider and orthopteran populations in a grassland ecosystem. Ecological Monographs, 41, 1–26.

van Rijn, P.C.J. & Tanigoshi, L.K. (1999a) The contribution of extrafloral nectar to survival and reproduction of the predatory mite *Iphesius degenerans* on *Ricinus communis*. Experimental and Applied Acarology, 23, 281–296.

van Rijn, P.C.J. & Tanigoshi, L.K. (1999b) Pollen as food for the predatory mites *Iphiseius degenerans* and *Neoseiulus cucumeris* (Acari: Phytoseiidae): Dietary range and life history. Experimental and Applied Acarology, 23, 785–802.

van Rijn, P.C.J., van Houten, Y.M., & Sabelis, M.W. (2002) How plants benefit from providing food to predators even when it is also edible to herbivores. Ecology, 83, 2664–2679.

Vander Wall, S.B., Kuhn, K.M. & Beck, M.J. (2005) Seed removal, seed predation, and secondary dispersal. Ecology, 86 (3), 801–806.

Vanhaelen, M., Vanhaelen-Fastre, R., Geerearts, J., & Wirthlin, T. (1979) *Cis*- and *trans*-octa-1, 5-dien-3-ol, new attractants to the cheese mite *Tyrophagous putrescentiae* (Schrank) (Acarina: Acaridae) identified in *Trichothecium roseum* (Fungi Imperfecti). Microbios, 23, 199–212.

Vanhaelen, M., Vanhaelen-Fastre, R., & Geeraerts, J. (1980) Occurrence in mushrooms (Homobasidiomycetes) of cis- and trans-octa-1, 5-dien-3-ol, attractants to the cheese mite, Tyrophagus putrescentiae (Schrank) (Acarina, Acaridae). Experientia, 36, 406–407.

Vansell, G.H. (1926) Buckeye poisoning of the honey bee, University of California Agriculture Experiment Station Circular 301, Berkeley, CA.

Vansell, G.H. & Watkins, W.G. (1933) A plant poisonous to adult bees. Journal of Economic Entomology, 26, 168–170.

Vansell, G.H. & Watkins, W.G. (1934) Adult bees found dying on spotted loco. Journal of Economic Entomology, 27, 635–637.

Vattala, H.D., Wratten, S.D., Phillips, C.B., & Wäckers, F.L. (2006) The influence of flower morphology and nectar quality on the longevity of a parasitoid biological control agent. Biological Control, 39, 179–185.

Vaughan, J. (1983) The use of seed proteins in taxonomy and phylogeny. In Seed Proteins (eds J. Daussant, J. Mosse, & J. Vaughan), pp. 135–153. Academic Press, London.

Veerman, A., Overmeer, W.P.J., van Zon, A.Q., de Boer, J.M., Waard, E.R., & Huisman, H.O. (1983) Vitamin A is essential for photoperiodic induction of diapause in an eyeless mite. Nature, 302, 248–249.

Vega, F.E. & Dowd, P.F. (2005). The role of yeasts as insect endosymbionts. In *Insect-Fungal Associations: Ecology and Evolution* (eds F.E. Vega & M. Blackwell), pp. 211–243. Oxford University Press, Oxford, UK.

Vega, F.E., Pava-Ripoll, M., Posada, F., & Buyer, J.S. (2005) Endophytic bacteria in *Coffea arabica* L. Journal of Basic Microbiology, 45, 371–380.

Venzon, M. & Carvalho, C.F. (1992) Biology of *Ceraeochrysa cubana* (Hagen, 1861) (Neuroptera: Chrysopidae) adults on different diets and temperatures. Ciencia e Practica, 16, 315–320.

Venzon, M., Rosado, M.C., Euzebio, D.E., Souza, B., & Schoereder, J.H. (2006) Suitability of leguminous cover crop pollens as food source for the green lacewing *Chrysoperla externa* (Hagen) (Neuroptera: Chrysopidae). Neotropical Entomology, 35, 371–376.

Verkerk, R.H.J., Leather, S.R., & Wright, D.J. (1998) The potential for manipulating crop-pest-natural enemy interactions for improved insect pest management. Bulletin of Entomological Research, 88, 493–501.

Villaneve, J., Thierry, D., Mamun, A.A., Lode, T., & Rat-Morris, E. (2005) The pollens consumed by common green lacewings *Chrysoperla* spp. (Neuroptera: Chrysopidae) in cabbage crop environment in western France. European Journal of Entomology, 102, 547–552.

Villanueva, R.T. & Childers, C.C. (2007) Development of Iphiseiodes quadripilis (Banks) (Acari: Phytoseiidae) on pollen or mite diets and predation on Aculops pelekassi (Keifer) (Acari: Eriophyidae) in the laboratory. Environmental Entomology, 36, 9–14.

Vinson, S.B. (1968) The distribution of an oil, carbohydrate, and protein food source to members of the imported fire ant colony. Journal of Economic Entomology, 61, 712–714.

Vinson, S.B. (1972) Imported fire ant feeding on *Papsalum* seeds. Annals of the Entomological Society of America, 65, 988.

Vinson, S.B. (1977) Behavioral chemicals in the augmentation of natural enemies. In Biological Control by Augmentation of Natural Enemies (eds R.L. Ridgway & S.B. Vinson), pp. 237–279. Plenum Press, New York.

Vinson, S.B. (1981) Habitat location. In Semiochemicals: Their Role in Pest Management (eds D.A. Nordlund, R.L. Jones, & W.J. Lewis), pp. 51–77. Wiley, New York.

Vinson, S.B., Harlan, D.P., & Hart, W.G. (1978) Response of the parasitoid Microterys flavus to the brown soft scale and its honeydew. Environmental Entomology, 7, 874–878.

Vogelei, A. & Greissl, R. (1989) Survival strategies of the crab spider *Thomisus onustus* Walckenaer 1806 (Chelicerata, Arachnida, Thomisidae). Oecologia, 80, 513–515.

Waage, J.K. & Ming, N.S. (1984) The reproductive strategy of a parasitic wasp, I. Optimal progeny and sex allocation in *Trichogramma evanescens*. Journal of Animal Ecology, 53, 401–415.

Wacht, S., Lunau, K., & Hansen, K. (1996) Optical and chemical stimuli control pollen feeding in the hoverfly *Eristalis tenax*. Entomologia Experimentalis et Applicata, 80, 50–53.

Wacht, S., Lunau, K., & Hansen, K. (2000) Chemosensory control of pollen ingestion in the hoverfly *Eristalis tenax* by labellar taste hairs. Journal of Comparative Physiology A, 186, 193–203.

Wäckers, F., Bruin, J., & van Rijn, P.C.J. (2005) Plant-Provided Food for Carnivorous Insects: A Protective Mutualism and Its Applications. Cambridge University Press, Cambridge.

Wäckers, F.L. (1994) The effect of food deprivation on the innate visual and olfactory preferences in the parasitoid *Cotesia rubecula*. Journal of Insect Physiology, 40, 641–649.

Wäckers, F.L. (1999) Gustatory response by the hymenopteran parasitoid Cotesia glomerata to a range of nectar and honeydew sugars. Journal of Chemical Ecology, 25, 2863–2877.
Wäckers, F.L. (2000) Do oligosaccharides reduce the suitability of honeydew for predators and parasitoids? A further facet to the function of insect-synthesized honeydew sugars. Oikos, 90, 197–201.
Wäckers, F.L. (2001) A comparison of nectar- and honeydew sugars with respect to their utilization by the hymenopteran parasitoid *Cotesia glomerata*. Journal of Insect Physiology, 47, 1077–1084.
Wäckers, F.L. (2003) The effect of food supplements on parasitoid-host dynamics. In International Symposium on Biological Control of Arthropods (ed R.G. Van Driesche), pp. 226–231. United States Department of Agriculture, Forest Service, Morgantown, WV/Honolulu, HA.
Wäckers, F.L. (2005) Suitability of (extra-)floral nectar, pollen, and honeydew as insect food sources. In Plant-Provided Food for Carnivorous Insects (eds F.L. Wäckers, P.C.J. van Rijn, & J. Bruin), pp. 17–74. Cambridge University Press, Cambridge.
Wäckers, F.L. & Bonifay, C. (2004) How to be sweet? Extrafloral nectar allocation by *Gossypium hirsutum* fits optimal defense theory predictions. Ecology, 85, 1512–1518.
Wäckers, F.L. & Stepphun, A. (2003) Characterizing nutritional state and food source use of parasitoids collected in fields with high and low nectar availability. IOBC WPRS Bulletin, 26, 209–214.
Wäckers, F.L., Bonifay, C., & Lewis, W.J. (2002) Conditioning of appetitive behavior in the hymenopteran parasitoid *Microplitis croceipes*. Entomologia Experimentalis et Applicata, 103, 135–138.
Wäckers, F.L., Lee, J.C., Heimpel, G.E., Winkler, K., & Wagenaar, R. (2006) Hymenopteran parasitoids synthesize "honeydew-specific" oligosaccharides. Functional Ecology, 20, 790–798.
Wäckers, F.L., Zuber, D., Wunderlin, R., & Keller, F. (2001) The effect of herbivory on temporal and spatial dynamics of foliar nectar production in cotton and castor. Annals of Botany, 87, 365–370.
Wada, A., Isobe, Y., Yamaguchi, S., Yamaoka, R., & Ozaki, M. (2001) Taste-enhancing effects of glycine on the sweetness of glucose: A gustatory aspect of symbiosis between the ant, *Camponotus japonicus*, and the larvae of the lycaenid butterfly, *Niphanda fusca*. Chemical Senses, 26, 983–992.
Waller, G.D. & Barker, R.J. (1979) Effects of dimethoate on honey bee colonies. Journal of Economic Entomology, 72, 549–551.
Waller, G.D., Erickson, B.J., Harvey, J., & Martin, J.H. (1984) Effects of dimethoate on honey bees (Hymenoptera: Apidae) when applied to flowering lemons. Journal of Economic Entomology, 77, 70–74.
Wallin, H., Chiverton, P.A., Ekbom, B.S., & Borg, A. (1992) Diet, fecundity and egg size in some polyphagous predatory carabid beetles. Entomologia Experimentalis et Applicata, 65, 129–140.
Walter, D.E. & O'Dowd, D.J. (1995) Life on the forest phylloplane: Hairs, little houses, and myriad mites. In Forest Canopies (eds M.D. Lowman & N.M. Nadkarni), pp. 325–351. Academic Press, New York.
Wang, Z.-Y., Sun, X.-F., Wang, F., Tang, K.-X., & Zhang, J.-R. (2005) Enhanced resistance of snowdrop lectin (*Galanthus nivalis* L. agglutinin)-expressing maize to Asian corn borer (*Ostrinia furnacalis* Guenée). Journal of Integrative Plant Biology, 47, 873–880.
Wanner, H., Gu, H., & Dorn, S. (2006) Nutritional value of floral nectar sources for flight in the parasitoid wasp, *Cotesia glomerata*. Physiological Entomology, 31, 127–133.
Watt, W.B., Hoch, P.C., & Mills, S.G. (1974) Nectar resource use by *Colias* butteflies. Oecologia, 14, 353–374.
Watson, S.J., Mauchline, A.L., Brown, V.K. & Froud-Williams, R.J. (2003) Post-dispersal losses of *Stellaria media* and *Polygonum aviculare* seeds in spring barley (*Hordeum vulgare*). Aspects of Applied Biology, 69, 203–208.
Watson, J.R. & Thompson, W.L. (1933) Food habits of *Leis conformis* Boisd. (Chinese ladybeetle). The Florida Entomologist, 17: 27–29.
Way, M.J. (1954) Studies on the association of the ant *Oecophylla longinoda* (Latr.) (Formicidae) with the scale insect *Saissetia zanzibarensis* Williams. Bulletin of Entomological Research, 45, 113–134.
Way, M.J. (1963) Mutualism between ants and honeydew-producing Homoptera. Annual Review of Entomology, 8, 307–344.

Webber, J.F. & Gibbs, J.N. (1989) Insect dissemination of fungal pathogens of trees. In Insect-Fungus Interactions (eds N. Wilding, N.M. Collins, P.M. Hammond, & J.F. Webber), pp. 161–194. Academic Press, London.

Webster, F.M. (1881) Notes upon the food of predaceous beetles. Bulletin of the Illinois State Laboratory of Natural History, 1, 149–152.

Webster, F.M. (1900) *Harpalus caliginosus* as a strawberry pest with notes on other phytophagous Carabidae. Canadian Entomologist, 32, 265–271.

Webster, F.M. (1903) Notes on the food of predaceous beetles. Bulletin of the Illinois State Laboratory of Natural History, 1, 162–166.

Wei, F., Ma, L.-Y., Cheng, X.-L., Lin, R.-C., Jin, W.-T., Khan, I.A., & Lu, J.-Q. (2005) Preparative HPLC for purification of four isomeric bioactive saponins from the seeds of *Aesculus chinensis*. Journal of Liquid Chromatography and Related Technologies, 28, 763–773.

Weigert, R.G. (1965) Energy dynamics of the grasshopper populations in old field and alfalfa field. Oikos, 16, 161–176.

Went, F.W., Wheeler, J., & Wheeler, G.C. (1972) Feeding and digestion in some ants (*Veromessor* and *Manica*). BioScience, 22, 82–88.

Weseloh, R.M. (1993) Manipulation of forest ant (Hymenoptera: Formicidae) abundance and resulting impact on gypsy moth (Lepidoptera: Lymantriidae) populations. Environmental Entomology, 22, 587–594.

Westerman, P.R., Borza, J.K., Andjelkovic, J., Liebman, M., & Danielson, B. (2008) Density-dependent predation of weed seeds in maize fields. Journal of Applied Ecology, in press.

Westerman, P.R., Hofman, A., Vet, L.E.M., & van der Werf, W. (2003a) Relative importance of vertebrates and invertebrates in epigeaic weed seed predation in organic cereal fields. Agriculture, Ecosystems and Environment, 95, 417–425.

Westerman, P.R., Liebman, M., Heggenstaller, A.H., & Forcella, F. (2006) Integrating measurements of seed availability and removal to estimate weed seed losses due to predation. Weed Science, 54, 566–574.

Westerman, P.R., Liebman, M., Menalled, F.D., Heggenstaller, A., Hartzler, R.G. & Dixon, P.M. (2005) Are many little hammers effective? Velvetleaf (*Abutilon theophrasti*) population dynamics in two- and four-year crop rotation systems. Weed Science, 53, 382–392.

Westerman, P.R., Wes, J.S., Kropff, M.J., & van der Werf, W. (2003b) Annual losses of weed seeds due to predation in organic cereal fields. Journal of Applied Ecology, 40, 824–836.

Westoby, M., French, K., Hughes, L., Rice, B., & Rodgerson, L. (1991) Why do more plant species use ants for dispersal on infertile compared with fertile soils? Australian Journal of Ecology, 16, 445–455.

Wetzel, J.M., Ohnishi, M., Fujita, T., Nakanishi, K., Naya, Y., Noda, H., & Sugiura, M. (1992) Diversity in steroidogenesis of symbiotic microorganisms from planthoppers. Journal of Chemical Ecology, 18, 2083–2094.

Wheeler, A.G., Jr. (1910) Ants. Columbia University Press, New York.

Wheeler, G.C. & Wheeler, J. (1953) The ant larvae of the subfamily Formicinae. Annals of the Entomological Society of America, 46, 126–171.

Wheeler, W.C. & Bailey, I.W. (1920) The feeding habits of Pseudomyrminae and other ants. Transactions of the American Philosophical Society, 22, 235–277.

White, A.J., Wratten, S.D., Berry, N.A., & Weigmann, U. (1995) Habitat manipulation to enhance biological control of *Brassica* pests by hover flies (Diptera: Syrphidae). Journal of Economic Entomology, 88, 1171–1176.

White, N.D. & Laing, J.E. (1977) Some aspects of the biology and a laboratory life table of the acarine predator *Zetzellia mali*. Canadian Entomologist, 109, 1275–1281.

Whitford, W.G. (1978) Foraging in seed-harvester ants, *Pogonomyrmex* spp. Ecology, 59, 185–189.

Whitford, W.G. & Ettershank, G. (1975) Factors affecting foraging activity in Chihuahuan desert harvester ants. Environmental Entomology, 4, 689–696.

Whitney, K.D. (2004) Experimental evidence that both parties benefit in a facultative plant-spider mutualism. Ecology, 85, 1642–1650.

Wickremasinghe, M.G.V. & van Emden, H.F. (1992) Reactions of adult female parasitoids, particularly *Aphidius rhopalosiphi,* to volatile chemical cues from the host plants of their aphid prey. Physiological Entomology, 17, 297–304.

Wiedenmann, R.N. & Smith, J.W.J. (1997) Attributes of natural enemies in ephemeral crop habitats. Biological Control, 10, 16–22.

Wilkinson, T.K. & Landis, D.A. (2005) Habitat diversification in biological control: The role of plant resources. In Plant-Provided Food for Carnivorous Insects (eds F.L. Wäckers, P.C.J. van Rijn, & J. Bruin), pp. 305–325. Cambridge University Press, Cambridge.

Willaman, J.J. & Schubert, B.G. (1961) Alkaloid-Bearing Plants and Their Contained Alkaloids U.S.D.A.A.R. Service Technical Bulletin No. 1234, 287 pp. United States Department of Agriculture: Agricultural Research Service.

Williams, D.F., Lofgren, C.S., & Lemire, A. (1980) A simple diet for rearing laboratory colonies of the red imported fire ant. Journal of Economic Entomology, 73, 176–177.

Williams, L.I., Roane, T.M., & Beach, J.P. (2005) Gustatory acceptance, longevity, and utilization of nectar and honeydew sugars by *Anaphes iole*, an egg parasitoid of *Lygus* bugs. In International Symposium on Biological Control of Arthropods (ed M.S. Hoddle), Vol. 2, pp. 409–419. United States Forest Service, Davos, Switzerland.

Williams, L. & Roane, T.M. (2007) Nutritional ecology of a parasitic wasp: Food source affects gustatory response, metabolic utilization, and survivorship. Journal of Insect Physiology, 53, 1262–1275.

Willmer, P.G. (1980) The effects of insect visitors on nectar constituents in temperate plants. Oecologia, 47, 270–277.

Wilson, E.O. (1978) Division of labor in fire ants based on physical castes (Hymenoptera: Formicidae: *Solenopsis*). Journal of the Kansas Entomological Society, 51, 615–636.

Wink, M. & Romer, P. (1986) Acquired toxicity – the advantages of specializing on alkaloid-rich lupins to Macrosiphon albifrons (Aphidae). Naturwissenschaften, 73, 210–212.

Wolcott, G.N. (1941) The establishment in Puerto Rico of *Larra americana* Suassure. Journal of Economic Entomology, 34, 53–56.

Wolcott, G.N. (1942) The requirements of parasites for more than hosts. Science, 96, 317–318.

Wolfenbarger, L.L., Naranjo, S.E., Lundgren, J.G., Bitzer, R.J., & Watrud, L.S. (2008) Bt crop effects on functional guilds of non-target arthropods: A meta-analysis. PLoS One, 3(5), e2118.

Wolff, A. & Debussche, M. (1999) Ants as seed dispersers in a Mediterranean old-field succession. Oikos, 84, 443–452.

Wolschin, F., Holldobler, B., Gross, R., & Zientz, E. (2004) Replication of the endosymbiotic bacterium *Blochmannia floridanus* Is correlated with the developmental and reproductive stages of its ant host. Applied and Environmental Microbiology, 70 (7), 4096–4102.

Wood, D.M. (1992) Tachinidae. In Manual of Nearctic Diptera (ed J.F. McAlpine), Vol. 2, pp. 1110–1269. Agriculture Canada, Minister of Supply and Services Canada, Ottowa, Canada.

Wood, H.B., Jr., Stromberg, V.L., Keresztesy, J.C., & Horning, E.C. (1954) Andromedotoxin. A potent hypotensive agent from *Rhododendron maximum*. Journal of the American Chemical Society, 76, 5689–5692.

Woodring, J., Wiedemann, R., Volkl, W., & Hoffmann, K.H. (2006) Oligosaccharide synthesis regulates gut osmolality in the ant-attended aphid *Metopeurum fuscoviride* but not in the unattended aphid, *Macrosiphoniella tanacetaria*. Journal of Applied Entomology, 131, 1–7.

Woolfolk, S.W. & Inglis, G.D. (2004) Microorganisms associated with field-collected Chrysoperla rufilabris (Neuroptera: Chrysopidae) adults with emphasis on yeast symbionts. Biological Control, 29, 155–168.

Woolfolk, S.W., Cohen, A.C., & Inglis, G.D. (2004) Morphology of the alimentary canal of Chrysoperla rufilabris (Neuroptera: Chrysopidae) adults in relation to microbial symbionts. Annals of the Entomological Society of America, 97, 796–808.

Wratten, S.D., van Emden, H.F., & Thomas, M.B. (1998) Within-field and border refugia for the enhancement of natural enemies. In Enhancing Biological Control: Habitat Management to Promote Natural Enemies of Agricultural Pests (eds C.H. Pickett & R.L. Bugg), pp. 375–403. University of California Press, Berkeley, CA.

Wratten, S.D., Berndt, L., Gurr, G.M., Tylianakis, J.M., Fernando, P., & Didham, R.K. (2003) Adding floral diversity to enhance parasitoid fitness and efficacy. In International Symposium on Biological Control of Arthropods (ed R.G. Van Driesche), pp. 211–214. United States Department of Agriculture, Forest Service, Morgantown, WV/Honolulu, HA.

Wu, J., Luo, X., Guo, H., Xiao, J., & Tian, Y. (2006) Transgenic cotton, expressing *Amaranthus caudatus* agglutinin, confers enhanced resistance to aphids. Plant Breeding, 125, 390–394.

Wyatt, I.J. (1970) The distribution of Myzus persicae (Sulz.) on year-round chrysanthemums. II. Winter season: The effect of parasitism by *Aphidius matricariae* Hal. Annals of Applied Biology, 65, 31–41.

Wykes, G.R. (1952) An investigation of the sugars present in the nectar of flowers of various species. New Phytologist, 51, 210–215.

Yahiro, K. (1990) A comparative morphology of the alimentary canal in the adults of ground-beetles (Coleoptera) I. Classification into the types. Esakia, Special Issue No. 1, 35–44.

Yano, E. (2006) Ecological considerations for biological control of aphids in protected culture. Population Ecology, 48, 333–339.

Yao, I. & Akimoto, S.-I. (2002) Flexibility in the composition and concentration of amino acids in honeydew of the drepanosiphid aphid *Tuberculatus quercicola*. Ecological Entomology, 27, 745–752.

Yazlovetsky, I.G. (2001) Features of the nutrition of Chrysopidae larvae and larval artificial diets. In Lacewings in the Crop Environment (eds P.K. McEwen, T.R. New, & A.E. Whittington), pp. 320–337. Cambridge University Press, Cambridge.

Yinon, U. (1969) Food consumption of the armored scale lady-beetle *Chilocorus bipustulatus* (Coccinellidae). Entomologia Experimentalis et Applicata, 12, 139–146.

Yokoyama, V.Y. (1978) Relation of seasonal changes in extrafloral nectar and foliar protein and arthropod populations in cotton. Environmental Entomology, 7, 799–802.

Yong, T.-H. (2003) Nectar-feeding by a predatory ambush bug (Heteroptera: Phymatidae) that hunts on flowers. Annals of the Entomological Society of America, 96, 643–651.

Yoshida, H., Hirakawa, Y., Murakami, C., Mizushina, Y., & Yamade, T. (2003) Variation in the content of tocopherols and distribution of fatty acids in soya bean seeds (*Glycine max* L.). Journal of Food Composition and Analysis, 16, 429–440.

Young, J.A. & Evans, R.A. (1973) Mucilaginous seed coats. Weed Science, 21, 52–54.

Yue, B. & Tsai, J.H. (1996) Development, survivorship, and reproduction of *Amblyseius largoensis* (Acari: Phytoseiidae) on selected plant pollens and temperatures. Environmental Entomology, 25, 488–494.

Zaher, M.A. & Shehata, K.K. (1971) Biological studies on the predator mite *Typhlodromus pyri* Sch. (Acarina: Phytoseiidae) with the effect of prey and non prey substances. Journal of Applied Entomology, 67, 389–394.

Zaher, M.A., Wafa, A.K., & Shehata, K.K. (1969) Life history of the predatory mite *Phytoseius plumifer* and the effect of nutrition on its biology (Acarina: Phytoseiidae). Entomologia Experimentalis et Applicata, 12, 383–388.

Zaher, M.A., Yousef, A.E.-T.A., & Kandil, M.M. (1981) Effect of food on the biology of *Cheletogenes ornatus* (C. & F.) (Acari: Prostigmata: Cheyletidae). Acarologia, 22, 361–366.

Zaitsev, V.F. (1982) Microstructure of the labella of the fly proboscis. I. Structure of the apparatus closing the pseudotracheae. Entomological Review, 61, 61–69.

Zandstra, B.H. & Motooka, P.S. (1978) Beneficial effects of weeds in pest management – a review. PANS, 24, 333–338.

Zangerl, A.R., Berenbaum, M.R., & Nitao, J.K. (1991) Parthenocarpic fruits in wild parsnip: Decoy defence against a specialist herbivore. Evolutionary Ecology, 5, 136–145.

Zemek, R. & Prenerova, E. (1997) Powdery mildew (Ascomycotina: Erysiphales) – an alternative food for the predatory mite *Typhlodromus pyri* Scheuten (Acari: Phytoseiidae). Experimental and Applied Acarology, 21, 405–414.

Zeng, F. & Cohen, A.C. (2000) Demonstration of amylase from the zoophytophagous anthocorid *Orius insidiosus*. Archives of Insect Biochemistry and Physiology, 44, 136–139.

Zettler, J.A. & Spira, T.P. (2001) Yellow jackets (*Vespula* spp.) disperse Trillium (spp.) seeds in eastern North America. American Midland Naturalist, 146, 444–446.

Zettler, J.A., Spira, T.P., & Allen, C.R. (2001) Ant-seed mutualisms: Can red imported fire ant sour the relationship? Biological Conservation, 101, 249–253.

Zhang, G., Zimmermann, O., & Hassan, S.A. (2004) Pollen as a source of food for egg parasitoids of the genus *Trichogramma* (Hymenoptera: Trichogrammatidae). Biocontrol Science and Technology, 14, 201–209.

Zhang, J., Drummond, F.A., Liebman, M., & Hartke, A. (1997) Phenology and dispersal of *Harpalus rufipes* DeGeer (Coleoptera: Carabidae) in agroecosystems in Maine. Journal of Agricultural Entomology, 14, 171–186.

Zhang, N.Y. & Li, Y. (1989) Rearing of the predacious mite, *Amblyseius fallacis* (Acari: Phytoseiidae) with plant pollen. (In Chinese). Chinese Journal of Biological Control, 5, 60–63.

Zhavoronkova, T.N. (1969) Certain structural peculiarities of the Carabidae (Coleoptera) in relation to their feeding habits. Entomological Reviews, 48, 462–471.

Zhimo, Z. & McMurtry, J.A. (1990) Development and reproduction in of three *Euseius* (Acari: Phytoseiidae) species in the presence and absence of supplementary foods. Experimental and Applied Acarology, 8, 233–242.

Zhou, W. & Wang, R. (1989) Rearing of *Orius sauteri* (Hem.: Anthocoridae) with natural and artificial diets. Chinese Journal of Biological Control, 5, 9–12.

Zientz, E., Beyaert, I., Gross, R., & Feldhaar, H. (2006) Relevance of the endosymbiosis of *Blochmannia floridanus* and carpenter ants at different stages of the life cycle of the host. Applied and Environmental Microbiology, 72(9), 6027–6033.

Zimina, L.V. (1957) Structure and function of the intestine of *Syrphus* (Diptera: Syrphidae). Zoologicheskii Zhurnal, 36, 1039–1043.

Znoiko, D.V. (1935) Die Larven des grossen Getreidelaufkafers (*Zabrus tenebrioides* Creutz.) und der Hirselaugkafers (*Harpalus calceatus* Duft.) (Coleoptera: Carabidae). Rev. Ent. U.S.S.R., 25, 232–238.

Zwahlen, C. & Andow, D.A. (2005) Field evidence for the exposure of ground beetles to Cry1Ab from transgenic corn. Environmental Biosafety Research, 4, 113–117.

Taxonomic Index

D

E

H

N

Subject Index

FSC
www.fsc.org
MIX
Papier | Fördert
gute Waldnutzung
FSC® C083411

Kolibri 360 GmbH
Ferdinand-Jühlke-Straße 7
99095 Erfurt, Deutschland
produktsicherheit@kolibri360.de